Measuring Elemental Impurities in Pharmaceuticals

A Practical Guide

PRACTICAL SPECTROSCOPY A SERIES

Infrared and Raman Spectroscopy (in three parts), *edited by Edward G. Brame, Jr. and Jeanette G. Grasselli*

X-Ray Spectrometry, *edited by H. K. Herglotz and L. S. Birks*

Mass Spectrometry (in two parts), *edited by Charles Merritt, Jr. and Charles N. McEwen*

Infrared and Raman Spectroscopy of Polymers, *H. W. Siesler and K. Holland-Moritz*

NMR Spectroscopy Techniques, *edited by Cecil Dybowski and Robert L. Lichter*

Infrared Microspectroscopy: Theory and Applications, *edited by Robert G. Messerschmidt and Matthew A. Harthcock*

Flow Injection Atomic Spectroscopy, *edited by Jose Luis Burguera*

Mass Spectrometry of Biological Materials, *edited by Charles N. McEwen and Barbara S. Larsen*

Field Desorption Mass Spectrometry, *László Prókai*

Chromatography/Fourier Transform Infrared Spectroscopy and Its Applications, *Robert White*

Modern NMR Techniques and Their Application in Chemistry, *edited by Alexander I. Popov and Klaas Hallenga*

Luminescence Techniques in Chemical and Biochemical Analysis, *edited by Willy R. G. Baeyens, Denis De Keukeleire, and Katherine Korkidis*

Handbook of Near-Infrared Analysis, edited by *Donald A. Burns and Emil W. Ciurczak*

Handbook of X-Ray Spectrometry: Methods and Techniques, *edited by René E. Van Grieken and Andrzej A. Markowicz*

Internal Reflection Spectroscopy: Theory and Applications, *edited by Francis M. Mirabella, Jr.*

Microscopic and Spectroscopic Imaging of the Chemical State, *edited by Michael D. Morris*

Mathematical Analysis of Spectral Orthogonality, *John H. Kalivas and Patrick M. Lang*

Laser Spectroscopy: Techniques and Applications, *E. Roland Menzel*

Practical Guide to Infrared Microspectroscopy, *edited by Howard J. Humecki*

Quantitative X-ray Spectrometry: Second Edition, *Ron Jenkins, R. W. Gould, and Dale Gedcke*

NMR Spectroscopy Techniques: Second Edition, Revised and Expanded, *edited by Martha D. Bruch*

Spectrophotometric Reactions, *Irena Nemcova, Ludmila Cermakova, and Jiri Gasparic*

Inorganic Mass Spectrometry: Fundamentals and Applications, *edited by Christopher M. Barshick, Douglas C. Duckworth, and David H. Smith*

Infrared and Raman Spectroscopy of Biological Materials, *edited by Hans-Ulrich Gremlich and Bing Yan*

Near-Infrared Applications In Biotechnology, *edited by Ramesh Raghavachari*

Ultrafast Infrared and Ramen Spectroscopy, *edited by M. D. Fayer*

Handbook of Near-Infrared Analysis: Second Edition, Revised and Expanded, *edited by Donald A. Burns and Emil W. Ciurczak*

Handbook of Raman Spectroscopy: From the Research Laboratory to the Process Line, *edited by Ian R. Lewis and Howell G. M. Edwards*

Handbook of X-Ray Spectrometry: Second Edition, Revised and Expanded, *edited by René E. Van Grieken and Andrzej A. Markowicz*

Ultraviolet Spectroscopy and UV Lasers, *edited by Prabhakar Misra and Mark A. Dubinskii*

Pharmaceutical and Medical Applications of Near-Infrared Spectroscopy, Second Edition, *Emil W. Ciurczak and James K. Drennen III*

Applied Electrospray Mass Spectrometry, *edited by Birendra N. Pramanik, A. K. Ganguly, and Michael L. Gross*

Practical Guide to ICP-MS, *edited by Robert Thomas*

NMR Spectroscopy of Biological Solids, *edited by A. Ramamoorthy*

Handbook of Near Infrared Analysis, Third Edition, *edited by Donald A. Burns and Emil W. Ciurczak*

Coherent Vibrational Dynamics, *edited by Guglielmo Lanzani, Giulio Cerullo, and Sandro De Silvestri*

Practical Guide to ICP-MS: A Tutorial for Beginners, Second Edition, *Robert Thomas*

Practical Guide to ICP-MS: A Tutorial for Beginners, Third Edition, *Robert Thomas*

Pharmaceutical and Medical Applications of Near-Infrared Spectroscopy, *Emil W. Ciurczak and Benoit Igne*

Measuring Elemental Impurities in Pharmaceuticals: A Practical Guide, *Robert Thomas*

Measuring Elemental Impurities in Pharmaceuticals

A Practical Guide

By

Robert Thomas

CRC Press is an imprint of the
Taylor & Francis Group, an **informa** business

CRC Press
Taylor & Francis Group
6000 Broken Sound Parkway NW, Suite 300
Boca Raton, FL 33487-2742

First issued in paperback 2021

CRC Press is an imprint of Taylor & Francis Group, an Informa business

ISBN 13: 978-1-138-19796-1 (hbk)
ISBN 13: 978-1-03-224089-3 (pbk)

DOI: 10.1201/b21952

Library of Congress Cataloging-in-Publication Data

Names: Thomas, Robert, 1949- author.
Title: Measuring elemental impurities in pharmaceuticals : a practical guide / Robert Thomas.
Other titles: Practical spectroscopy ; v. 40.
Description: Boca Raton : Taylor & Francis, 2018. | Series: Practical spectroscopy ; [v. 40] | Includes bibliographical references and index.
Identifiers: LCCN 2017038198 | ISBN 9781138197961 (hardback : alk. paper)
Subjects: | MESH: Pharmaceutical Preparations--analysis | Mass Spectrometry--methods | Drug Contamination | Metals, Heavy | Pharmaceutical Preparations--standards | Laboratory Manuals
Classification: LCC RM301 | NLM QV 25 | DDC 615.1--dc23
LC record available at https://lccn.loc.gov/2017038198

Visit the Taylor & Francis Web site at
http://www.taylorandfrancis.com

and the CRC Press Web site at
http://www.crcpress.com

Publisher's Note
The publisher has gone to great lengths to ensure the quality of this reprint but points out that some imperfections in the original copies may be apparent.

Contents

Foreword

In 2004, the FDA initiated a laboratory study of lead in widely used drugs and dietary supplements. At that time little was known about elemental impurities in pharmaceuticals, and the study's goal was to perform a risk-based survey of lead concentrations in prescription and over-the-counter drugs and dietary supplements. This study was published in 2007 and concluded that, though permissible levels of lead in pharmaceuticals at the time were high, products evaluated in the study had low levels of lead. During this same time, the United States Pharmacopeia (USP) began in earnest to modernize methods of heavy metals analysis.

In 2008, the USP engaged the Institute of Medicine to develop a workshop that brought together leading experts in metals analysis and toxicology and stakeholders from the pharmaceutical industry to plot a course for modernization of USP <231>. The workshop led to the development of draft chapters <232> and <233> and ultimately to the formation of the International Conference on Harmonization of Technical Requirements for Human Drugs (ICH) Q3D Expert Working Group on Elemental Impurities in Pharmaceuticals. Historically, metal levels in pharmaceuticals were based on the capabilities of the wet chemical methods described in USP <231>. USP <232> was intended to replace <231> with permitted daily exposure levels that were based on assessments of elemental toxicological data. The permitted daily exposure approach has already been established for residual catalysts by the European Medicines Agency. Q3D continued the toxicological assessment of elements of concern and emphasized a science-based and risk-based approach to control of elemental impurities in pharmaceuticals. Risk assessment was at the heart of elemental impurity control in Q3D, and informed the selection of elemental impurities of concern in the components of drug products.

Notably, the USP was at the Q3D table, along with Pharm Europa and the Japanese Pharmacopeia. Subsequent to the adoption of Q3D by regulatory authorities in the ICH regions, Ph. Eur. Adopted Q3D as its standard for control of elemental impurities, and shortly thereafter, the USP Elemental Impurities Expert Panel agreed to adopt many of the features of Q3D, including the classification of elements and the promotion of risk-based assessment of elemental impurities in drug products. The ultimate harmonization of the pharmacopeias with ICH provides a model for the development of common standards for technical requirements for drugs. This process, which included numerous fora for stakeholder input, resulted in adoption of a single standard for control of elemental impurities among both regulators and standard setting organizations. Regulatory authorities from non-ICH regions participated in the implementation phase of ICH Q3D and provided input into both the elaboration of examples to implementation and development of regulatory expectations for pharmaceutical manufacturers.

Over the course of these deliberations, several important surveys of elemental impurities in drug products and excipients were published. On several occasions I have summarized these surveys by noting that there are no surprises. Excipients and active pharmaceutical ingredients (APIs) that are highly refined prior to use (e.g., cellulose and synthetic excipients such as povidone) have very low levels of elemental impurities. Certain mined excipients that are purified prior to use (e.g., calcium carbonate, talc, titanium dioxide) exhibit relatively elevated levels of certain elemental impurities. Whether or not these elevated levels result in pharmaceuticals that exceed permitted daily exposures (PDEs) of elemental impurities depends largely on the total mass of the excipient in a daily dose of the finished drug product, as well as the route of administration. Importantly, certain rare elements (Q3D class 2B elements) were found only to be present in finished products when they were used as catalysts in late stages of the manufacturing process. These observations can inform the risk assessment of elemental impurities in pharmaceutical products.

Notwithstanding the above considerations, risk assessments as well as control of relatively high risk elemental impurities will ultimately require laboratory measurements. USP <233> provides some guidance on the use of inductively-coupled plasma (ICP) based techniques for analysis of elemental impurities, focusing on acceptable approaches to demonstrating the suitability of an ICP-OES or ICP-MS method for analysis of elemental impurities in pharmaceutical products. But these methods are commonly performed by experts with extensive experience in their operation and capabilities. Analytical chemists that are new to this area will need more detailed guidance to navigate the idiosyncrasies of ICP-MS and ICP-OES. Sample digestion and preparation, selection of internal standards, correction for interferences, identification of common instrumental concerns, preparation of suitable standards for a breadth of elements, and isotope selection for analysis are among the choices that the analyst will face.

Rob's book provides invaluable expert advice on these and other considerations that must be addressed during development of robust and accurate methods for both the risk assessment phase and the quantitative analysis phase of control of elemental impurities in pharmaceutical products. It begins with a detailed examination of modern instrumentation for analysis using ICP atomization and excitation, including collision and reaction cells. Understanding instrumental factors that govern electronic excitation and ionization of elements is critical to trouble-shooting and optimizing methods for risk assessment and quantitation of elemental impurities in pharmaceuticals. Chapters covering common interferences and method of quantitation follow. The book includes chapters covering sample digestion and preparation, which are critical for accurate and reliable assessment of elemental impurities in pharmaceutical components and finished products, as well as chapters on elemental speciation, which is often critical for analysis of dietary supplements. Analytical chemists who are tasked with developing spectrochemical methods for evaluation of elemental impurities in pharmaceutical will benefit considerably from this book, which provides detailed information on the key success factors for applying ICP-OES and ICP-MS to problems in pharmaceutical analysis.

John Kauffman, PhD, MBA

Preface

I cannot believe that it has been 14 years since I published the first edition of my textbook, *Practical Guide to Inductively Coupled Plasma Mass Spectrometry*, and 3 years since the third edition was launched at the Pittsburgh conference (PittCon) in 2014. What was originally intended as a series of tutorials on the basic principles of inductively coupled plasma mass spectrometry (ICP-MS) for *Spectroscopy* magazine in 2001 quickly grew into a textbook focusing on the practical benefits of the technique. With more than 5000 copies of the English version sold, and almost 3000 copies of the Chinese (Mandarin) book in print, I'm very much honored that the book has gained the reputation as being the reference book of choice for novices and beginners to the technique all over the world. Sales of the book have exceeded my wildest expectations. Of course, it helps when it is "recommended reading" for a PittCon ICP-MS short course that I teach every year on how to select an ICP-MS. It also helps when your book is displayed at 15 different vendors' booths at the PittCon every year. But there is no question in my mind that the major reason for its success has been that it presents ICP-MS in a way that is very easy to understand for beginners, and also shows the practical benefits of the technique for carrying out routine trace element analysis.

APPLICATION LANDSCAPE

During the time the book has been in print, the application landscape has slowly changed. When the technique was first commercialized in 1983, it was mainly the environmental and geological communities that were the first to realize its benefits. Then as its capabilities became better known, and its performance improved, other application areas, like clinical, toxicological, and semiconductor markets, embraced the technique. Today, ICP-MS, with all its performance and productivity enhancement tools, is the most dominant trace element technique, and besides the traditional application fields mentioned, it is now being utilized in other exciting areas, including nanoparticle research, oil exploration, and pharmaceutical manufacturing. As a result, with a street price tag for a single-quadrupole-based ICP-MS in the order of $100,000, it is being purchased for applications that were previously being carried out by inductively coupled plasma optical emission spectrometry (ICP-OES) and atomic absorption.

PHARMACEUTICAL ANALYSIS

One of the newer market segments that have realized the benefits of ICP-MS is the pharmaceutical and nutraceutical application areas. Recent regulations on heavy metals testing have required the industry to monitor a suite of elemental impurities in pharmaceutical raw materials, drug products, and dietary supplements. These guidelines, which are described in the new United States Pharmacopeia (USP), chapters <232>, <2232>, and <233>, are being implemented in January 2018. Similar methodology has also been approved by the International Conference on Harmonisation of Technical

Requirements for Registration of Pharmaceuticals for Human Use (ICH), a consortium of global pharmaceutical industries, including the European Pharmacopeia (Ph. Eu.) under the umbrella of the European Medicine Association (EMA), the Japanese Pharmacopeia (JP), and the USP through the Q3D, Step 4 guidelines, which has been applied to new drug products since June 2016 and will go into effect for over-the-counter (OTC) and existing prescription drugs in January 2018 (Step 5).

NEW REGULATIONS

These new directives have been driven by the fact that even though the risk factors for heavy metal contamination have changed dramatically, standard methods for their testing and control have changed little for more than 100 years, and as a result, most heavy metal limits have little basis in toxicology. This standard method, described in Chapter <231> of the USP's National Formulary, has been in existence since 1908. Unfortunately, the test involving precipitation of the metal sulfides using thioacetamide, an extremely toxic chemical, is also known to be unreliable and prone to errors. As a result, the USP has spent the past ten years investigating the use of plasma-based techniques for this analysis, particularly inductively coupled plasma spectrometry (ICP-MS), which has far greater specificity, is more sensitive, and is applicable to a wider range of metals of interest.

Based on these investigations, three brand new USP general chapters, <232>, <233>, and <2232>, were eventually approved in early 2013. Chapters <232> and <233> specify toxicity limits and analytical procedures for elemental impurities in pharmaceutical products, whereas Chapter <2232> deals with elemental contaminants in neutraceuticals and dietary supplements. Stakeholders will be required to fully implement these new USP directives and ICH Q3D step 4 guidelines in January 2018.

MY INVOLVEMENT WITH USP THROUGH THE ACS

In June 2016, the 11th edition of the American Chemical Society's (ACS) book *Specifications and Procedures for Reagent Chemicals* was officially published, which was the culmination of 7 years of work for 25 dedicated analytical chemists and their affiliated organizations, who volunteer their time and resources by serving on this ACS committee. This new compendium reflected new methodology, updated procedures, and more stringent specifications with regard to chemical reagents used for analytical testing purposes

It was through my involvement with the ACS Committee on Reagent Chemicals, which I've served on for the past 17 years, that I became involved with the USP. Because our book is used by the pharmaceutical community (and other similar organizations where analytical testing is carried out, such as ASTM and the U.S. Environmental Protection Agency), we have worked closely with the USP to ensure our methods on heavy metals testing align with theirs. In fact, the general notices and requirements in the USP National Formulary book of compendial standards state, "Unless otherwise specified, reagents conforming to the specifications set forth in the current edition of Reagent Chemicals published by the American Chemical Society (ACS) shall be used."

I took a lead on the elemental impurities task force and, in consultation with other committee members, wrote the new plasma-based spectrochemical methods in the ACS book, which has replaced a similar heavy metals test in USP <Chapter 231>. In addition, for the past 5 years I have been invited to give talks on the topic at workshops, expositions, and webinars, and also have taught a PittCon short course on the implementation of the new USP chapters on elemental impurities. For that reason, I have become very familiar with the elemental impurity demands of the pharmaceutical manufacturing community.

MY INCENTIVE FOR WRITING THE BOOK

It was mainly the feedback I was getting from people who were attending my workshops and short courses that gave me the incentive to write this textbook. It was clear that there was a great deal of misunderstanding and confusion regarding the new USP chapters. This was further complicated by the fact that these were people who were not familiar with atomic spectroscopic testing methods, so they did not have a solid grounding in trace element analysis and/or sample preparation techniques. For that reason, a large component of my short course addressed the fundamental principles of both ICP-OES and ICP-MS. It became very clear to me that these operators needed a great deal of support and hand-holding if they were going to fully understand these techniques and be able to generate high-quality data for the pharmaceutical materials they would be required to test. This became my major driving force to put together a reference book focusing on the analytical methodology, instrumental techniques, and sample preparation procedures used for testing elemental impurities in drug products.

It's also important to emphasize that manufacturing facilities of pharmaceutical and nutraceutical companies have never really been required to use plasma spectrochemical-based techniques before. ICP-MS in particular has been considered more applicable to the demands of the drug research and development process. For that reason, it has primarily been used by research and development groups who have analytical chemists with a high level of expertise. With the approval of these three USP methods and ICH guidelines, pharmaceutical production facilities will have to either invest in this new technology or send it out to a contract lab for testing. Unfortunately, if it's done in-house, they could be asking technician-level people to operate this equipment who have little or no experience with using sophisticated analytical instrumentation.

CONTRIBUTIONS

When I began the process of putting the book chapters together, I realized that I did not have the experience of working in or for the pharmaceutical industry, contract laboratory, or regulatory agency. My background is in atomic spectroscopic techniques, having worked in trace element analysis for more than 40 years—12 years as an analytical chemist in the metallurgical industry and 24 years in sales, marketing, application, and product development, for a manufacturer of ICP-MS instrumentation. In addition, for the past 15 years, I have run my own consulting company, specializing in trace element analysis and atomic spectroscopic techniques. Additionally, I have authored almost 100 technical papers and articles, including a 15-part tutorial series on ICP-MS, and published 3 ICP-MS textbooks. I am currently the editor and contributor to the Atomic Perspectives column in *Spectroscopy* magazine. So to give this book added credibility in the pharmaceutical community, I needed to get contributions from experts in this field. For that reason, I reached out to my colleagues who were experts in their respective disciplines and asked them to contribute to my book. I would therefore like to acknowledge the following people.

John Kauffman, PhD, MBA

Dr. Kauffman earned a BS in chemistry from the University of Oregon, Eugene, and a PhD in physical chemistry from the University of Illinois, Champaign-Urbana. He was professor of analytical chemistry at the University of Missouri, Columbia, from 1991 to 2004 and developed an internationally recognized research program studying fast chemical events with ultrafast laser spectroscopy. John joined the Food and Drug Administration (FDA) Division of Pharmaceutical Analysis in 2004 to evaluate applications of process analytical technologies, such as near-infrared spectroscopy and Raman chemical imaging, to pharmaceutical manufacturing. From 2007 to 2010, he led a team of scientists in the development of rapid spectroscopic screening technologies for pharmaceutical surveillance. He served as deputy director and director of the division for nearly 4 years, and directed a group of highly skilled analytical chemists who performed regulatory studies and analytical chemistry research to support pharmaceutical product quality.

As part of his research, John began evaluating elemental impurity levels in pharmaceuticals in 2004. He served as FDA liaison to the USP expert panel on elemental impurities (< 232> and < 233>) from 2008 until 2016. In 2010, John was appointed the FDA topic lead to the ICH Q3D Elemental Impurities Expert Working Group, and served as Step 3 rapporteur of the Expert Working Group from 2013 until its completion in fall of 2014. He was also rapporteur of the Q3D Implementation Working Group until the completion of its work in 2016. In November 2016, John joined 3M as the quality lab supervisor at its Columbia, Missouri, manufacturing facility. He has published more than 80 papers on his research and has lectured extensively on spectroscopy, chemometrics, and elemental impurities in pharmaceuticals.

I asked John to write the foreword to my book. His message is self-explanatory, but he talks about the implementation of these new directives and how there is a critical need for support material to help the pharmaceutical industry through this challenging period. And in particular to understand that a combination of analytical spectrochemistry and risk assessment approach, will be the key to managing the elemental impurity workload for pharmaceutical manufacturers. This message is consistent with what he has been saying in public forums for the past 7 years. I very much appreciate his complimentary words, and his endorsement will be invaluable to the credibility and success of my book.

Tony DeStefano, PhD

Tony comes from a 31-year career at Procter & Gamble, which focused primarily on analytical and bioanalytical chemistry. After retiring from Procter & Gamble, he spent 5 years at the USP, where he was senior vice president of General Chapters and Healthcare Quality Standards. His team was responsible for reviewing and updating existing general chapters and writing new ones, including the new elemental impurities chapters, <232> and <233>. He was also the USP representative to the ICH Q3D working group from its inception in 2009 until he left the USP in February 2013. Tony holds a doctorate from Cornell University, Ithaca, New York.

I've asked Tony to contribute a chapter on an overview of the new elemental impurities, because he spent 5 years at the USP during the critical time that the organization was developing the new general chapters and, with the establishment of the ICH Q3D working group, aligning USP thinking with that of the ICH Q3D guidelines and, in particular, setting the permissible daily exposure (PDE) levels for the different elements and drug delivery categories, based on up-to-date global toxicological data. In addition, Tony and other key USP staff, working through the USP expert committees and panels, and in conjunction with representatives from the European and Japanese Pharmacopeias through the Pharmacopeial Discussion Group, made the decision to separate the new elemental impurities' methodology into two separate chapters, Chapter <232>, "Elemental Impurities: Limits," and Chapter <233>, "Elemental Impurities: Procedures." Thus, ICH Q3D focused on the elements and their PDEs, while the decisions with regard to the measurement technologies were left to the individual pharmacopeias. Tony is currently consulting in the areas of bioanalytical, analytical, and compendial chemistry.

Jonathan Sims, CChem, MRSC

Jon is a consultant in trace element analysis as applied to the pharmaceutical industry. He previously worked as analytical manager at the pharmaceutical giant GlaxoSmithKline in the United Kingdom, where he helped develop and implement strategies for ICH Q3D, USP <232>, and pioneered automation with the use of inductively coupled plasma spectrometries. His experience in developing automated analytical methodologies and his depth of knowledge on the benefits of risk assessment have been especially helpful to me with my PittCon short courses to educate users on the best way to implement USP chapters <232> and <233> and ICH Q3D Step 4 guidelines.

In addition, his extensive experience in the good manufacturing practices environment is an invaluable resource to help users with their validation requirements. Jon has recently been hired by PerkinElmer Inc. as their technical advisor to enhance and support comprehensive solutions for the new USP regulations on elemental impurities in pharmaceutical materials and contaminants in dietary supplements. He holds a BSc (honors) in chemistry from Loughborough University, United Kingdom.

I asked Jon to contribute a chapter on risk assessment, which will be a critically important aspect for companies that have a good understanding of their raw materials and manufacturing processes. Rather than routinely testing drug products against a broad specification for elemental impurities, which may cause delays in product delivery, correct use of the risk assessment process will ensure that targeted and appropriate testing of materials is performed where control is needed, to ensure that the control strategy is appropriate and does not impact the product quality or patient safety. Also, in preparation for inspection by the FDA, or your local regulatory agency if outside the United States, Jon has written a chapter on inspection readiness, which offers tips on how best to prepare and present your elemental impurity data for inspection.

Maura Rury, PhD
Maura is a Portfolio Marketing Manager for organic products in the Discovery and Analytical Solutions Division of PerkinElmer Inc. In this role, Maura is responsible for developing marketing strategy to support workflow solutions for applications in environmental health and life science market segments. Her prior roles include product management and technical applications at Thermo Fisher Scientific, supporting atomic absorption and inductively coupled plasma-based instruments for a wide range of applications. This provided her with a significant amount of experience with atomic spectroscopy applications in specific markets including: food and beverage, pharmaceutical, environmental, industrial, and clinical. Maura holds a PhD in analytical chemistry from the University of Massachusetts, Amherst, under the tutelage of professors Julian Tyson and Ramon Barnes.

I asked Maura to put together a chapter on ICP-OES, where she has extensive knowledge in the fundamental principles, as well as developing robust methods for various applications of the technique. She has published widely on atomic spectroscopic applications and has contributed a number of articles on plasma spectrochemistry to my Atomic Perspectives column in Spectroscopy magazine. Maura was also responsible for coordinating a six-city pharmaceutical workshop in the spring of 2017, where I was one of the guest speakers. This was a hands-on, 2-day workshop that put users in front of the appropriate plasma spectrochemical techniques, sample preparation equipment, and productivity enhancement tools to give them a better understanding of what is required to follow the new USP/ICH guidelines.

SO HOW SHOULD THE BOOK BE USED?

Let me say up front that this is not a reference book to explain all the historical and background information about testing for elemental impurities in the pharmaceutical industry. There are enough references in the public domain, which talk about the history of heavy metals testing in the pharmaceutical industry over the past 100 years, including the process of replacing the old sulfide-based heavy metals test with the new USP and ICH guidelines. However, Chapter 2, written by Tony Destefano is a really good overview of the past 20 years, with lots of references to explain how we got to this point today and how the new methods are guiding elemental impurities regulations for the global pharmaceutical industry.

My objective for this book was to give analytical chemists and technicians in the pharmaceutical manufacturing and quality control and assurance communities a better understanding of the plasma-based analytical techniques that are being recommended to carry out the determination of elemental impurities. The book primarily focuses on USP Chapter <232> and ICH Q3D Step 4 guidelines and, to a lesser extent, on Chapter <2232>, because both chapters refer to the use of Chapter <233> to monitor the elemental impurity PDE levels.

My goal is that it should also leverage the success of my previous atomic spectroscopy publications and textbooks, including the fundamental principles and practical benefits of both ICP-OES and ICP-MS, in a reader-friendly format that a novice in the pharmaceutical and nutraceutical industries will find easy to understand. It walks a user through both techniques, including easy-to-read chapters on hardware components, calibration and measurement protocols, typical interferences, routine maintenance, and troubleshooting procedures. These chapters should give the operator the tools to develop robust and bulletproof methods for pharmaceutical-type samples, and also help them to optimize sample preparation and instrumental method development to improve detection capability for the various drug delivery categories. And if plasma spectrochemical techniques alone are not suitable for the elemental impurities or drug delivery method, there is a detailed chapter on sampling accessories to improve performance and maximize productivity. The user should find this chapter very useful, particularly if there is a requirement to determine speciated forms of arsenic or mercury in the drug compound.

As mentioned previously, there are also excellent contributions on risk assessment, which very much could be a viable approach for pharmaceutical companies that have a good handle on their sources of raw material and manufacturing processes. On a similar theme, to help pharma companies prepare for compliance, we include a brief chapter called "Regulatory Inspection Readiness," which talks about how best to present your data in preparation for regulatory inspection.

I'm hoping that the pharma community will also find the other chapters useful, including the first chapter of the book, about my work on the ACS Reagent Chemicals Committee and, in particular, the process for the alignment of ACS plasma-based heavy metals testing with the new USP chapters and ICH guidelines for the determination of elemental impurities. The process of selecting the optimum spectrochemical technique for reagent chemicals in this chapter then lines up with a chapter later in the book by comparing the four major atomic spectroscopic techniques for the determination of elemental impurities in drug products and materials. And on a similar theme, I have written a chapter on presenting a set of evaluation guidelines to pharmaceutical companies that are thinking about investing in ICP-MS instrumentation.

There is no question that the new USP directives and ICH guidelines represent a wonderful opportunity to put together a focused book on this topic of determination of elemental impurities in pharmaceutical raw materials and drug compounds, which will hopefully be a very useful training tool, not only for novices and inexperienced users of plasma spectrochemistry, but also for supervisors and senior management who want to better understand the analytical issues. So, let me end my introduction by wishing you the best of luck with your laboratory endeavors. I'm hoping my book will be placed next to the analytical instrumentation to use as a reference source for all your colleagues—or better still, convince them that they should have their own copy!

Thanks for your support.

Robert Thomas, CSci, CChem, FRSC
Scientific Solutions: Serving the Needs of the Trace Element Analysis User Community

Acknowledgment

I would like to take this opportunity once again to thank some of the people and organizations that have helped me put this book together. First, I would like to thank Dr. Ramon Barnes, professor emeritus of chemistry at the University of Massachusetts and director of the Research Institute for Analytical Chemistry in Amherst, Massachusetts, and the driving force behind the winter conference on plasma spectrochemistry. He has written the foreword of all three editions of my book and has been very active in promoting it at the conference every other year in the United States. His endorsement has been invaluable, and there's no question it has helped in the success of the book around the world.

Second, I would also like to thank the editorial staff of *Spectroscopy* magazine over the past 15+ years, who gave me the opportunity to write a monthly tutorial on ICP-MS back in 2001. This was most definitely the spark I needed to start my first ICP-MS book project in 2004, and subsequently two more in 2009 and 2013. They also allowed me to use many of the figures from the series, together with material from other ICP-MS articles I wrote for the magazine. In particular, I'd like to mention the current editorial team, including Laura Bush, editorial director; Meg L'Heureux, managing editor; and Steve Brown, technical editor. My close working relationship with them over the past 5+ years has led to my quarterly Atomic Perspectives column on the fundamentals and applications of atomic spectroscopy.

Third, I would like to thank all the manufacturers of atomic spectroscopy instrumentation, ancillary equipment, sampling accessories, consumables, calibration standards, chemical reagents, and high-purity gases, who supplied me with the information, data, drawings, figures and schematics, and so forth, and particularly their willingness over the past 15 years to display my *Practical Guide to ICP-MS* at their Pittsburgh conference exhibition booths. There are way too many people to name, but at last year's exposition in Chicago, we had 17 vendors showing the book. This alone has made a huge difference to the visibility of the book, and its success would not have been possible without their help. In fact, one vendor purchased 100 copies of the book, which they are using as a training tool for their customers. We are hoping more vendors will follow suit with my new book, particularly as there will be many new and novice users of plasma spectrochemical techniques in the pharmaceutical and nutraceutical industries.

Finally, I would also like to thank the four contributors to my book, John Kauffman, Tony DeStefano, Jonathan Sims, and Maura Rury, who are all experts in their own fields of work. When I began the process of putting the book together, I realized that even though I had expertise in elemental analysis and plasma spectrochemistry, I did not have experience working in the pharmaceutical industry. For that reason, I reached out to them for contributions to my book. You will find more information about how they contributed, together with their professions backgrounds, in the book's preface.

Author

Robert J. Thomas has worked in the field of trace element analysis for more than 40 years, including 24 years for an inductively coupled plasma mass spectrometry manufacturer and 15 years as a principal of his own consulting company. He has served on the American Chemical Society's Reagent Chemical Committee for the past 17 years as leader of the elemental impurities task force, where he has worked closely with the United States Pharmacopeia to align heavy metals testing procedures in reagent chemicals with those of pharmaceutical materials. He has authored almost 100 publications on trace element analysis and written 3 textbooks on ICP-MS and related topics, including this new book, which focuses on the new global directives on elemental impurities in pharmaceutical materials. He is currently the editor of and a frequent contributor to the Atomic Perspectives column in *Spectroscopy* magazine. He has an advanced degree in analytical chemistry from the University of Wales, Cardiff, and is a Fellow of the Royal Society of Chemistry and a chartered chemist.

1 Testing for Heavy Metals
An ACS Perspective

My relationship with the United States Pharmacopeia (USP) and the topic of heavy metals testing began when I got involved with the American Chemical Society's (ACS) Committee on Analytical Reagents (CAR) in 2001. This led to me working closely with the USP as the ACS began the process of aligning its elemental impurities testing procedures in ACS-grade reagent chemicals using plasma spectrochemical techniques with those of USP chapters <232> and <233>. There's no doubt that my work with the other dedicated committee members to publish three editions of the ACS's *Reagent Chemicals: Specifications and Procedures for Reagents and Standard-Grade Reference Materials* over the past 15 years gave me the motivation to compile this textbook. For that reason, I would like to begin my book by giving an overview of the work we carried out on the committee, which should give the reader a better understanding of the role of *Reagent Chemicals* for general laboratory testing procedures, and discuss some of the updated methods being adopted. In particular, I focus on the process of replacing the old ACS sulfide precipitation heavy metals test for reagent chemicals, and discuss some of the important decisions we had to make to implement plasma-based spectrochemical methodologies in the 11th edition of the ACS book.

INTRODUCTION

In June 2016, the 11th edition of the ACS's *Reagent Chemicals* was officially published, which was the culmination of 7 years' work for 25 dedicated analytical chemists and their affiliated organizations, who volunteer their time and resources by serving on this ACS committee.[1] This new compendium reflected new methodology, updated procedures, and more stringent specifications with regard to chemical reagents used for analytical testing purposes.

The process of updating procedures and setting new specifications is time-consuming and can sometimes take years to complete. By carrying out various testing protocols, including sample preparation and spike recovery procedures, it ensures that new methods are rugged and robust and will stand up to scrutiny, wherever the book is used around the world. Many standards organizations and federal agencies that set guidelines, specifications, and/or analytical testing methods—including the USP[2] and the U.S. Environmental Protection Agency (EPA)[3] —require the use of ACS-grade reagent chemicals in many of their test procedures. For that reason, it has become the *de facto* reference book worldwide for the chemicals used in high-purity laboratory applications. This chapter gives an overview of the new book, and in particular focuses on the new plasma spectrochemical-based methodology for carrying out the determination of a suite of heavy metals in reagent chemicals.

WHAT IS A REAGENT CHEMICAL?

The specifications in the ACS book are intended to serve for reagent chemicals and standard-grade reference materials to be used in precise analytical work of a general nature. The term *reagent-grade chemical* implies that it is a substance of sufficient purity to be used in most chemical analyses or reactions. Standard-grade reference materials are suitable for the preparation of analytical standards used for a variety of applications, including instrument calibration, quality control, analyte identification, method performance, and other applications requiring high-purity materials. It is recognized that there may be special uses for reagents and standard-grade reference materials, which may need to conform to more rigorous specifications. Therefore, where necessary, some of the specifications include requirements and tests for certain specialized uses. However, it is impossible to include specifications for all such uses, and thus there may be occasions when it will be necessary for the analyst to further purify reagents known to have special purity requirements for certain uses.

HISTORICAL PERSPECTIVE

There is a very proud tradition in the ACS CAR, which dates back more than 100 years. It evolved from the Committee on the Purity of Chemical Reagents, which was established in 1903. Analysts at that time were concerned with the quality of reagents available and by the discrepancies between labels and the actual purity of the materials. The committee's role in resolving these issues expanded rapidly after its 1921 publication of specifications for ammonium hydroxide and for hydrochloric, nitric, and sulfuric acids. Specifications appeared initially in *Industrial & Engineering Chemistry*, and later in its Analytical Edition. In 1941, the existing specifications were reprinted in a single pamphlet. Revisions and new specifications were later gathered into a book, the 1950 edition of *Reagent Chemicals*, and new editions appeared regularly thereafter.

The commonplace introduction of instrumentation into analytical laboratories, beginning in the late 1950s, resulted in dramatic improvements in the sensitivity and accuracy of analytical measurements. As a result, the specifications for reagent chemicals and the tests measuring their purity were improved so that the methods would be as accurate and cost-effective as possible. In many cases, these improved methods were able to detect impurities previously found to be absent. The eighth edition, which became official in 1993, substantially changed and updated the general procedures and attempted to make the book easier to read. The ninth edition, which became official in 2001, eliminated and simplified some of the tedious classical procedures for trace analysis and added instrumental methods where possible. The 10th edition continued the trend of removing and replacing more of these outdated methods of analysis, and included the use of plasma-based spectrochemical techniques for the very first time.[4]

ELEVENTH EDITION

With the publication of the 11th edition, some of the many highlights include all the additional "supplements" posted online since the publication of the 10th edition in 2009, removal of some obsolete test methods, clearer instructions for many of the existing ones, and the introduction of many new methods. Overall, the safety, accuracy, and ease of use in specifications for approximately 70 of the 430 listed reagents have been improved, and 7 new reagents have been added. There are also numerous minor changes, such as incorporation of the International Union of Pure and Applied Chemistry (IUPAC), recalculation and redefinition of some atomic weights in 2011. In particular, there are several changes and additions worth noting:

- The old heavy metal sulfide precipitation test method is still valid, but has been replaced with inductively coupled plasma optical emission spectroscopy (ICP-OES) for more than 50 reagent chemicals.
- The use of inductively coupled plasma mass spectroscopy (ICP-MS) is highly recommended for analyzing many of the ultra-high-purity acids and chemicals.
- Atomic absorption (AA) techniques are replaced with plasma-based ones, as long as method validation is carried out.
- Polarography for measuring carbonyl impurities is replaced with gas chromatography–mass spectrometry (GC-MS).
- New methodologies introduced for the first time include liquid chromatography–mass spectrometry (LC-MS) and headspace gas chromatography (HS-GC) analysis.
- The green chemistry initiative is continued, which incorporates fewer toxic chemicals in existing test methodologies.

Previous improvements from the 10th edition were intended to make the book easier to use, including a CAS number index, a separate index for the standard-grade reference materials, complete assay calculations with titer values, an updated table of atomic weights, frequently used mathematical equations, a quick reference page on how to read a monograph, division of the book into parts, and a detailed table of contents for each section. An example of a typical reagent chemical monograph from the 11th edition is shown in Figure 1.1.

Each specification is then accompanied with a detailed description of the tests required to carry out the measurement of each of the analytes listed in the monograph. Where more information is required, the monograph directs the reader to the relevant page in the book describing the test in greater detail. An example of this would be replacement of the sulfide precipitation colorimetric test for heavy metals with a brand new section on optical plasma spectrochemistry.

HEAVY METALS TEST

One of the most significant new methods in the 11th edition has been the replacement of the old sulfide precipitation colorimetric method for heavy metals in reagent chemicals. This test is very similar to the method described in the USP and National

Magnesium nitrate hexahydrate

$Mg(NO_3)_2 \cdot 6H_2O$ Formula Wt 256.40 CAS No. 13446-18-9

General description

Typical appearance: Colorless or white solid

Applications: Preparation of standards

Change in state (approximate): Melting point, 89°C

Aqueous solubility: 125 g in 100 mL at 25°C

Specifications

Specification	
Assay	98.0–102.0% $Mg(NO_3)_2 \cdot 6H_2O$
pH of a 5% solution at 25.0°C	5.0–8.2
	Maximum allowable
Insoluble matter	0.005%
Chloride (Cl)	0.001%
Phosphate (PO_4)	5 ppm
Sulfate (SO_4)	0.005%
Ammonium (NH_4)	0.003%
Barium (Ba)	0.005%
Calcium (Ca)	0.01%
Manganese (Mn)	5 ppm
Potassium (K)	0.005%
Sodium (Na)	0.005%
Strontium (Sr)	0.005%
Heavy metals (by ICP–OES)	5 ppm
Iron (Fe)	5 ppm

FIGURE 1.1 Typical specification from the 11th edition of the ACS's *Reagent Chemicals.*

Formulary (NF) General Chapter <231>, the 100-year-old test for heavy metals in pharmaceutical materials,[5] which is a colorimetric test based on precipitation of the metal sulfide in a sample, and comparing it to a lead standard. This USP method is being replaced by two new chapters: Chapter <232>, which defines the elemental impurity limits,[6] and Chapter <233>, which describes the plasma-based analytical procedures and validation protocols to meet those limits[7] The ACS Committee on Reagent Chemicals had many discussions and meetings with USP personnel during the drafting and writing of the new plasma-based methods. So let's take a closer look at the traditional ACS sulfide precipitation method and the new plasma spectrochemical techniques that are replacing it.

SULFIDE COLORIMETRIC AND PRECIPITATION PROCEDURE

Colorimetric analytical methods have been used in analytical chemistry for more than 100 years and are based on measuring color changes of solutions that arise from specific chemical interactions with the analyte elements. The traditional colorimetric test for heavy metals in chemical reagents or pharmaceutical matrices is based on a chemical reaction of the heavy metal with a sulfide solution, and compared with a standard prepared from a stock lead solution. It relies on the ability of heavy elements, such as lead, mercury, bismuth, arsenic, antimony, tin, cadmium, silver, copper, and molybdenum, to react with the sulfide at a pH of 3–4 to produce a precipitate of the metallic sulfide that is then compared with a lead standard solution. It is used to demonstrate that the metallic impurities colored by sulfide ions under the specific test conditions do not exceed a predefined limit.

One of the many drawbacks of this approach is the assumption that formation of the sulfides in the sample is very similar to the formation of the lead standard solution and is not affected by the sample matrix. However, since many metals behave very differently, the method requires that visual comparison be performed very quickly after the precipitate has formed. Unfortunately, analysts can differ in their interpretation of the color change, so different analysts may not consistently read the sample and standard solutions correctly each time.

Another limitation of the technique is that the sample preparation procedure for some chemicals involves ashing at high temperature and acid dissolution of the sample residue, which is prone to sample losses, particularly for volatile elements like mercury. The loss of metals is also matrix dependent, and because the procedures are time-consuming and labor-intensive, recoveries can vary significantly among different analysts.

For all these reasons, the 11th edition of the book has started the process of replacing the sulfide precipitation test for heavy metals in a number of reagent chemicals with ICP-OES. The first suite will include those where the maximum allowable concentrations are at an appropriate level for the technique and where the spectral and matrix-induced interferences are well understood and easily compensated. In addition, the use of ICP-MS is encouraged for the analysis of ultratrace reagents and acids where ICP-OES is not suitable based on its inferior detection capability. However, it is important to emphasize that the use of the older, more traditional techniques will still be allowed for many chemicals, because in many parts of the world, modern analytical techniques are either not available or can be extremely cost-prohibitive to purchase.

WHICH PLASMA-BASED SPECTROCHEMICAL TECHNIQUE IS THE BEST ONE TO USE?

OES using ICP as the excitation source is widely recognized as an extremely rapid technique to carry out multielement analysis in various sample matrices.[8] Whereas AA is used to determine small numbers of elements in solution, ICP-OES is utilized for multielement determinations or when high sample throughput is required. This technique is applicable to determine ultratrace elemental levels in many of the reagent chemicals specified in the book, including Ag, As, Bi, Cd, Cu, Hg, Mo, Pb, Sb, and Sn, and all the elements that can be determined by AA (Ca, Ba, Na, K, Li, Mn, and Fe).The first phase of the changeover to the plasma emission technique, which has occurred in the 11th edition of the book, will lead to the eventual elimination of both the sulfide precipitation method and AA spectrometry.

In order to achieve the specifications, many of the reagents can simply be dissolved with deionized water, acidified with 5% nitric acid, made up to volume, and aspirated directly into the instrument's sample introduction system. However, for analytical reagents that are more difficult to get into solution, an acid dissolution procedure using hot plate or microwave digestion might be required. For extremely low ultratrace specifications, such as in high-purity mineral acids, the samples may need to be preconcentrated by evaporation under clean air conditions and made up to volume using 5% nitric acid, prior to analysis by ICP-OES, or analyzed directly by ICP-MS. Additionally, if the matrix of the reagent chemical is spectrally complex, such as iron, chromium, tungsten, and molybdenum salts, interference-free analyte emission lines may be difficult to find. If this is the case, the use of ICP-MS is strongly recommended, as long as the method protocol can be validated.[9]

It's also important to point out that in the ICP-based ACS procedure, specific wavelengths are suggested, because of the strong possibility that the reagent chemical matrix will produce a spectral- or matrix-derived spectral interference on one or more of the suite of analyte elements. This is very different from the USP methodology for elemental impurities in pharmaceutical and nutraceutical materials, which provides generic procedures and relies on a comprehensive set of validation protocols to ensure the quality of the data.

OPTIMUM TECHNIQUE BASED ON DETECTION LIMITS, SENSITIVITY, AND LINEAR DYNAMIC RANGE

To decide which technique is most suitable, limits of detection, sensitivity, and linear dynamic range should be investigated and established for each individual analyte line or wavelength on the particular instrument being used. A list of typical detection limits of all the atomic spectroscopy (AS) techniques is given in the book to make an assessment of whether the technique of choice is suitable for the element being determined at the specification defined in the particular reagent chemical being tested. Table 1.1 gives the AS detection limits in parts per billion (ppb) of the most common heavy metals.[10]

The required detection capability will also depend on the sample preparation technique used. For example, the weight of the sample and the final volume will impact the concentration of the heavy metals in solution, as shown in Table 1.2.

TABLE 1.1
AS Detection Limits (ppb) of the Most Common Heavy Metals

Element	FAAS	CVAA/HGAA	ETA	ICP-OES	ICP-MS
Ag	1.5	Not suitable	0.005	0.6	0.00009
As	150	0.03	0.05	1	0.0004
Bi	30	0.03	0.05	1	0.00002
Cd	0.8	Not suitable	0.002	0.1	0.00007
Cu	1.5	Not suitable	0.014	0.4	0.0002
Hg	300	0.009	0.6	1	0.001
Mo	45	Not suitable	0.03	0.5	0.00008
Pb	15	Not suitable	0.05	1	0.00004
Sb	45	0.15	0.05	2	0.0002
Sn	150	Not suitable	0.1	2	0.0002

Note: FAA = flame atomic absorption, HGAA = hydride generation atomic absorption, ETA = electrothermal atomization.

TABLE 1.2
Elemental Concentrations in Solution Based on the Specification and the Sample Preparation Used to Get the Sample into Solution

	Sample Weight/Final Volume			
Specification of heavy metals (by ICP-OES)	0.5 g/100 mL	1.0 g/100 mL	2.0 g/100 mL	5.0 g/100 mL
	Elemental Concentration in Solution (ppm)			
1 ppm (0.0001%)	0.005	0.01	0.02	0.05
5 ppm (0.0005%)	0.025	0.05	0.10	0.25
10 ppm (0.001%)	0.050	0.10	0.20	0.50
100 ppm (0.01%)	0.50	1.0	2.0	5.0
1000 ppm (0.10%)	1.00	10.0	20.0	50.0

Consideration should also be given to the linear dynamic range of the particular wavelength being selected for the analysis. Although this will vary slightly depending on the type of instrument and viewing configuration used, most ICP-OES systems on the market are capable of achieving linear dynamic ranges of approximately five orders of magnitude. The book also emphasizes that the manufacturer's application-specific methodology should be followed and, if needed, with reference literature in the public domain for sensitivity, detection limits, and linear dynamic range of the emission lines being used for the analysis.

APPROACHES TO REDUCING INTERFERENCES

In general, plasma-based spectrochemical techniques are mostly prone to spectral-type overlaps and matrix suppression effects produced by other matrix components in the sample. These types of interferences must be fully investigated in the method development stage, before accurate multielement determinations can be carried out. Special attention must also be paid to the possibility of external contamination during sample preparation and analysis, especially if ultratrace determinations are being carried out. These external contamination sources can not only affect the accuracy of analyte recoveries, but also impact the detection capability by generating unforeseen spectral and matrix-induced interferences.

The use of an internal standard is usually required, as it compensates for the physical and matrix interferences resulting from differences in sample transport efficiencies and/or excitation conditions between the sample matrix and calibration solutions. The choice of internal standard should be left to the analyst, as the optimum selection will depend on a number of different factors. However, it is very important that the internal standard element chosen is not present in the samples. It is also desirable that the excitation or ionization potential of the internal standard is similar to that of the analytes.

Shifts in background intensity levels from recombination effects and/or molecular band spectral contributions may be corrected by the use of an appropriate background correction technique. For example, in ICP-OES, direct spectral overlaps are best addressed by selecting alternative wavelengths, whereas in ICP-MS, mathematical corrections are often used, because suitable alternative masses are not always easy to find. For these reasons, spectral interference studies should be conducted on all new matrices to determine whether interference correction factors should be applied to concentrations obtained from certain spectral line intensities to minimize biases. It is therefore highly recommended that the spectral areas around the analyte emission lines should be examined to assess whether background correction techniques are needed, or even if alternative analyte lines or masses should be selected. For example, if a fixed-channel, polychromator-type spectrometer is being used for the analysis by ICP-OES, it is recommended that interelement correction factors be applied, based on the expected interfering elements present in the samples. The user may also choose multiple wavelengths if the spectrometer has the capability, to help verify that emission line selection is optimized for the particular reagent chemical being analyzed. If the spectral complexity is too difficult to overcome using ICP-OES, ICP-MS is probably a better approach to use, as there are less direct overlaps to compensate for and many of them can now be alleviated using collision/reaction cell technology.

CALIBRATION

Multielement standard solutions should be prepared on the day of analysis by dilution of a suitable stock standard solution. They should be matrix matched to the sample solution, which is typically 5% nitric acid (HNO_3). It's also important that all stock solutions should be traceable to an internationally recognized standards organization, such as the National Institute of Standards and Technology (NIST). The elemental components should be selected based on the analytical requirements and the chemistry of the individual elements to provide stable solutions for the required concentration ranges. Analysis of the samples should be carried out using matrix-matched calibration standards and a blank. The number and range of calibration standards should be selected based on the reagent specification and the sample preparation dilution factor. However, two standards—one at the lower end and one at the higher end of the calibration range—and a reagent blank are fairly typical to use.

INSTRUMENTATION

No specific procedures are given in the book for the choice of what instrumentation to use except for recommended emission lines, since operating conditions will vary depending on instrument design, or technique being used. However, whatever instrument is being used, an ICP-OES with sufficient resolution to separate the relevant analyte emission lines of interest is required for this analysis. It is important that the optimum plasma conditions and sample introduction parameters are used for samples being analyzed. To ensure that the instrument is achieving adequate sensitivity, the user should also check its performance specifications on a routine basis. If in doubt, refer to the instrument operator's manual for guidance. The instrument should have the ability to monitor one or more wavelengths for each analyte to ensure the optimum one is being used.

MEASUREMENT PROCEDURE AND DATA INTEGRITY

To measure the heavy metal impurities in the reagent chemical, the weight of sample suggested in the individual monograph is dissolved in approximately 50 mL of water in a 100 mL volumetric flask. The sample is then acidified with 5 mL of high-purity nitric acid. If mercury is one of the analytes, add 0.1 mL of 1000 ppm gold standard solution, and dilute to 100 mL with water, giving a final concentration of 1 ppm gold in solution. It is well recognized that the addition of low concentrations of gold to form a stable amalgam is important in the determination of mercury. However, it is strongly recommended that samples and standards be analyzed immediately after preparation to optimize the stabilization process. For more information, refer to the EPA document on mercury preservation techniques.[11] It's also worth mentioning that methodology for determining mercury by the cold vapor atomic absorption (CVAA) technique is also given in the book. So if lower levels of mercury need to be quantified, the ICP-OES method can be replaced by either a dedicated mercury analyzer or a mercury generation system coupled to a stand-alone AA instrument.

To ensure the integrity of the ICP-OES data, the sample is divided into two. One aliquot is analyzed by ICP-OES for Ag, As, Bi, Cd, Cu, Hg, Mo, Pb, Sb, and Sn. The second aliquot is spiked with an appropriate concentration of a suitable multielemental standard and analyzed. The spike recoveries are calculated based on the results from solutions with and without the spike and a 5% nitric acid blank, and results reported for each element in the unspiked aliquot, together with a method detection limit, based on statistical analysis of replicate measurements of the blank. The data is considered acceptable if spike recoveries of 75%–125% are achieved. The sum of the results of the individual analytes should be less than the heavy metals specification defined in the monograph.

FINAL THOUGHTS

The 11th edition of the ACS's *Reagent Chemicals* represents the transition occurring in the technical book business where print editions are being enhanced or in some cases replaced by digital forms. There is no question that it has been the most difficult to upgrade and improve in the modern era. Validation of methods according to current best practices has dramatically increased the method development time and cost, which has created new challenges for all the volunteer committee members to accomplish this important work in a timely manner.

With regard to heavy metals testing, this current edition of the book has focused more on ICP-OES, because it is by far the most common trace multielement technique being used by laboratories that utilize ACS-grade reagent chemicals for testing purposes. Future editions of the book will expand its use to include more and more chemicals and analytes, where appropriate. However, this could change as ICP-MS becomes less expensive, and as a result will probably be used by more and more routine labs, and new users will get more comfortable with operating the technique. As this happens, future editions of *Reagent Chemicals* will reflect the growing trend in the acceptance of ICP-MS and update its procedures accordingly.

ACKNOWLEDGMENTS

The author would like to acknowledge the assistance and guidance of members of the ACS Committee on Reagent Chemicals in writing this chapter.

REFERENCES

1. American Chemical Society. *Reagent Chemicals: Specifications and Procedures for Reagents and Standard-Grade Reference Materials.* 11th ed. Oxford: Oxford University Press, 2016.
2. United States Pharmacopeia (USP) and the National Formulary (NF) online (USP-NF). http://www.usp.org/usp-nf.
3. Environmental Protection Agency. SW-846: Methods, 6020B, Update V—Revision 2, Section 7.1, Reagents and Standards. July 2014. http://www3.epa.gov/epawaste/hazard/testmethods/sw846/pdfs/6020b.pdf
4. American Chemical Society. *Reagent Chemicals: Specifications and Procedures for Reagents and Standard-Grade Reference Materials.* 10th ed. Oxford: Oxford University Press, 2006.
5. United States Pharmacopeia. General Chapter <231>: Heavy metals test in USP National Formulary (NF): Second supplement to USP 38–NF 33. December 2015. http://www.usp.org/usp-nf/notices/general-chapter-heavy-metals-and-affected-monographs-and-general-chapters.
6. United States Pharmacopeia. General Chapter <232>: Elemental impurities in pharmaceutical materials—Limits: Second supplement to USP 38–NF 33. December 2015. http://www.usp.org/chemical-medicines/key-issues-elemental-impurities.
7. United States Pharmacopeia. General Chapter <233>: Elemental impurities in pharmaceutical materials—Procedures: Second supplement to USP 38–NF 33. December 2015. http://www.usp.org/chemical-medicines/key-issues-elemental-impurities.
8. J. Nolte. *ICP Emission Spectrometry: A Practical Guide.* Weinheim, Germany: Wiley-VCH, 2003.
9. R. Thomas. *Practical Guide to ICP-MS: A Tutorial for Beginners.* Boca Raton, FL: CRC Press, 2014.
10. PerkinElmer Inc. *Atomic Spectroscopy—A Guide to Selecting the Appropriate Technique and System.* Waltham, MA: PerkinElmer, 2013.
11. Environmental Protection Agency. Mercury preservation techniques. https://www.inorganicventures.com/sites/default/files/mercury_preservation_techniques.pdf.

2 Elemental Impurities in Pharmaceuticals

An Overview

Heavy metal (now elemental impurity) contamination of pharmaceutical components (i.e., drug substances, excipients, and drug products) has been a cause for concern for the pharmaceutical industry for many years. The concern was sufficiently great in the early part of the twentieth century that a test for heavy metals was introduced into the United States Pharmacopeia (USP) for the first time in 1908. While the pharmaceutical industry has come a long way in the century since 1908, until recently the same could not be said for heavy metals testing, since the same basic testing USP General Chapter <231>, "Heavy Metals," remained essentially unchallenged as the compendial test for heavy metals. Further, the test had global support, since it was harmonized across the International Conference on Harmonisation of Technical Requirements for Registration of Pharmaceuticals for Human Use (ICH) regions (United States, Europe, and Japan) and accepted as the test by which heavy metals were to be determined. Much has changed in the way the pharmaceutical industry has come to think about measuring heavy metals in its components and products in the time since the early 2000s. This chapter, compiled by Tony DeStefano (former vice president of General Chapters at USP—see his biography in the Preface), reviews some of the history of heavy metals testing and discusses ways forward as the thinking about heavy metal contamination and testing for it has evolved in light of modern analytical technology and a risk-based approach for assessing the impact of heavy metals on patient safety.

BACKGROUND

Heavy metals contamination was a serious concern in the early part of the twentieth century. Lead, for example, was an issue since it was used in the production of solder. Without the level of good manufacturing practice control that is present today, iron and other metals from manufacturing systems were potential contaminants. In addition, mined excipients that carried with them elements such as lead, cadmium, mercury, and arsenic posed potential contamination issues. For parenteral products, containers and container closure systems posed an additional risk. In total, there was, for many formulations, a considerable risk of heavy metal contamination, and far fewer ways to test for and control the levels of impurities than we have available today.

As those who have made it through a laboratory course in qualitative inorganic chemistry are painfully aware, the first step in the identification of an inorganic element is typically its precipitation from solution as its sulfide (e.g., PbS and HgS). This concept formed the basis for the testing in USP General Chapter <231>. In principle, the metals in solution are precipitated as their respective sulfides, and the intensity of the dark-colored precipitate is visually compared against that of a precipitate of lead sulfide prepared from a lead salt solution with a concentration at the level of the allowed specification. Lead sulfide made the best comparator since its precipitation from solution was quantitative and it produced a black precipitate. In addition, lead was a known human toxicant and assumed to be the most toxic of the metals likely to be present, so comparison with this standard as the maximum permitted for the sum of all the heavy metals present seemed reasonable.

Unfortunately, the use of <231> testing as a practical way to accurately assess the levels of heavy metals in pharmaceutical drug substances, excipients, and drug products is confounded by a number of practical issues.

- Its utility was limited to elements that quantitatively form dark sulfide precipitates from a solution of their salts. As described in <231>, these elements include lead, mercury, bismuth, arsenic, antimony, tin, cadmium, silver, copper, and molybdenum. Components of stainless steel, such as chromium and nickel, are not included, nor are common catalysts such as palladium or platinum.
- It is a screening test. It is not element specific, and cannot tell one element from another if more than one is present in the precipitate.
- Sensitivity varies by element, since not all of the elements form precipitates as dark as that of the lead sulfide comparator, and since the colloidal nature of the precipitates for each element varies.
- Large sample sizes are needed. The <231> tests typically need 1–2 g of sample, making the test impractical in some cases.
- There are safety issues with the reagents. The sulfide precipitation of the metals requires generation of the sulfide ion in solution. In the USP chapter, sulfide ion is generated *in situ* using the combination of thioacetamide and glycerin base. This procedure generates hydrogen sulfide, which is toxic and must be handled carefully. In addition, thioacetamide is a Class 2B carcinogen.
- Reproducibility is quite difficult and varies by analyst. There is an assumption made that the mechanism and form of the precipitates of all the elements are essentially the same as that of lead sulfide. However, many of the elements of interest form colloids, so the formations and forms of the precipitates are dependent on the nature of the sample matrix and the time between precipitate formation and visualization. In addition, the comparison is in part dependent on the visual acuity and expertise of the analyst, and it is difficult for an analyst to accurately and consistently compare the sample and standard solutions every time.
- The test relies on being able to get the metals of interest into solution so that they can be precipitated as the sulfide salt. Many excipients are not able to be dissolved in water, making USP <231> Method I impractical. Method II, the most commonly used, relies on ashing at 500°C–600°C and acid

dissolution of the sample residue. This procedure is very prone to losses, from both the nature of the procedures and the nature of the matrices, which make reproducibility across analysts difficult.

- Finally, this is a visual acuity test, whose limits were based on what an analyst could reasonably be expected to see (typically corresponding to 10–20 ppm of lead in solution). The levels were not toxicologically based and do not take into account the toxicity of any individual element.

Although the problems with USP <231> were well known for many years, the test remained the gold standard for heavy metals testing for decades. In large part, this was due to the lack of analytical testing capable of generating accurate and precise numbers and in part due to the lack of accepted permissible daily exposures (PDEs) of the elements so that toxicologically based safety decisions could be made.

BEGINNINGS OF CHANGE

As modern technology advanced, the need for improvement in the measurement of heavy metals in pharmaceuticals was recognized and called out. The first time this was broached in the USP was in the Stimuli to the Revision Process article by Katherine Blake.[1] In this article, it was noted that often 50% of the metal of interest was lost during the ash process, and that there was essentially zero recovery for mercury, one of the more toxic and common elements of interest.

In 2000, Dr. Taibang Wang noted this in his publication in the *Journal of Pharmaceutical and Biomedical Analysis*:[2] "Although still widely accepted and used in the pharmaceutical industry, these methods based on the intensity of the color of sulfide precipitation are non-specific, labor intensive, and more often than hoped, yield low recoveries or no recoveries at all." In 2003, Dr. Wang made similar comments as he proposed inductively coupled plasma mass spectrometry (ICP-MS) technology to replace both the USP <231>, "Heavy Metals," and USP <281>, "Residue on Ignition," chapters.[3]

In 2004, Nancy Lewen and coworkers[4] directly compared the recoveries of 14 different elements using USP <231> Method II and ICP-MS. As can be seen in Figure 2.1, consistent with Dr. Blake's observations, many of recoveries using <231> Method II are on the order of 50%, with recoveries of less than 5% observed for Se, Sn, Sb, and Ru and no recovery observed for Hg. At about the same time, Schenkenberger and Lewen proposed in the *Pharmacopeial Forum* Stimuli to the Revision Process article the use of inductively coupled plasma optical emission spectroscopy (ICP-OES) as an alternative to the heavy metals test.[5]

From a regulatory perspective, the European Medicines Agency (EMA) began developing a guideline on the control of residual catalysts in pharmaceuticals with the goal of establishing limits based on toxicological safety assessments of common catalytic elements as early as 1998. This guideline was officially implemented for new drug products in 2008.[6] The guideline introduced the concept of mass-based Permissable Daily Exposures (PDEs) to establish permissible exposures in drug products rather than concentration limits in drug substances. The PDE for each element in the guideline was established based on an assessment of the toxicological data on the individual metal.

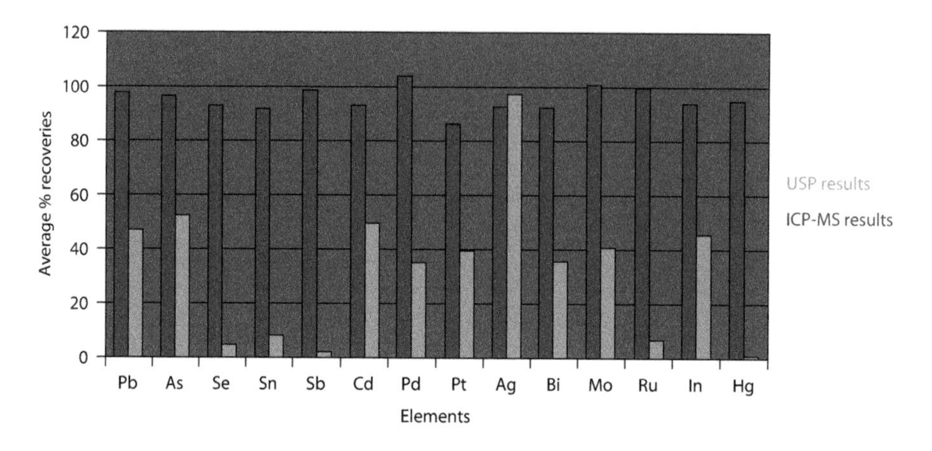

FIGURE 2.1 Comparisons between ICP-MS and <231> Method II. (From Lewen, N., et al., *J. Pharm. Biomed. Anal.*, 35, 739–752, 2004.)

DEVELOPMENT OF NEW LIMITS AND PROCEDURE GENERAL CHAPTERS

Beginning in earnest in the 2005–2010 revision cycle, the USP initiated a series of workshops and stakeholder forums for the purpose of revising General Chapter <231>. In 2008, the USP commissioned the U.S. Institute of Medicine (now the National Academy of Medicine) to organize a workshop with analytical and toxicology experts in metals analysis to evaluate current elemental toxicology thinking regarding potential metals to be controlled, and the capabilities of modern methods of elemental analysis. Later in 2008, the USP proposed to replace <231> with two chapters: <232>, "Elemental Impurities—Limits," which would establish safety-based limits on elemental impurities in pharmaceutical products, and <233>, "Elemental Impurities—Procedures," which would establish appropriate criteria for methods for elemental analysis.[7]

This Stimuli to the Revision Process article, entitled "General Chapter on Inorganic Impurities: Heavy Metals," was published in *Pharmacopeial Forum* (34 (5), 2008) by the USP Ad Hoc Advisory Panel on Inorganic Impurities and Heavy Metals and USP staff. For commonly used catalysts, it drew on the recommendations of an EMA catalyst guideline,[6] so its basic approach was similar to the European approach. The new USP approach differed from the approach in <231> in several fundamental ways.

- It proposed the control of potentially 31 elements at toxicologically relevant levels. (This was pared to 15 elements at one point by the USP toxicology expert panel, but is now 24 based on ICH deliberations and discussions between the USP and ICH toxicology groups.) The control was risk based, with the elements of interest selected based on a combination of their toxicities in the route of administration of interest (oral, parenteral, or inhalational) and the potential of the element to be present in the article.

- The limits were based on the drug product. There are no elemental impurity specifications for individual components unless specified in a monograph. The drug product components and manufacturing processes (e.g., excipients, drug substances, container closure systems, or manufacturing conditions) that go into the production of the drug product are critical to the overall level of the elemental impurity in the drug product, but without knowledge of such things as the dosage form, route of administration, and daily dose, establishing a specification for individual drug product components is not possible.
- Unlike <231>, the methodology of testing was separated from the elemental impurities and their limits. This allowed General Chapter <232> to focus strictly on the risk-based safety of the individual metals. Separately, how to make these measurements in ways that satisfied compendial requirements wase left to General Chapter <233>, "Elemental Impurities—Procedures."

Even with all its faults, one of the advantages of USP <231> was that it was for the most part harmonized within the ICH. There was a strong desire within the pharmaceutical industry, and within the pharmacopeias, to remain harmonized, at least with respect to the metals of interest and their toxicologically based limits. With that goal in mind, a concept paper and business plan were written, and in 2009, the ICH initiated the Q3D Expert Working Group on Elemental Impurities in Pharmaceutical Products with the intention of harmonizing technical requirements for elemental impurities in pharmaceutical products across three regions: Europe, Japan, and the United States.[8] As with the EMA guideline and the USP chapters, the Q3D Expert Working Group tried to set maximum PDEs for elemental impurities in pharmaceutical products based on an assessment of existing toxicological data for the oral, parenteral, and inhalation routes of administration. Q3D reached Step 2 of the ICH process in June 2013, and the guideline was published for public review and comment. Q3D reached Step 4 in November 2014.

USP and the other ICH pharmacopeias were active participants as observers of the ICH process, and indeed, the ICH Expert Working Group began its work with the background provided by USP. To minimize delays in developing official general chapters, the USP process for moving general chapters forward went on in parallel with the ICH process. The USP general chapters, both <232> and <233>, were published several times in this time frame to obtain real-time industry feedback. Both General Chapters <232> and <233> became official in February 2013. The USP Expert Panel on Elemental Impurities worked closely with the ICH Expert Working Group to resolve their differences and aligned General Chapter <232> with Q3D, with minor exceptions, with the version of General Chapter <232>, "Elemental Impurities—Limits," published in *Pharmacopeial Forum* (42 (2), 2016) and official with the first supplement of USP 40–National Formulary (NF) 35 on August 1, 2017. The Food and Drug Administration (FDA) has published the ICH document with relatively minor regional modifications,[9] and subsequently published a draft guidance in June 2016.[10]

The ICH Expert Working Group recommended implementation of its guideline 36 months after publication for new drug products. Currently, for existing drug products, the FDA expects compliance with the specifications of <232> as of June 2016, and for new drug products, consistent with ICH as of January 2018. Since the USP can only have one standard at a time, its official implementation date, via the General Notices, is January 1, 2018, for both General Chapters <232> and <233>. It is anticipated that at that time, General Chapter <231> will be omitted from the USP-NF. Since the general chapters have been official since February 2013 and the FDA had agreed to early implementation of the chapters if desired, manufacturers have had the option of using these chapters for some time. Regarding General Chapter <233>, "Elemental Impurities—Procedures," it was published in USP 38-NF 33, second supplement, official December 1, 2015, and <2232>, "Elemental Contaminants in Dietary Supplements," was published in USP 36-NF 31, official December 1, 2013.

GENERAL CHAPTER <232>, "ELEMENTAL IMPURITIES—LIMITS": OVERVIEW

The goal of General Chapter <232> is to provide specifications for the elemental impurities that are considered toxic if present above their PDE levels in pharmaceutical drug products. The scope of the chapter is limited to small- and large-molecule drug products. Dietary supplements, which are regulated via the Dietary Supplement Health and Education Act of 1994 (DSHEA), rather than the Food, Drug and Cosmetic Act, are discussed in General Chapter <2232>, since dietary supplement general chapters occupy the <2000>–<2999> section of the USP-NF. In addition to dietary supplements, also excluded from the chapter are

- Radiopharmaceuticals
- Articles intended only for veterinary use
- Vaccines
- Cells (cell therapy) and cell metabolites
- DNA products
- Allergenic extracts
- Cells, whole blood, cellular blood components, or blood derivatives, including plasma and plasma derivatives
- Products based on genes
- Tissue (tissue engineering)
- Total parental nutrition (TPN)
- Elements intentionally added for therapeutic benefit

TPN products deserve a special note. While ICH (and in like manner, <232>) excludes them from consideration, the FDA, in its current guidance, notes that it plans to include them in its risk assessment. Thus, for TPN products regulated by FDA, an elemental impurity risk assessment is currently an expectation.

APPLICATION OF LIMITS AND RISK ASSESSMENT OF EXCIPIENTS AND DRUG SUBSTANCES

As discussed earlier, it is important to remember that the limits apply to the drug product, not to excipients or drug substances, unless a limit is specified in the drug substance or excipient monograph. Of course, manufacturers of drug products will need information regarding the content of elemental impurities in drug substances and excipients in order to meet the criteria of the chapter. Being able to provide elemental impurity information may prove to be a competitive advantage to a drug substance or excipient supplier, or it may prove to not be worthwhile for the supplier to provide, depending on many factors. If the drug product manufacturer cannot obtain the information from the material supplier, testing of the material or some form of documented risk assessment will likely be necessary. Drug product manufacturers do not need to retest or reevaluate the material if the supplier does provide sufficient information; they may use data from qualified suppliers of excipients and drug substances (or risk assessments). If the drug product manufacturer does receive a risk assessment from a drug substance or excipient supplier, it is important to ensure that the risk assessments were done using Table 2 in <232> (Table 5.1 in ICH Q3D). In addition to the elements identified in Table 2, any other elements inherent in the material (e.g., naturally sourced materials) must be considered in the risk assessment.

SPECIATION

Speciation in our context is defined as the determination of oxidation state, organic complex, or a combination of both. In terms of pharmaceuticals and dietary supplements, arsenic and mercury are realistically the only elements of concern.

The arsenic limits were established based on the inorganic form. Inorganic arsenic is the most toxic form, and in fact, some organic arsenic compounds are considered nutrients. Organic arsenic compounds are very seldom found in small- or large-molecule pharmaceuticals, but can be quite common in dietary supplements. If organic arsenic is suspected, arsenic may be measured using a total arsenic test assuming all the arsenic is in the inorganic form. If the limit is exceeded using this procedure, it may be possible to show via a procedure that separately quantifies the inorganic and organic forms that the inorganic form meets the specification.

Mercury limits are based on the inorganic 2+ oxidation state. This is the only form found in small- or large-molecule pharmaceuticals. Since methyl mercury is not observed for these products, the limit was based on the inorganic mercuric form. Methyl mercury is a metabolite of inorganic mercury and can be found in fish-derived products (e.g., salmon calcitonin) and other aquatic products, such as kelp. If methyl mercury has been identified as an issue, it is handled in the individual monograph. For fish-derived products in dietary supplements, methyl mercury can be much more of an issue. The toxic level for methyl mercury (2 μg/day) is markedly lower than that for the mercuric form (30 μg/day). Thus, if methyl mercury is suspected, it must be measured using a technique that separates and quantifies it separately from the mercuric form. Once it is established that methyl mercury is present, if it is below the 2 μg/day PDE limit, a total mercury procedure may be performed to ensure that the mercuric form is also within its allowed specification.

ELEMENTAL IMPURITY CLASSES

As discussed previously, the General Chapter <232> was intended to be a risk-based chapter. Here, risk can be defined as the amount of harm a substance can cause, scaled (attenuated or enhanced) by the likelihood of its presence. Thus, a ubiquitous, highly toxic species (e.g., Pb) is high risk, while a less toxic, rarer element (e.g., Cu) is considered of lower risk. Given this, the elements of concern in the general chapter were divided into three classes based on their toxicity (PDE) and likelihood of occurrence in the drug product. Table 2.1 contains the elements of concern, their class, and their route-dependent PDEs, discussed below.

Class 1: The elements arsenic (As), cadmium (Cd), mercury (Hg), and lead (Pb) are human toxicants that have limited or no use in the manufacture of pharmaceuticals.

TABLE 2.1
Permitted Daily Exposures by Route of Administration for Elemental Impurities (USP <232> Table 1)

Element	Class	Oral PDE (μg/day)	Parenteral PDE (μg/day)	Inhalational PDE (μg/day)
Cd	1	5	2	2
Pb	1	5	5	5
As	1	15	15	2
Hg	1	30	3	1
Co	2A	50	5	3
V	2A	100	10	1
Ni	2A	200	20	5
Tl	2B	8	8	8
Au	2B	100	100	1
Pd	2B	100	10	1
Ir	2B	100	10	1
Os	2B	100	10	1
Rh	2B	100	10	1
Ru	2B	100	10	1
Se	2B	150	80	130
Ag	2B	150	10	7
Pt	2B	100	10	1
Li	3	550	250	25
Sb	3	1200	90	20
Ba	3	1400	700	300
Mo	3	3000	1500	10
Cu	3	3000	300	30
Sn	3	6000	600	60
Cr	3	11000	1100	3

They are also ubiquitous in nature. These four elements should be evaluated in all risk assessments.

- Class 2: The Class 2 elements are route-dependent human toxicants. They were further divided into subclasses 2A and 2B based on their relative likelihood of occurrence in the drug product.
    - Class 2A elements (Ni, V, and Cr) have a relatively high probability of occurrence in the drug product, since they are found in material such as stainless steel or other materials used in the manufacturing process. Thus, along with the Class 1 elements, these should be evaluated in all risk assessments.
    - Class 2B elements have a reduced probability of occurrence in the drug product since they are typically found in low abundance in nature and have a low potential to be co-isolated with other materials. As a result, they can be excluded from the risk assessment unless they are intentionally added during the manufacture of drug substances, excipients, or other components of the drug product.
- Class 3: These are elements with relatively low toxicities by the oral route of administration. However, these are route-dependent toxicants, and due to their route-specific toxicity, some (Li, Cu, and Sb) need to be considered in the risk assessment of parenteral products and all need to be considered in the risk assessment of inhalational products.

The elements of interest and how they should be considered in a drug product risk assessment are shown in Table 2.2. Regardless of toxicity level, if an element is intentionally added to any of the components of the drug product, it must be considered in the risk assessment.

Editor's Note: The PDE limits defined in Chapter <2232>, "Elemental Contaminants in Dietary Supplements," are restricted to the "big four" heavy metal contaminants, Pb, As, Cd, and Hg. They are discussed in greater detail in Chapter 23, which offers guidance to which analytical technique is the best to use for the different sample types, based on the instrumental limit of quantitation (LOQ) and the target concentration/J-value of the elemental impurity or contaminant. As with the other elemental impurities chapters, Chapter <2232> is implemented via the USP General Notices Section 5.60.30 rather than referenced in each monograph to which it applies. Thus, it applies to all dietary supplement product monographs. Compliance with a dietary supplement monograph is voluntary unless the product or ingredient is labeled as being USP compliant.

EXPOSURE AND ROUTE OF ADMINISTRATION

As discussed, ICH and USP placed the elements of interest into four classes (1, 2A, 2B, and 3) based on toxicity and likelihood of occurrence in the drug product. This grouping was done in part to help focus the risk assessment on those elements that are most toxic and also have a reasonable probability of being included in the drug product.

TABLE 2.2
Elements to Be Considered in the Risk Assessment (USP <232> Table 2)

			Not Intentionally Added		
Element	**Class**	**Intentionally Added (All Routes)**	**Oral**	**Parenteral**	**Inhalational**
Cd	1	Yes	Yes	Yes	Yes
Pb	1	Yes	Yes	Yes	Yes
As	1	Yes	Yes	Yes	Yes
Hg	1	Yes	Yes	Yes	Yes
Co	2A	Yes	Yes	Yes	Yes
V	2A	Yes	Yes	Yes	Yes
Ni	2A	Yes	Yes	Yes	Yes
Tl	2B	Yes	No	No	No
Au	2B	Yes	No	No	No
Pd	2B	Yes	No	No	No
Ir	2B	Yes	No	No	No
Os	2B	Yes	No	No	No
Rh	2B	Yes	No	No	No
Ru	2B	Yes	No	No	No
Se	2B	Yes	No	No	No
Ag	2B	Yes	No	No	No
Pt	2B	Yes	No	No	No
Li	3	Yes	No	Yes	Yes
Sb	3	Yes	No	Yes	Yes
Ba	3	Yes	No	No	Yes
Mo	3	Yes	No	No	Yes
Cu	3	Yes	No	Yes	Yes
Sn	3	Yes	No	No	Yes
Cr	3	Yes	No	No	Yes

When determining the PDE, the toxicologists considered the impact of bioavailability (extent of exposure) in their risk assessments. Thus, bioavailability has been factored into the PDE for each of the elemental impurities of interest for each of the three routes of administration. In addition to bioavailability, which of course can vary greatly, depending on the route of administration, other parameters were also established prior to calculating the PDEs. Thus, these limits are based on chronic exposure (70-year exposure), determined for a patient weighing 50 kg, and based on a drug product taken by a patient according to the route of administration and label instructions. The ICH Q3D document[8] contains a detailed explanation of the derivation of each of the limits, including the studies and factors considered in each evaluation.

OPTIONS FOR DEMONSTRATING COMPLIANCE

There are three options for demonstrating compliance to the specifications outlined in Chapter <232>. (Editor's note: These are similar but slightly different in

Chapter <2232>.) These are the drug product option, the summation option, and the individual component option.

Drug Product Option

The results obtained from the analysis of a typical dosage unit, scaled to a maximum daily dose, are compared with the daily dose PDE.

Specification: The measured amount of each impurity is not more than (NMT) the daily dose PDE, unless otherwise stated in the monograph.

Daily dose PDE ≥ Measured value (µg/g) × Maximum daily dose (g/day)

Summation Option

Separately, add the amounts of each elemental impurity (in µg/g) present in each of the components of the drug product.

Specification: The result of the summation of each impurity is NMT the daily dose PDE, unless otherwise stated in the monograph.

$$\text{Daily dose PDE} \geq \left[\sum_{1}^{M}(C_M \times W_M)\right] \times D_D$$

where:

Σ = Sum of

M = Each ingredient used to manufacture a dose unit

C_M = Element concentration in component (drug substance or excipient) (µg/g)

W_M = Weight of component in dosage unit (g/dosage unit)

D_D = Number of units in the maximum daily dose (unit/day)

- Ensure no additional elemental impurities can be inadvertently added via the manufacturing process or container closure system.

Individual Component Option

- For drugs with a daily dose of NMT 10 g, if all drug substance and excipient components meet the limits in USP Table 3, then these components may be used in any proportion. No further calculation is necessary.
- Be sure to consider the manufacturing process and container closure system.
- For doses other than 10 g/day, the equation below may be used (ICH Option 2a).

Concentration (µg/g) = PDE (µg/day)/Daily amount of drug product (g/day)

If all components are NMT the concentration calculated, then these components may be used in any proportion.

USP <232> Table 3 is shown in Table 2.3.

Table 2.3 is simply Table 2.1 (the per-gram PDE) divided by 10. The concept is that if the total daily dose is 10 g, then any of these elements can be in the drug

TABLE 2.3
Permitted Concentrations of Elemental Impurities per Individual Component (10 g) Option

Element	Class	Oral PDE (µg/day)	Parenteral PDE (µg/day)	Inhalational PDE (µg/day)
Cd	1	0.5	0.2	0.2
Pb	1	0.5	0.5	0.5
As	1	1.5	1.5	0.2
Hg	1	3	0.3	0.1
Co	2A	5	0.5	0.3
V	2A	10	1	0.1
Ni	2A	20	2	0.5
Tl	2B	0.8	0.8	0.8
Au	2B	10	10	0.1
Pd	2B	10	1	0.1
Ir	2B	10	1	0.1
Os	2B	10	1	0.1
Rh	2B	10	1	0.1
Ru	2B	10	1	0.1
Se	2B	15	8	13
Ag	2B	15	1	0.7
Pt	2B	10	1	0.1
Li	3	55	25	2.5
Sb	3	120	9	2
Ba	3	140	70	30
Mo	3	300	150	1
Cu	3	300	30	3
Sn	3	6000	60	6
Cr	3	1100	110	0.3

product at these levels without fear of going over the limit. These limits are, however, quite difficult to achieve for some mined excipients; thus, this option is not always a practical alternative.

JUSTIFICATION FOR LEVELS HIGHER THAN THE PDE

Levels of elemental impurities higher than an established PDE may be acceptable in certain cases. These cases could include

- Intermittent dosing
- Short-term dosing (i.e., 30 days or less)
- Specific indications (e.g., life threatening, unmet medical needs, and rare diseases)

These exceptions are not discussed in <232>, which is strictly a specification. However, the ICH and FDA guidances provide examples of justifying an increased level of an elemental impurity using a subfactor approach of a modifying factor.[8,9] Other approaches may also be used to justify an increased level. Any proposed level higher than an established PDE should be justified on a case-by-case basis and, of course, higher-than-PDE levels subject to regulatory review.

RISK ASSESSMENT AND REGULATORY CONSIDERATIONS

The concept of quality by design (QbD) requires that quality be built into a process rather than tested in. In the case of ensuring that a drug product will continue to meet its elemental impurity specifications for the life cycle of the product, a robust risk assessment is essential. The risk assessment both assesses the current state of affairs relative to elemental impurities and provides a scientific, risk-based assurance that the levels of elemental impurities will remain in control. As discussed in the EMA's "Implementation Strategy of ICH Q3D Guideline,"[11] "analytical data, without a risk assessment, will not be sufficient to justify to omit a specification for an element."

The risk assessment process begins with identifying known and potential sources of elemental impurities that may find their way into the drug product. Then, for each potential impurity, evaluate its presence in the drug product by determining the observed or predicted level of the impurity and comparing it with the established PDE. The findings should be summarized and documented as part of this process. One needs to show that either the controls are in place that will ensure that the elemental impurity levels are, and will remain, below the control threshold (see the following) or if there are elemental impurities that could be above the control threshold, that additional controls are in place to ensure the PDE or agreed-upon specification is not exceeded.

With regard to deciding whether there is a need for long-term monitoring of specific elemental impurities, ICH discusses the use of a control threshold of 30% of the established PDE in the drug product to determine if additional controls are warranted. If the total elemental impurity level from all sources in the drug product is expected to be consistently less than 30% of the PDE, then additional controls are not required, provided the applicant has demonstrated adequate controls. Otherwise, controls should be established to ensure that the elemental impurity level does not exceed the PDE in the drug product.

Variability is important in deciding whether the control threshold concept should be used to make the case that long-term testing is not needed. Variability occurs in all aspects of the process, including the analytical method, the elemental impurity level in the specific sources, and the elemental impurity levels in the drug product. Variability is especially important in the cases where mined excipients, or other materials where levels of elemental impurities are difficult to manage, are present in the drug product.

The FDA notes in its draft guidance that manufacturers should establish that the analytical procedures used during risk assessments possess characteristics (e.g., accuracy, precision, and specificity) such that the manufacturers can be reasonably certain (e.g., at the 95% confidence level) that the measurements can be relied on to decide whether to include routine testing of materials in the control strategy. This decision depends on whether the amounts of the elemental impurities in the materials are consistently below control thresholds. Thus, the analytical procedures should be validated with this goal in mind.

With regard to the data required at the time of submission, the FDA draft guidance advises that in the absence of other justification, the level and variability of an elemental impurity can be established by providing the data from three representative production scale lots or six representative pilot scale lots of the component or components or drug product. It is further noted that for some components that have inherent variability (e.g., mined excipients), additional data may be needed to apply the control threshold. Note: Chapter 25 gives more details on risk assessment.

GENERAL CHAPTER <233>, "ELEMENTAL IMPURITIES—PROCEDURES": OVERVIEW

The ICH Q3D document and USP General Chapter <232> purposefully omit discussions regarding how to test for whether a drug product meets its elemental impurities specifications. It was decided that the details of testing and its accompanying validation should be left to the pharmacopeias. In the case of the USP, the testing and validation requirements are provided in General Chapter <233>, "Elemental Impurities—Procedures."

The USP realized that as is the case with General Chapter <467>, "Residual Solvents," one or two methods are not likely to work for the many monographs that may require testing. Thus, General Chapter <233> provides two procedures as potential starting points, which applies to the measurement of elemental impurities in Chapter <232>, as well as the four heavy metal contaminants in Chapter <2232>.

Any procedure that meets the validation requirements of the chapter is acceptable for use. It is important to note that for the purposes of this chapter, the procedures provided in the chapter must be verified via the validation procedures provided for alternative procedures. That is, verification equals validation for the procedures provided in the general chapter. In addition, it should be noted that the chapter provides minimum validation acceptance criteria. Individual drug products may need further validation. It is also an expectation that the validation requirements of ICH Q2(R1), or USP <1225>, be met in conjunction with the USP requirements. The dietary supplements chapter, <2232>, also refers to Chapter <233> for all but methyl mercury and speciation issues, since speciation is not covered in this chapter.

USP <233> COMPENDIAL PROCEDURES 1 AND 2

There are two procedures described in General Chapter <233>. Procedure 1 can be used for elemental impurities generally amenable to detection by inductively coupled plasma atomic (optical) emission spectroscopy (ICP-AES or ICP-OES). Procedure 2 can be used for elemental impurities generally amenable to detection by ICP-MS. Before initial use, the analyst should verify that the procedure is appropriate for the instrument and sample used by meeting the alternative procedure validation requirements. The procedures must be validated using the criteria for validation of alternative procedures. In addition, the procedures contain a system suitability criterion. The drift should be compared from results obtained from two calibration standard solutions prepared at 1.5J and 0.5J (see Sample Preperation) with the target elements before and after the analysis of the sample solution. The suitability criteria are that the drift for the 15.J standard must be NMT 20% for each target element.

Sample Preparation

There are four primary sample preparation procedures.

- Neat: Used for liquids or alternative procedures that allow for examination of unsolvated samples.
- Direct aqueous solution: Used when the sample is soluble in an aqueous solvent.
- Direct organic solution: Used when the sample is soluble in an organic solvent.
- Indirect solution: Used when a material is not directly soluble in aqueous or organic solvents. Solubilization is typically accomplished via digestion of the sample using closed-vessel digestion.

Any of these are acceptable if no specific procedure is indicated in the individual monograph.

When preparing blanks, they should be spiked with the same target elements and, where possible, the same spiking solutions. Standard solutions may contain multiple target elements. It is recommended that all liquids be weighed to minimize errors associated with pipetting. When deciding on target elements (elements of concern), consider any element with the potential of being present in the material under test. Include Class 1 and 2A when testing is done to demonstrate compliance for any of the routes of administration, and Class 3 elements as appropriate, and include any elements that may be added through material processing or storage. Note: More detailed chapters on sample preparation and contamination issues are given in Chapters 18 and 19.

USP <233> J-Value Parameter

The validation sections of General Chapter <233> discuss most of the target element concentration in terms of the parameter J. The J-value is defined as the concentration (w/w) of the elements of interest at the target limit, appropriately diluted to the working range of the instrument (detection limit to the end of the linear part of the calibration curve). Here the target limit is the acceptance value for the elemental impurity being evaluated. Exceeding the target limit indicates the material under test exceeds the acceptable value. As an example of the calculation of the J-value, consider the case of the target element Pb, to be evaluated in an oral dosage form with the maximum daily dose of 10 g/day via ICP-MS. From Table 2.3, the target limit concentration for Pb is 0.5 μg/g. So if the linear dynamic range for Pb by ICP-MS is 0.001 ng/mL to 1.0 μg/mL, the sample will require a dilution factor of at least 100:1 to ensure that analysis occurs in the linear dynamic range of the instrument. However, for practical purposes, it is best to keep the total dissolved solids (TDSs) below 0.2%

in ICP-MS to minimize buildup of solid material on the interface cones. So a more realistic dilution factor for ICP-MS is 500:1. Thus, in this example of a target limit of 0.5 μg/g for Pb and a sample preparation or dilution of 0.2 g in 100 mL,

J-value for Pb = (0.5 μg/g) (0.2 g/100 mL) = 0.001 μg/mL = 1 ng/mL

To meet the suitability criteria, a calibration is then made up of two standards:

$$\text{Standard 1} = 1.5\text{J}, \text{ Standard 2} = 0.5\text{J}$$

So for Pb, that's 1.5 ng/mL (Standard 1) and 0.5 ng/mL (Standard 2).

Calibration drift is measured by comparing results for Standard 1 before and after the analysis of all the sample solutions under test. The drift should be <20% for each target element.

It should also be remembered that the J-value is the parameter used by the analyst, since units of J correspond to concentrations compatible with the instrument used for the analysis. It is therefore critically important to report results back in units of value to the customer (e.g., micrograms per gram of material or micrograms per tablet).

Alternative Procedure Validation: USP <233> Requirements

As discussed previously, even if the compendial procedures one or two work as written, they need to be validated using the criteria in this section, and in USP <1225> or ICH Q2(R1) as appropriate. If the specified compendial procedures don't meet the needs of a specific application, alternative procedures may be developed, and they too must be validated and shown to be acceptable. The level of validation required depends on whether a limit test or quantitative determination is specified in the monograph. Any alternative procedure that has been validated and meets the acceptance criteria that follow is considered suitable for use. As a general rule, the suitability of the method must be determined by conducting studies with the material under test supplemented with known concentrations of each Target Element of interest at the appropriate acceptance limit concentration. The material of interest under test must be spiked before any sample preparation steps are performed.

Limit Procedures: Detectability

Both instrumental tests and noninstrumental tests are permitted, if they meet the validation requirements. Acceptance criteria will, of course, differ depending on whether an instrumental or noninstrumental test is used. The detectability test is done with the following solutions:

Standard solution: A preparation of reference materials for the target elements at the target concentration.

Spiked sample solution 1: Solution of sample under test, spiked with reference materials for the target elements at the target concentration, solubilized or digested as described in sample preparation.

Spiked sample solution 2: As per spiked sample solution 1, spiked at 80% of the target concentration.

Unspiked sample solution: Solution of sample under test, solubilized or digested as described in sample preparation

Acceptance Criteria: Noninstrumental Procedures

Spiked sample solution 1 provides a signal or intensity equivalent to or greater than the standard solution.

Spiked sample solution 2 must provide a signal or intensity less than that of spiked sample solution 1.

The signal from each spiked sample solution is not less than (NLT) the unspiked sample solution determination.

Thus, the acceptance criteria for noninstrumental procedures say that the sample should not attenuate the elemental impurity signal relative to a similarly prepared standard solution, that the procedure must be able to distinguish a sample prepared at 100% of the target limit from one prepared at 80%, and that a spiked sample solution must provide a signal NLT that of an unspiked sample solution.

Acceptance Criteria: Instrumental Procedures

The average of three replicate measurements of spiked sample solution 1 is within ±15% of the average value obtained for the replicate measurement of the standard solution.

The average value of the replicate measurements of spiked sample solution 2 must provide a signal intensity that is less than that of the standard solution.

It is important to correct the values obtained for each of the spiked solutions using the unspiked sample solution.

Limit Procedures: Precision (Repeatability) and Specificity

Noninstrumental measurement precision is demonstrated by meeting the detectability requirement. For instrumental procedures, six independent samples of the material under test, spiked with appropriate reference materials for the target elements at their target concentrations, are analyzed. The acceptance criteria state that the relative standard deviation (RSD) is NMT 20% for each target element.

The specificity requirement is consistent with USP General Chapter <1225>. The procedure must be able to unequivocally assess each target element in the presence of components that may be expected to be present, including other target elements and matrix components. For a noninstrumental test, this is often difficult unless only one element is being assessed.

Quantitative Testing Validation Requirements

The quantitative testing validation requirements are summarized in Table 2.4.

Several points should be noted regarding the tests and acceptance criteria. First, the accuracy acceptance criteria are asymmetric. This is done because a recovery that is too high is a manufacturer issue, whereas a recovery that is too low can be a

TABLE 2.4
Summary of Quantitative Testing Validation Requirements

Parameter	Test	Acceptance Criteria
Accuracy	Comparison of spiked samples with standards at 0.5–1.5 J	70%–150%
Precision (repeatability)	Analysis of 6 individual sample preps spiked at 1.0 J	RSD $\leq$ 20% ($n = 6$)
Precision (intermediate precision)	Repeatability test performed by separate analyst, different system, different day (only one required)	RSD $\leq$ 25% ($n = 12$)
Specificity	As required	Accuracy and precision criteria met
LOQ, range, linearity	Accuracy	Accuracy criteria met

patient issue. Thus, more latitude is provided at the high end of the recovery, since patient safety is not an issue if recoveries run high and the PDEs are not exceeded. A tighter requirement is imposed for lower recoveries to protect against lots being approved that in reality have levels of elemental impurities that exceed the limits.

With regard to intermediate precision, it is noted that many laboratories may only have one instrument or one analyst. Thus, the requirement is that the test be repeated either on a separate date, by a separate analyst, or on a separate instrument. Only one of these is required, although as many as possible are encouraged in assessing the true precision of the procedure.

FINAL THOUGHTS

Elemental impurities testing has come a long way since 1908. The elements of interest and their levels are toxicologically based, and the testing provides the ability to accurately and precisely quantify the elements of interest at the required levels for each drug product. The risk assessment process ensures that the drug product has elemental impurity levels that are acceptable now and throughout the life cycle of the product. In addition, the elements, limits, and validated testing requirements in the USP-NF provide a mechanism by which any questionable material (e.g., potentially adulterated) can be rapidly and accurately tested for authenticity. The end result is a set of general chapters that add value to the quality assessment of the drug product and help protect the drug product supply chain.

It is also strongly recommended that the person who is carrying out the elemental impurity measurements should become familiar with the chapters on the fundamental principles of ICP-OES and/or ICP-MS to have a better understanding of the technique they are using and help them to become more comfortable with the terminology used in both USP chapters <232> and <233>.

There are, in addition, a number of other USP general chapters that serve as excellent references and background information. These include, in addition to the currently official versions of <232>, "Elemental Impurities—Limits"; <233>,

"Elemental Impurities—Procedures"; and <2232>, "Elemental Contaminants in Dietary Supplements," General Chapters <730>, "Plasma Spectroscopy"; <467>, "Residual Solvents"; <1225>, "Validation of Compendial Procedures"; and <1151>, "Pharmaceutical Dosage Forms." It is recommended that the reader check the current version of the USP-NF and/or the USP elemental impurities information web pages (http://www.usp.org/chemical-medicines/key-issues-elemental-impurities) to ensure that the latest updates to these chapters are considered.

REFERENCES

1. K. Blake. *Pharmacopeial Forum*, 21(6), 1632–1637, 1995.
2. T. Wang. *Journal of Pharmaceutical and Biomedical Analysis*, 23, 867–890, 2000.
3. T. Wang, J. Wu, X. Jia, X. Bu, I. Santos, and R. S. Egan. An atomic spectroscopic method as an alternative to both USP heavy metal <231> and USP residue on ignition <281>. Stimuli to the revision process. *Pharmacopeial Forum*, 29(4), 1328–1336, 2003.
4. N. Lewen, S. Mathew, M. Schenkenberger, and T. Raglione. *Journal of Pharmaceutical and Biomedical Analysis*, 35, 739–752, 2004.
5. M. Schenkenberger and N. Lewen. Inductively coupled plasma–optical emission spectroscopy as an alternative to the heavy metals test. Stimuli to the revision process. *Pharmacopeial Forum*, 30(6), 2271–2274, 2004.
6. European Medicines Agency, Committee for Medicinal Products for Human Use. Guideline on the specification limits for residues of metal catalysts or metal reagents. EMEA/CHMP/SWP/4446/20007. 2008.
7. USP Ad Hoc Advisory Panel on Inorganic Impurities and Heavy Metals and USP Staff. General chapter on inorganic impurities: Heavy metals. *Pharmacopeial Forum*, 34(5), 1345–1347, 2008.
8. International Conference on Harmonisation. http://www.ich.org/products/guidelines/quality/article/quality-guidelines.html.
9. Food and Drug Administration. http://www.fda.gov/downloads/drugs/guidancecomplianceregulatoryinformation/guidances/ucm371025.pdf.
10. Food and Drug Administration. https://www.fda.gov/downloads/Drugs/Guidances/UCM509432.pdf.
11. Implementation strategy of ICH Q3D guideline. http://www.ema.europa.eu/docs/en_GB/document_library/Scientific_guideline/2017/03/WC500222768.pdf.

3 An Overview of ICP *Mass Spectrometry*

Inductively coupled plasma mass spectrometry (ICP-MS) not only offers extremely low detection limits in the sub-parts-per-trillion (ppt) range, but also enables quantitation at the high parts per million (ppm) level. This unique capability makes the technique very attractive compared with other trace metal techniques, such as electrothermal atomization (ETA), which is limited to determinations at the trace level, or flame atomic absorption (FAA) and inductively coupled plasma optical emission spectrometry (ICP-OES), which are traditionally used for the detection of higher concentrations. In this chapter, we present an overview of ICP-MS and explain how its characteristic low detection limits are achieved.

ICP-MS is undoubtedly the fastest-growing trace element technique available today. Since its commercialization in 1983, approximately 20,000 systems have been installed worldwide for many varied and diverse applications. The most common ones, which represent approximately 80% of the ICP-MS analysis being carried out today, include environmental, geological, semiconductor, biomedical, and nuclear market segments. However, over the past few years, with new heavy metal regulations being implemented, pharmaceutical applications are rapidly catching up. There is no question that the major reason for its unparalleled growth is its ability to carry out rapid multielement determinations at the ultratrace level. Even though it can broadly determine the same suite of elements as other atomic spectroscopic techniques, such as FAA, ETA, and ICP-OES, ICP-MS has clear advantages in its multielement characteristics, speed of analysis, detection limits, and isotopic capability. Figure 3.1 shows approximate detection limits of all the elements that can be detected by ICP-MS, together with their isotopic abundance.

PRINCIPLES OF OPERATION

There are a number of different ICP-MS designs available today that share many similar components, such as nebulizer, spray chamber, plasma torch, interface cones, vacuum chamber, ion optics, mass analyzer, and detector. However, the engineering design and implementation of these components can vary significantly from one instrument to another. Instrument hardware is described in greater detail in the subsequent chapters on the basic principles of the technique. So let us begin here by giving an overview of the principles of operation of ICP-MS. Figure 3.2 shows the basic components that make up an ICP-MS system. The sample, which usually must be in a liquid form, is pumped at 1 mL/min, usually with a peristaltic pump into a nebulizer, where it is converted into a fine aerosol with argon gas at about 1 L/min. The fine droplets of the aerosol, which represent only 1%–2% of the sample, are separated from larger droplets by means of a spray chamber. The fine aerosol then

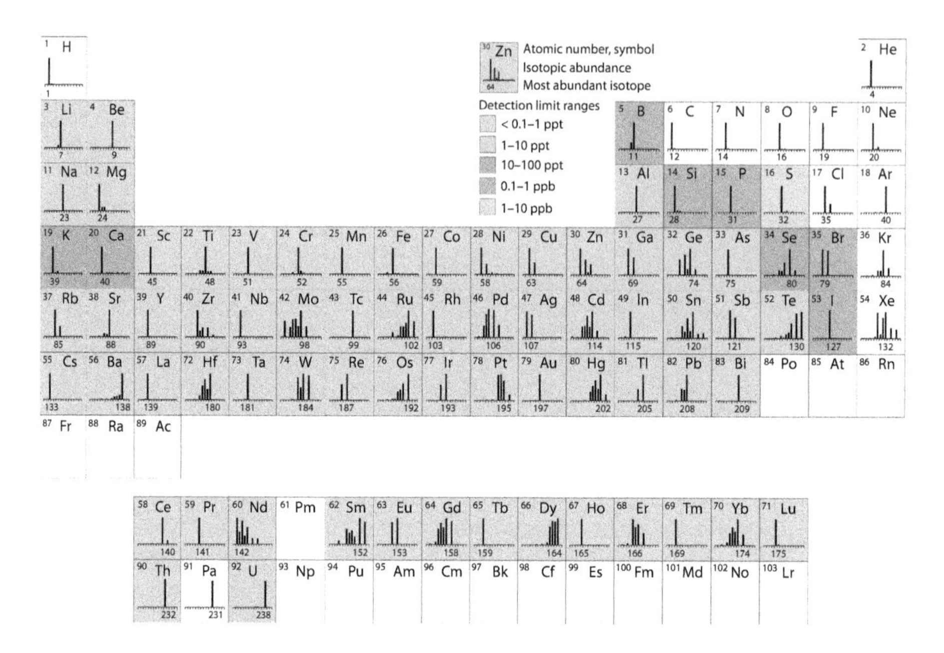

FIGURE 3.1 Approximate detection capability of ICP-MS, together with elemental isotropic abundances. (Reproduced with permission from PerkinElmer Inc., Waltham, MA, 2013. All rights reserved.)

emerges from the exit tube of the spray chamber and is transported into the plasma torch via a sample injector.

It is important to differentiate between the roles of the plasma torch in ICP-MS compared with ICP-OES. The plasma is formed in exactly the same way, by the interaction of an intense magnetic field (produced by radio frequency [RF] passing through a copper coil) on a tangential flow of gas (normally argon), at about 15 L/min, through a concentric quartz tube (torch). This has the effect of ionizing

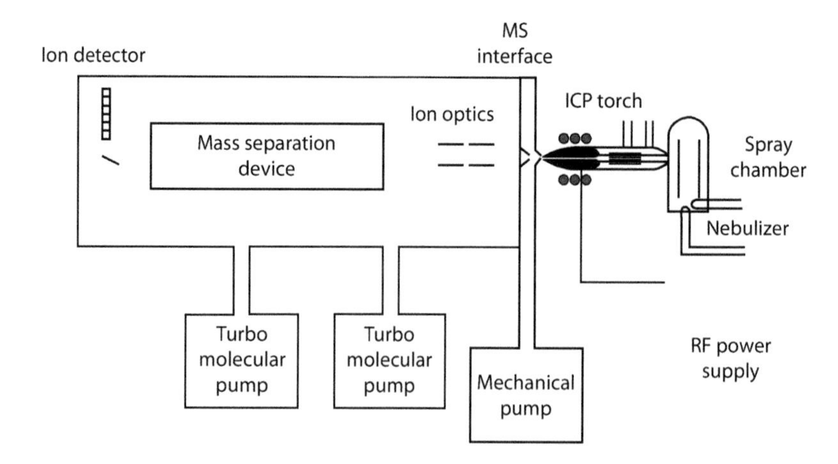

FIGURE 3.2 Basic instrumental components of an ICP mass spectrometer.

the gas, which, when seeded with a source of electrons from a high-voltage spark, forms a very high-temperature plasma discharge (~10,000 K) at the open end of the tube. However, this is where the similarity ends. In ICP-OES, the plasma, which is normally vertical (but can be horizontal with dual-view designs), is used to generate photons of light by the excitation of electrons of a ground-state atom to a higher energy level. When the electrons "fall" back to the ground state, wavelength-specific photons are emitted that are characteristic of the element of interest. In ICP-MS, the plasma torch, which is positioned horizontally, is used to generate positively charged ions and not photons. In fact, every attempt is made to stop the photons from reaching the detector because they have the potential to increase signal noise. It is the production and detection of large quantities of these ions that give ICP-MS its characteristic low-parts-per-trillion detection capability—about three to four orders of magnitude lower than ICP-OES.

Once the ions are produced in the plasma, they are directed into the mass spectrometer via the interface region, which is maintained at a vacuum of 1–2 torr with a mechanical roughing pump. This interface region consists of two or three metallic cones (depending on the design), called the sampler and a skimmer cone, each with a small orifice (0.6–1.2 mm) to allow the ions to pass through to the ion optics, where they are guided into the mass separation device.

The interface region is one of the most critical areas of an ICP mass spectrometer, because the ions must be transported efficiently and with electrical integrity from the plasma, which is at atmospheric pressure (760 torr), to the mass spectrometer analyzer region, which is at approximately 10^{-6} torr. Unfortunately, there is the likelihood of capacitive coupling between the RF coil and the plasma, producing a potential difference of a few hundred volts. If this is not eliminated, an electrical discharge (called a secondary discharge or pinch effect) between the plasma and the sampler cone occurs. This discharge increases the formation of interfering species and also dramatically affects the kinetic energy of the ions entering the mass spectrometer, making optimization of the ion optics very erratic and unpredictable. For this reason, it is absolutely critical that the secondary charge be eliminated by grounding the RF coil. There have been a number of different approaches used over the years to achieve this, including a grounding strap between the coil and the interface, balancing the oscillator inside the RF generator circuitry, a grounded shield or plate between the coil and the plasma torch, or the use of a double interlaced coil where RF fields go in opposing directions. They all work differently but basically achieve a similar result, which is to reduce or eliminate the secondary discharge.

Once the ions have been successfully extracted from the interface region, they are directed into the main vacuum chamber by a series of electrostatic lenses, called ion optics. The operating vacuum in this region is maintained at about 10^{-3} torr with a turbomolecular pump. There are many different designs of the ion optic region, but they serve the same function, which is to electrostatically focus the ion beam toward the mass separation device, while stopping photons, particulates, and neutral species from reaching the detector.

The ion beam containing all the analyte and matrix ions exits the ion optics and now passes into the heart of the mass spectrometer—the mass separation device, which is kept at an operating vacuum of approximately 10^{-6} torr with a second turbomolecular pump. There are many different mass separation devices, all with their strengths and weaknesses. Three of the most common types are discussed in this book—quadrupole, magnetic sector, and time-of-flight technology—but they basically serve the same purpose, which is to allow analyte ions of a particular mass-to-charge ratio through to the detector and to filter out (reject) all the nonanalyte, interfering, and matrix ions. Depending on the design of the mass spectrometer, this is either a scanning process where the ions arrive at the detector in a sequential manner, or a simultaneous process where the ions are either sampled or detected at the same time. Most quadrupole instruments nowadays are also sold with collision/reaction cells or interfaces. This technology offers a novel way of minimizing polyatomic spectral interferences by bleeding a gas into the cell or interface and using ion–molecule collision/reaction mechanisms to reduce the impact of the ionic interference.

The final process is to convert the ions into an electrical signal with an ion detector. The most common design used today is called a discrete dynode detector, which contains a series of metal dynodes along the length of the detector. In this design, when the ions emerge from the mass filter, they impinge on the first dynode and are converted into electrons. As the electrons are attracted to the next dynode, electron multiplication takes place, which results in a very high stream of electrons emerging from the final dynode. This electronic signal is then processed by the data handling system in the conventional way and converted into an analyte concentration using ICP-MS calibration standards. Most detection systems can handle up to nine orders of dynamic range, which means they can be used to analyze samples from low- or sub-parts-per-trillion levels up to a few hundred parts per million, depending on the analyte mass.

It is important to emphasize that because of the enormous interest in the technique, most ICP-MS instrument companies have very active research and development programs in place, in order to get an edge in a very competitive marketplace. This is obviously very good for the consumer, because not only does it drive down instrument prices, but also the performance, applicability, usability, and flexibility of the technique are being improved at a dramatic rate. Although this is extremely beneficial for the ICP-MS user community, it can pose a problem for a textbook writer who is attempting to present a snapshot of instrument hardware and software components at a particular moment in time. Hopefully, I have struck the right balance in not only presenting the fundamental principles of ICP-MS, but also making the reader aware of what the technique is capable of achieving and where new developments might be taking it, particularly with regard to pharmaceutical analysis.

4 Principles of Ion Formation

Chapter 4 gives a brief overview of the fundamental principles of ion formation in inductively coupled plasma mass spectrometry (ICP-MS)—the use of a high-temperature argon plasma to generate positively charged ions. The highly energized argon ions that make up the plasma discharge are used to first produce analyte ground-state atoms from the dried sample aerosol, and then to interact with the atoms to remove one or more electrons and generate positively charged ions, which are then steered into the mass spectrometer for detection and measurement.

In ICP-MS, the sample, which is usually in liquid form, is delivered into the sample introduction system, comprising a spray chamber and nebulizer. It emerges as an aerosol, where it eventually finds its way, via a sample injector, into the base of the plasma. As it travels through the different heating zones of the plasma torch, it is dried, vaporized, atomized, and ionized. During this time, the sample is transformed from a liquid aerosol to solid particles, and then into a gas. When it finally arrives at the analytical zone of the plasma, at approximately 6000–7000 K, it exists as ground-state atoms and ions, representing the elemental composition of the sample. The excitation of the outer electrons of a ground-state atom to produce wavelength-specific photons of light is the fundamental basis of atomic emission. However, there is also enough energy in the plasma to remove one or more electrons from its orbital to generate a free ion. The energy available in an argon plasma is ~15.8 eV, which is high enough to ionize most of the elements in the periodic table (the majority have first ionization potentials on the order of 4–12 eV). It is the generation, transportation, and detection of significant numbers of positively charged ions that give ICP-MS its characteristic ultratrace detection capabilities. It is also important to mention that although ICP-MS is predominantly used for the detection of positive ions, negative ions are also produced in the plasma. However, because the extraction and transportation of negative ions is different from that of positive ions, most commercial instruments are not designed to measure them. The process of the generation of positively charged ions in the plasma is conceptually shown in greater detail in Figure 4.1.

ION FORMATION

The actual process of conversion of a neutral ground-state atom to a positively charged ion is shown in Figure 4.2 and Figure 4.3. Figure 4.2 shows a very simplistic view of the chromium atom Cr^0, consisting of a nucleus with 24 protons (p^+) and 28 neutrons (n), surrounded by 24 orbiting electrons (e^-). It must be emphasized that this is not meant to be an accurate representation of the electron's shells and

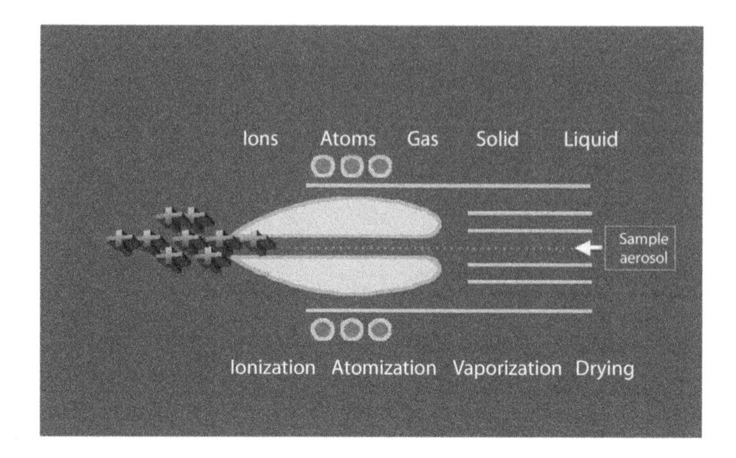

FIGURE 4.1 Generation of positively charged ions in the plasma.

subshells, but just a conceptual explanation for the purpose of clarity. From this, we can conclude that the atomic number of chromium is 24 (number of protons) and its atomic mass is 52 (number of protons + neutrons).

If energy is then applied to the chromium ground-state atom in the form of heat from a plasma discharge, one or more orbiting electrons will be stripped off the outer shell. This will result in only 23 (or less) electrons left orbiting the nucleus. Because the atom has lost a negative charge (e^-) but still has 24 protons (p^+) in the nucleus, it is converted into an ion with a net positive charge. It still has an atomic mass of 52

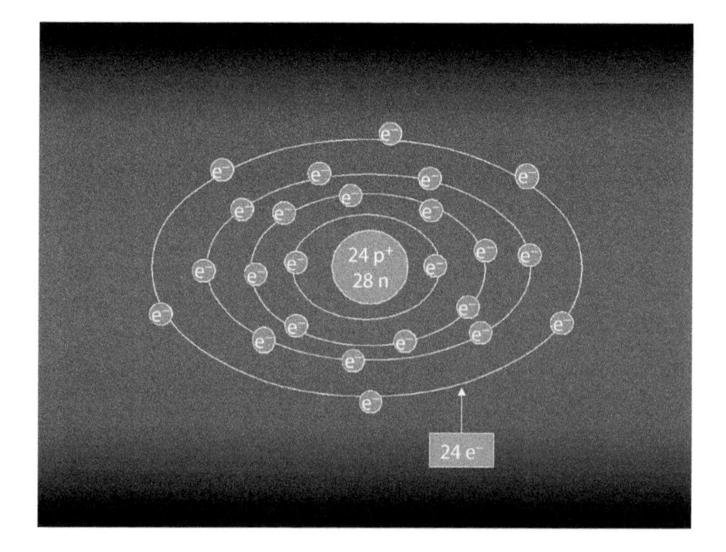

FIGURE 4.2 Simplified schematic of a chromium ground-state atom (Cr^0).

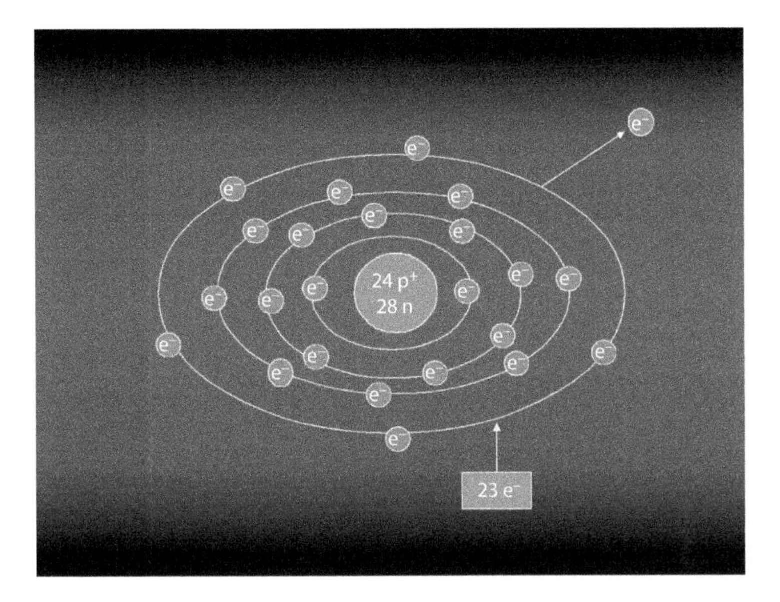

FIGURE 4.3 Conversion of a chromium ground-state atom (Cr^0) to an ion (Cr^+).

and an atomic number of 24, but is now a positively charged ion and not a neutral ground-state atom. This process is shown in Figure 4.3.

NATURAL ISOTOPES

This is a very basic look at the process, because most elements occur in more than one form (isotope). In fact, chromium has four naturally occurring isotopes, which means that the chromium atom exists in four different forms, all with the same atomic number of 24 (number of protons), but with different atomic masses (numbers of neutrons).

To make this a little easier to understand, let us take a closer look at an element such as copper, which only has two different isotopes—one with an atomic mass of 63 (^{63}Cu) and another with an atomic mass of 65 (^{65}Cu). They both have the same number of protons and electrons, but differ in the number of neutrons in the nucleus. The natural abundances of ^{63}Cu and ^{65}Cu are 69.1% and 30.9%, respectively, which gives copper a nominal atomic mass of 63.55—the value you see for copper in atomic weight reference tables. Details of the atomic structure of the two copper isotopes are shown in Table 4.1.

When a sample containing naturally occurring copper is introduced into the plasma, two different ions of copper, $^{63}Cu^+$ and $^{65}Cu^+$, are produced that generate two different masses—one at mass 63 and the other at mass 65. This can be seen in Figure 4.4, which is an actual ICP-MS spectral scan of a sample containing copper, showing a peak for the $^{63}Cu^+$ ion on the left, which is 69.17% abundant, and a peak for $^{65}Cu^+$ at 30.83% abundance on the right.

TABLE 4.1
Breakdown of the Atomic Structure of Copper Isotopes

	^{63}Cu		^{65}Cu
Number of protons (p^+)	29		29
Number electrons (e^-)	29		29
Number of neutrons (n)	34		36
Atomic mass (p^+ + n)	63		65
Atomic number (p^+)	29		29
Natural abundance	69.17%		30.83%
Nominal atomic weight		63.55[a]	

[a] The nominal atomic weight of copper is calculated using the formula 0.6917n (^{63}Cu) + 0.3083n (^{65}Cu) + p^+ and referenced to the atomic weight of carbon.

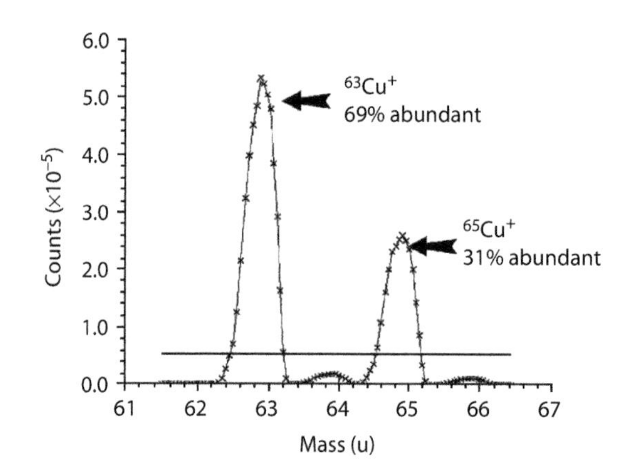

FIGURE 4.4 Mass spectra of the two copper isotopes—$^{63}Cu^+$ and $^{65}Cu^+$.

You can also see small peaks for two Zn isotopes at mass 64 ($^{64}Zn^+$) and mass 66 ($^{66}Zn^+$). (Zn has a total of five isotopes at masses 64, 66, 67, 68, and 70.) In fact, most elements have at least two or three isotopes, and many elements, including zinc and lead, have four or more isotopes. Figure 4.5 is a chart showing the relative abundance of the naturally occurring isotopes of all the elements.

Isotope		%		%		%
1	H	99.985				
2	H	0.015				
3			He	0.000137		
4			He	99.999863		
5						
6					Li	7.5
7					Li	92.5
8						
9	Be	100				
10			B	19.9		
11			B	80.1		
12					C	98.90
13					C	1.10
14	N	99.643				
15	N	0.366				
16			O	99.762		
17			O	0.038		
18			O	0.200		
19					F	100
20	Ne	90.48				
21	Ne	0.27				
22	Ne	9.25				
23			Na	100		
24					Mg	78.99
25					Mg	10.00
26					Mg	11.01
27	Al	100				
28			Si	92.23		
29			Si	4.67		
30			Si	3.10		
31					P	100
32	S	95.02				
33	S	0.75				
34	S	4.21				
35			Cl	75.77		
36	S	0.02			Ar	0.337
37			Cl	24.23		
38					Ar	0.063
39	K	93.2581				
40	K	0.0117	Ca	96.941	Ar	99.600
41	K	6.7302				
42			Ca	0.647		
43			Ca	0.135		
44			Ca	2.086		
45					Sc	100
46	Ti	8.0	Ca	0.004		
47	Ti	7.3				
48	Ti	73.8	Ca	0.187		
49	Ti	5.5				
50	Ti	5.4	V	0.250	Cr	4.345
51			V	99.750		
52					Cr	83.789
53					Cr	9.501
54	Fe	5.8			Cr	2.365
55			Mn	100		
56	Fe	91.72				
57	Fe	2.2				
58	Fe	0.28			Ni	68.077
59			Co	100		
60					Ni	26.223

Isotope		%		%		%
61					Ni	1.140
62					Ni	3.634
63	Cu	69.17				
64			Zn	48.6	Ni	0.926
65	Cu	30.83				
66			Zn	27.9		
67			Zn	4.1		
68			Zn	18.8		
69					Ga	60.108
70	Ge	21.23	Zn	0.6		
71					Ga	39.892
72	Ge	27.66				
73	Ge	7.73				
74	Ge	35.94	Se	0.89		
75					As	100
76	Ge	7.44	Se	9.36		
77			Se	7.63		
78	Kr	0.35	Se	23.78		
79					Br	50.69
80	Kr	2.25	Se	49.61		
81					Br	49.31
82	Kr	11.6	Se	8.73		
83	Kr	11.5				
84	Kr	57.0	Sr	0.56		
85					Rb	72.165
86	Kr	17.3	Sr	9.86		
87			Sr	7.00	Rb	27.835
88			Sr	82.58		
89					Y	100
90	Zr	51.45				
91	Zr	11.22				
92	Zr	17.15	Mo	14.84		
93					Nb	100
94	Zr	17.38	Mo	9.25		
95			Mo	15.92		
96	Zr	2.80	Mo	16.68	Ru	5.52
97			Mo	9.55		
98			Mo	24.13	Ru	1.88
99					Ru	12.7
100			Mo	9.63	Ru	12.6
101					Ru	17.0
102	Pd	1.02			Ru	31.6
103			Rh	100		
104	Pd	11.14			Ru	18.7
105	Pd	22.33				
106	Pd	27.33	Cd	1.25		
107					Ag	51.839
108	Pd	26.46	Cd	0.89		
109					Ag	48.161
110	Pd	11.72	Cd	12.49		
111			Cd	12.80		
112	Sn	0.97	Cd	24.13		
113			Cd	12.22	In	4.3
114	Sn	0.65	Cd	28.73		
115	Sn	0.34			In	95.7
116	Sn	14.53	Cd	7.49		
117	Sn	7.68				
118	Sn	24.23				
119	Sn	8.59				
120	Sn	32.59	Te	0.096		

Isotope		%		%		%
121					Sb	57.36
122	Sn	4.63	Te	2.603		
123			Te	0.908	Sb	42.64
124	Sn	5.79	Te	4.816	Xe	0.10
125			Te	7.139		
126			Te	18.95	Xe	0.09
127	I	100				
128			Te	31.69	Xe	1.91
129					Xe	26.4
130	Ba	0.106	Te	33.80	Xe	4.1
131					Xe	21.2
132	Ba	0.101			Xe	26.9
133			Cs	100		
134	Ba	2.417			Xe	10.4
135	Ba	6.592				
136	Ba	7.854	Ce	0.19	Xe	8.9
137	Ba	11.23				
138	Ba	71.70	Ce	0.25	La	0.0902
139					La	99.9098
140			Ce	88.48		
141					Pr	100
142	Nd	27.13	Ce	11.08		
143	Nd	12.18				
144	Nd	23.80	Sm	3.1		
145	Nd	8.30				
146	Nd	17.19				
147			Sm	15.0		
148	Nd	5.76	Sm	11.3		
149			Sm	13.8		
150	Nd	5.64	Sm	7.4		
151					Eu	47.8
152	Gd	0.20	Sm	26.7		
153					Eu	52.2
154	Gd	2.18	Sm	22.7		
155	Gd	14.80				
156	Gd	20.47	Dy	0.06		
157	Gd	15.65				
158	Gd	24.84	Dy	0.10		
159					Tb	100
160	Gd	21.86	Dy	2.34		
161			Dy	18.9		
162	Er	0.14	Dy	25.5		
163			Dy	24.9		
164	Er	1.61	Dy	28.2		
165					Ho	100
166	Er	33.6				
167	Er	22.95				
168	Er	26.8	Yb	0.13		
169					Tm	100
170	Er	14.9	Yb	3.05		
171			Yb	14.3		
172			Yb	21.9		
173			Yb	16.12		
174			Yb	31.8	Hf	0.162
175	Lu	97.41				
176	Lu	2.59	Yb	12.7	Hf	5.206
177					Hf	18.606
178					Hf	27.297
179					Hf	13.629
180	Ta	0.012	W	0.13	Hf	35.100

Isotope		%		%		%
181	Ta	99.988				
182			W	26.3		
183			W	14.3		
184	Os	0.02	W	30.67		
185					Re	37.40
186	Os	1.58	W	28.6		
187	Os	1.6			Re	62.60
188	Os	13.3				
189	Os	16.1				
190	Os	26.4			Pt	0.01
191			Ir	37.3		
192	Os	41.0			Pt	0.79
193			Ir	62.7		
194					Pt	32.9
195					Pt	33.8
196	Hg	0.15			Pt	25.3
197			Au	100		
198	Hg	9.97			Pt	7.2
199	Hg	16.87				
200	Hg	23.10				
201	Hg	13.18				
202	Hg	29.86				
203					Tl	29.524
204	Hg	6.87	Pb	1.4		
205					Tl	70.476
206			Pb	24.1		
207			Pb	22.1		
208			Pb	52.4		
209	Bi	100				
210						
211						
212						
213						
214						
215						
216						
217						
218						
219						
220						
221						
222						
223						
224						
225						
226						
227						
228						
229						
230						
231	Pa	100				
232	Th	100				
233						
234	U	0.0055				
235	U	0.7200				
236						
237						
238	U	99.2745				

FIGURE 4.5 Relative abundance of the naturally occurring isotopes of the elements. (From International Union of Pure and Applied Chemistry, *Pure Appl. Chem.*, 75(6), 683–799, 2003.)

5 Sample Introduction

Chapter 5 examines one of the most critical areas of the inductively coupled plasma mass spectrometry (ICP-MS) instrument—the sample introduction system. It discusses the basic principles of converting a liquid into a fine-droplet aerosol suitable for ionization in the plasma, and presents an overview of the different types of commercially available nebulizers and spray chambers. Although this chapter briefly touches on some of the newer sampling components introduced in the past few years, such as microflow nebulizers and aerosol dilution systems, the new advancements in desolvating nebulizers, chilled spray chambers, on-line chemistry approaches, autodilution and autocalibration, intelligent autosamplers, and productivity enhance systems are specifically described in Chapter 20.

The majority of current ICP-MS applications involve the analysis of liquid samples. Even though the technique has been adapted over the years to handle solids and slurries, it was developed in the early 1980s primarily to analyze solutions. There are many different ways of introducing a liquid into an ICP mass spectrometer, but they all basically achieve the same result, which is to generate a fine aerosol of the sample so that it can be efficiently ionized in the plasma discharge. The sample introduction area has been called the Achilles' heel of ICP-MS, because it is considered the weakest component of the instrument. Only about 2% of the sample finds its way into the plasma, depending on the matrix and method of introducing the sample.[1] Although there has recently been significant innovation in this area, particularly in instrument-specific components custom built by third-party vendors, the fundamental design of a traditional ICP-MS sample introduction system has not dramatically changed since the technique was first introduced in 1983.

Before I discuss the mechanics of aerosol generation in greater detail, let us look at the basic components of a sample introduction system. Figure 5.1 shows the location of the sample introduction area relative to the rest of the ICP mass spectrometer, whereas Figure 5.2 represents a more detailed view showing the individual components.

The traditional way of introducing a liquid sample into an analytical plasma can be considered as two separate events: aerosol generation using a nebulizer and droplet selection using a spray chamber.[2]

AEROSOL GENERATION

As mentioned previously, the main function of the sample introduction system is to generate a fine aerosol of the sample. It achieves this with a nebulizer and a spray chamber. The sample is normally pumped at about 1 mL/min via a peristaltic or syringe pump into the nebulizer. A peristaltic pump is a small pump with lots of minirollers that all rotate at the same speed. The constant motion and pressure of the rollers on the pump tubing feeds the sample through to the nebulizer. A syringe

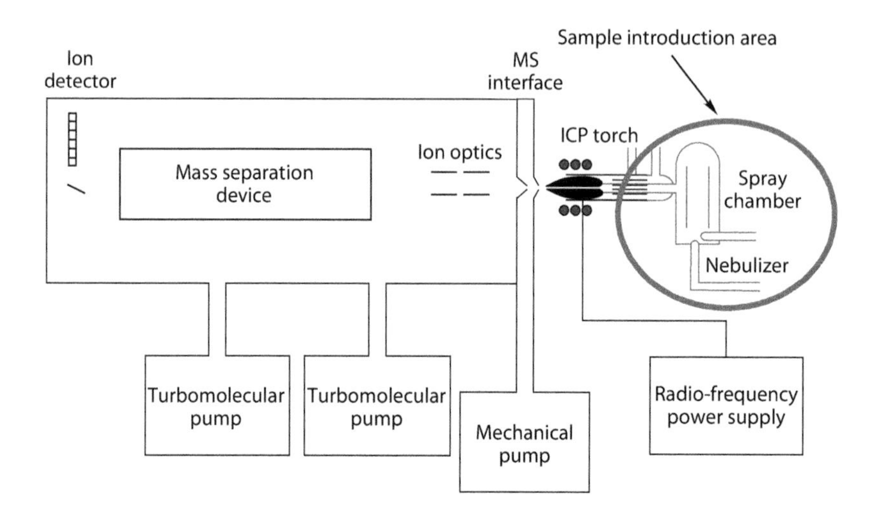

FIGURE 5.1 Location of the ICP-MS sample introduction area.

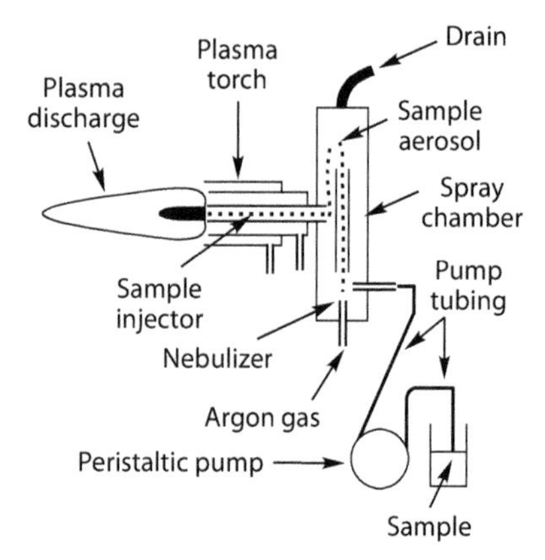

FIGURE 5.2 More detailed view of the ICP-MS sample introduction area.

pump delivers the sample via a pneumatic piston, which is typically two to three times faster than a peristaltic pump, which means that not only is sample throughput increased, but also faster rinse-out is achieved, and therefore there is less carryover to the next sample. The benefits of using a syringe-type pump for autodilution and the online addition of internal standards are well recognized in ICP-MS and will be discussed in greater detail in Chapter 20. However, in its basic configuration, a syringe pump with switching valve, it significantly improves precision by eliminating the pulsations of a peristaltic pump, allowing shorter measurement times to achieve the same performance levels. The stability of a continuous ICP-MS signal with a syringe pump compared with a peristaltic pump is shown in Figure 5.3.

Once the sample enters the nebulizer, the liquid is then broken up into a fine aerosol by the pneumatic action of a flow of gas (~1 L/min) "smashing" the liquid

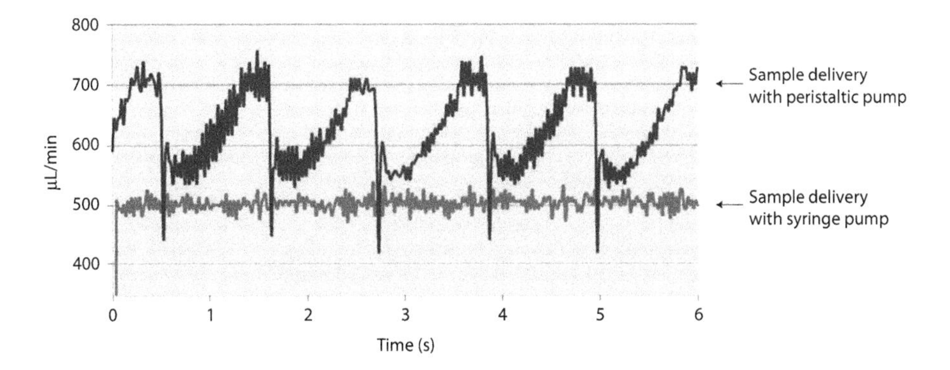

FIGURE 5.3 Stability of a continuous ICP-MS signal with a syringe pump compared with a peristaltic pump.

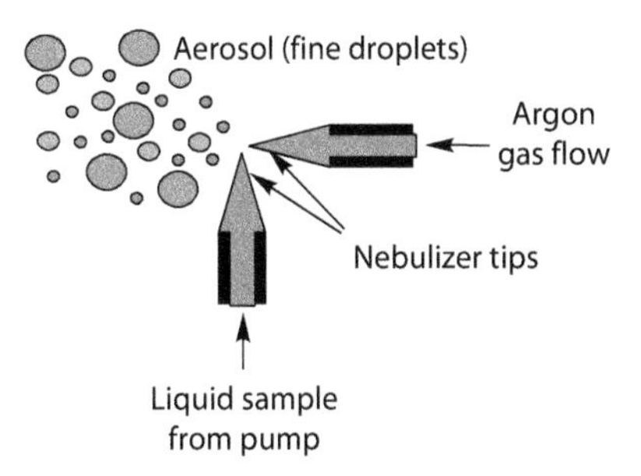

FIGURE 5.4 Conceptual representation of aerosol generation using a cross-flow nebulizer.

into tiny droplets, very similar to the spray mechanism in a can of deodorant. It should be noted that although pumping the sample is the most common approach to introducing the sample, some pneumatic designs, such as concentric nebulizers, do not require a pump, because they rely on the natural "Venturi effect" of the positive pressure of the nebulizer gas to suck the sample through the tubing. Solution nebulization is conceptually represented in Figure 5.4, which shows aerosol generation using a cross-flow-designed nebulizer.

DROPLET SELECTION

Because the plasma discharge is not very efficient at dissociating large droplets, the function of the spray chamber is primarily to allow only the small droplets to enter the plasma. Its secondary purpose is to smooth out pulses that occur during the nebulization process, mainly from the peristaltic pump. Spray chambers are discussed in greater detail later in this chapter, but the most common type is the double-pass design, where the aerosol from the nebulizer is directed into a central tube running the entire length of the chamber. The droplets then travel the length of this tube, where the large droplets (>10 μm diameter) will fall out by gravity and exit through the drain tube at the end of the spray chamber. The fine droplets (<10 μm diameter) then

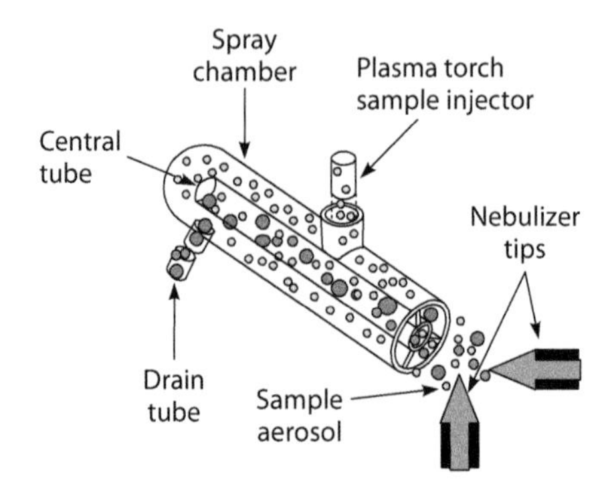

FIGURE 5.5 Simplified representation of the separation of large and fine droplets in a double-pass spray chamber.

pass between the outer wall and the central tube, where they eventually emerge from the spray chamber and are transported into the sample injector of the plasma torch.[3] Although there are many different designs available, the spray chamber's main function is to allow only the smallest droplets into the plasma for dissociation, atomization, and finally, ionization of the sample's elemental components. A simplified schematic of this process using a double-pass-designed spray chamber is shown in Figure 5.5.

Let us now look at the most common nebulizers and spray chamber designs used in ICP-MS. We cannot cover every conceivable design, because over the past few years there has been a huge demand for application-specific solutions, which has generated a number of third-party manufacturers that sell sample introduction components directly to ICP-MS users.

NEBULIZERS

By far the most common design used for ICP-MS is the pneumatic nebulizer, which uses mechanical forces of a gas flow (normally argon at a pressure of 20–30 psi) to generate the sample aerosol. Some of the most popular designs of pneumatic nebulizers include the concentric, microconcentric, microflow, and cross-flow. They are usually made from glass, but other nebulizer materials, such as various kinds of polymers and plastics, are becoming more popular, particularly for highly corrosive samples and specialized applications.

It should be emphasized at this point that nebulizers designed for use with ICP-OES are far from ideal for use with ICP-MS. This is the result of a limitation in the quantity of total dissolved solids (TDSs) that can be put into the ICP-MS interface area. Because the orifice sizes of the sampler and skimmer cones used in ICP-MS are so small (~0.6–1.2 mm), the matrix components must generally be kept below 0.2%, although higher concentrations of some matrices can be tolerated (refer to Chapter 5).[4] This means that general-purpose ICP-OES nebulizers that are designed to aspirate 1%–2% dissolved solids, or high-solids nebulizers, such as the Babbington, V-groove, or cone-spray, which are designed to handle up to 20% dissolved solids, are not ideally suited to analyzing solutions by ICP-MS.

Some researchers have attempted to analyze slurries by ICP-MS using this approach. However, this is not recommended for high-throughput, routine work because of the potential of blocking the interface cones, but as long as the particle size of the slurry is kept below 10 μm in diameter, some success has been achieved using these types of nebulizers.[5] Additionally, there are researchers who are attempting to characterize engineered nanoparticles by using a technique called single-particle (SP) ICP-MS. This is in its early stages, but it is a very exciting development, which uses a novel way to separate out individual nanoparticles in suspension, typically in environmental samples, and then detecting them by optimizing the measurement electronics of the ICP-MS detection system.[6] This technique will be described in greater detail later in the book.

The most common of the pneumatic nebulizers used in commercial ICP mass spectrometers are the concentric and cross-flow design types. The concentric design is the most widely used nebulizer for clean samples, whereas the cross-flow is generally more tolerant to samples containing higher solids and particulate matter. However, recent advances in the concentric design have allowed for the aspiration of these types of samples.

Concentric Design

In traditional concentric nebulization, a solution is introduced through a fine-bore capillary tube, where it comes into contact with a rapidly moving flow of argon gas at a pressure of approximately 30–50 psi. The high-speed gas and the lower-pressure sample combine to create a Venturi effect, which results in the sample being sucked through to the end of the capillary, where it is broken up into a fine-droplet aerosol. Most concentric nebulizers being used today are manufactured from borosilicate glass or quartz. However, polymer-based materials are now being used for applications that require corrosion resistance. Typical sample flow rates for a standard concentric nebulizer are on the order of 1–3 mL/min, although lower flows can be used to accommodate more volatile sample matrices, such as organic solvents. A schematic of a glass concentric nebulizer with the different parts labeled is shown in Figure 5.6, and the aerosol generated by the nebulization process is shown in Figure 5.7.

The standard concentric pneumatic nebulizer will give excellent sensitivity and stability, particularly with clean solutions. However, the narrow capillary can be plagued by blockage problems, especially if heavier matrix samples are being aspirated. For that reason, manufacturers of concentric nebulizers offer modifications to the basic design utilizing different size capillary tubing and recessed tips to allow aspiration of samples with higher dissolved solids and particulate matter. There are even specially designed concentric nebulizers with a smaller-bore input capillary to significantly reduce the dead volume for better coupling of a high-performance liquid chromatography (HPLC) system to the ICP-MS when carrying out trace element speciation studies.

Cross-Flow Design

For the routine analysis of samples that contain a heavier matrix, or maybe small amounts of undissolved matter, the cross-flow design is the more rugged one. With

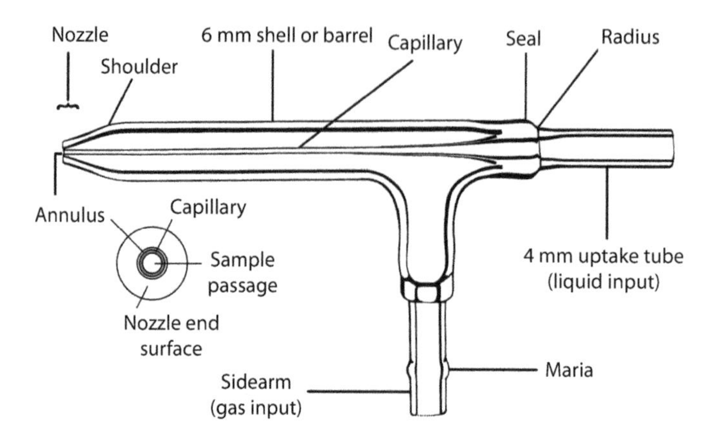

FIGURE 5.6 Schematic of a glass concentric nebulizer. (Courtesy of Meinhard Glass Products, a part of Elemental Scientific Inc., Omaha, NE.)

FIGURE 5.7 Aerosol generated by a concentric nebulizer. (Courtesy of Meinhard Glass Products, a part of Elemental Scientific Inc., Omaha, NE.)

this nebulizer, the argon gas is directed at right angles to the tip of a capillary tube, in contrast to the concentric design, where the gas flow is parallel to the capillary. The solution is either drawn up through the capillary tube via the pressure created by the high-speed gas flow or, as is most common with cross-flow nebulizers, fed through the tube with a peristaltic pump. In either case, contact between the high-speed gas and the liquid stream causes the liquid to break up into an aerosol. Cross-flow nebulizers are generally not as efficient as concentric nebulizers at creating the very small droplets needed for ionization in the plasma. However, the larger-diameter liquid capillary and longer distance between liquid and gas injectors reduces the potential for clogging problems. Many analysts feel that the small penalty to be paid in analytical sensitivity and precision with cross-flow nebulizers compared with the concentric design is compensated by the fact that they are better suited for high-throughput, routine applications. In addition, they are typically manufactured from plastic materials, which make them far more rugged than a glass concentric nebulizer. A cross section of a cross-flow nebulizer is shown in Figure 5.8.

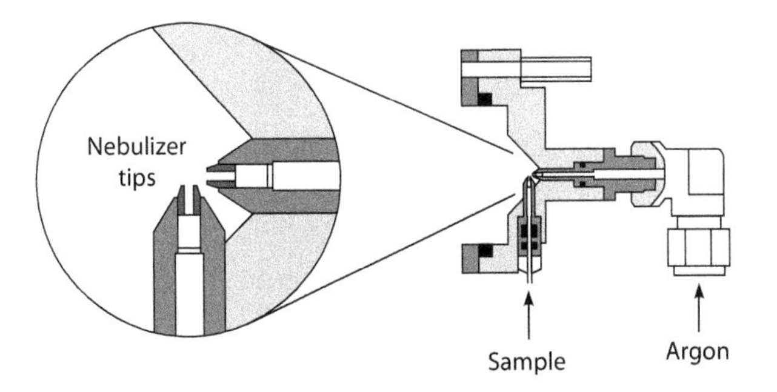

FIGURE 5.8 Schematic of a cross-flow nebulizer. (Copyright © 2013 PerkinElmer Inc., Waltham, MA. All rights reserved.)

Microflow Design

More recently, microflow or high-efficiency nebulizers have been designed for ICP-MS to operate at much lower sample flows. Whereas conventional nebulizers have a sample uptake rate of about 1 mL/min, microflow or high-efficiency nebulizers typically run at less than 0.1 mL/min. They are based on the concentric principle, but usually operate at higher gas pressure to accommodate the lower sample flow rates. The extremely low uptake rate makes them ideal for applications where sample volume is limited or where the sample or analyte is prone to sample introduction memory effects. The additional benefit of this design is that it produces an aerosol with smaller droplets, and as a result, it is generally more efficient than a conventional concentric nebulizer.

These nebulizers and their components are typically constructed from polymer materials, such as polytetrafluoroethylene (PTFE), perfluoroalkoxy (PFA), or polyvinylfluoride (PVF), although some designs are available in borosilicate glass or quartz. The excellent corrosion resistance of the polymer nebulizers means they have naturally low blank levels. This characteristic, together with their ability to handle small sample volumes found in applications such as vapor phase decomposition (VPD), makes them an ideal choice for semiconductor laboratories that are carrying out ultratrace element analysis.[7,8] A microflow concentric nebulizer made from PFA is shown in Figure 5.9, and a typical spray pattern of the nebulization process is shown in Figure 5.10.

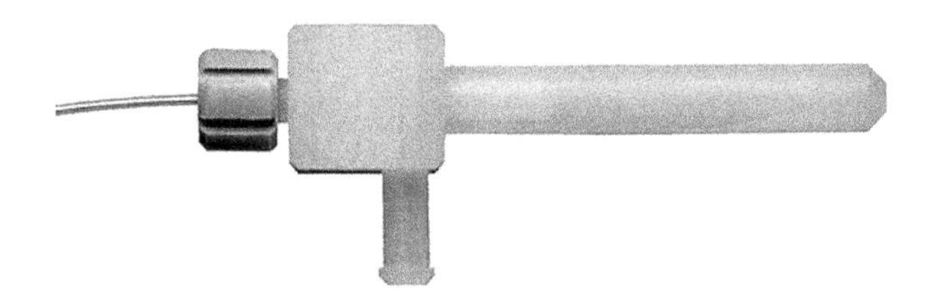

FIGURE 5.9 OpalMist™ microflow concentric nebulizer made from PFA. (Courtesy of Glass Expansion Inc., Pocasset, MA.)

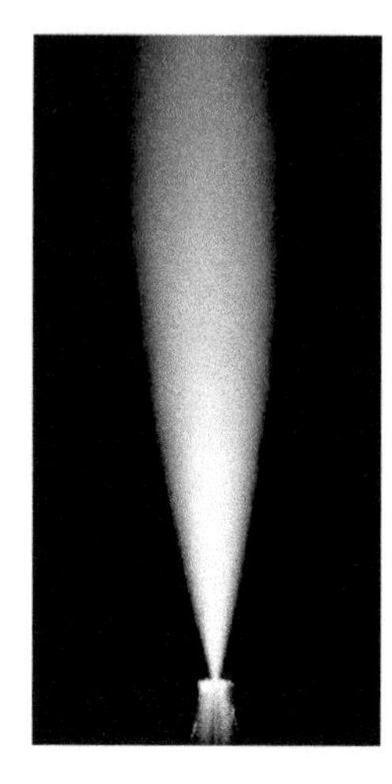

FIGURE 5.10 Spray pattern of a PFA microflow concentric nebulizer. (Courtesy of Elemental Scientific Inc., Omaha, NE.)

The disadvantage of microconcentric nebulizers is that they use an extremely fine capillary, which makes them not very tolerant to high concentrations of dissolved solids or suspended particles. Their high efficiency also means that most of the sample makes it into the plasma, and as a result can cause more severe matrix suppression problems. In addition, the higher level of matrix components entering the interface has the potential to cause cone blockage problems over extended periods of operation. For these reasons, they have been found to be most applicable for the analysis of aqueous-type samples or samples containing low levels of dissolved solids.

One of the application areas in which high-efficiency nebulizers are well suited is the handling of extremely small volumes being eluted from an HPLC or flow injection analyzer (FIA) system into an ICP-MS for doing speciation or microsampling work. The analysis of discrete sample volumes encountered in these types of applications allows for detection limits equivalent to those of a standard concentric nebulizer, while consuming 10–20 times less sample.

SPRAY CHAMBERS

Let us now turn our attention to spray chambers. There are basically two designs that are used in today's commercial ICP-MS instrumentation: double-pass and cyclonic spray chambers. The double-pass is by far the most common, with the cyclonic type rapidly gaining in popularity. As mentioned earlier, the function of the spray chamber is to reject the larger aerosol droplets and also to smooth out nebulization pulses produced by the peristaltic pump, if it is used. In addition, some ICP-MS spray chambers are externally cooled for thermal stability of the sample and to reduce the amount of solvent going into the plasma. This can have a number of beneficial effects, depending on the application, but the main advantages are to reduce oxide species, minimize signal drift, and reduce the solvent loading on the plasma, particularly when aspirating volatile organic solvents.

Double-Pass Spray Chamber

By far the most common design of the double-pass spray chamber is the Scott design, which selects the small droplets by directing the aerosol into a central tube. The larger droplets emerge from the tube, and exit the spray chamber via a drain tube. The liquid in the drain tube is kept at positive pressure (usually by way of a loop), which forces the small droplets back between the outer wall and the central tube, and emerges from the spray chamber into the sample injector of the plasma torch. Double-pass spray chambers come in a variety of shapes, sizes, and materials, and are generally considered the most rugged design for routine use. Figure 5.11 shows a Scott double-pass spray chamber made of a polysulfide-type material, coupled to a cross-flow nebulizer.

Cyclonic Spray Chamber

The cyclonic spray chamber operates by centrifugal force. Droplets are discriminated according to their size by means of a vortex produced by the tangential flow of the sample aerosol and argon gas inside the chamber. Smaller droplets are carried with the gas stream into the ICP-MS, whereas the larger droplets impinge on the walls and fall out through the drain. It is generally accepted that a cyclonic spray chamber has a higher sampling efficiency, which for clean samples translates into higher sensitivity and lower detection limits. However, the droplet size distribution appears to be different from a double-pass design, and for certain types of samples can give slightly inferior precision. Beres and coworkers published a very useful study describing the capabilities of a cyclonic spray chamber.[9] Figure 5.12 shows a cyclonic spray chamber connected to a concentric nebulizer.

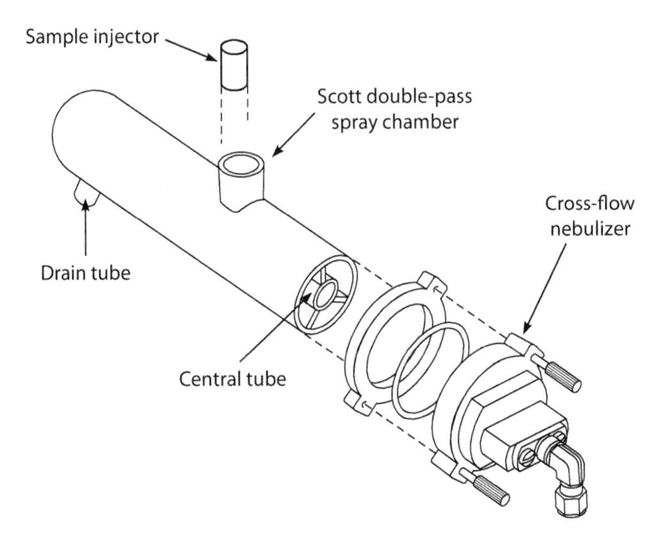

FIGURE 5.11 Scott double-pass spray chamber with cross-flow nebulizer. (Copyright © 2013 PerkinElmer Inc., Waltham, MA. All rights reserved. [https://www.agilent.com/cs/library/technicaloverviews/public/5989-7737EN.pdf])

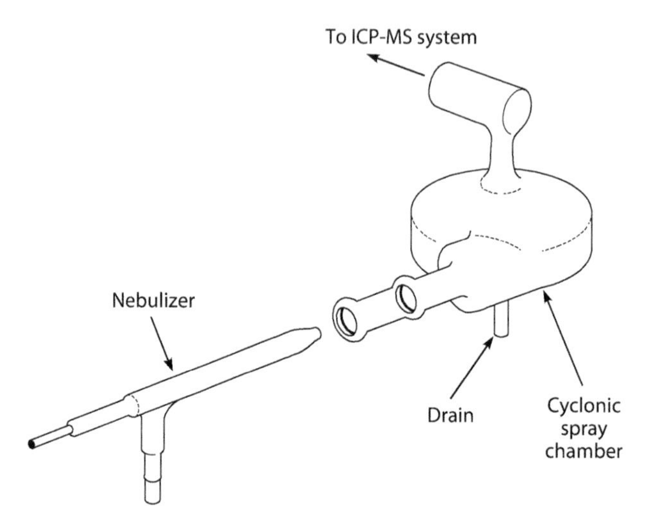

FIGURE 5.12 Cyclonic spray chamber (shown with a concentric nebulizer). (From Beres, S. A., et al., *Atomic Spectrosc.*, 15(2), 96–99, 1994.)

FIGURE 5.13 Low-flow Cinnabar, a water-cooled cyclonic spray chamber for use with a microflow concentric nebulizer. (Courtesy of Glass Expansion Inc., Pocasset, MA.)

The cyclonic spray chamber is growing in popularity, particularly as its potential is getting realized in more and more application areas. Just as there is a wide selection of nebulizers available for different applications, there is also a wide choice of customized cyclonic spray chambers, manufactured from glass, quartz, and different polymer materials. Depending on the application being carried out, modifications to the cyclonic design are available for low sample flows, high dissolved solids, fast sample washout, corrosion resistance, and organic solvents. Figure 5.13 shows one of the many variations of a cyclonic spray chamber, the jacketed Cinnabar™, which is a water-cooled borosilicate glass spray chamber optimized for aspirating small sample volumes with a microflow concentric nebulizer.

It is worth emphasizing that cooling the spray chamber is generally beneficial in ICP-MS, because it reduces the solvent loading on the plasma. This has three major benefits. First, because very little plasma energy is wasted vaporizing the solvent, more is available to excite and ionize the analytes. Second, if there is less water being delivered to the plasma, there is less chance of forming oxide and hydroxide species, which can potentially interfere with other analytes. Finally, if the spray chamber is kept at a constant temperature, it leads to better long-term signal stability, especially if there are environmental temperature changes over the time period of the analysis. For these reasons, some manufacturers supply cooled spray chambers as the standard, whereas others offer the capability as an option. There is also a wide variety of cooled and chilled spray chambers available from third-party vendors.

AEROSOL DILUTION

To address the limitation of the TDS capability of ICP-MS, some vendors offer an aerosol dilution system, which introduces a flow of argon gas between the nebulizer and the torch to carry out aerosol dilution of the sample. This has the effect of reducing the sample's solvent loading on the plasma, so it can tolerate much higher TDSs levels than the <0.2%, which is typical for most ICP-MS instrumentation. However, it's important to emphasize that the dilution is done after the nebulizer, so care must be taken in selecting the optimum nebulizer if the sample contains high levels of dissolved solids. The principles of aerosol dilution are shown in Figure 5.14. Some of the benefits of this novel type of dilution include

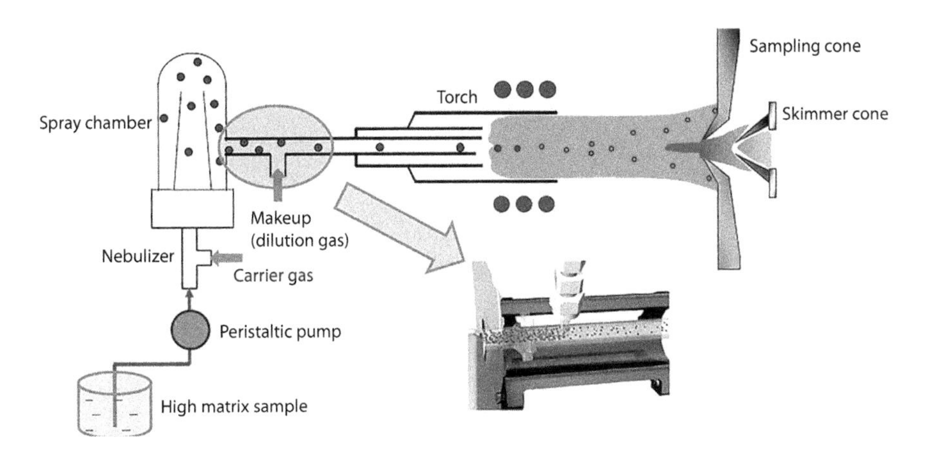

FIGURE 5.14 Principle of aerosol dilution. (From "Today's agilent—New atomic spectroscopy solutions for environmental laboratories," online webinar. [https://www.agilent.com/cs/library/technicaloverviews/public/5989-7737EN.pdf])

- Enables the direct analysis of samples containing medium to high percentage levels of dissolve solids, assuming the nebulizer can handle them
- Significantly improves plasma robustness compared with conventional sample introduction methods
- Because less solvent or matrix is entering the plasma, reduces oxide interferences to very low levels, providing better accuracy and more stable sampling conditions
- Eliminates the need for conventional liquid dilution of high matrix samples prior to analysis, which has several disadvantages, including increased risk of sample contamination, dilution errors, and sample prep time

FINAL THOUGHTS

There are many other nonstandard sample introduction devices, such as laser ablation, ultrasonic nebulizers, desolvation devices, direct injection nebulizers, flow injection systems, enhanced productivity systems, autodilution, and online chemistry techniques, which are not described in this chapter. However, because they are becoming more and more important, particularly as ICP-MS users are demanding higher performance, more productivity, and greater flexibility, they are covered in greater detail in Chapter 20.

REFERENCES

1. R. A. Browner and A. W. Boorn. *Analytical Chemistry*, 56, 786–798A, 1984.
2. B. L. Sharp. *Analytical Atomic Spectrometry*, 3, 613, 1980.
3. L. C. Bates and J. W. Olesik. *Journal of Analytical Atomic Spectrometry*, 5(3), 239, 1990.
4. R. S. Houk. *Analytical Chemistry*, 56, 97A, 1986.
5. J. G. Williams, A. L. Gray, P. Norman, and L. Ebdon. *Journal of Analytical Atomic Spectrometry*, 2, 469–472, 1987.
6. J. Ranville, K. Neubauer, and R. Thomas. *Spectroscopy*, 27(8), 20–27, 2012.
7. E. Debrah, S. A. Beres, T. J. Gluodennis, R. J. Thomas, and E. R. Denoyer. *Atomic Spectroscopy*, 16(7), 197–202, 1995.
8. R. A. Aleksejczyk and D. Gibilisco. Micro, September 1997.
9. S. A. Beres, P. H. Bruckner, and E. R. Denoyer. *Atomic Spectroscopy*, 15(2), 96–99, 1994.

6 Plasma Source

Chapter 6 takes a look at the region of the ICP-MS where the ions are generated—the plasma discharge. It gives a brief historical perspective of some of the common analytical plasmas used over the years and discusses the components used to create the inductively coupled plasma (ICP). It then goes on to explain the fundamental principles of formation of a plasma discharge and how it is used to convert the sample aerosol into a stream of positively charged ions of low kinetic energy required by the ion-focusing system and the mass spectrometer.

ICPs are by far the most common type of plasma sources used in today's commercial ICP optical emission spectroscopy (ICP-OES) and ICP mass spectroscopy (ICP-MS) instrumentation. However, it was not always that way. In the early days, when researchers were attempting to find the ideal plasma source to use for spectrometric studies, it was not clear which approach would prove to be the most successful. In addition to ICPs, some of the other novel plasma sources developed were direct current plasmas (DCPs) and microwave-induced plasmas (MIPs). Before I go on to describe the ICP, let us first take a closer look at these other two excitation sources.

A DCP is formed when a gas (usually argon) is introduced into a high current flowing between two or three electrodes. Ionization of the gas produces a Y-shaped plasma. Unfortunately, early DCP instrumentation was prone to interference effects and also had some usability and reliability problems. For these reasons, the technique never became widely accepted by the analytical community.[1] However, its one major benefit was that it could aspirate high dissolved or suspended solids because there was no restrictive sample injector for the solid material to block. This feature alone made it very attractive for some laboratories, and once the initial limitations of DCPs were better understood, the technique became more accepted. Limitations in the DCP approach led to the development of electrodeless plasma, of which the MIP was the simplest form. MIP technology has mainly been used as an ion source for mass spectrometry (MS),[2] and also as emission-based detectors for gas chromatography. It is only recently that that the technology has advanced and been viewed as a possible alternative to the ICP for elemental analysis.

An MIP basically consists of a quartz tube surrounded by a microwave waveguide or cavity. Microwaves produced from a magnetron fill the cavity and cause the electrons in the plasma support gas to oscillate. The oscillating electrons collide with other atoms in the flowing gas to create and maintain a high-temperature plasma. As in the ICPs, a high-voltage spark is needed to create the initial electrons to create the plasma, which achieves a temperature of approximately 5000 K.

The limiting factor to their use was that with the low power and high frequency of the MIP, it was very difficult to maintain the stability of the plasma when aspirating liquid samples containing high levels of dissolved solids. Various attempts had been made over the years to couple desolvation techniques to the MIP, but only managed to achieve limited success. However, an MIP–atomic emission spectroscopy (AES) system using nitrogen gas was recently developed that appears to have overcome many of the limitations of the earlier designs and is achieving detection limits better than those of flame atomic absorption and only slightly inferior to those of ICP-OES for many elements.[3]

Because of the limitations of the DCP and MIP approaches, ICPs became the dominant area of research for both optical emission and mass spectrometric studies. As early as 1964, Greenfield and coworkers reported that an atmospheric pressure ICP coupled with OES could be used for elemental analysis.[4] Although crude by today's standards, it showed the enormous possibilities of the ICP as an excitation source and most definitely opened the door in the early 1980s to the even more exciting potential of using the ICP to generate ions.[5]

PLASMA TORCH

Before we take a look at the fundamental principles behind the creation of an ICP used in ICP-MS, let us take a look at the basic components used to generate the source—a plasma torch, radio-frequency (RF) coil, and power supply. Figure 6.1 shows their proximity compared with the rest of the instrument, and Figure 6.2 is a more detailed view of the plasma torch and RF coil relative to the MS interface.

The plasma torch consists of three concentric tubes, which are normally made from quartz. In Figure 6.2, these are shown as the outer tube, middle tube, and sample injector. The torch either can be one piece, in which all three tubes are connected, or can employ a demountable design in which the tubes and the sample injector are separate. The gas (usually argon) that is used to form the plasma (plasma gas) is passed between the outer and middle tubes at a flow rate of ~12–17 L/min. A second gas flow (auxiliary gas) passes between the middle tube and the sample injector at

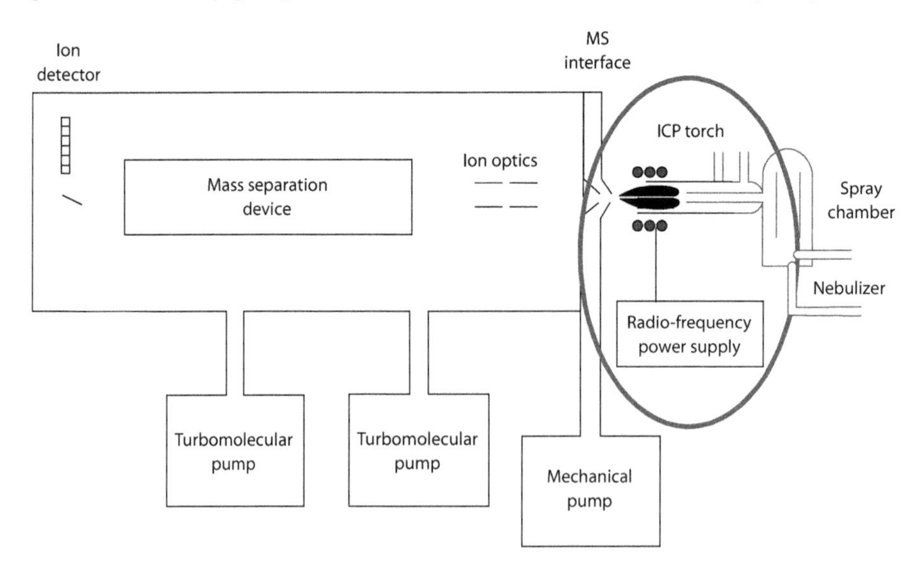

FIGURE 6.1 ICP-MS system showing location of the plasma torch and RF power supply.

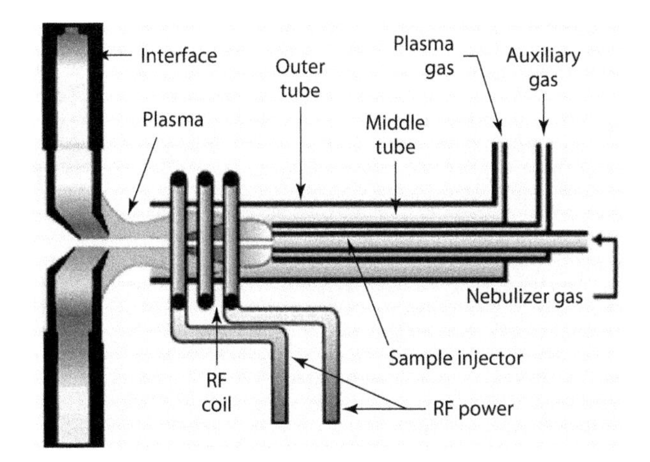

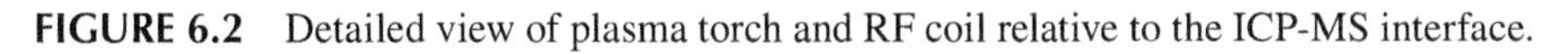
FIGURE 6.2 Detailed view of plasma torch and RF coil relative to the ICP-MS interface.

~1 L/min and is used to change the position of the base of the plasma relative to the tube and the injector. A third gas flow (nebulizer gas), also at ~1 L/min, brings the sample, in the form of a fine-droplet aerosol, from the sample introduction system and physically punches a channel through the center of the plasma. The sample injector is often made from other materials besides quartz, such as alumina, platinum, and sapphire—if highly corrosive materials need to be analyzed. It is worth mentioning that although argon is the most suitable gas to use for all three flows, there are analytical benefits in using other gas mixtures, especially in the nebulizer flow.[6] The plasma torch is mounted horizontally and positioned centrally in the RF coil, approximately 10–20 mm from the interface. This can be seen in Figure 6.3, which shows a photograph of a plasma torch mounted in an instrument.

Ceramic components are also available for most ICP-MS torches. The outer, inner, and sample injector tubes are normally made of quartz, but for some applications, it is beneficial to consider using an alternative material like ceramic. And if a demountable torch is being used, any or all of the tubes can be replaced. Some of the applications that might benefit from a ceramic torch include

- When silicon is one of the analytes, because quartz outer tubes often produce high-Si background signals
- For the analysis of fusion mixtures or samples with high levels of dissolved solids, which might cause devitrification of quartz tubes
- For the analysis of organic-based samples where quartz outer tubes often suffer from a short lifetime

A fully demountable ceramic torch is shown in Figure 6.4.

It's also worth emphasizing that the coil used in an ICP-MS plasma is slightly different from the one used in ICP-OES, the reason being that in a plasma discharge, there is a potential difference of a few hundred volts produced by capacitive coupling between the RF coil and the plasma. In an ICP mass spectrometer, this would result in a secondary discharge between the plasma and the interface cone, which can

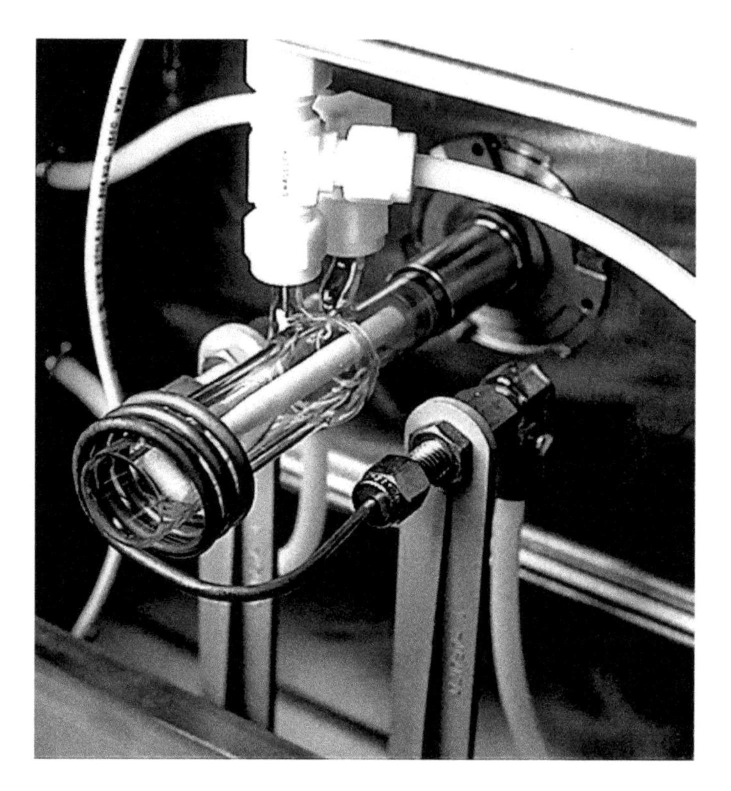

FIGURE 6.3 Photograph of a plasma torch mounted in an instrument. (Copyright © 2013 PerkinElmer Inc., Waltham, MA. All rights reserved.)

FIGURE 6.4 Fully demountable ceramic torch, showing the outer and inner tubes, together with the base and sample injector. (Courtesy of Glass Expansion Inc., Pocasset, MA.)

negatively affect the performance of the instrument. To compensate for this, the coil must be grounded to keep the interface region as close to zero potential as possible. The full implications of this are discussed in greater detail in the next chapter.

FORMATION OF AN ICP DISCHARGE

Let us now discuss the mechanism of formation of the plasma discharge in greater detail. First, a tangential (spiral) flow of argon gas is directed between the outer and middle tube of a quartz torch. A load coil (usually copper) surrounds the top end of the torch and is connected to an RF generator. When RF power (typically 750–1500 W, depending on the sample) is applied to the load coil, an alternating current oscillates within the coil at a rate corresponding to the frequency of the generator. In most ICP generators, this frequency is either 27 or 40 MHz (commonly known as megahertz or million cycles per second). This RF oscillation of the current in the coil causes an intense electromagnetic field to be created in the area at the top of the torch. With argon gas flowing through the torch, a high-voltage spark is applied to the gas, causing some electrons to be stripped from their argon atoms. These electrons, which are caught up and accelerated in the magnetic field, then collide with other argon atoms, stripping off still more electrons. This collision-induced ionization of the argon continues in a chain reaction, breaking down the gas into argon atoms, argon ions, and electrons, forming what is known as an ICP discharge. The ICP discharge is then sustained within the torch and load coil as RF energy is continually transferred to it through the inductive coupling process. The amount of energy required to generate argon ions in this process is on the order of 15.8 eV (first ionization potential), which is enough energy to ionize the majority of the elements in the periodic table. The sample aerosol is then introduced into the plasma through a third tube called the sample injector. The entire process is conceptually shown in Figure 6.5.[7]

FUNCTION OF THE RF GENERATOR

Although the principles of an RF power supply have not changed since the work of Greenfield, the components have become significantly smaller. Some of the early generators that used nitrogen or air required 5–10 kW of power to sustain the plasma discharge—and literally took up half the room. Most of today's generators use solid-state electronic components, which means that vacuum power amplifier tubes are no longer required. This makes modern instruments significantly smaller and, because vacuum tubes were notoriously unreliable and unstable, far more suitable for routine operation.

As mentioned previously, two frequencies have typically been used for ICP RF generators—27 and 40 MHz. These frequencies have been set aside specifically for RF applications of this kind, so that they will not interfere with other communication-based frequencies. There has been much debate over the years as to which frequency gives the best performance.[8,9] I think it is fair to say that although there have been a number of studies, no frequency appears to give a significant analytical advantage over the other. In fact, of all the commercially available ICP-MS systems, there seems to be roughly an equal number of 27 and 40 MHz generators.

The more important consideration is the coupling efficiency of the RF generator to the coil. The majority of modern solid-state RF generators are on the order

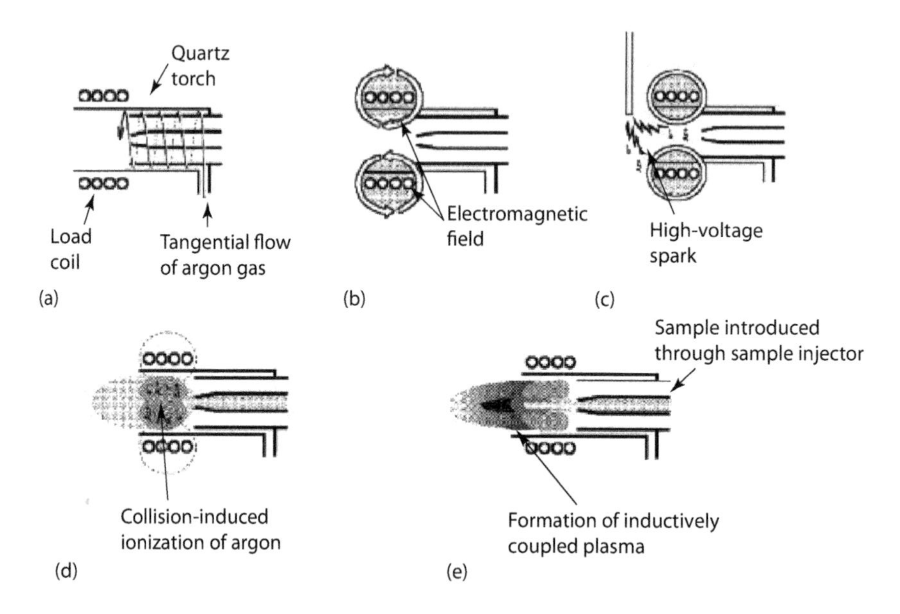

FIGURE 6.5 Schematic of an ICP torch and load coil showing how the ICP is formed. (a) A tangential flow of argon gas is passed between the outer and middle tube of the quartz torch. (b) RF power is applied to the load coil, producing an intense electromagnetic field. (c) A high-voltage spark produces free electrons. (d) Free electrons are accelerated by the RF field, causing collisions and ionization of the argon gas. (e) The ICP is formed at the open end of the quartz torch. The sample is introduced into the plasma via the sample injector. (From Boss, C. B., and Fredeen, K. J., *Concepts, Instrumentation and Techniques in Inductively Coupled Plasma Optical Emission Spectrometry*, 2nd ed., PerkinElmer Inc., Waltham, MA, 1997.)

of 70%–75% efficient, which means that 70%–75% of the delivered power actually makes it into the plasma. This was not always the case, and some of the older vacuum-tube-designed generators were notoriously inefficient, with some of them experiencing more than a 50% power loss. Another important criterion to consider is the way the matching network compensates for changes in impedance (a material's resistance to the flow of an electric current) produced by the sample's matrix components or differences in solvent volatility, or both. In earlier-designed crystal-controlled generators, this was usually done with servo-driven capacitors. They worked very well with most sample types but, because they were mechanical devices, struggled to compensate for very rapid impedance changes produced by some samples. As a result, it was fairly easy to extinguish the plasma, particularly when aspirating volatile organic solvents.

These problems were partially overcome by the use of free-running RF generators, in which the matching network was based on electronic tuning of small changes in frequency brought about by the sample solvent or matrix components, or both. The major benefit of this approach was that compensation for impedance changes was virtually instantaneous, because there were no moving parts. This allowed for the successful analysis of many sample types, which would most probably have extinguished the plasma of a crystal-controlled generator. However, because of improvements in electronic components over the years, the more recent crystal-controlled generators appear to be as responsive as free-running designs.

However, it should be mentioned that the recent development of a novel 34 MHz free-running-designed RF generator using solid-state electronics has enhanced the capability of ICP-MS to analyze some real-world samples, particularly when using cool plasma conditions (see Chapter 16 on interference reduction). This new design, which is based on an air-cooled plasma load coil, allows the matching network electronics to rapidly respond to changes in the plasma impedance produced by different sampling conditions and sample matrices, while still maintaining low plasma potential at the interface region. This technology appears to offer some real benefits over traditional 27 and 40 MHz generators for some applications.[10]

IONIZATION OF THE SAMPLE

To better understand what happens to the sample on its journey through the plasma source, it is important to understand the different heating zones within the discharge. Figure 6.6 shows a cross-sectional representation of the discharge along with the approximate temperatures for different regions of the plasma.

As mentioned previously, the sample aerosol enters the injector via the spray chamber. When it exits the sample injector, it is moving at such a velocity that it physically punches a hole through the center of the plasma discharge. It then goes through a number of physical changes, starting at the preheating zone and continuing through the radiation zone, before it eventually becomes a positively charged ion in the analytical zone. To explain this in a very simplified way, let us assume that the element exists as a trace metal salt in solution. The first step that takes place is desolvation of the droplet. With the water molecules stripped away, it then becomes a very small solid particle. As the sample moves further into the plasma, the solid particle changes first into gaseous form and then into a ground-state atom. The final process of conversion of an atom to an ion is achieved mainly by collisions of energetic argon electrons (and to a lesser extent by argon ions) with the ground-state atom.[11] The ion then emerges from the plasma and is directed into the interface of the mass

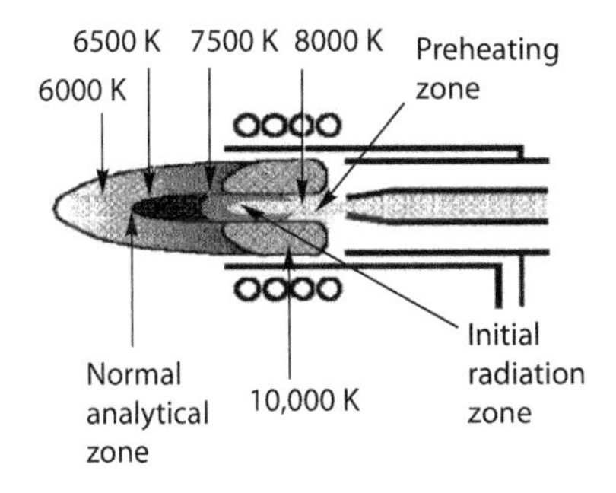

FIGURE 6.6 Different temperature zones in the plasma. (From Boss, C. B., and Fredeen, K. J., *Concepts, Instrumentation and Techniques in Inductively Coupled Plasma Optical Emission Spectrometry*, 2nd ed., PerkinElmer Inc., Waltham, MA, 1997.)

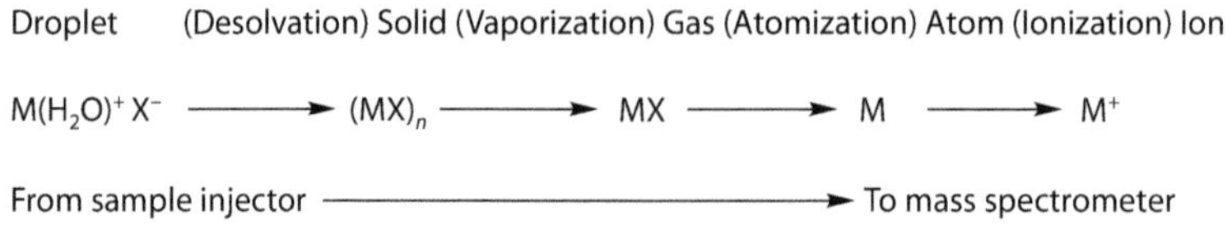

Droplet (Desolvation) Solid (Vaporization) Gas (Atomization) Atom (Ionization) Ion

$M(H_2O)^+ X^- \longrightarrow (MX)_n \longrightarrow MX \longrightarrow M \longrightarrow M^+$

From sample injector ⟶ To mass spectrometer

FIGURE 6.7 Mechanism of conversion of a droplet to a positive ion in the ICP.

spectrometer (for details on the mechanisms of ion generation, refer to Chapter 4). This process of conversion of droplets into ions is represented in Figure 6.7.

REFERENCES

1. L. Gray. *Analyst*, 100, 289–299, 1975.
2. D. J. Douglas and J. B. French. *Analytical Chemistry*, 53, 37–41, 1981.
3. R. J. Thomas. Emerging technology trends in atomic spectroscopy are solving real-world application problems. *Spectroscopy*, 29(3), 42–51, 2014.
4. S. Greenfield, I. L. Jones, and C. T. Berry. *Analyst*, 89, 713–720, 1964.
5. R. S. Houk, V. A. Fassel, and H. J. Svec. *Dynamic Mass Spectrometry*, 6, 234, 1981.
6. J. W. Lam and J. W. McLaren. *Journal of Analytical Atomic Spectrometry*, 5, 419–424, 1990.
7. C. B. Boss and K. J. Fredeen. *Concepts, Instrumentation and Techniques in Inductively Coupled Plasma Optical Emission Spectrometry*. 2nd ed. Waltham, MA: Perkin Elmer Inc., 1997.
8. K. E. Jarvis, P. Mason, T. Platzner, and J. G. Williams. *Journal of Analytical Atomic Spectrometry*, 13, 689–696, 1998.
9. G. H. Vickers, D. A. Wilson, and G. M. Hieftje. *Journal of Analytical Atomic Spectrometry*, 4, 749–754, 1989.
10. K. Neubaeur. The use of ion-molecule chemistry combined with optimized plasma conditions to meet SEMI tier C guidelines for elemental impurities in semiconductor grade hydrochloric acid by ICP-MS. *Spectroscopy*, October 2017.
11. T. Hasegawa and H. Haraguchi. *ICPs in Analytical Atomic Spectrometry*, ed. A. Montasser and D. W. Golightly. 2nd ed. New York: VCH, 1992.

7 Interface Region

Chapter 7 takes a look at the inductively coupled plasma mass spectrometry (ICP-MS) interface region, which is probably the most critical area of the entire ICP-MS system. It gave the early pioneers of the technique the most problems to overcome. Although we take all the benefits of ICP-MS for granted, the process of taking a liquid sample, generating an aerosol that is suitable for ionization in the plasma, and then sampling a representative number of analyte ions, transporting them through the interface, focusing them via the ion optics into the mass spectrometer, and finally ending up with detection and conversion to an electronic signal is not a trivial task. Each part of the journey has its own unique problems to overcome, but probably the most challenging is the extraction of the ions from the plasma into the mass spectrometer.

The role of the interface region, which is shown in Figure 7.1, is to transport the ions efficiently, consistently, and with electrical integrity from the plasma, which is at atmospheric pressure (760 torr), to the mass spectrometer analyzer region at approximately 10^{-6} torr.

This is first achieved by directing the ions into the interface region. The interface consists of two or three metallic cones (depending on the design) with very small orifices, which are maintained at a vacuum of ~1–2 torr with a mechanical roughing pump. After the ions are generated in the plasma, they pass into the first cone, known as the sampler cone, which has an orifice of 0.8–1.2 mm inner diameter (i.d.). From there, they travel a short distance to the skimmer cone, which is generally smaller and more pointed than the sampler cone. The skimmer also has a much smaller orifice (typically 0.4–0.8 mm i.d.) than the sampler cone. In some designs, there is a third cone called the hyper skimmer cone, which is used to reduce the vacuum into smaller steps and provide less dispersion of the ion beam. Whether the system uses two cones or incorporates a triple-cone interface, they are usually made of nickel, but can be made of other materials, such as platinum, which is far more tolerant to corrosive liquids. To reduce the effects of high-temperature plasma on the cones, the interface housing is water cooled and made from a material that dissipates heat easily, such as copper or aluminum. The ions then emerge from the skimmer cone, where they are directed through the ion optics and, finally, guided into the mass separation device. Figure 7.2 shows the interface region in greater detail, and Figure 7.3 shows a close-up of a platinum sampler cone on the left and a platinum skimmer cone on the right.

It should be noted that for most sample matrices, it is desirable to keep the total dissolved solids (TDSs) below 0.2%, because of the possibility of deposition of the matrix components around the sampler cone orifice. This is not such a serious problem with short-term use, but it can lead to long-term signal instability if the instrument is being run for extended periods of time. The TDS levels can be higher (0.5%–1%) when analyzing a matrix that forms a volatile oxide, such as sodium chloride, because once deposited on the cones, the volatile sodium oxide tends to

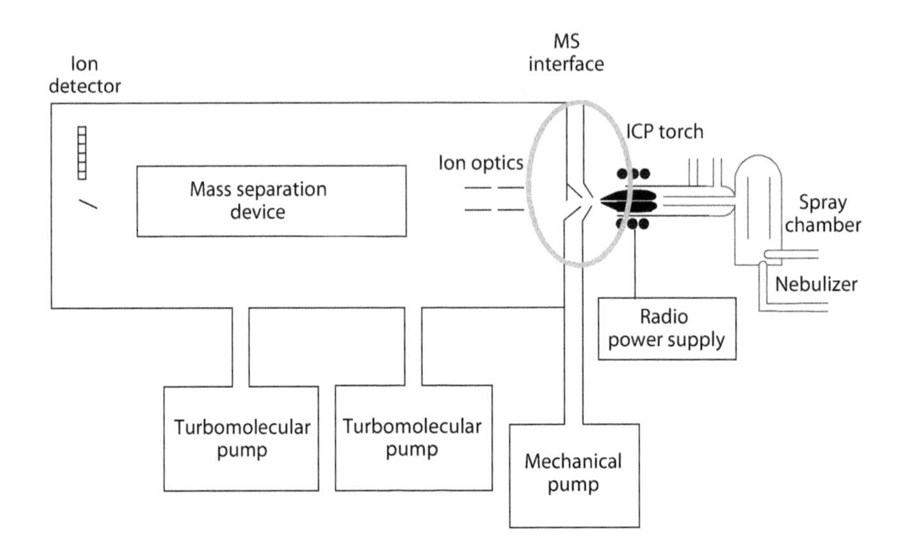

FIGURE 7.1 Schematic of an ICP mass spectrometer showing the proximity of the interface region.

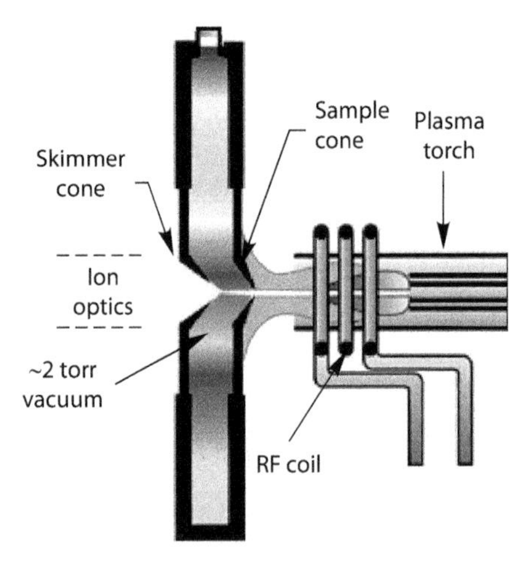

FIGURE 7.2 Detailed view of the interface region.

revaporize without forming a significant layer that could potentially affect the flow through the cone orifice. in fact, some researchers have reported running a 1:1 dilution of seawater (1.5% NaCl) for extended periods of time with good stability and no significant cone blockage—by careful optimization of the plasma RF power, sampling depth, and extraction lens voltage.[1]

More recently, a novel technique for the handling of high-matrix samples has been developed that introduces a makeup gas between the spray chamber and the torch. This technique, which has been termed "aerosol dilution," does not require

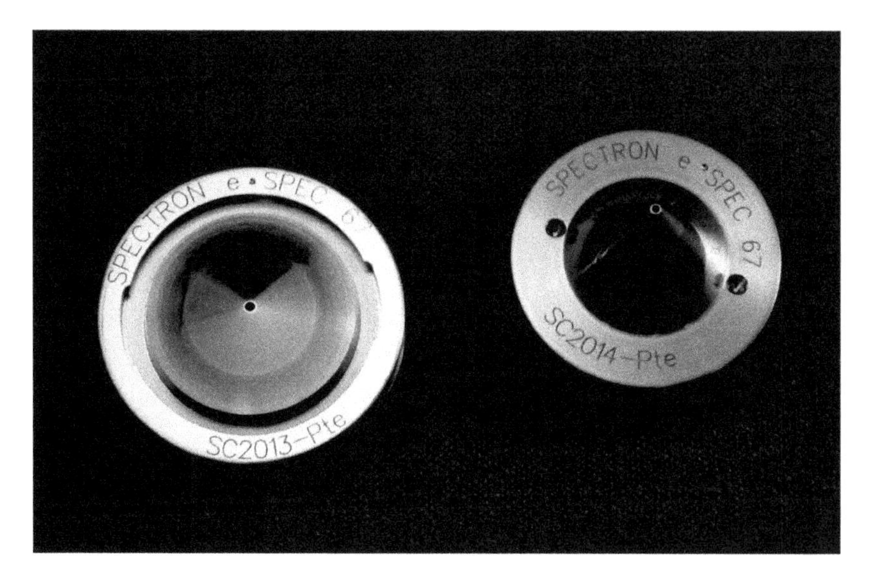

FIGURE 7.3 Close-up of a platinum sampler cone (left) and a platinum skimmer cone (right). (Courtesy of Spectron Inc., Ventura, CA.)

the introduction of more water to the system and protects the mass spectrometer components from the high levels of matrix components in the sample. This technique enables the ICP-MS system to directly aspirate 2%–3% TDSs, with the added benefit of reducing oxide levels, because the sample is not being diluted with water in the traditional way. This technique was described in greater detail in Chapter 5 on sample introduction.

CAPACITIVE COUPLING

The coupling of the plasma to the mass spectrometer proved to be very problematic during the early development of ICP-MS because of an undesirable electrostatic (capacitive) coupling between the voltage on the load coil and the plasma discharge, producing a potential difference of 100–200 V. Although this potential is a physical characteristic of all ICP discharges, it was more serious in an ICP mass spectrometer, because the capacitive coupling created an electrical discharge between the plasma and the sampler cone. This discharge, commonly called the "pinch effect" or secondary discharge, shows itself as arcing in the region where the plasma is in contact with the sampler cone.[2] This is shown in a simplified manner in Figure 7.4.

If not addressed, this arcing can cause all kinds of problems, including an increase in doubly charged interfering species, a wide kinetic energy spread of sampled ions, the formation of ions generated from the sampler cone, and a decreased orifice lifetime. These were all problems reported by many of the early researchers into the technique.[3,4] In fact, because the arcing increased with sampler cone orifice size, the source of the secondary discharge was originally thought to be the result of an electro-gas-dynamic effect, which produced an increase in electron density at the orifice.[5] After many experiments, it was eventually realized that the secondary

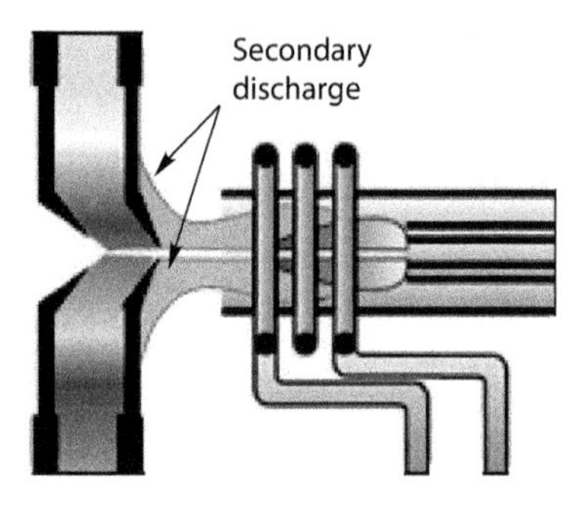

FIGURE 7.4 Interface showing area affected by a secondary discharge.

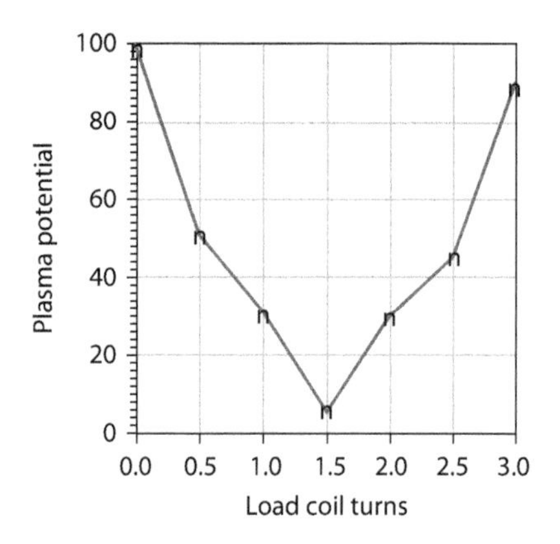

FIGURE 7.5 Reduction in plasma potential as the load coil is grounded at different positions (turns) along its length.[6]

discharge was a result of electrostatic coupling of the load coil to the plasma. The problem was first eliminated by grounding the induction coil at the center, which had the effect of reducing the RF potential to a few volts. This can be seen in Figure 7.5, which is taken from one of the early papers and shows the reduction in plasma potential as the coil is grounded at different positions (turns) along its length.[6]

This work has since been supported by other researchers who carried out Langmuir probe measurements, the results indicating that plasma potential was lowest with a center-tapped coil, as opposed to the grounding being elsewhere on the coil.[7,8] In today's instrumentation, the "grounding" is implemented in a number of different ways, depending on the design of the interface. Some of the most popular designs include balancing the oscillator inside the circuitry of the RF generator,[9] positioning a grounded shield or plate between the coil and the plasma torch,[10] and using two interlaced coils where the RF fields go in opposite directions.[11] They all work differently, but many experts believe that the center-tapped coil and the interlaced coil achieve the lowest plasma potential compared with the other designs. However, they all appear to work equally well when it comes to using cool plasma conditions requiring higher RF power and lower nebulizer gas flow. Further details about cool and cold plasma technology can be found in Chapter 16 on reducing interferences.

ION KINETIC ENERGY

The impact of a secondary discharge cannot be overemphasized with respect to its effect on the kinetic energy of the ions being sampled. It is well documented that the energy spread of the ions entering the mass spectrometer must be as low as possible to ensure they can all be focused efficiently and with full electrical integrity by the ion optics and the mass separation device. When the ions emerge from the argon plasma, they will all have different kinetic energies, depending on their mass-to-charge ratio. Their velocities should all be similar, because they are controlled by rapid expansion of the bulk plasma, which will be neutral as long as it is maintained at zero potential. As the ion beam passes through the sampler cone into the skimmer cone or cones, expansion will take place, but its composition and integrity will be maintained, assuming the plasma is neutral. This can be seen in Figure 7.6.

Electrodynamic forces do not play a role as the ions enter the sampler or the skimmer, because the distance over which the ions exert an influence on one another (known as the Debye length) is small (typically 10^{-3} to 10^{-4} mm) compared with the diameter of the orifice (0.5–1.0 mm),[6] as shown in Figure 7.7.

It is therefore clear that maintaining a neutral plasma is of paramount importance to guarantee the electrical integrity of the ion beam as it passes through the interface region. If a secondary discharge is present, the electrical characteristics of the plasma change, which will affect the kinetic energy of the ions differently, depending on their mass-to-charge ratio. If the plasma is at zero potential, the ion energy spread is on the order of 5–10 eV. However, if a secondary discharge is present, it results in a much wider spread of ion energies entering the mass spectrometer (typically 20–40 eV), which makes ion focusing far more complicated.[6]

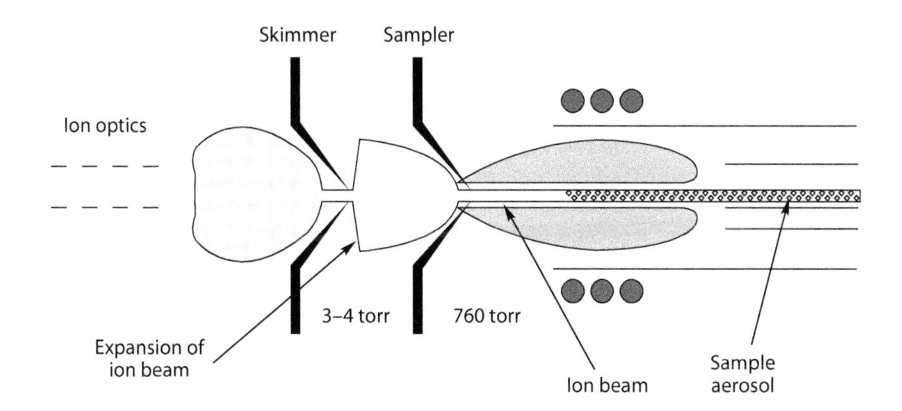

FIGURE 7.6 The composition of the ion beam is maintained as it passes through the interface, a neutral plasma being assumed.

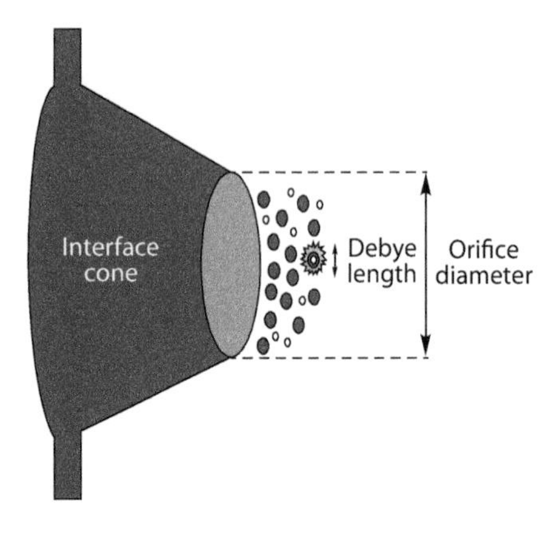

FIGURE 7.7 Electrodynamic forces do not affect the composition of the ion beam entering the sampler or the skimmer cone.

BENEFITS OF A WELL-DESIGNED INTERFACE

The benefits of a well-designed interface are not readily obvious if simple aqueous samples are being analyzed using only one set of operating conditions. However, it becomes more apparent when many different sample types are being handled, requiring different operating parameters. The design of the interface is really put to the test when plasma conditions need to be changed, when the sample matrix changes, or when ICP-MS is being used to analyze solid materials. Analytical scenarios such as these have the potential to induce a secondary discharge, change the kinetic energy of the ions entering the mass spectrometer, and affect the tuning of the ion optics. It is therefore critical that the interface grounding mechanism be able to handle these types of real-world analytical situations, including the following:

- *Using cool plasma conditions:* Although not utilized so much (but still useful for some applications) since the development of collision/reaction cells and interfaces, all instruments today have the ability to use cool plasma conditions. By reducing RF power to 500–700 W and increasing nebulizer gas flow to 1.0–1.3 L/min, the plasma temperature is lowered, which reduces argon-based polyatomic interferences such as $^{40}Ar^{16}O^+$, $^{40}Ar^+$, and $^{38}ArH^+$ in the determination of difficult elements such as $^{56}Fe^+$, $^{40}Ca^+$, and $^{39}K^+$. Such dramatic deviations from normal operating conditions (~1000 W, 0.8 L/min) will affect the electrical characteristics of the plasma. (Note: For some applications, it can be beneficial to combine cool or cold plasma conditions with collision/reaction cell technology to reduce some polyatomic spectral interferences.)
- *Running organic solvents:* Analyzing oil or organic-based samples requires a chilled spray chamber or a membrane desolvation system to reduce the solvent loading on the plasma. In addition, higher RF power (~1300–1500 W) and lower nebulizer gas flow (~0.4–0.8 L/min) are required to dissociate the organic components in the sample. A reduction in the amount of

solvent entering the plasma, combined with higher power and lower nebulizer gas flow, translates into a hotter plasma, and a change in its ionization mechanism.

- *Optimizing conditions for low oxides:* The formation of oxide species can be problematic in some sample types. For example, in geochemical applications it is quite common to sacrifice sensitivity by lowering the nebulizer gas flow and increasing the RF power to reduce the formation of rare earth oxides, which can spectrally interfere with the determination of other analytes. Unfortunately, these conditions will change the electrical characteristics of the plasma, which can induce a secondary discharge.
- *Using sampling accessories:* Sampling accessories such as membrane desolvators and laser ablation systems are being used more routinely to improve performance and productivity and enhance the flexibility of ICP-MS. Some of these sampling devices, such as laser ablation or membrane desolvation, generate a "dry" sample aerosol, which requires completely different operating conditions compared with conventional "wet" plasma. An aerosol that contains no solvent can have a dramatic effect on the ionization conditions in the plasma.

FINAL THOUGHTS

Even though most modern ICP-MS interfaces have been designed to minimize the effects of the secondary discharge, it should not be taken for granted that they can all handle changes in operating conditions and matrix components with the same ease. The most noticeable problems that have been reported include spectral peaks of the cone material appearing in the blank, erosion and discoloration of the sampling cones, widely different optimum plasma conditions (neb flow and RF power) for different masses, and frequent retuning of the ion optics.[12,13] Chapter 27 on how best to evaluate ICP-MS instrumentation goes into this subject in greater detail, but there is no question that the plasma discharge, interface region, and ion optics have to be designed in concert to ensure that the instrument can handle a wide range of operating conditions and sample types. For the best way to clean interface cones that may have been impacted by long-term aspiration of matrix components or discolored by acid erosion, refer to Spectron Inc.[14]

REFERENCES

1. M. Plantz and S. Elliott. Application Note ICP-MS 17. Palo Alto, CA: Varian Instruments, 1998.
2. A. L. Gray and A. R. Date. *Analyst*, 108, 1033, 1983.
3. R. S. Houk, V. A. Fassel, and H. J. Svec. *Dynamic Mass Spectrometry*, 6, 234, 1981.
4. A. R. Date and A. L. Gray. *Analyst*, 106, 1255, 1981.
5. A. L. Gray and A. R. Date. *Dynamic Mass Spectrometry*, 6, 252, 1981.
6. D. J. Douglas and J. B. French. *Spectrochimica Acta*, 41B(3), 197, 1986.
7. A. L. Gray, R. S. Houk, and J. G. Williams. *Journal of Analytical Atomic Spectrometry*, 2, 13–20, 1987.

8. R. S. Houk, J. K. Schoer, and J. S. Crain. *Journal of Analytical Atomic Spectrometry*, 2, 283–286, 1987.
9. S. D. Tanner. *Journal of Analytical Atomic Spectrometry*, 10, 905, 1995.
10. K. Sakata and K. Kawabata. *Spectrochimica Acta*, 49B, 1027, 1994.
11. S. Georgitus and M. Plantz. Winter Conference on Plasma Spectrochemistry, FP4, Fort Lauderdale, FL, 1996.
12. D. J. Douglas. *Canadian Journal of Spectroscopy*, 34, 2, 1989.
13. J. E. Fulford and D. J. Douglas. *Applied Spectroscopy*, 40, 7, 1986.
14. Spectron Inc. Cone care and maintenance. Application Note. http://www.spectronus.com/uploadcache/1253135846 Cone_Cleaning_Final_909.pdf.

8 Ion-Focusing System

Chapter 8 takes a detailed look at the inductively coupled plasma mass spectrometry (ICP-MS) ion-focusing system—a crucial area of the ICP mass spectrometer—where the ion beam is focused before it enters the mass analyzer. Sometimes known as the ion optics, it comprises one or more ion lens components, which electrostatically steer the analyte ions in an axial (straight) or orthogonal (right-angled) direction from the interface region into the mass separation device. The strength of a well-designed ion-focusing system is its ability to produce a flat signal response over the entire mass range, low background levels, good detection limits, and stable signals in real-world sample matrices.

Although the detection capability of ICP-MS is generally recognized as being superior to that of any of the other atomic spectroscopic techniques, it is probably most susceptible to the sample's matrix components. The inherent problem lies in the fact that ICP-MS is relatively inefficient—out of a million ions generated in the plasma, only a few ions actually reach the detector. One of the main contributing factors to the low efficiency is the higher concentration of matrix elements compared with the analyte, which has the effect of defocusing the ions and altering the transmission characteristics of the ion beam. This is sometimes referred to as a space charge effect, and can be particularly severe when the matrix ions are of a heavier mass than the analyte ions.[1] The role of the ion-focusing system is therefore to transport the maximum number of analyte ions from the interface region to the mass separation device, while rejecting as many of the matrix components and non-analyte-based species as possible. Let us now discuss this process in greater detail.

ROLE OF THE ION OPTICS

The ion optics, shown in Figure 8.1, are positioned between the skimmer cone (or cones) and the mass separation device. They typically consist of one or more electrostatically controlled lens components, maintained at a vacuum of approximately 10^{-3} torr with a turbomolecular pump.

They are not traditional optics that we associate with ICP emission or atomic absorption, but are made up of a series of metallic plates, barrels, or ion mirrors, which have a voltage placed on them. A recent commercial design implements a quadrupole deflector, which turns the ion beam at right angles into the mass spectrometer. Whatever the design, the function of the ion optic system is to take ions from the hostile environment of the plasma at atmospheric pressure via the interface cones and steer them into the mass analyzer, which is under high vacuum. The nonionic species, such as particulates, neutral species, and photons, are prevented from reaching the detector by using some kind of physical barrier, positioning the mass analyzer off axis relative to the ion beam, or electrostatically bending the ions by 90° into the mass analyzer.

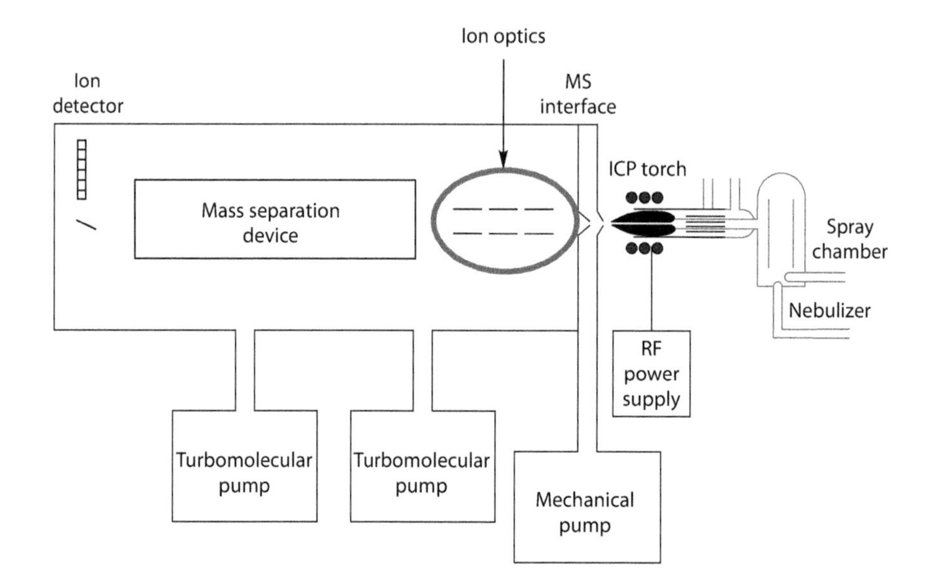

FIGURE 8.1 Position of ion optics relative to the plasma torch and interface region. RF, radio frequency.

As mentioned in previous chapters, the plasma discharge and interface region have to be designed in concert with the ion optics. It is absolutely critical that the composition and electrical integrity of the ion beam be maintained as it enters the ion optics. For this reason, it is essential that the plasma be at zero potential to ensure that the magnitude and spread of ion energies are as low as possible.[2]

A secondary, but also very important, role of the ion optic system is to stop particulates, neutral species, and photons from getting through to the mass analyzer and the detector. These species cause signal instability and contribute to background levels, which ultimately affect the performance of the system. For example, if photons or neutral species reach the detector, they will elevate the noise of the background and therefore degrade detection capability. In addition, if particulates from the matrix penetrate further into the mass spectrometer region, they have the potential to deposit on lens components and, in extreme cases, get into the mass analyzer. In the short term, this will cause signal instability, and in the long term, it will increase the frequency of cleaning and routine maintenance.

There are basically four different approaches of reducing the chances of these undesirable species entering the mass spectrometer. The first method is to place a grounded metal stop (disk) behind the skimmer cone. This stop allows the ion beam to move around it and physically block the particulates, photons, and neutral species from traveling "downstream."[3] Although implemented very successfully in earlier instruments, this design has not been utilized for a number of years, mainly because it can have a negative impact on instrument sensitivity, by reducing the number of ions reaching the detector. The second approach is to set the mass analyzer off axis to the ion lens system (in some systems this is called a chicane design). The positively charged ions are then steered with the lens components into the mass analyzer, while the photons, neutrals, and nonionic species are ejected out of the ion beam.[4]

The third development is to deflect the ion beam 90° with a "hollow" ion mirror.[5] This allows the photons, neutrals, and solid particles to pass through, whereas the ions are deflected at right angles into an off-axis mass analyzer that incorporates curved fringe rod technology.[6] The fourth and most recent approach is to deflect the ion beam emerging from the plasma by 90°. This has the effect of changing direction and focusing the ion beam into the mass spectrometer, while allowing the neutral species, photons, and particulate matter to go straight through and be ejected.

It is also worth mentioning that some lens systems incorporate an extraction lens after the skimmer cone to electrostatically "pull" the ions from the interface region. This has the benefit of improving the transmission and detection limits of the low-mass elements (which tend to be pushed out of the ion beam by the heavier elements), resulting in a more uniform response across the full mass range. In an attempt to reduce these space charge effects, some older designs have utilized lens components to accelerate the ions downstream. Unfortunately, this can have the effect of degrading the resolving power and abundance sensitivity (ability to differentiate an analyte peak from the wing of an interference) of the instrument, because of the much higher kinetic energy of the accelerated ions as they enter the mass analyzer.[7]

DYNAMICS OF ION FLOW

To fully understand the role of the ion optics in ICP-MS, it is important to get an appreciation of the dynamics of ion flow from the plasma through the interface region into the mass spectrometer. When the ions generated in the plasma emerge from the skimmer cone (or cones), there is a rapid expansion of the ion beam as the pressure is reduced from 760 torr (atmospheric pressure) to approximately 10^{-3} to 10^{-4} torr in the lens chamber with a turbomolecular pump. The composition of the ion beam immediately behind the cone is the same as that in front of the cone because the expansion at this stage is controlled by normal gas dynamics and not by electrodynamics. One of the main reasons for this is that in the ion-sampling process, the Debye length (the distance over which ions exert influence on one another) is small compared with the orifice diameter of the sampler or skimmer cone. Consequently, there is little electrical interaction between the ion beam and the cone, and relatively little interaction between the individual ions in the beam. In this way, the compositional integrity of the ion beam is maintained throughout the interface region.[8] With the rapid drop in pressure in the lens chamber, electrons diffuse out of the ion beam. Because of the small size of the electrons relative to the positively charged ions, the electrons diffuse further from the beam than the ions, resulting in an ion beam with a net positive charge. This is represented schematically in Figure 8.2.

The generation of a positively charged ion beam is the first stage in the charge separation process. Unfortunately, the net positive charge of the ion beam means that there is now a natural tendency for the ions to repel each other. If nothing is done to compensate for this, ions of higher mass-to-charge ratio will dominate the center of the ion beam and force the lighter ions to the outside. The degree of loss will depend on the kinetic energy of the ions—those with high kinetic energy (high-mass elements) will be transmitted in preference to ions with medium (midmass elements) or low (low-mass elements) kinetic energy. This is shown in Figure 8.3.

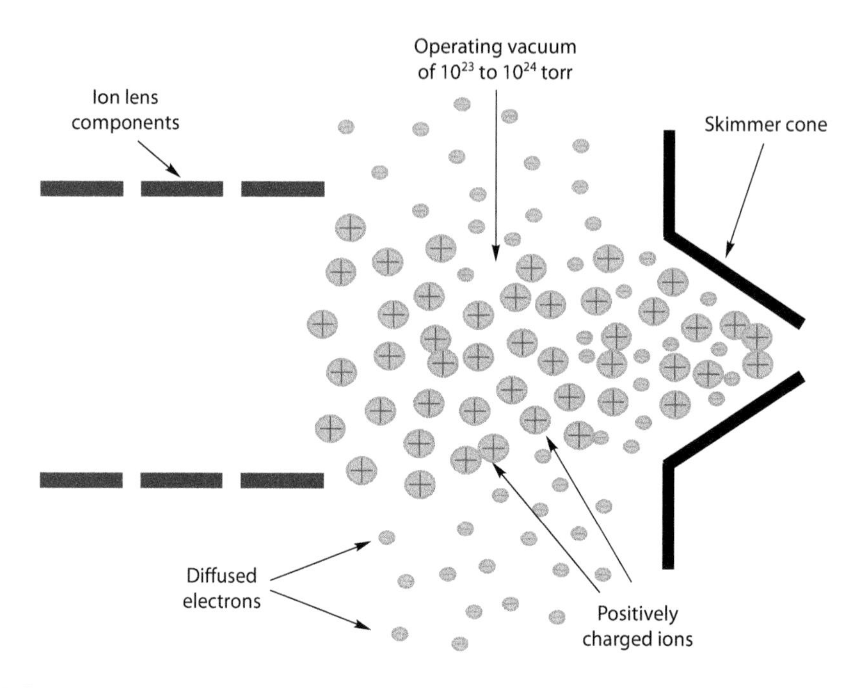

FIGURE 8.2 An extreme pressure drop in the ion optic chamber produces diffusion of electrons, resulting in a positively charged ion beam.

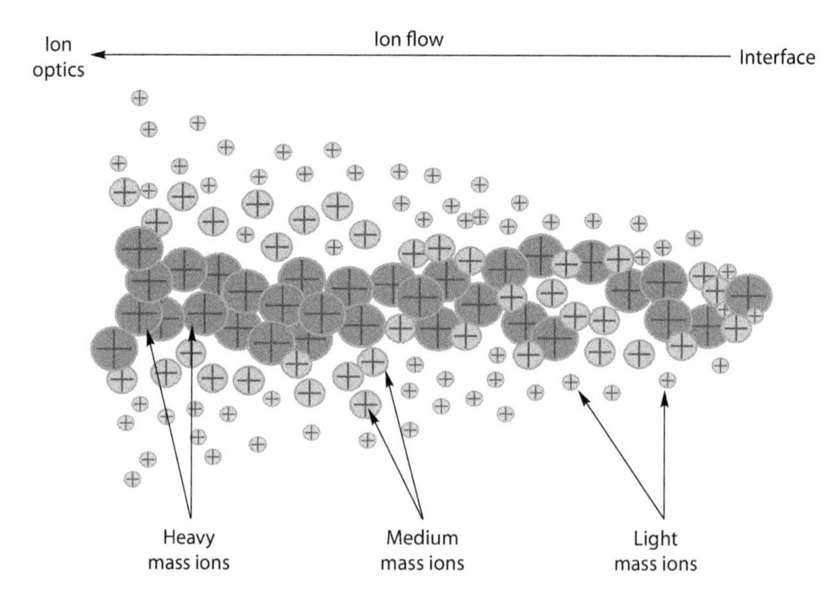

FIGURE 8.3 The degree of ion repulsion will depend on the kinetic energy of the ions—those with high kinetic energy (heavy masses) will be transmitted in preference to those with medium (medium masses) or low (light masses) kinetic energy.

The second stage of charge separation therefore consists of electrostatically steering the ions of interest back into the center of the ion beam using the ion lens system. It should be emphasized that this is only possible if the interface is kept at zero potential, which ensures a neutral gas dynamic flow through the interface, maintaining the compositional integrity of the ion beam. It also guarantees that the average ion energy and energy spread of each ion entering the lens systems are at levels optimum for mass separation. If the interface region is not grounded correctly, stray capacitance will generate a discharge between the plasma and sampler cone and increase the kinetic energy of the ion beam, making it very difficult to optimize the ion lens voltages (refer to Chapter 7 for details).

COMMERCIAL ION OPTIC DESIGNS

Over the years, there have been many different ion optic designs. Although they have their own individual characteristics, they perform the same basic function of allowing the maximum number of analyte ions through to the mass analyzer, while at the same time rejecting the undesirable matrix- and solvent-based ions. The oldest and most mature design of ion optics in use today consists of several lens components, all of which have a specific role to play in the transmission of the analyte ions with a minimum of mass discrimination. With these multicomponent lens systems, the voltage can be optimized on every lens of the ion optics to achieve the desired ion specificity. This type of lens configuration has been used in commercial instrumentation for almost 30 years and has proved to be very durable. One of its main benefits is that it produces a uniform response across the mass range with very low background levels, particularly when combined with an off-axis mass analyzer.[9] A schematic of a commercially available multicomponent lens systems is shown in Figure 8.4.

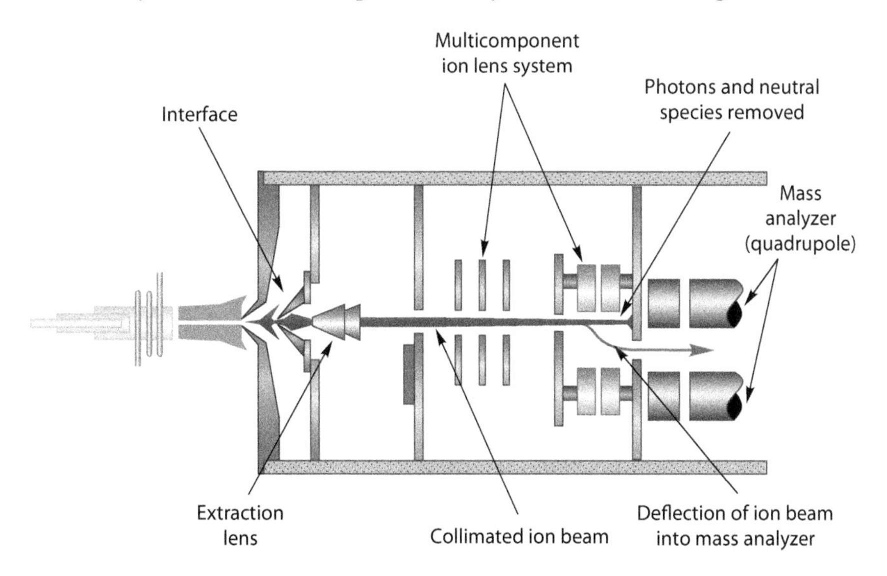

FIGURE 8.4 Schematic of a multicomponent lens system. (From Kishi, Y., *Agilent Technologies Application Journal*, August 1997.)

It should be emphasized that because of the interactive nature of parameters that affect the signal response, the more complex the lens system, the more the variables that have to be optimized. For this reason, if many different sample types are being analyzed, extensive lens optimization procedures have to be carried out for each matrix or group of elements. This is not such a major problem, because most of the lens voltages are computer controlled and methods can be stored for every new sample scenario. However, it could be a factor if the instrument is being used for the routine analysis of many diverse sample types, all requiring different lens settings.

Another well-established approach was the use of a cylinder lens, combined with a grounded stop—positioned just inside the skimmer cone. With this design, the voltage is dynamically ramped "on the fly," in concert with the mass scan of the analyzer. The benefit is that the optimum lens voltage is placed on every mass in a multielement run to allow the maximum number of analyte ions through, while keeping the matrix ions down to an absolute minimum.[10] This design is typically used in conjunction with a grounded stop to act as a physical barrier to reduce the chances that particulates, neutral species, and photons will reach the mass analyzer and detector. Although this design does not generate such a uniform mass response across the full range as an off-axis multilens system with an extraction lens, it appears to offer better long-term stability with real-world samples. It works well for many sample types, but is most effective when low-mass elements are being determined in the presence of high-mass matrix elements.

A more recent variation of the single-lens approach uses a right-angled cylinder lens. This is a very simple design that utilizes a single, fixed voltage ion lens to eliminate particulates and photons from reaching the detector. This is done by deflecting the positive ion beam 90°, thus allowing the neutral species to go straight through and be pumped out of the mass spectrometer. The benefits of this design are its simplicity, reduced maintenance, and very low background levels, as well as it being easy to clean. This lens system is unique to one instrument design and is used in conjunction with a low-mass cutoff collision/reaction cell. A photograph of this ion optic lens system is shown in Figure 8.5.

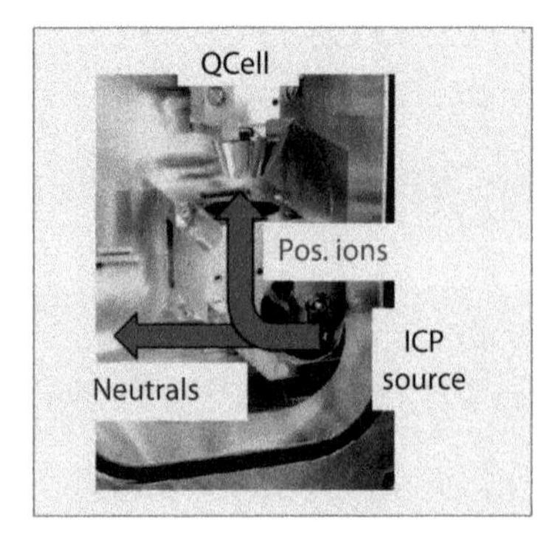

FIGURE 8.5 Right-angled cylinder lens showing how it deflects the positive ion beam 90° into the collision/reaction cell, while allowing the neutral species to go straight through.

Another design in ion-focusing optics utilizes a parabolic electrostatic field created with an ion mirror to reflect and refocus the ion beam at 90° to the ion source.[5] This ion mirror incorporates a hollow structure, which allows photons, neutrals, and solid particles to pass through it, while allowing ions to be reflected at right angles into the mass analyzer. The major benefit of this design is the very efficient way the ions are refocused, offering the capability of extremely high sensitivity across the mass range, with very little sacrifice in oxide performance. In addition, there is very little contamination of the ion optics, because a vacuum pump sits behind the ion mirror to immediately remove these particles before they have a chance to penetrate further into the mass spectrometer. Removing these undesirable species and photons before they reach the detector, in addition to incorporating curved fringe rods prior to an off-axis mass analyzer, means that background levels are very low. Figure 8.6 shows a schematic of a quadrupole-based ICP-MS that utilizes a 90° ion optic design.[6,11]

The most novel commercial development in ion optic design uses a quadrupole ion deflector (QID), which utilizes a miniaturized quadrupole. This novel filtering technology bends the ion beam 90°, focusing ions of a specified mass into the mass separation device, while discarding all neutral species, photons, and particulates into the turbo pump. The major benefit of removing these nonionic species means they won't be deposited on component surfaces in the mass analyzer region, which significantly minimizes drift and ensures good signal stability, even when running the most challenging sample matrices. This QID approach to ion filtering is shown in Figure 8.7.

It is also worth emphasizing that a number of ICP-MS systems offer what is called a high-sensitivity option. All these work slightly differently but share similar components. By using a combination of slightly different cone geometry, higher vacuum at the interface, one or more extraction lenses, and slightly modified ion optic design, they offer up to 10 times the sensitivity of a traditional interface. However, in some

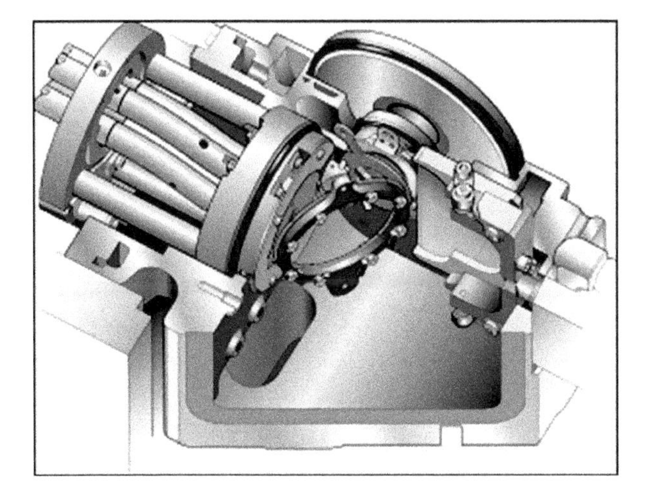

FIGURE 8.6 A 90° ion optic design used with curved fringe rods and an off-axis quadrupole mass analyzer. (Courtesy of Analytik Jena, Jena, Germany.)

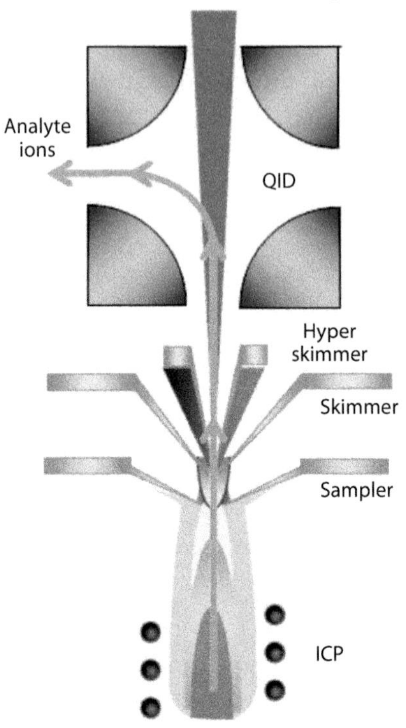

FIGURE 8.7 A novel commercial development in ion optic design utilizes a QID. (Copyright © 2013 PerkinElmer Inc., Waltham, MA. All rights reserved.)

systems, this increased sensitivity sometimes comes with slightly worse stability and an increase in background levels, particularly for samples with a heavy matrix. To get around this, these kinds of samples typically need to be diluted before analysis—which has somewhat limited their applicability to real-world samples with high dissolved solids.[12] However, they have found a use in non-liquid-based applications in which high sensitivity is crucial—for example, in the analysis of small spots on the surface of a geological specimen using laser ablation ICP-MS. For this application, the instrument must offer high sensitivity, because a single laser pulse is often used to ablate very small amounts of the sample, which is then swept into the ICP-MS for analysis.

The importance of the ion-focusing system cannot be overemphasized, because it has a direct bearing on the number of ions that find their way to the mass analyzer. In addition to affecting background levels and instrument response across the entire mass range, it has a huge impact on both long- and short-term signal stability, especially in real-world samples. However, there are many different ways of achieving this. It is almost irrelevant whether the design of the ion optics is based on a multi-component lens system, a radio-frequency multipole guide, or a right-angled deflection approach using an ion mirror or a quadrupole. The most important consideration when evaluating any ion lens system is not the actual design but its ability to perform well with real sample matrices.

REFERENCES

1. J. A. Olivares and R. S. Houk. *Analytical Chemistry*, 58, 20, 1986.
2. D. J. Douglas and J. B. French. *Spectrochimica Acta*, 41B(3), 197, 1986.
3. S. D. Tanner, L. M. Cousins, and D. J. Douglas. *Applied Spectroscopy*, 48, 1367, 1994.
4. D. Potter. *American Lab*, July 1994.
5. I. Kalinitchenko. Ion optical system for a mass spectrometer. Patent 750860, 1999.
6. S. Elliott, M. Plantz, and L. Kalinitchenko. Presented at Pittsburgh Conference, Orlando, FL, 2003, Paper 1360–8.
7. P. Turner. Presented at 2nd International Conference on Plasma Source Mass Spec, Durham, UK, 1990.
8. S. D. Tanner, D. J. Douglas, and J. B. French. *Applied Spectroscopy*, 48, 1373, 1994.
9. Y. Kishi. *Agilent Technologies Application Journal*, August 1997.
10. E. R. Denoyer, D. Jacques, E. Debrah, and S. D. Tanner. *Atomic Spectroscopy*, 16(1), 1, 1995.
11. I. Kalinitchenko. Mass spectrometer including a quadrupole mass analyzer arrangement. Patent applied for—WO 01/91159 A1.
12. B. C. Gibson. Presented at Surrey International Conference on ICP-MS, London, 1994.

9 Mass Analyzers
Quadrupole Technology

Chapters 9–12 deal with the heart of the inductively coupled plasma mass spectrometry (ICP-MS) system—the mass separation device. Sometimes called the mass analyzer, it is the region of the ICP mass spectrometer that separates the ions according to their mass-to-charge ratio. This selection process is achieved in a number of different ways, depending on the mass separation device, but they all have one common goal, which is to separate the ions of interest from all other nonanalyte, matrix, solvent, and argon-based ions. Quadrupole mass filters are described in this chapter, followed by magnetic sector systems in Chapter 10, time-of-flight mass spectrometers in Chapter 11, and finally, collision/reaction cell and interface technology in Chapter 12.

Although ICP-MS was commercialized in 1983, the first 10 years of its development utilized a traditional quadrupole mass analyzer to separate the ions of interest. These worked exceptionally well for most applications, but proved to have limitations when determining difficult elements or dealing with more complex sample matrices. This led to the development of alternative mass separation devices that allowed ICP-MS to be used for applications that were previously beyond the capabilities of quadrupole-based technology. Before we discuss these different mass spectrometers in greater detail, let us take a look at the proximity of the mass analyzer in relation to the ion optics and detector. Figure 9.1 shows this in greater detail.

As can be seen, the mass analyzer is positioned between the ion optics and detector, and it is maintained at a vacuum of approximately 10^{-6} torr, with an additional turbomolecular pump for the one that is used for the lens chamber. Assuming the ions are emerging from the ion optics at the optimum kinetic energy, they are ready to be separated according to their mass-to-charge ratio (m/z) by the mass analyzer. There are basically three different kinds of commercially available mass analyzers: quadrupole mass filters, double-focusing magnetic sectors, and time-of-flight mass spectrometers. It should be noted that although collision/reaction cell and interface technology has been given its own chapter in this book, it is not utilized on its own to carry out mass separation. It is typically used in conjunction with a quadrupole to reduce the impact of polyatomic spectral interferences before the ions are passed into the mass analyzer for separation. They are very powerful enhancements to the technique, but they are not considered primary separation devices on their own.

They all have their own strengths and weaknesses, which will be discussed in greater detail in the next four chapters. Let us first begin with the most common type of mass separation device used in ICP-MS—the quadrupole mass filter.

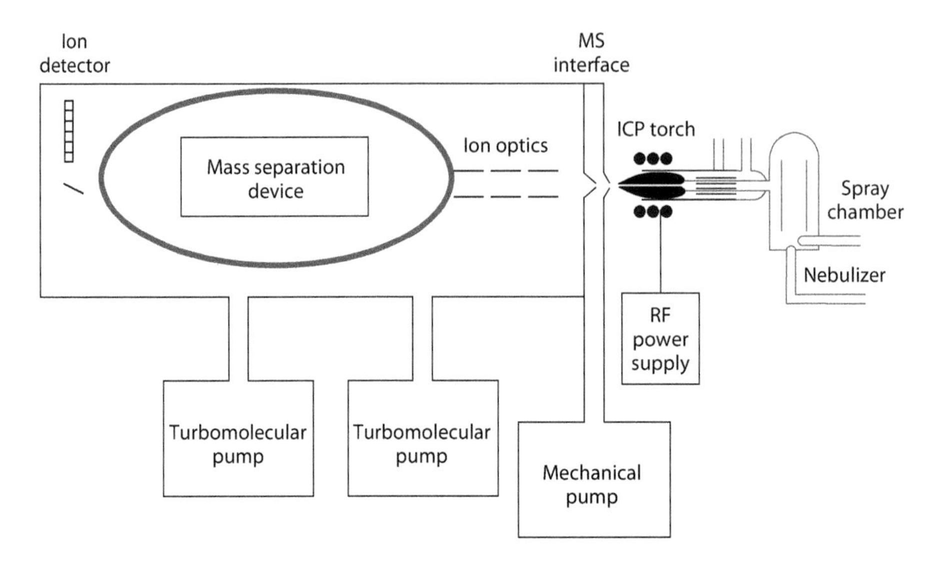

FIGURE 9.1 The mass separation device is positioned between the ion optics and the detector.

QUADRUPOLE TECHNOLOGY

Developed in the early 1980s for ICP-MS, quadrupole-based systems represent approximately 90% of all ICP mass spectrometers used today. This design was the first to be commercialized, and as a result, today's quadrupole ICP-MS technology is considered a very mature, routine trace element technique. A quadrupole usually consists of four cylindrical or hyperbolic metallic rods of the same length and diameter, although in one design of collision/reaction cell a quadrupole with flat sides is used (this "flatapole" design is described in greater detail in Chapter 12 on collision/reaction cells). Quadrupole rods are typically made of stainless steel or molybdenum and sometimes coated with a ceramic coating for corrosion resistance. When used for ICP-MS, they are typically 15–25 cm in length, about 1 cm in diameter, and operate at a frequency of 2–3 MHz. Figure 9.2 shows a photograph of a quadrupole system mounted in its housing.

BASIC PRINCIPLES OF OPERATION

A quadrupole operates by placing both a direct current (DC) field and a time-dependent alternating current (AC) of radio frequency on opposite pairs of the four rods. By selecting the optimum AC/DC ratio on each pair of rods, ions of a selected mass are allowed to pass through the rods to the detector, whereas the others are unstable and ejected from the quadrupole. Figure 9.3 shows this in greater detail.

In this simplified example, the analyte ion (black) and four other ions (gray) have arrived at the entrance to the four rods of the quadrupole. When a particular AC/DC potential is applied to the rods, the positive or negative bias on the rods will electrostatically steer the analyte ion of interest down the middle of the four rods to

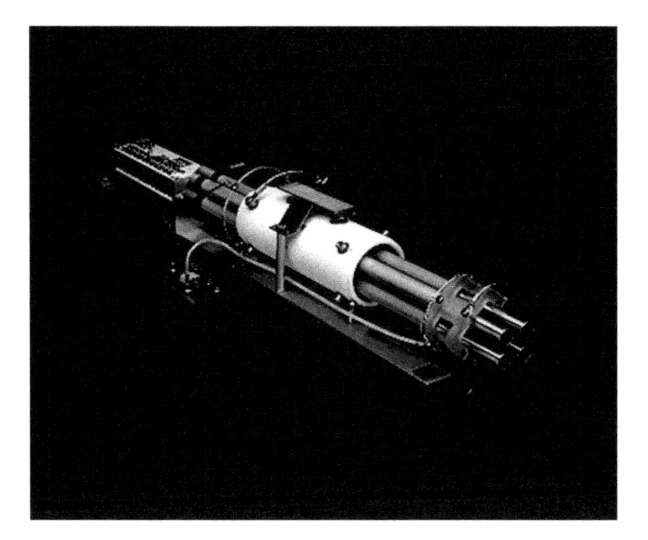

FIGURE 9.2 Photograph of a quadrupole system mounted in its housing. (Copyright © 2013 PerkinElmer Inc., Waltham, MA. All rights reserved.)

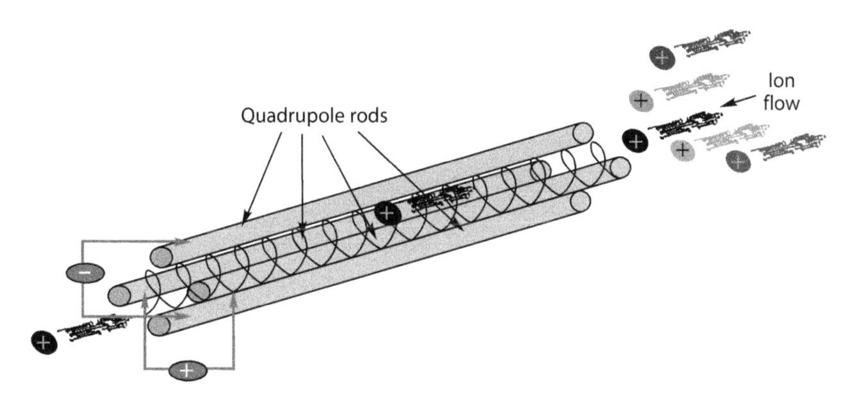

FIGURE 9.3 Schematic representation showing principles of mass separation using a quadrupole mass filter.

the end, where it will emerge and be converted to an electrical pulse by the detector. The other ions of different *m*/*z* values will be unstable, pass through the spaces between the rods, and be ejected from the quadrupole. This scanning process is then repeated for another analyte with a completely different mass-to-charge ratio until all the analytes in a multielement analysis have been measured. The process for the detection of one particular mass in a multielement run is represented in Figure 9.4.

It shows a $^{63}Cu^+$ ion emerging from the quadrupole and being converted to an electrical pulse by the detector. As the AC/DC voltage of the quadrupole—corresponding to $^{63}Cu^+$—is repeatedly scanned, the ions are stored and counted by a multichannel analyzer as electrical pulses. This multichannel data acquisition system typically has 20 channels per mass. As the electrical pulses are counted in each

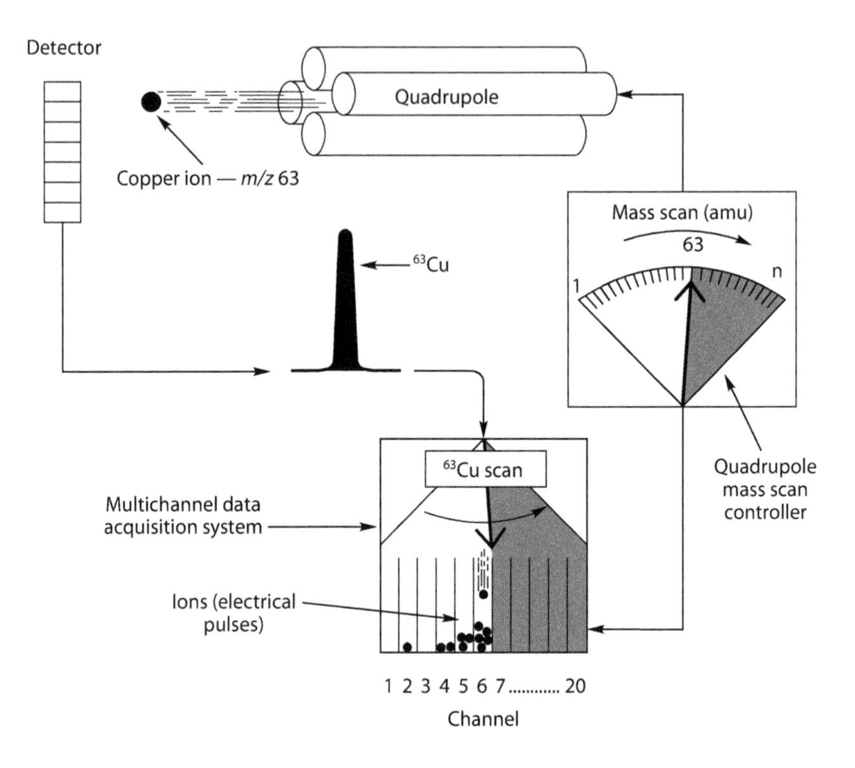

FIGURE 9.4 Profiles of different masses are built up using a multichannel data acquisition system. (Copyright © 2013 PerkinElmer Inc., Waltham, MA. All rights reserved.)

channel, a profile of the mass is built up over the 20 channels, corresponding to the spectral peak of $^{63}Cu^{+}$. In a multielement run, repeated scans are made over the entire suite of analyte masses, as opposed to just one mass represented in this example.

Quadrupole scan rates are typically on the order of 2500–5000 amu per second (depending on the commercial design) and can cover the entire mass range of 0–300 amu in about 1/10 of a second. However, real-world analytical speed and sample throughput is much slower than this, and in practice, 25 elements can be determined in duplicate with good precision in 1–2 min, depending on the analytical requirements.

QUADRUPOLE PERFORMANCE CRITERIA

There are two very important performance specifications of a mass analyzer that govern its ability to separate an analyte peak from a spectral interference. The first is the resolving power (R), which in traditional mass spectrometry is represented by the equation $R = m/\Delta m$, where m is the nominal mass at which the peak occurs and Δm is the mass difference between two resolved peaks.[1] However, for quadrupole technology, the term *resolution* is more commonly used and is normally defined as the width of a peak at 10% of its height. The second specification is abundance sensitivity, which is the signal contribution of the tail of an adjacent peak at one mass lower and one mass higher than the analyte peak.[2] Even though they are somewhat related and both define the quality of a quadrupole, the abundance sensitivity is probably the most critical. If a quadrupole has good resolution but poor abundance sensitivity, it will often prohibit the measurement of an ultratrace analyte peak next to a major interfering mass.

Resolution

Let us now discuss this area in greater detail. The ability to separate different masses with a quadrupole is determined by a combination of factors, including shape, diameter, and length of the rods; frequency of quadrupole power supply; operating vacuum; applied radio-frequency (RF)/DC voltages; and the motion and kinetic energy of the ions entering and exiting the quadrupole. All these factors will have a direct impact on the stability of the ions as they travel down the middle of the rods, and therefore the quadrupole's ability to separate ions with differing *m/z* values. This is represented in Figure 9.5, which shows a simplified version of the Mathieu mass stability plot of two separate masses (A and B) entering the quadrupole at the same time.[3]

Any of the RF/DC conditions shown under the peak on the left will only allow mass A to pass through the quadrupole, whereas any combination of RF/DC voltages under the peak on the right will only allow mass B to pass through the quadrupole. If the slope of the RF/DC scan rate is steep, represented by the top line (high resolution), the spectral peaks will be narrow and masses A and B will be well separated. However, if the slope of the scan is shallow, represented by the middle line (low resolution), the spectral peaks will be wide and masses A and B will not be well separated. On the other hand, if the slope of the scan is too shallow, represented by the lower line (inadequate resolution), the peaks will overlap each other and both masses A and B will pass through the quadrupole without being separated. Theoretically, the resolution of a quadrupole mass filter can be varied between 0.3 and 3.0 amu,

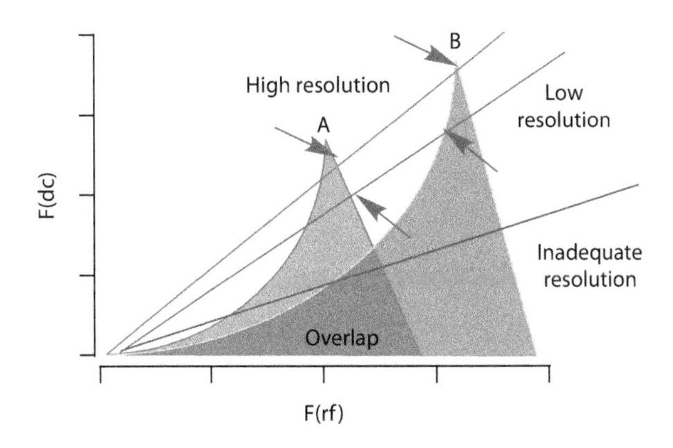

FIGURE 9.5 Simplified Mathieu stability diagram of a quadrupole mass filter, showing separation of two different masses. A and B. (From Dawson, P. H., ed., *Quadrupole Mass Spectrometry and Its Applications*, Elsevier, Amsterdam, 1976; reissued by AIP Press, Woodbury, NY, 1995.)

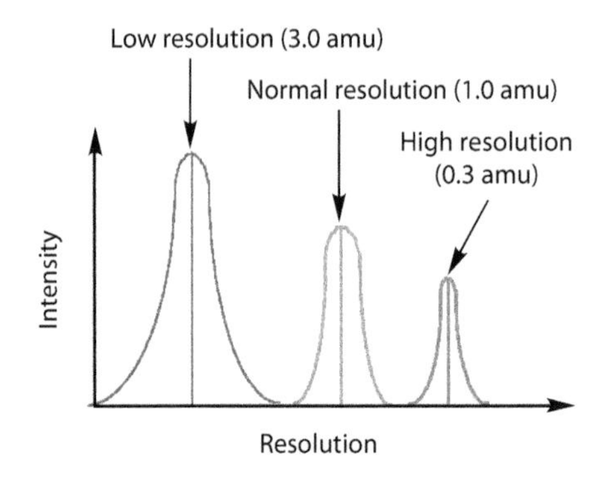

FIGURE 9.6 Sensitivity comparison of a quadrupole operated at 3.0, 1.0, and 0.3 amu resolutions.

but is normally kept at 0.7–1.0 amu for most applications. However, improved resolution is always accompanied by a sacrifice in sensitivity, as seen in Figure 9.6, which shows a comparison of the same mass at resolutions of 3.0, 1.0, and 0.3 amu.

It can be seen that the peak height at 3.0 amu is much larger than that at 0.3 amu, but as expected, it is also much wider. This would prohibit using a resolution of 3.0 amu with spectrally complex samples. Conversely, the peak width at 0.3 amu is very narrow, but the sensitivity is low. For this reason, a compromise between peak width and sensitivity normally has to be reached, depending on the application. This can clearly be seen in Figure 9.7, which shows a spectral overlay of two copper isotopes—$^{63}Cu^+$ and $^{65}Cu^+$—at resolution settings of 0.70 and 0.50 amu. In practice, the quadrupole is normally operated at a resolution of 0.7–1.0 amu for the majority of applications.

It is worth mentioning that most quadrupoles are operated in the first stability region, where resolving power is typically on the order of 500–600. If the quadrupole is operated in the second or third stability regions, resolving powers of 4000[4] and 9000,[5] respectively, can be achieved. However, improving resolution using this approach has resulted in a significant loss of signal. Although there are ways of improving sensitivity, other problems have been encountered. As a result, to date there are no commercial quadrupole instruments available that use higher-stability regions.

Some instruments can vary the peak width "on the fly," which means that the resolution can be changed between 3.0 and 0.3 amu for every analyte in a multielement run. Although this appears to offer some benefits, in reality they are few and far between, and for the vast majority of applications it is adequate to use the same resolution setting for every analyte. Even though quadrupoles can be operated at a higher resolution (in the first stability region), up to now the slight improvement has not been shown to be of practical benefit for most routine applications.

Abundance Sensitivity

It can be seen in Figure 9.7 that the tail of the spectral peaks drops off more rapidly at the high-mass end of the peak compared with the low-mass end. The overall peak

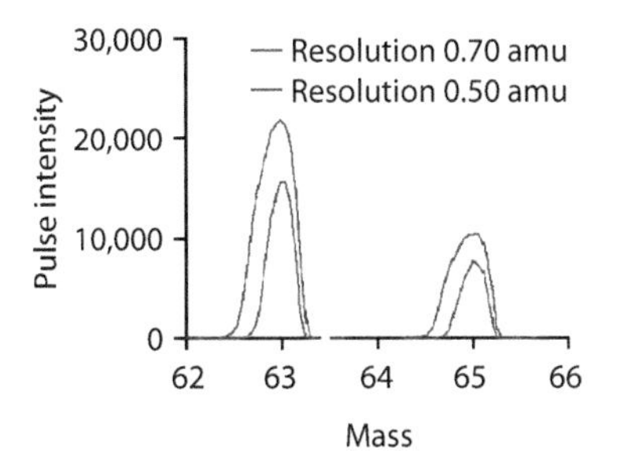

FIGURE 9.7 Sensitivity comparison of two copper isotopes—$^{63}Cu^+$ and $^{65}Cu^+$— at resolution settings of 0.70 and 0.50 amu.

shape, particularly its low-mass and high-mass tail, is determined by the abundance sensitivity of the quadrupole, which is impacted by a combination of factors, including the design of the rods, frequency of the power supply, and operating vacuum.[6] Even though they are all important, probably the biggest impact on abundance sensitivity is the motion and kinetic energy of the ions as they enter and exit the quadrupole. If the Mathieu stability plot in Figure 9.5 is examined, it can be seen that the stability boundaries of each mass are less defined (not so sharp) on the low-mass side compared with the high-mass side.[3] As a result, the characteristic of ion motion at the low-mass boundary is different from that at the high-mass boundary and is therefore reflected in poorer abundance sensitivity at the low-mass side than at the high-mass side. The velocity, and therefore the kinetic energy, of the ions entering the quadrupole will affect the ion motion and, as a result, will have a direct impact on the abundance sensitivity. For that reason, factors that affect the kinetic energy of the ions, such as high plasma potential and the use of lenses to accelerate the ion beam, could have a negative effect on the instrument's abundance sensitivity.[7]

These are the fundamental reasons why the peak shape is not symmetrical with a quadrupole and explains why there is always a pronounced shoulder at the low-mass side of the peak compared with the high-mass side—as represented in Figure 9.8, which shows the theoretical peak shape of a nominal mass M.

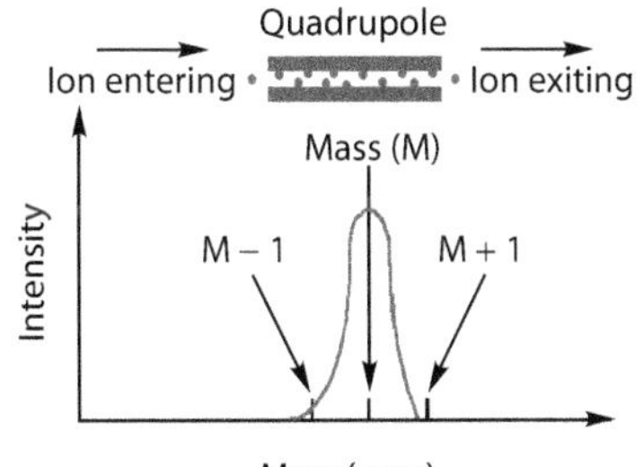

FIGURE 9.8 Ions entering the quadrupole are slowed down by the filtering process and produce peaks with a pronounced tail or shoulder at the low-mass end (M – 1) compared with the high-mass end (M + 1).

It can be seen that the shape of the peak at one mass lower (M – 1) is slightly different from that of the other side of the peak at one mass higher (M + 1) than mass M. For this reason, the abundance sensitivity specification for all quadrupoles is always worse on the low-mass side than the high-mass side and is typically 1×10^{-6} at M – 1 and 1×10^{-7} at M + 1. In other words, an interfering peak of 1 million counts per second (Mcps) at M – 1 would produce a background of 1 cps at M, whereas it would take an interference of 10 million cps at M + 1 to produce a background of 1 cps at M.

Benefit of Good Abundance Sensitivity

An example of the importance of abundance sensitivity is shown in Figure 9.9. Figure 9.9a is a spectral scan of 50 ppm of the doubly charged europium ion, $^{151}Eu^{++}$, at 75.5 amu (a doubly charged ion is one with two positive charges, as opposed to a normal singly charged positive ion, and exhibits an *m*/*z* peak at half its mass). It can be seen that the intensity of the peak is so great that its tail overlaps the adjacent mass at 75 amu, which is the only available mass for the determination of arsenic. This is highlighted in Figure 9.9b, which shows an expanded view of the tail of the $^{151}Eu^{++}$, together with a scan of 1 ppb As at mass 75. It can be seen very clearly that the $^{75}As^{+}$ signal lies on the sloping tail of the $^{151}Eu^{++}$ peak. Measurement on a sloping background similar to this would result in a significant degradation in the arsenic detection limit, particularly as the element is monoisotopic and no alternative mass is available. In this particular example, a slightly higher resolution setting was also used (0.5 amu instead of 0.7 amu) to enhance the separation of the arsenic peak from the europium peak, but it nevertheless still emphasizes the importance of good abundance sensitivity in ICP-MS.

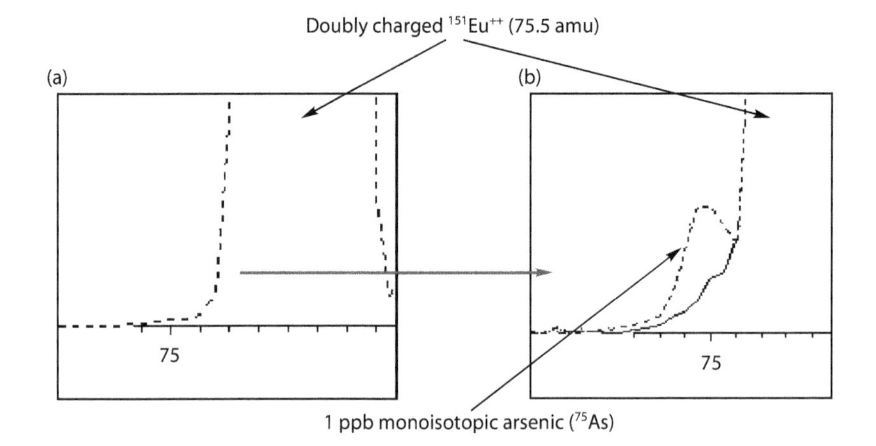

FIGURE 9.9 A low abundance sensitivity specification is critical to minimize spectral interferences, as shown by (a), which represents a spectral scan of 50 ppb $^{151}Eu^{++}$ at 75.5 amu, and (b), which shows how the tail of the $^{151}Eu^{++}$ elevates the spectral background of 1 ppb As at mass 75.

There are many different designs of quadrupole used in ICP-MS, all made from different materials with varied dimensions, shape, and physical characteristics. In addition, they are all maintained at a slightly different vacuum chamber pressure and operate at different frequencies. Theoretically, these hyperbolic rods should generate a better hyperbolic (elliptical) field than cylindrical rods, resulting in higher transmission of ions at higher resolution. It also tells us that a higher operating frequency means a higher rate of oscillation—and therefore separation—of the ions as they travel down the quadrupole. Finally, it is very well accepted that a higher vacuum produces fewer collisions between gas molecules and ions, resulting in a narrower spread in kinetic energy of the ions, and therefore a reduction in the tail at the low-mass side of a peak. Given all these theoretical differences, in reality the practical capabilities of most modern quadrupoles used in ICP-MS are very similar. However, there are some subtle differences in each instrument's measurement protocol and the software's approach to peak quantitation. This is such an important area that it will be discussed in greater detail in a Chapter 14.

REFERENCES

1. F. Adams, R. Gijbels, and R. Van Grieken. *Inorganic Mass Spectrometry*. New York: John Wiley & Sons, 1988.
2. E. Montasser, ed. *Inductively Coupled Plasma Mass Spectrometry*. Berlin: Wiley-VCH, 1998.
3. P. H. Dawson, ed. *Quadrupole Mass Spectrometry and Its Applications*. Amsterdam: Elsevier, 1976. Reissued by AIP Press, Woodbury, NY, 1995.
4. Z. Du, T. N. Olney, and D. J. Douglas. *Journal of American Society of Mass Spectrometry*, 8, 1230–1236, 1997.
5. P. H. Dawson and Y. Binqi. International Journal of Mass Spectrometry, 56, 25, 1984.
6. D. Potter. Agilent Technologies Application Note, January 1996, 228–349.
7. E. R. Denoyer, D. Jacques, E. Debrah, and S. D. Tanner. *Atomic Spectroscopy*, 16(1), 1, 1995.

10 Mass Analyzers *Double-Focusing Magnetic Sector Technology*

Although quadrupole mass analyzers represent approximately 90% of all inductively coupled plasma mass spectrometry (ICP-MS) systems installed worldwide, limitations in their resolving power have led to the development of high-resolution spectrometers based on the double-focusing magnetic sector design. In this chapter, we take a detailed look at this very powerful mass separation device, which has found its niche in solving challenging application problems that require excellent detection capability, exceptional resolving power, and very high precision.

As discussed in Chapter 9, a quadrupole-based ICP-MS system typically offers a resolution of 0.7–1.0 amu. This is quite adequate for most routine applications, but has proved to be inadequate for many elements that are prone to argon-, solvent-, and/or sample-based spectral interferences. These limitations in quadrupoles drove researchers in the direction of traditional high-resolution magnetic sector technology to improve quantitation by resolving the analyte mass away from the spectral interference.[1] These ICP-MS instruments that were first commercialized in the late 1980s offered resolving power of up to 10,000, compared with that of a quadrupole, which was on the order of ~300. This dramatic improvement in resolving power allowed difficult elements, such as Fe, K, As, V, and Cr, to be determined with relative ease, even in complex sample matrices.

MAGNETIC SECTOR MASS SPECTROSCOPY: A HISTORICAL PERSPECTIVE

Mass spectrometers, using separation based on velocity focusing[2,3] and magnetic deflection,[4,5] were first developed more than 80 years ago, primarily to investigate isotopic abundances and calculate atomic weights. Even though these designs were combined into one instrument in the 1930s to improve both sensitivity and resolving power,[6,7] they were still considered rather bulky and expensive to build. For that reason, in the late 1930s and 1940s, magnetic field technology, and in particular the small radius sector design of Nier,[8] became the preferred method of mass separation. Because Nier was a physicist, most of the early work carried out with this design was used for isotope studies in the disciplines of earth and planetary sciences. However, it was the oil industry that accelerated the commercialization of MS, because of its demand for the fast and reliable analysis of complex hydrocarbons in oil refineries.

Once scanning magnetic sector technology became the most accepted approach for high-resolution mass separation in the 1940s, the challenges that lay ahead for mass spectroscopists were in the design of the ionization source—especially as the technique was being used more and more for the analysis of solids. The gas discharge ion source that was developed for gases and high-vapor-pressure liquids proved to be inadequate for most solid materials. For this reason, one of the first successful methods of ionizing solids was carried out using the hot anode method,[9] where the previously dissolved material was deposited onto a strip of platinum foil and evaporated by passing an electric current through it. Unfortunately, although there were variations of this approach that all worked reasonably well, the main drawback of a thermal evaporation technique was selective ionization. In other words, because of the different volatilities of the elements, it could not be guaranteed that the ion beam properly represented the compositional integrity of the sample.

It was finally the work carried out by Dempster in 1946,[10] using a vacuum spark discharge and a high-frequency, high-voltage spark, that led researchers to believe that it could be applied to sample electrodes and used as a general-purpose source for the analysis of solids. The breakthrough came in 1954 with the development of the first modern spark source mass spectrometer (SSMS) based on the Mattauch–Herzog mass spectrometer design, which separated the ions in the same flat plane, so they could be detected by a linear detector, such as a photographic plate.[11] Using this design, Hannay and Ahearn showed that it was possible to determine sub-parts-per-million impurity levels directly in a solid material.[12]

Over the years, as a result of a demand for more stable ionization sources, lower detection capability, and higher precision, researchers were led in the direction of other techniques, such as secondary ion mass spectrometry (SIMS),[13] ion microprobe mass spectrometry (IMMS),[14] and laser-induced mass spectrometry (LIMS).[15] Although they are considered somewhat complementary to SSMS, they all had their own strengths and weaknesses, depending on the analytical objectives for the solid material being analyzed. However, it should be emphasized that these techniques were predominantly used for microanalysis because only a very small area of the sample is vaporized. This meant that it could only provide meaningful analytical data for the bulk material if the sample was sufficiently homogeneous. For that reason, other ionization sources that sampled a much larger area, such as the glow discharge, became a lot more practical for the bulk analysis of solids by MS.[16]

USE OF MAGNETIC SECTOR TECHNOLOGY FOR ICP-MS

Even though magnetic sector technology was the most common mass separation device for the analysis of inorganic compounds using traditional ion sources, it lost out to quadrupole technology when ICP-MS was first developed in the early 1980s. However, it was not until the mid- to late 1980s that the analytical community realized that quadrupole ICP-MS suffered from serious limitations in its ability to resolve troublesome polyatomic spectral interferences. In addition, because quadrupoles were scanning devices, the technique was not suitable for applications that required high precision, such as isotope ratio measurements. As a result, analytical chemists began to look at double-focusing magnetic sector technology to alleviate these kinds of limitations.

Initially, this technology was found to be unsuitable as a separation device for an ICP because of the high voltage required to accelerate the ions into the mass analyzer. This high potential at the interface region dramatically changed the energy of the ions entering the mass spectrometer, and therefore made it very difficult to steer the ions through the ion optics and still maintain a narrow spread of ion kinetic energies. For this reason, basic changes had to be made to the ion acceleration mechanism for magnetic sector technology to be successfully used as a separation device for ICP-MS. This was a significant challenge when magnetic sector systems were first developed in the late 1980s.

However, by the early 1990s, this problem was solved by moving the high-voltage components away from the plasma and positioning the interface closer to the mass spectrometer. Modern instrumentation has typically been based on two different approaches, the "standard" and "reverse" Nier–Johnson geometry. Both of these designs, which use the same basic principles, consist of two analyzers: a traditional electromagnet and an electrostatic analyzer (ESA). In the standard (sometimes called forward) design, the ESA is positioned before the magnet, and in the reverse design it is positioned after the magnet. A schematic of the reverse Nier–Johnson spectrometer is shown in Figure 10.1.

PRINCIPLES OF OPERATION OF MAGNETIC SECTOR TECHNOLOGY

The original concept of magnetic sector technology was to scan over a large mass range by varying the magnetic field over time with a fixed acceleration voltage. During a small window in time, which was dependent on the resolution chosen, ions

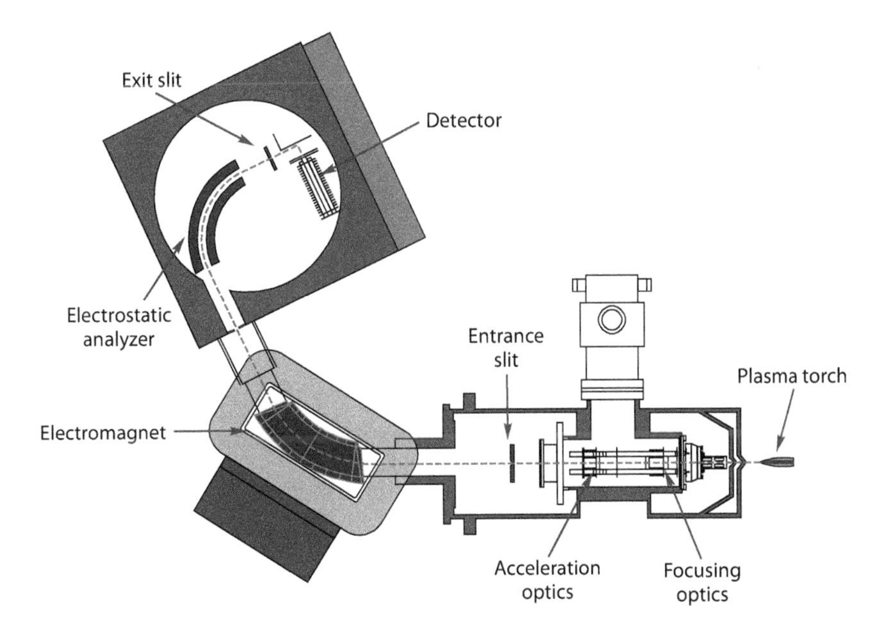

FIGURE 10.1 Schematic of a reverse Nier–Johnson double-focusing magnetic sector mass spectrometer. (From Geismann, U., and Greb, U., *Fresenius J. Anal. Chem.*, 350, 186–193, 1994.)

of a particular mass-to-charge ratio are swept past the exit slit to produce the characteristic flat-top peaks. As the resolution of a magnetic sector instrument is independent of mass, ion signals, particularly at low mass, are far apart. Unfortunately, this results in a relatively long time being spent scanning and settling the magnet. This was not such a major problem for qualitative analysis or mass spectral fingerprinting of unknown compounds, but proved to be impractical for rapid trace element analysis, where you had to scan to individual masses, slow down, settle the magnet, stop, take measurements, and then scan to the next mass.

However, by using the double-focusing approach, the ions are sampled from the plasma in a conventional manner and then accelerated in the ion optic region to a few kilovolts before they enter the mass analyzer. The magnetic field, which is dispersive with respect to ion energy and mass, then focuses all the ions with diverging angles of motion from the entrance slit. The ESA, which is only dispersive with respect to ion energy, then focuses all the ions onto the exit slit, where the detector is positioned. If the energy dispersions of the magnet and ESA are equal in magnitude but opposite in direction, they will focus both ion angles (first focusing) and ion energies (second or double focusing) when combined together. Changing the electric field in the opposite direction during the cycle time of the magnet (in terms of the mass passing the exit slit) has the effect of "freezing" the mass for detection. Then, as soon as a certain magnetic field strength is passed, the electric field is set to its original value and the next mass is "frozen." The voltage is varied on a per-mass basis, allowing the operator to scan only the mass peaks of interest rather than the full mass range.[17,18]

It should be pointed out that although this approach represents an enormous time savings over traditional magnet scanning technology, it is still slower than quadrupole-based instruments. The inherent problem lies in the fact that a quadrupole can be electronically scanned faster than a magnet. Typical speeds for a full mass scan (0–250 amu) of a magnet are on the order of 200 ms compared with 50–100 ms for a quadrupole. In addition, it takes much longer for a magnet to slow down, settle, and stop to take measurements—typically, 20 ms compared with <1 ms for a quadrupole. So, even though in practice the electric scan dramatically reduces the overall analysis time, modern double-focusing magnetic sector ICP-MS systems are still slower than state-of-the-art quadrupole instruments, which make them less than ideal for fast, high-throughput multielement applications or the characterization of rapid transient peaks.

Resolving Power

As mentioned previously, most commercial magnetic sector ICP-MS systems offer up to 10,000 resolving power (10% valley definition), which is high enough to resolve the majority of spectral interferences. It is worth emphasizing that resolving power is represented by the equation $R = m/\Delta m$, where m is the nominal mass at which the peak occurs and Δm is the mass difference between two resolved peaks.[19] In a quadrupole, the resolution is selected by changing the ratio of the radio-frequency (RF) and direct current (DC) voltages on the quadrupole rods.

However, because a double-focusing magnetic sector instrument involves focusing ion angles and ion energies, mass resolution is achieved by using two mechanical slits—one at the entrance to the mass spectrometer and another at the exit, prior to

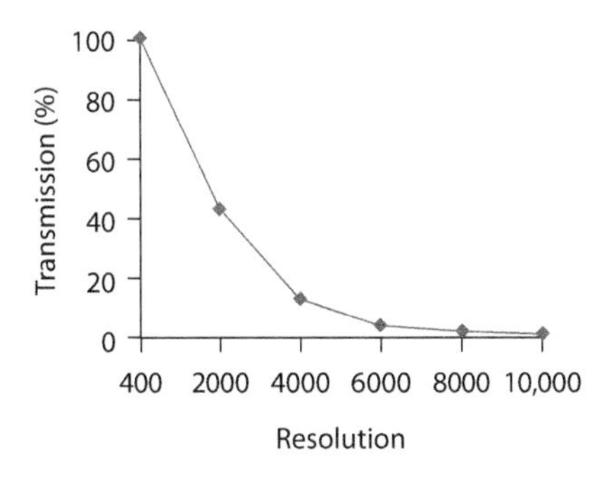

FIGURE 10.2 Ion transmission with a magnetic sector instrument decreases as the resolution increases.

the detector. Varying resolution is achieved by scanning the magnetic field under different entrance and exit slit width conditions. Similar to optical systems, low resolution is achieved by using wide slits, whereas high resolution is achieved with narrow slits. Varying the width of both the entrance and exit slits effectively changes the operating resolution.

However, it should be emphasized that, similar to optical spectrometry, as the resolution is increased, the transmission decreases. So even though extremely high resolution is available, detection limits will be compromised under these conditions. This can be seen in Figure 10.2, which shows a plot of resolution against ion transmission.

It can be seen that a resolving power of 400 produces 100% transmission, but at a resolving power of 10,000, only ~2% is achievable. This dramatic loss in sensitivity could be an issue if low detection limits are required in spectrally complex samples that require the highest possible resolution. However, spectral demands of this nature are not very common. Table 10.1 shows the resolution required to resolve

TABLE 10.1
Resolution Required to Separate Some Common Polyatomic Interferences from a Selected Group of Isotopes

Isotope	Matrix	Interference	Resolution	Transmission (%)
$^{39}K^+$	H_2O	$^{38}ArH^+$	5,570	6
$^{40}Ca^+$	H_2O	$^{40}Ar^+$	199,800	0
$^{44}Ca^+$	HNO_3	$^{14}N^{14}N\,^{16}O^+$	970	80
$^{56}Fe^+$	H_2O	$^{40}Ar^{16}O^+$	2,504	18
$^{31}P^+$	H_2O	$^{15}N^{16}O^+$	1,460	53
$^{34}S^+$	H_2O	$^{16}O^{18}O^+$	1,300	65
$^{75}As^+$	HCl	$^{40}Ar^{35}Cl^+$	7,725	2
$^{51}V^+$	HCl	$^{35}Cl^{16}O^+$	2,572	18
$^{64}Zn^+$	H_2SO_4	$^{32}S^{16}O^{16}O^+$	1,950	42
$^{24}Mg^+$	Organics	$^{12}C^{12}C^+$	1,600	50
$^{52}Cr^+$	Organics	$^{40}Ar^{12}C^+$	2,370	20
$^{55}Mn^+$	HNO_3	$^{40}Ar^{15}N^+$	2,300	20

fairly common polyatomic interferences from a selected group of elemental isotopes, together with the achievable ion transmission.

Figure 10.3 is a comparison between a quadrupole and a magnetic sector instrument of one of the most common polyatomic interferences, $^{40}Ar^{16}O^+$ on $^{56}Fe^+$, which requires a resolution of 2504 to separate the peaks. Figure 10.3a shows a spectral scan of $^{56}Fe^+$ using a quadrupole instrument. What it does not show is the massive polyatomic interference $^{40}Ar^{16}O^+$ (produced by oxygen ions from the water combining with argon ions from the plasma) completely overlapping the $^{56}Fe^+$. It shows very

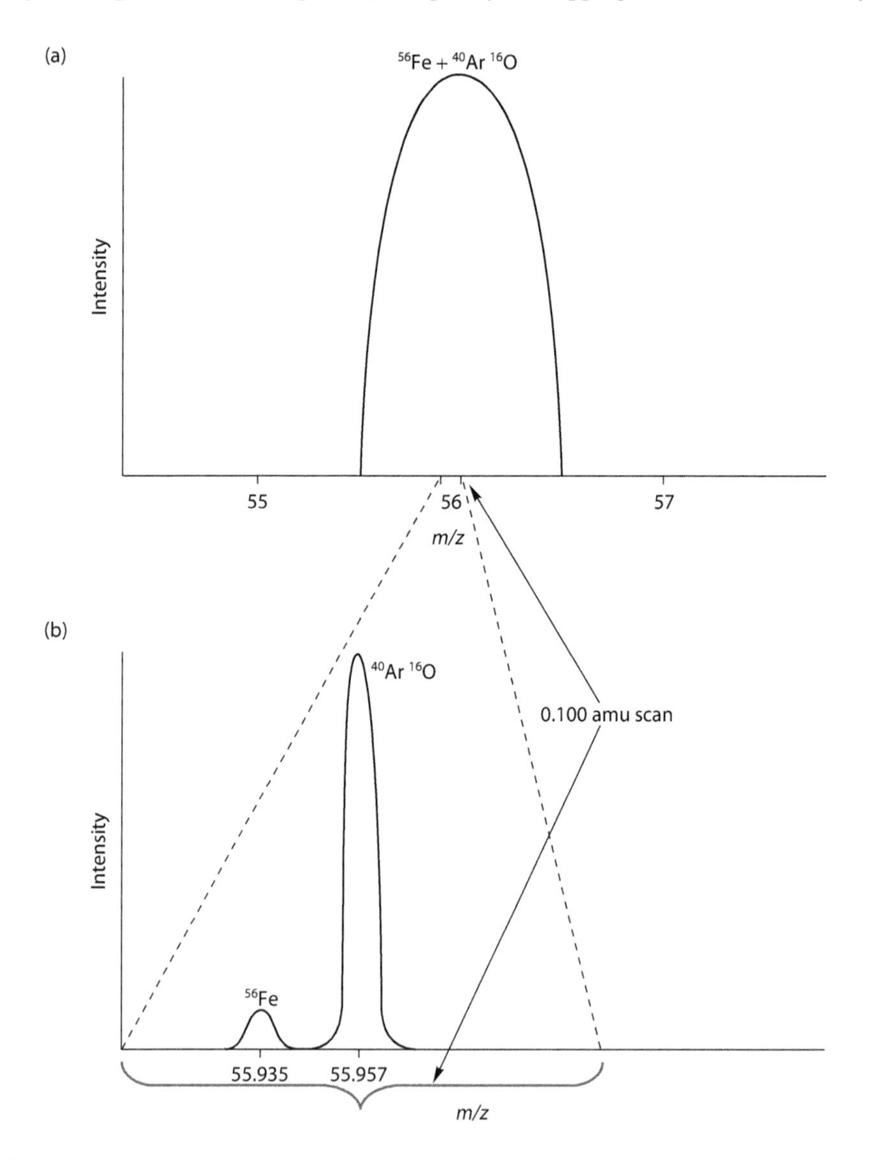

FIGURE 10.3 Comparison of resolution between (a) a quadrupole and (b) a magnetic sector instrument for the polyatomic interference of $^{40}Ar^{16}O^+$ on $^{56}Fe^+$. (From Greb, U., and Rottman, L., *Labor Praxis*, August 1994.)

clearly that these two masses are irresolvable with a quadrupole. If that same spectral scan is carried out on a magnetic sector–type instrument, the result is the scan shown in Figure 10.3b.[20] It should be pointed out that in order to see the spectral scan on the same scale, it is necessary to examine a much smaller range. For this reason, a 0.100 amu window was taken, as indicated by the dotted lines.

OTHER BENEFITS OF MAGNETIC SECTOR INSTRUMENTS

Besides high resolving power, another attractive feature of magnetic sector instruments is their very high sensitivity combined with extremely low background levels. High ion transmission in low-resolution mode translates into sensitivity specifications of up to 1 billion counts per second (Bcps) per parts per million, whereas background levels resulting from extremely low dark current noise are typically 0.1–0.2 counts per second (cps). This compares to typical sensitivity levels of 100 million counts per second (Mcps) and background levels of <5 cps for a quadrupole technology, although the newer instruments are now capable of generating in excess of 200 Mcps and up to 500 Mcps when optimized for high sensitivity, with BG (Background) levels in the order of 1 cps. However, because background levels are usually significantly lower on a magnetic sector system, detection limits, especially for high-mass elements such as uranium, where high resolution is generally not required, are typically 5–10 times better than those for a quadrupole-based instrument.

Besides good detection capability, another of the recognized benefits of the magnetic sector approach is its ability to quantitate with excellent precision. Measurement of the characteristically flat-topped spectral peaks translates directly into high-precision data. As a result, in the low-resolution mode relative standard deviation (RSD) values of 0.01%–0.05% are fairly common, which makes it an ideal approach for carrying out high-precision isotope ratio work.[21] Although precision is usually degraded as resolution is increased, modern instrumentation with high-speed electronics and low-mass bias are still capable of precision values of <0.1% RSD in medium- or high-resolution mode.[22]

The demand for ultra-high-precision data, particularly in the field of geochemistry, has led to the development of instruments dedicated to isotope ratio analysis. These are based on the double-focusing magnetic sector design, but instead of using just one detector, these instruments use multiple detectors. Often referred to as multicollector systems, they offer the capability of detecting and measuring multiple ion signals at exactly the same time. As a result of this simultaneous measurement approach, they are recognized as producing extremely low isotope ratio precision.[23]

SIMULTANEOUS MEASUREMENT APPROACH USING ONE DETECTOR

The scanning limitations of a single-detector magnetic sector instrument have been addressed in the past 10 years by the development of spectrometers based on Mattauch–Herzog geometry and simultaneous measurement of the ions using linear plane array detectors.[24-27] The early designs were limited in their applicability to the entire mass spectrum and the use of an ICP excitation source, but they eventually led to the development of a commercially available instrument in 2010.[28] This particular instrument is made up of an entrance slit, ESA, energy slit, and permanent magnet. In this design, the magnetic field focuses all of the ions onto a flat linear focal plane, without having to adjust the voltage of the mass analyzer or the strength of the magnetic field. In this way, no scanning is involved, and as a result, all the ions generated can be collected on a solid-state ion detector for the simultaneous detection of all the elemental masses in the sample, similar to using a charge-coupled device (CCD) or charge injection device (CID) for optical emission work. The principles of this design are exemplified in Figure 10.4. The ion detector has more than 4800 channels, which is enough to record all 210 isotopes of the 75 elements that can be determined by ICP-MS, using an average of 20 channels per isotope. Every channel is a combination of two detector arrays with different signal amplifiers. In this way, it is possible to cover six orders of dynamic range plus an additional three orders by optimizing the measurement times. More details will be given about this type of detector in Chapter 13.

The major benefit of this approach is that no scanning of the magnet is required. For this reason, it is ideal for any application that benefits from a rapid simultaneous measurement of the analyte masses. Some of these applications include

- Analysis that requires high precision, such as isotope ratio and dilution studies, or the real-time measurement of internal standards
- Multielement analysis of a fast transient event, such as laser ablation or chromatographic separation techniques coupled with ICP-MS
- Applications where rapid speed of analysis is important, such as in high-throughput environmental contract labs
- Because the complete mass spectrum is collected every time, any mass can be interrogated after the analysis to check for any unexpected interferences.

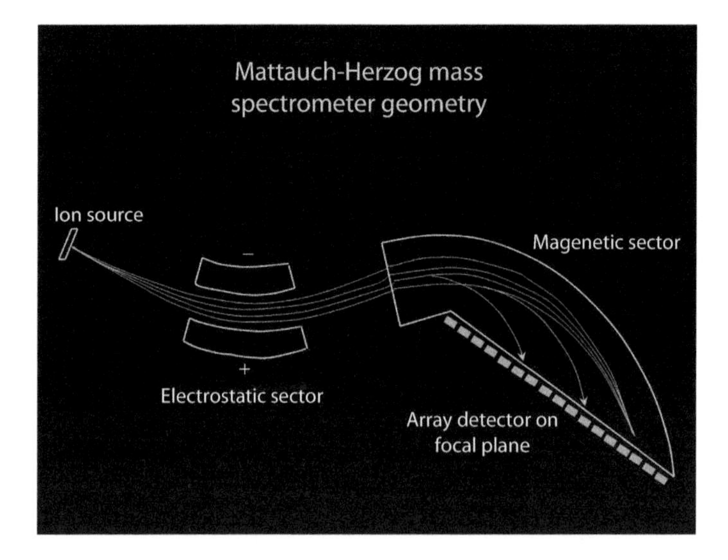

FIGURE 10.4 The principles of the Mattauch–Herzog double-focusing magnet sector mass spectrometer with a simultaneous detector plane.

However, it should be emphasized that this type of magnetic sector mass spectrometer does not offer the high resolving power of a traditional Nier–Johnson scanning system, which is typically in the order of 10,000. The resolution of this design of a Mattauch–Herzog mass spectrometer is only slightly better than that of a quadrupole system, which limits its use for the analysis of spectrally complex sample matrices.

FINAL THOUGHTS

There is no question that magnetic sector ICP-MS instruments are no longer novel analytical techniques. They have proved themselves to be a valuable addition to the trace element tool box, particularly for challenging applications that require good detection capability, exceptional resolving power, and very high precision. Even though they are not trying to compete with quadrupole instruments when it comes to rapid, high-sample-throughput applications, the measurement speed of modern systems, particularly the newer Mattauch–Herzog simultaneous detection approach, has opened up the technique to brand new application areas, which were traditionally beyond the scope of earlier designs of this technology.

REFERENCES

1. N. Bradshaw, E. F. H. Hall, and N. E. Sanderson. *Journal of Analytical Atomic Spectrometry*, 4, 801–803, 1989
2. F. W. Aston. *Philosophical Magazine*, 38, 707, 1919.
3. J. L. Costa. *Annals of Physics*, 4, 425, 1925.
4. A. J. Dempster. *Physical Review*, 11, 316, 1918.
5. W. F. G. Swann. *Journal of the Franklin Institute*, 210, 751, 1930.
6. A. J. Dempster. *Proceedings of the American Philosophical Society*, 75, 755, 1935.
7. K. T. Bainbridge and E. B. Jordan. *Physical Review*, 50, 282, 1936.
8. A. O. Nier. *Review of Scientific Instruments*, 11, 252, 1940.
9. G. P. Thomson. *Philosophical Magazine*, 42, 857, 1921.
10. A. J. Dempster. MDDC 370. Washington, DC: U.S. Department of Commerce, 1946.
11. J. Mattauch and R. Herzog. *Zeitschrift für Physik*, 89, 786, 1934.
12. N. B. Hannay and A. J. Ahearn. *Analytical Chemistry*, 26, 1056, 1954.
13. R. E. Honig. *Journal of Applied Physics*, 29, 549, 1958.
14. R. Castaing and G. Slodzian. *Journal of Microscopy*, 1, 395, 1962.
15. R. E. Honig and J. R. Wolston. *Applied Physics Letters*, 2, 138, 1963.
16. J. W. Coburn. *Review of Scientific Instruments*, 41, 1219, 1970.
17. R. Hutton, A. Walsh, D. Milton, and J. Cantle. *ChemSA*, 17, 213–215, 1991.
18. U. Geismann and U. Greb. *Fresenius' Journal of Analytical Chemistry*, 350, 186–193, 1994.
19. F. Adams, R. Gijbels, and R. Van Grieken. *Inorganic Mass Spectrometry*. New York: John Wiley & Sons, 1988.
20. U. Greb and L. Rottman. *Labor Praxis*, August 1994.
21. F. Vanhaecke, L. Moens, R. Dams, and R. Taylor. *Analytical Chemistry*, 68, 567, 1996.
22. M. Hamester, D. Wiederin, J. Willis, W. Keri, and C. B. Douthitt. *Fresenius' Journal of Analytical Chemistry*, 364, 495–497, 1999.
23. J. Walder and P. A. Freeman. *Journal of Analytical Atomic Spectrometry*, 7, 571, 1992.

24. J. H. Barnes, R. P. Sperline, M. B. Denton, C. J. Barinaga, D. W. Koppenaal, E. T. Young, and G. M. Hieftje. *Analytical Chemistry*, 74(20), 5327–5332, 2002.
25. J. H. Barnes, G. D. Schilling, R. P. Sperline, M. B. Denton, E. T. Young, C. J. Barinaga, D. W. Koppenaal, and G. M. Hieftje. *Analytical Chemistry*, 76(9), 2531–2536, 2004.
26. G. D. Schilling, F. J. Andrade, J. H. Barnes, R. P. Sperline, M. B. Denton, C. J. Barinaga, D. W. Koppenaal, and G. M. Hieftje. *Analytical Chemistry*, 78(13), 4319–4325, 2006.
27. G. D. Schilling, S. J. Ray, A. A. Rubinshtein, J. A. Felton, R. P. Sperline, M. B. Denton, C. J. Barinaga, D. W. Koppenaal, and G. M. Hieftje. *Analytical Chemistry*, 81(13), 5467–5473, 2009.
28. Spectro Analytical Instruments. A new era in mass spectrometry. http://www.spectro.com/pages/e/p010402tab_overview.htm.

11 Mass Analyzers
Time-of-Flight Technology

Let us turn our attention to time-of-flight (TOF) technology in this chapter. Although the first TOF mass spectrometer was first described in the literature in the late 1940s,[1] it has taken more than 50 years to adapt it for use with a commercial inductively coupled plasma (ICP) mass spectrometer. The interest in TOF ICP mass spectrometry (MS) instrumentation has come about because of its ability to sample all ions generated in the plasma at exactly the same time, which is ideally suited for multielement determinations of rapid transient signals, high-precision isotope ratio analysis, fast data acquisition for high-throughput workloads, and the rapid semiquantitative fingerprinting of unknown samples.

BASIC PRINCIPLES OF TOF TECHNOLOGY

The simultaneous nature of sampling ions in TOF offers distinct advantages over traditional scanning (sequential) quadrupole technology for ICP-MS applications in which large amounts of data need to be captured in a short span of time. To understand the benefits of this mass separation device, let us first take a look at its fundamental principles. All TOF mass spectrometers are based on the same principle: the kinetic energy (KE) of an ion is directly proportional to its mass (m) and velocity (V). This can be represented by the following equation:

$$KE = \frac{1}{2}mV^2$$

Therefore, if a population of ions—all with different masses—is given the same KE by an accelerating voltage (U), the velocities of the ions will all be different, depending on their masses. This principle is then used to separate ions of different mass-to-charge ratios (m/e) in the time (t) domain over a fixed flight path distance (D), represented by the following equation:

$$\frac{m}{e} = \frac{2Ut^2}{D^2}$$

This is schematically shown in Figure 11.1, which shows three ions of different mass-to-charge ratios being accelerated into a "flight tube" and arriving at the detector at different times. It can be seen that, depending on their velocities, the lightest ion arrives first, followed by the medium-mass ion, and finally the heaviest one.

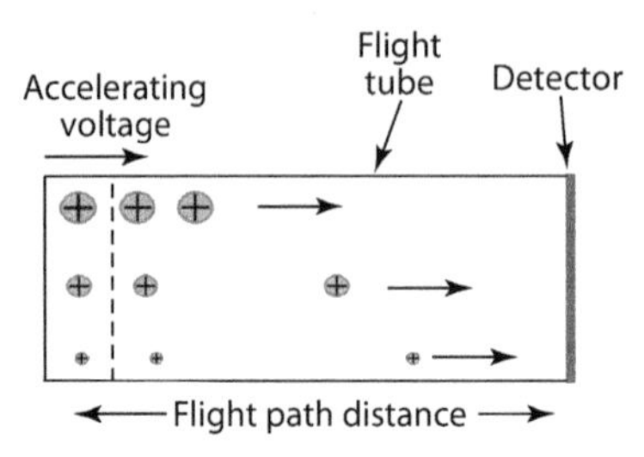

FIGURE 11.1 Principles of ion detection using TOF technology, showing separation of three different masses in the time domain.

Using flight tubes of 1 m in length, even the heaviest ions typically take less than 50 μs to reach the detector. This translates into approximately 20,000 mass spectra per second—three orders of magnitude faster than the sequential scanning mode of a quadrupole system.

COMMERCIAL DESIGNS

Even though this process sounds fairly straightforward, it is not a trivial task to sample the ions in a simultaneous manner from a continuous source of ions being generated in the plasma discharge. When TOF ICP-MS technology was first commercialized, there were basically two different sampling approaches used—the orthogonal design,[2] in which the flight tube is positioned at right angles to the sampled ion beam, and the axial design,[3] in which the flight tube is along the same axis as the ion beam. However, the only approach that is commercially available today is the orthogonal design.[4] The axial design was discontinued about 10 years ago. For the purpose of this chapter, I describe both approaches, but the reader should be aware that the only commercially available design today is the orthogonal approach.

In both designs, all ions that contribute to the mass spectrum are sampled through the interface cones, but instead of being focused into the mass filter in the conventional way, packets (groups) of ions are electrostatically injected into the flight tube at exactly the same time. With the orthogonal approach, an accelerating potential is applied at right angles to the continuous ion beam from the plasma source. The ion beam is then "chopped" by using a pulsed voltage supply coupled to the orthogonal accelerator to provide repetitive voltage "slices" at a frequency of a few kilohertz. The "sliced" packets of ions, which are typically tall and thin in cross section (in the vertical plane), are then allowed to "drift" into the flight tube, where the ions are temporally resolved according to their differing velocities. This is shown schematically in Figure 11.2.

The axial approach is similar in design to the orthogonal approach, except that an accelerating potential is applied axially (in the same axis) to the incoming ion beam as it enters the extraction region. Because the ions are in the same plane as the detector, the beam has to be modulated using an electrode grid to repel the "gated" packet of ions into the flight tube. This kind of modulation generates an ion packet that is long and thin in cross section (in the horizontal plane). The different masses are then

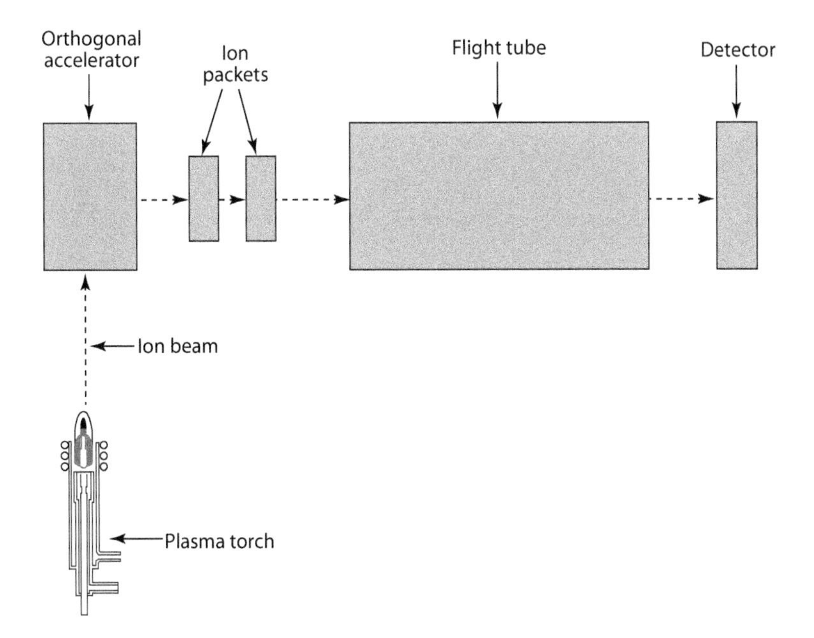

FIGURE 11.2 Schematic of an orthogonal acceleration TOF analyzer. (From GBC Scientific, Technical Note 001-0877-00, February 1998.)[4]

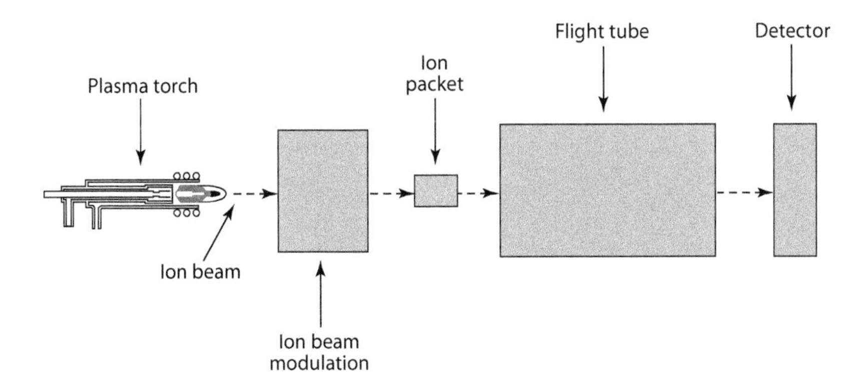

FIGURE 11.3 Schematic of an axial acceleration TOF analyzer. (From GBC Scientific, Technical Note 001-0877-00, February 1998.)[4]

resolved in the time domain in a manner similar to that of the orthogonal design. An on-axis TOF system is schematically shown in Figure 11.3.

Figures 11.2 and 11.3 offer a rather simplified explanation of the TOF principles of operation. In practice, there are many complex ion-focusing components in a TOF mass analyzer that ensure that the maximum number of analyte ions reach the detector, and also that undesired photons, neutral species, and interferences are ejected from the ion beam. Some of these components are seen in Figure 11.4, which shows a more detailed view of a commercial orthogonal TOF ICP-MS system.

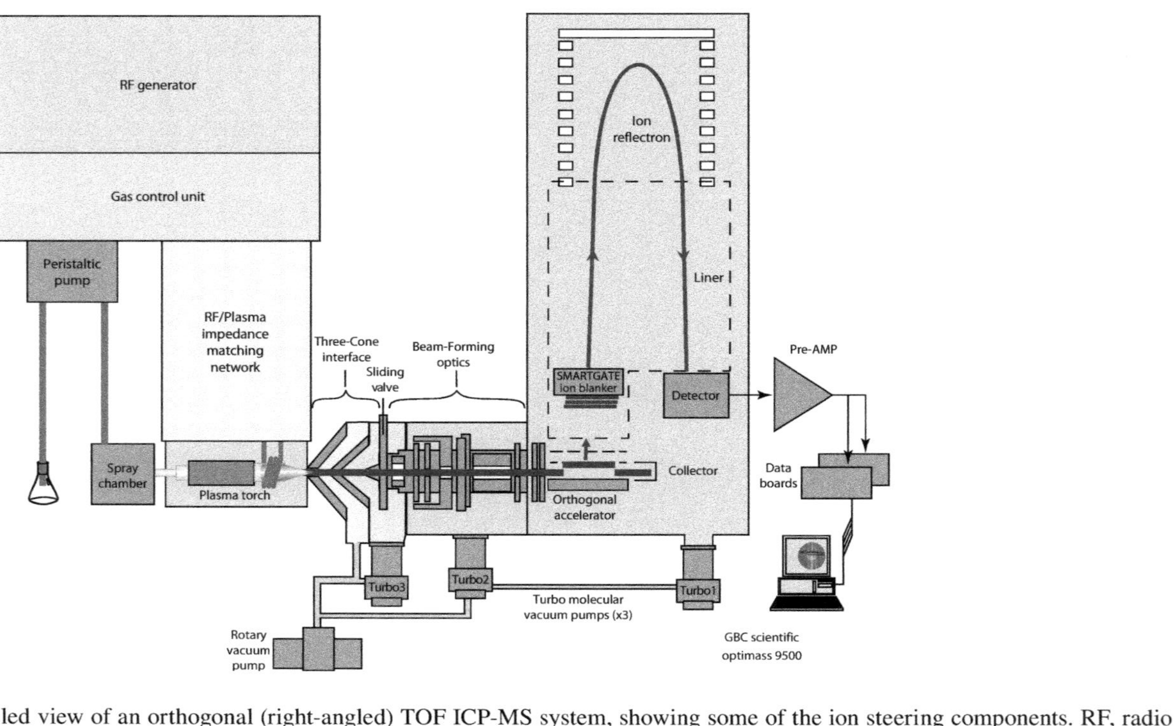

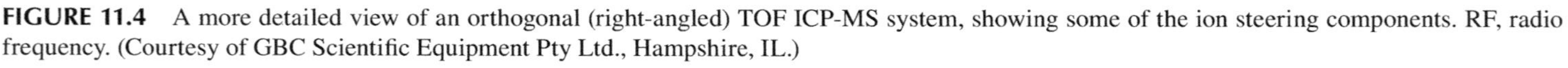
FIGURE 11.4 A more detailed view of an orthogonal (right-angled) TOF ICP-MS system, showing some of the ion steering components. RF, radio frequency. (Courtesy of GBC Scientific Equipment Pty Ltd., Hampshire, IL.)

It can be seen in this design that an orthogonal accelerator is used to inject packets of ions at right angles from the ion beam emerging from the MS interface. They are then directed toward an ion blanker, where unwanted ions are rejected from the flight path by deflection plates. The packets of ions are then directed into an ion reflectron, where they do a U-turn and are deflected back 180°, where they are detected by a channel electron multiplier or discrete dynode detector. The reflectron, which is a type of ion mirror, functions as an energy compensation device so that different ions of the same mass arrive at the detector at the same time.

DIFFERENCES BETWEEN ORTHOGONAL AND ON-AXIS TOF

Although there are real benefits of using TOF over quadrupole technology for some ICP-MS applications, there are also subtle differences in the capabilities of each type of TOF design. Some of these differences include

- Sensitivity: The axial approach tends to produce higher ion transmission because the steering components are in the same plane as the ion generation system (plasma) and the detector. This means that the direction and magnitude of greatest energy dispersion is along the axis of the flight tube. In addition, when ions are extracted orthogonally, the energy dispersion can produce angular divergence of the ion beam, resulting in poor transmission efficiency. However, on the basis of current evidence, the sensitivity of both TOF designs is generally an order of magnitude lower than that of the latest commercial quadrupole instruments.
- Background levels: The on-axis design tends to generate higher background levels because neutral species and photons stand a better chance of reaching the detector. This results in background levels on the order of 20–50 counts per second (cps)—approximately one order of magnitude higher than the orthogonal design. However, because the ion beam in the axial design has a smaller cross section, a smaller detector can be used, which generally has better noise characteristics. In comparison, most commercial quadrupole instruments offer background levels of 1–10 cps, depending on the design.
- Duty cycle: This is usually defined as the fraction (percentage) of extracted ions that actually make it into the mass analyzer. Unfortunately, with a TOF ICP mass spectrometer, which has to use "pulsed" ion packets from a continuous source of ions generated in the plasma, this process is relatively inefficient. It should also be emphasized that even though the ions are sampled at the same time, detection is not simultaneous because of different masses arriving at the detector at different times. The difference between the sampling mechanisms of the orthogonal and axial TOF designs translates into subtle differences in their duty cycles.
    - With the orthogonal design, duty cycle is defined by the width of the extracted ion packets, which are typically tall and thin in cross section, as shown in Figure 11.2. In comparison, the duty cycle of an axial design is defined by the length of the extracted ion packet, which is typically wide and thin in cross section, as shown in Figure 11.3. The duty cycle

can be improved by changing the cross-sectional area of the ion packet, but depending on the design, it is generally at the expense of resolution. However, this is not a major issue, because TOF instruments are generally not used for high-resolution ICP-MS applications. In practice, the duty cycles for both orthogonal and axial designs are on the order of 15%–20%.
- Resolution: The resolution of the orthogonal approach is slightly better because of its two-stage extraction and acceleration mechanism. Because a pulse of voltage pushes the ions from the extraction area into the acceleration region, the major energy dispersion lies along the axis of ion generation. For this reason, the energy spread is relatively small in the direction of extraction compared with the spread with the axial approach, resulting in better resolution. However, the resolving power of commercial TOF ICP-MS systems is typically on the order of 500–2000, depending on the mass region, which makes them inadequate to resolve many of the problematic polyatomic species encountered in ICP-MS. In comparison, commercial high-resolution systems based on the double-focusing magnetic sector design offer resolving power up to 10,000, whereas commercial quadrupoles achieve 300–400.
- Mass bias: This is also known as mass discrimination and is the degree to which ion transport efficiency varies with mass. All instruments show some degree of mass bias, which is usually compensated for by measuring the difference between the theoretical and observed ratios of two different isotopes of the same element. In TOF, the velocity (energy) of the initial ion beam will affect the instrument's mass bias characteristics. In theory, it should be less with the axial design because the extracted ion packets do not have any velocity in a direction perpendicular to the axis of the flight tube, which could potentially impact their transport efficiency.

BENEFITS OF TOF TECHNOLOGY FOR ICP-MS

It should be emphasized that these performance differences between the two designs are subtle and should not detract from the overall benefits of the TOF approach for ICP-MS. As mentioned earlier, a scanning device, such as a quadrupole, can only detect one mass at a time, which means that there is always a compromise between the number of elements, detection limits, precision, and overall measurement time. However, with the TOF approach, the ions are sampled at the same moment in time, which means that multielement data can be collected with no significant deterioration in quality. The ability of a TOF system to capture a full mass spectrum, approximately three orders of magnitude faster than a quadrupole, translates into three major benefits—multielement determinations in a fast transient peak; improved precision, especially for isotope ratioing techniques; and rapid data acquisition for carrying out qualitative or semiquantitative scans. Let us look at these in greater detail.

RAPID TRANSIENT PEAK ANALYSIS

Probably, the most exciting potential for TOF ICP-MS is in the multielement analysis of a rapid transient signal generated by sampling accessories such as laser ablation

(LA),[5] electrothermal vaporization (ETV),[6] and flow injection systems.[7] Even though a scanning quadrupole can be used for this type of analysis, it struggles to produce high-quality multielement data when the transient peak lasts only a few seconds. The simultaneous nature of TOF instrumentation makes it ideally suited for this type of analysis, because the entire mass range can be collected in less than 50 μs. In particular, when used with an ETV system, the high acquisition speed of TOF can help to reduce matrix-based spectral overlaps by resolving them from the analyte masses in the temperature domain.[7] There is no question that TOF technology is ideally suited (probably more than any other design of ICP mass spectrometer) for the analysis of transient peaks.

Improved Precision

To better understand how TOF technology can help improve precision in ICP-MS, it is important to know the major sources of instability. The most common source of noise in ICP-MS is the flicker noise associated with the sample introduction process (peristaltic pump pulsations, nebulization mechanisms, plasma fluctuations, etc.) and the shot noise derived from photons, electrons, and ions hitting the detector. Shot noise is based on counting statistics and is directly proportional to the square root of the signal. It therefore follows that as the signal intensity gets larger, the shot noise has less of an impact on the precision (percent relative standard deviation [% RSD]) of the signal. This means that at high ion counts, the most dominant source of imprecision in ICP-MS is derived from the flicker noise generated in the sample introduction area.

One of the most effective ways to reduce instability produced by flicker noise is to use a technique called internal standardization, where the analyte signal is compared and ratioed to the signal of an internal standard element (usually of similar mass or ionization characteristics) that is spiked into the sample. Even though a quadrupole-based system can do an adequate job of compensating for these signal fluctuations, it is ultimately limited by its inability to measure the internal standard at precisely the same time as the analyte isotope. So, in order to compensate for sample introduction– and plasma-based noise and achieve high precision, the analyte and internal standard isotopes need to be sampled and measured simultaneously. For this reason, the design of a TOF mass analyzer is perfect for the simultaneous internal standardization required for high-precision work. It therefore follows that TOF is also well suited for high-precision isotope ratio analysis, where its simultaneous nature of measurement is capable of achieving precision values close to the theoretical limits of counting statistics. Also, unlike a scanning quadrupole–based system, it can measure ratios for as many isotopes or isotopic pairs as needed—all with excellent precision.[8]

Rapid Data Acquisition

As with a scanning ICP-OES system, the speed of a quadrupole ICP mass spectrometer is limited by its scanning rate. To determine 10 elements in duplicate with good precision and detection limits, an integration time of 3 s per mass is normally required. When overhead scanning and settling times are added for each mass and replicate, this translates into approximately 2 min per sample. With a TOF system, the same analysis would take significantly less time, because all the data are captured simultaneously. In fact, detection limit levels in a TOF instrument are typically achieved within a 10–30 s integration time, which translates into a 5- to 10-fold improvement in data acquisition time over a quadrupole instrument. The added benefit of a TOF instrument is that the speed of analysis is not impacted by the number of analytes being determined: it would not matter if the method contained 10 or 70 elements—the measurement time would be virtually the same. However, there is one point that must be stressed. A large portion of the overall analysis time is taken up for flushing an old sample out of and pumping a new sample into the sample introduction system. This can be as much as 2 min per sample for real-world matrices. So, when this is taken into account, the difference between the sample throughput of a quadrupole and that of a TOF ICP mass spectrometer is not so evident.

Another benefit of the fast acquisition time is that qualitative or semiquantitative analysis is relatively seamless compared with scanning quadrupole technology, because every multielement scan contains data for every mass. This also makes spectral identification much easier by comparing the spectral fingerprint of unknown samples against a known reference standard. This is particularly useful for forensic work, where the evidence is often an extremely small sample.

HIGH-SPEED MULTIELEMENTAL IMAGING

A more recent example of the benefits of TOF technology for the characterization of rapid transients is in the field of high-speed multielement imaging using LA. Since its first appearance in the mid-1980s, laser ablation inductively coupled plasma mass spectrometry (LA-ICP-MS) has established itself as a routine method for the quantitation of trace elements in solid samples. In recent times, LA-ICP-MS has further become an important tool for elemental imaging applied in geological, biological, and medical research studies. To date, the most common imaging approach is to run the laser in continuous-scan mode, which involves firing the laser continuously with a specific repetition rate (usually 1–10 Hz) while slowly moving the stage with the mounted sample underneath the laser beam. The image is then constructed by ablating parallel lines on the sample surface. The ablated aerosol is washed out of the airtight ablation chamber (cell) to the ICP with a continuous flow of inert gas.

The washout times for conventional ablation cells are on the order of 0.5–30 s. After ionization in the ICP, the ablation signal is measured in a time-resolved manner, most commonly using a quadrupole or a sector field mass spectrometer. These instruments operate sequentially, resulting in the measurement of only one isotope at a time. For the analysis of multiple isotopes or elements, sequential peak hopping must be performed, which, depending on the number of elements of interest, can become very restrictive and time-consuming. The conventional approach to LA imaging is associated with drawbacks on spatial resolution and speed of analysis, which is usually limited to 1–2 pixels per second. The recent advent of fast-washout ablation cells and high-speed TOF ICP-MS allowed some of these limitations to be overcome by enabling spot-resolved imaging.

In order to capitalize on the benefits of fast-washout ablation cells, a mass spectrometer capable of extremely fast and simultaneous multielement data acquisition is required, and in particular its ability to handle rapid transient signals from these cells. The TOF ICP-MS is well suited for the shot-resolved imaging approach due to its simultaneous and fast multielement analysis, because the entire mass spectrum is measured simultaneously with each ion package extraction. Different isotopes from this package are then separated based on their TOF traversing of the region from extraction to detector, which is directly related to their mass-to-charge ratio. Thus, in contrast to conventional sequential mass spectrometers, it is not necessary to predefine or limit a range of isotopes for analysis. Moreover, the TOF ICP-MS can record complete mass spectra at a rate of 30 μs, which amounts to a maximum of ~33,000 spectra per second. This allows short transient signals from single laser pulses to be temporally resolved with sufficient sampling density (e.g., up to ~1000 spectra per 30 ms pulse). In practice, however, multiple spectra are integrated to improve counting statistics and to simplify data processing. Due to this simultaneous measurement of complete mass spectra, TOF ICP-MS can have a higher total ion utilization than sequential mass spectrometers, and as a result, a greater proportion of the ablated sample is actually used for analysis. This work has been presented in a recent publication by Bussweiler and coworkers.[9]

FINAL THOUGHTS

There is no question that TOF ICP-MS, with its rapid, simultaneous mode of measurement, excels at multielement applications that generate fast transient signals, such as LA. It offers excellent precision, particularly for isotope ratioing techniques, and also has the potential for very fast data acquisition. As mentioned previously, only the orthogonal (right-angled) design is currently available on a commercial basis. However, this should not detract from its overall capabilities. So although TOF ICP-MS is a fairly recent development compared with quadrupole technology, it definitely should be considered an option if the application demands it.

REFERENCES

1. E. Cameron and D. F. Eggers. *Review of Scientific Instruments*, 19(9), 605, 1948.
2. D. P. Myers, G. Li, P. Yang, and G. M. Hieftje. *Journal of American Society of Mass Spectrometry*, 5, 1008–1016, 1994.
3. D. P. Myers. *Elemental Mass Spectroscopy*. Presented at 12th Asilomar Conference on Mass Spectrometry, Pacific Grove, CA, September 20–24, 1996.
4. GBC Scientific. Technical Note 001-0877-00. February 1998.
5. P. Mahoney, G. Li, and G. M. Hieftje. *Journal of American Society of Mass Spectrometry*, 11, 401–406, 1996.
6. GBC Scientific. Technical Note 001-0876-00. GBC Scientific. February 1998.
7. R. E. Sturgeon, J. W. H. Lam, and A. Saint. *Journal of Analytical Atomic Spectrometry*, 15, 607–616, 2000.

8. F. Vanhaecke, L. Moens, R. Dams, L. Allen, and S. Georgitis. *Analytical Chemistry*, 71, 3297, 1999.
9. Y. Bussweiler, O. Borovinskaya, and M. Tanner. Laser ablation and inductively coupled plasma time-of-flight mass spectrometry—A powerful combination for high-speed multi-elemental imaging on the micrometer-scale. *Spectroscopy Magazine*, 32(5), 14–22, 2017.

12 Mass Analyzers *Collision/Reaction Cell and Interface Technology*

The detection capability for some elements using traditional quadrupole mass analyzer technology is severely compromised because of the formation of polyatomic spectral interferences generated by a combination of argon, solvent, and matrix-derived ions. Although there are ways to minimize these interferences, including correction equations, cool plasma technology, and matrix separation, they cannot be completely eliminated. However, a novel approach using collision/reaction cell (CRC) and collision/reaction interface (CRI) technology (more recently referred to as the integrated Collision Reaction Cell [iCRC] design) has been developed that significantly reduces the formation of many of these harmful species before they enter the mass analyzer. This chapter takes a detailed look at this very powerful technique and the exciting potential it has to offer.

There are a small number of elements that are recognized as having poor detection limits by inductively coupled plasma mass spectrometry (ICP-MS). These are predominantly elements that suffer from major spectral interferences generated by ions derived from the plasma gas, matrix components, or solvent or acid used in the sample preparation. Examples of these interferences include the following:

$^{40}Ar^{16}O^+$ in the determination of $^{56}Fe^+$
$^{38}ArH^+$ in the determination of $^{39}K^+$
$^{40}Ar^+$ in the determination of $^{40}Ca^+$
$^{40}Ar^{40}Ar^+$ in the determination of $^{80}Se^+$
$^{40}Ar^{35}Cl^+$ in the determination of $^{75}As^+$
$^{40}Ar^{12}C^+$ in the determination of $^{52}Cr^+$
$^{35}Cl^{16}O^+$ in the determination of $^{51}V^+$

The cold or cool plasma approach, which uses a lower temperature to reduce the formation of the argon-based interferences, has been a very effective way to get around some of these problems.[1] However, this approach can sometimes be difficult to optimize, is only suitable for a few of the interferences, and is susceptible to more severe matrix effects. Also, it can be time-consuming to change back and forth between normal and cool plasma conditions. These limitations and the desire to improve performance have led to the commercialization of CRCs, CRIs, and iCRCs. Their designs were based on the early work of Rowan and Houk, who used Xe and CH_4 in the late 1980s to reduce the formation of ArO^+ and Ar_2^+ species in the determination of Fe and Se with a modified tandem mass spectrometer.[2] This research was investigated further by Koppenaal and coworkers in 1994, who carried out studies using an ion trap for the determination of Fe, V, As, and Se in a 2% hydrochloric acid matrix.[3] However, it was not until 1996 that studies describing the coupling of a CRC with a traditional quadrupole ICP mass spectrometer were published. Eiden and coworkers experimented using hydrogen as a collision gas,[4] whereas Turner and coworkers based their investigations on using helium gas.[5] These studies and the work of other groups at the time[6,7] proved to be the basis for modern collision and reaction cells and interfaces that are commercially available today.

Let's take a look at the fundamental principles of CRCs, CRIs, and iCRCs.

BASIC PRINCIPLES OF COLLISION/REACTION CELLS

With all CRCs, ions enter the interface in the normal manner, and then are directed into a CRC positioned prior to the analyzer quadrupole. A collision/reaction gas (e.g., helium, hydrogen, ammonia, or oxygen, depending on the design) is then bled via an inlet aperture into the cell containing a multipole (a quadrupole, hexapole, or octapole), usually operated in the radio-frequency (RF)-only mode. The RF-only field does not separate the masses as a traditional quadrupole mass analyzer does, but instead, it has the effect of focusing the ions, which then collide and react with molecules of the collision/reaction gas. By a number of different ion–molecule collision and reaction mechanisms, either polyatomic interfering ions such as $^{40}Ar^+$, $^{40}Ar^{16}O^+$, and $^{38}ArH^+$ will be converted to harmless noninterfering species, or the analyte will be converted to another ion that is not interfered with. This process is exemplified by the equation below, which shows the use of hydrogen as a reaction gas to reduce the $^{40}Ar^+$ interference in the determination of $^{40}Ca^+$.

$$H_2 + {}^{40}Ar^+ = Ar + H_2^+$$

$$H_2 + {}^{40}Ca^+ = {}^{40}Ca^+ + H_2 \text{ (no reaction)}$$

It can be seen that the hydrogen molecule interacts with the argon interference to form atomic argon and the harmless H_2^+ ion. However, there is no interaction between the hydrogen and the calcium. As a result, the $^{40}Ca^+$ ions, free of the argon interference, emerge from the CRC through the exit aperture where they are directed toward the quadrupole analyzer for normal mass separation. Other gases are better suited to reduce the $^{40}Ar^+$ interference, but this process at least demonstrates the principles of the reaction mechanisms in a CRC. The layout of a typical CRC within the instrument is shown in Figure 12.1.

The equation above shows an example of an ion–molecule reaction using the process of charge transfer. By the transfer of a positive charge from the argon ion to the hydrogen molecule, an innocuous neutral Ar atom is formed, which is invisible to the mass analyzer. There are many other reaction and collisional mechanisms that can take place in the cell, depending on the nature of the analyte ion, the interfering species, the reaction/collision gas, and the type of multipole used. Other possible mechanisms that can occur in the cell, in addition to charge transfer, include the following:

- *Proton transfer*—The interfering polyatomic species gives up a proton, which is then transferred to the reaction gas molecule to form a neutral atom.

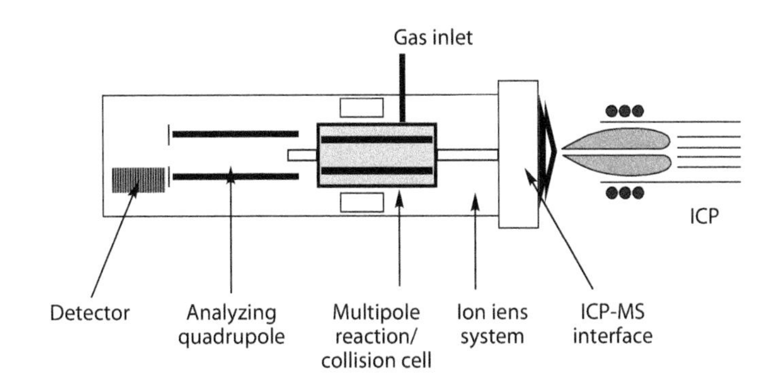

FIGURE 12.1 Layout of a typical CRC instrument.

- *Hydrogen atom transfer*—A hydrogen atom is transferred to the interfering ion, which is converted to an ion at one mass higher.
- *Molecular association reactions*—An interfering ion associates with a neutral species (atom or molecule) to form a molecular ion.
- *Collisional fragmentation*—The polyatomic ion is broken apart or fragmented by the process of multiple collisions with the gaseous atoms.
- *Collisional retardation*—The gas atoms or molecules undergo multiple collisions with the polyatomic interfering ion in order to retard or lower its kinetic energy. Because the interfering ion has a larger cross-sectional area than the analyte ion, it undergoes more collisions and, as a result, can be separated or discriminated from the analyte ion based on their kinetic energy differences.
- *Collisional focusing*—Analyte ions lose energy as they collide with the gaseous molecules and, depending on the molecular weight of the gas, will either enhance ion transmission as the ions migrate toward the central axis of the cell or decrease sensitivity if ion scattering takes place.

The CRI, which will be discussed later in this chapter, uses a slightly different principle to remove the interfering ions. It does not use a pressurized cell before the mass analyzer, but instead injects a reaction/collision gas directly into the aperture of interface skimmer cone. The injection of the collision/reaction into this region of the ion beam produces collisions between the argon gas and the injected gas molecules, and as a result, argon-based polyatomic interferences are destroyed or removed before they are extracted into the ion optics.

DIFFERENT COLLISION/REACTION CELL APPROACHES

All these possible interactions between ions and molecules indicate that many complex secondary reactions and collisions can take place, which generate undesirable interfering species. If these species are not eliminated or rejected, they could potentially lead to additional spectral interferences. There are basically two different approaches used to reduce the formation of polyatomic interferences and discriminate the products of these unwanted side reactions from the analyte ion. They are

- **Collision mechanisms** using nonreactive gases and kinetic energy discrimination (KED)[8]
- **Reaction mechanisms** using highly reactive gases and discrimination by selective bandpass mass filtering[9]

The major differences between the two approaches are how the gaseous molecules interact with the interfering species and what type of multipole is used in the cell. These dictate whether it is an ion–molecule collision or reaction mechanism taking place. Let us take a closer look at each process because there are distinct differences in the way the interference is rejected and separated from the analyte ion.

Collisional Mechanisms Using Nonreactive Gases and Kinetic Energy Discrimination

The collisional mechanisms approach was adapted from collision-induced dissociation (CID) technology, which was first used in the early to mid-1990s in the study of organic molecules using tandem mass spectrometry. The basic principle relies on using a nonreactive gas in a hexapole collision cell to stimulate ion–molecule collisions. The more collision-induced daughter species that are generated, the better the chance of identifying the structure of the parent molecule.[10,11] However, this very desirable CID characteristic for identifying and quantifying biomolecules was a disadvantage in inorganic mass spectrometry, where uncontrolled secondary reactions are generally something to be avoided. The limitation restricted the use of hexapole-based collision cells in ICP-MS to inert gases, such as helium, or low-reactivity gases, such as hydrogen, because of the potential to form undesirable reaction by-products, which could spectrally interfere with other analytes. Unfortunately, higher-order multipoles have little control over these secondary reactions because their stability boundaries are very diffuse and not well defineds like a quadrupole. As a result, they do not provide adequate mass separation capabilities to suppress the unwanted secondary reactions. Thus, the need is to rely mainly on collisional mechanisms and a process called *kinetic energy discrimination* to distinguish the interfering ions and by-product species from the analyte ions. So, what is KED?

KED relies on the principle of separating ions depending on their different ion energies. As ions enter the interface region, they all have differing kinetic energies based on the ionization process in the plasma and their mass-to-charge ratio. When the analyte and plasma or matrix-based interfering ions (sometimes referred to as *precursor ions*) enter the pressurized cell, they undergo multiple collisions with the collision gas. Because the collisional cross-sectional area of the precursor ions and other collision-induced by-product ions are usually larger than the analyte ion, they will undergo more collisions. This has the effect of lowering the kinetic energy of these interfering species compared with the analyte ion. If the collision cell rod offset potential is set slightly more negative than the mass filter potential, the polyatomic ions with lower kinetic energy are rejected or discriminated by the potential

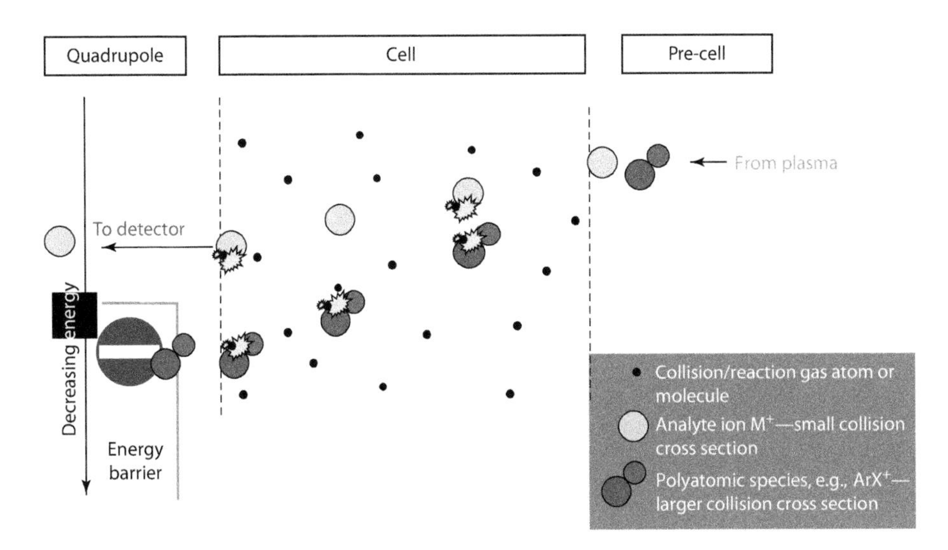

FIGURE 12.2 Principles of KED. (Courtesy of Thermo Scientific, Waltham, MA.)

energy barrier at the cell exit. On the other hand, the analyte ions, which have a higher kinetic energy, are transmitted to the mass analyzer. Figure 12.2 shows the principles of KED.

In addition to the kinetic energy generated by the ionization process, the spread of ion energies will also be dictated by the efficiency of the RF-grounding mechanism (see Chapter 5). Therefore, for KED to work properly, the ion energy spread of ions generated in the plasma must be as narrow as possible to ensure that there is very little overlap between the analyte and the polyatomic interfering ion as they enter the mass analyzer. This means that it is absolutely critical for the RF-grounding mechanism to guarantee a low potential at the interface. If this is not the case, and there is a secondary discharge between the plasma and the interface, it will increase the ion energy spread of ions entering the collision cell and make it extremely difficult to separate the polyatomic interfering ion from the analyte of interest based on their kinetic energy difference. The relevance of having a narrow spread of ion energies is shown in Figure 12.3.

It can be seen that the ion energy spread of the analyte ion (gray peak) and a polyatomic ion (black peak) is very similar as they enter the collision cell. This allows the collision process and KED system to easily separate the ions as they exit the cell. If the ion energy spread is larger, there would be more of an overlap as the ions enter the mass analyzer, and therefore compromise the detection limit for that analyte.

KED using helium as the collision gas works very well when the interfering polyatomic ion is physically larger than the analyte ion. This is exemplified in Figure 12.4, which shows helium flow optimization plots for six elements in 1:10 diluted seawater.[12]

It can be seen that the signal intensities for the analytes—Cr, V, Co, Ni, Cu, and As—are all at a maximum, whereas their respective matrix, argon, and solvent-based polyatomic interferences are all at a minimum at a similar helium flow rate of

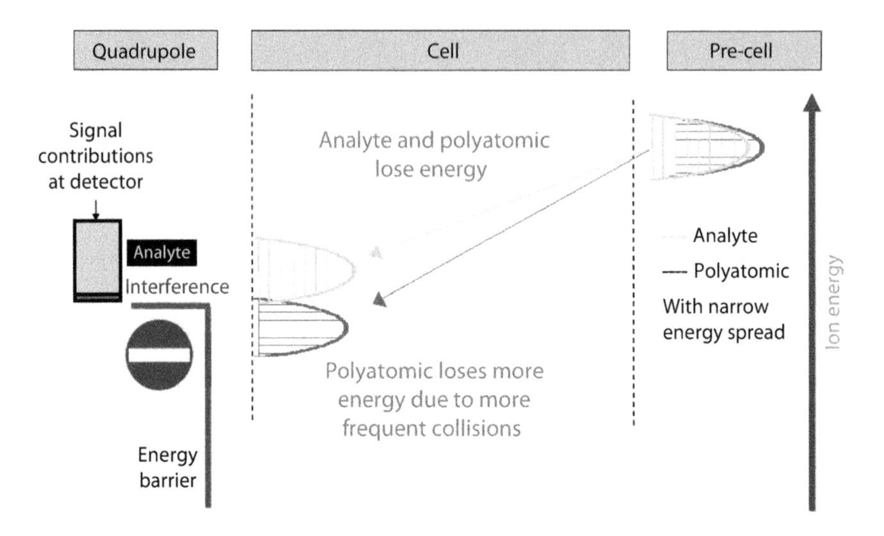

FIGURE 12.3 For optimum separation of an analyte ion from the interfering species, it is important to have a narrow spread of ion energies entering the collision cell. (Courtesy of Thermo Scientific, Waltham, MA.)

5–6 mL/min. All the analytes show very good sensitivity because the KED process has allowed the analytes to be efficiently separated from their respective polyatomic interfering ions. The additional benefit of using helium is that it is inert, and even if it is not being used in an interference reduction mode, it can have a beneficial effect on the other elements in a multielement run by increasing sensitivity via the process of collisional focusing. This makes the use of helium and KED very useful for both quantitative and semiquantitative multielement analysis using one set of tuning conditions.

For the KED process to work efficiently, there must be a distinct difference between the kinetic energy of the analyte and that of the interfering ion. In most cases where the polyatomic ion is large, this is not a problem. However, in many cases where the interfering and analyte ions have a similar physical size (cross section), the process requires an extremely large number of collisions, which will have an impact not only on the attenuation of the interfering ion, but also on the analyte ion. This means that for some situations, especially if the requirement is for ultra-trace detection capability, the collisional process with KED is not enough to reduce the interference to acceptable levels. That is why most collision cells also have basic reaction mechanism capabilities, which allows low-reactivity gases, such as hydrogen, and in some cases small amounts of highly reactive gases, such as ammonia or oxygen mixed with helium, to be used.[13] But it should be pointed out that even though this initiates a basic reaction mechanism, rejection of the reaction by-product ions is still handled through the process of KED, which in some applications might not offer the most efficient way of reducing the interfering species. Using the collision cell as a basic reaction cell with low-reactivity gases is described later in this chapter.

One further point to keep in mind is that a KED-based cell relies on interactions of the interfering ion with an inert or low-reactivity gas, such that it can be separated

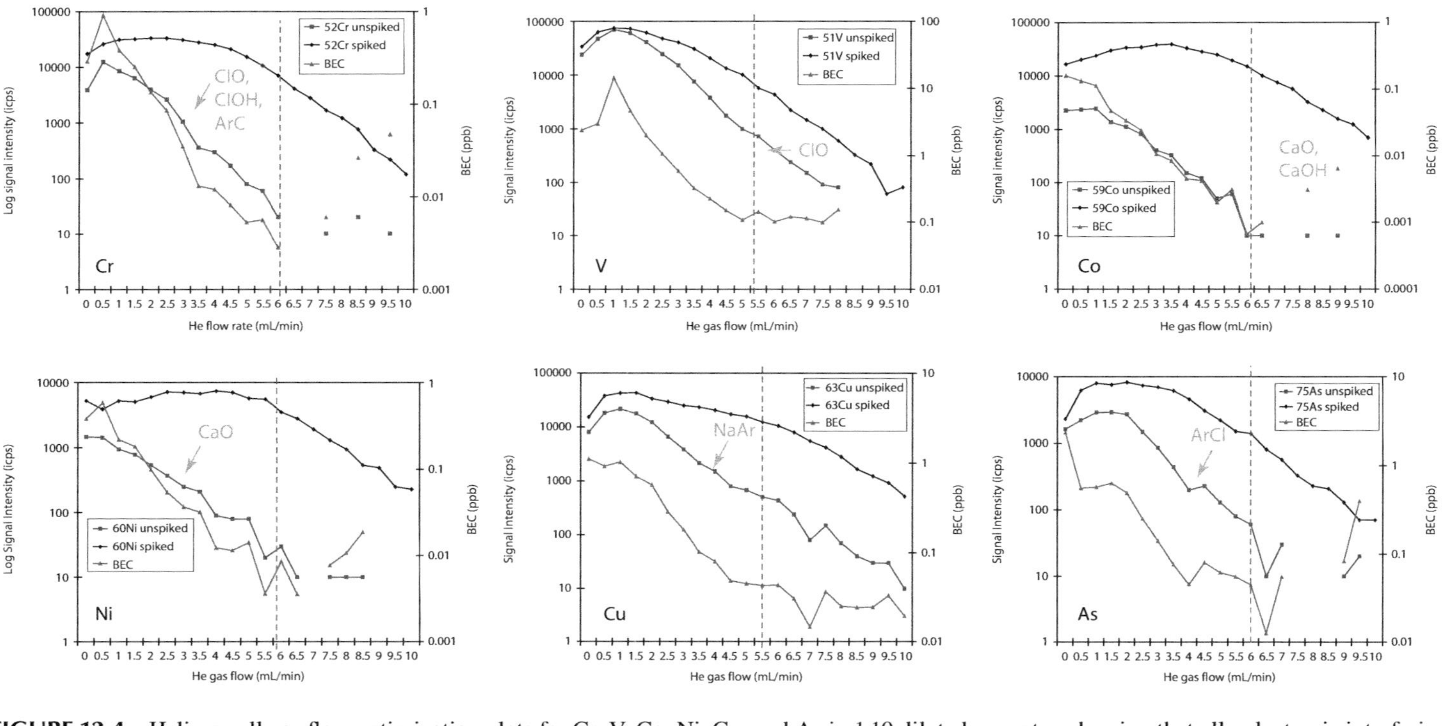

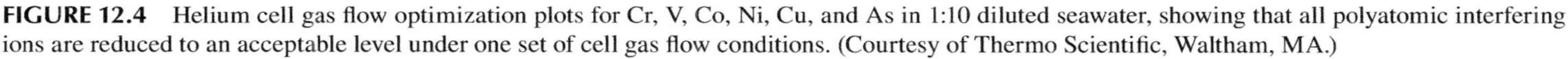

FIGURE 12.4 Helium cell gas flow optimization plots for Cr, V, Co, Ni, Cu, and As in 1:10 diluted seawater, showing that all polyatomic interfering ions are reduced to an acceptable level under one set of cell gas flow conditions. (Courtesy of Thermo Scientific, Waltham, MA.)

from the analyte based on their differences in kinetic energy. If the gas contains impurities such as organic compounds or water vapor, the impurity could be the dominant reaction or collision pathway, as opposed to the predicted collision/reaction with bulk gas. In addition, other unexpected ion–molecule reactions can readily occur if there are chemical impurities in the gas. This could also pose a secondary problem because of the formation of unexpected cluster ions, such as metal oxide and hydroxide species, which have the potential to interfere with other analyte ions. Fortunately, many of these new ions formed in the cell as a result of reactions with the impurities have low energy and are adequately handled by KED. However, depending on the level of the impurity, some of the ions formed have higher energies and are therefore too high to be attenuated by the KED process, which could negatively impact the performance of the interference reduction process. For this reason, it is strongly advised that the highest-purity collision/reaction gases be used. If this is not an option, it is recommended that a gas purifier (getter) system be placed in the gas line to cleanse the collision/reaction gas of impurities such as H_2O, O_2, CO_2, CO, or hydrocarbons. If you want to learn more about this subject, Yamada and coworkers published a very interesting paper describing the effects of cell–gas impurities and KED in an octapole-based collision cell.[14]

Let us now go on to discuss the other major way of interference rejection in a CRC using highly reactive gases and mass (bandpass) filtering discrimination.

Reaction Mechanisms with Highly Reactive Gases and Discrimination by Selective Bandpass Mass Filtering

Another way of rejecting polyatomic interfering ions and the products of secondary reactions/collisions is to discriminate them by mass. As mentioned previously, higher-order multipoles cannot be used for efficient mass discrimination because the stability boundaries are diffuse and sequential secondary reactions cannot be easily intercepted. The only way this can be done is to utilize a quadrupole (instead of a hexapole or octapole) inside the reaction/collision cell and use it as a selective bandpass (mass) filter. There are a number of commercial designs using this approach, so let's take a look at them in greater detail in order to get a better understanding of how they work and how they differ.

Dynamic Reaction Cell

The first commercial instrument to use this approach was called *dynamic reaction cell* (DRC) *technology*.[15] Similar in appearance to the hexapole and octapole CRCs, the DRC is a pressurized multipole positioned prior to the analyzer quadrupole. However, this is where the similarity ends. In DRC technology, a quadrupole is used instead of a hexapole or octapole. A highly reactive gas, such as ammonia, oxygen, or methane, is bled into the cell, which is a catalyst for ion–molecule chemistry to take place. By a number of different reaction mechanisms, the gaseous molecules react with the interfering ions to convert them into either an innocuous species different from the analyte mass or a harmless neutral species. The analyte mass then emerges from the DRC free of its interference and is steered into the analyzer quadrupole for conventional mass separation.

The advantage of using a quadrupole in the reaction cell is that the stability regions are much better defined than higher-order multipoles, so it is relatively straightforward to operate the quadrupole inside the reaction cell as a mass or bandpass filter and not just as an ion-focusing guide. Therefore, by careful optimization of the quadrupole electrical fields, unwanted reactions between the gas and the sample matrix or solvent, which could potentially lead to new interferences, are prevented. It means that every time an analyte and interfering ions enter the DRC, the bandpass of the quadrupole can be optimized for that specific problem and then changed on the fly for the next one. This is shown schematically in Figure 12.5, where an analyte ion $^{56}Fe^+$ and an isobaric interference $^{40}Ar^{16}O^+$ enter the DRC. As can be seen, the reaction gas NH_3 picks up a positive charge from the $^{40}Ar^{16}O^+$ ion to form atomic oxygen, argon, and a positive NH_3 ion (this is known as a "charge transfer reaction"). There is no reaction between the $^{56}Fe^+$ and the NH_3, as predicted by thermodynamic reaction kinetics. The quadrupole's electrical field is then set to allow the transmission of the analyte ion $^{56}Fe^+$ to the analyzer quadrupole, free of the problematic isobaric interference, $^{40}Ar^{16}O^+$. In addition, the NH_3^+ is prevented from reacting further to produce a new interfering ion.

The practical benefit of using highly reactive gases is that they increase the number of ion–molecule reactions taking place inside the cell, which results in a faster, more efficient removal of the interfering species. Of course, they will also generate more side reactions, which, if not prevented, will lead to new polyatomic ions being formed and could possibly interfere with other analyte masses. However, the quadrupole reaction cell is well characterized by well-defined stability boundaries. So, by careful selection of bandpass parameters, ions outside the mass/charge (*m*/*z*) stability

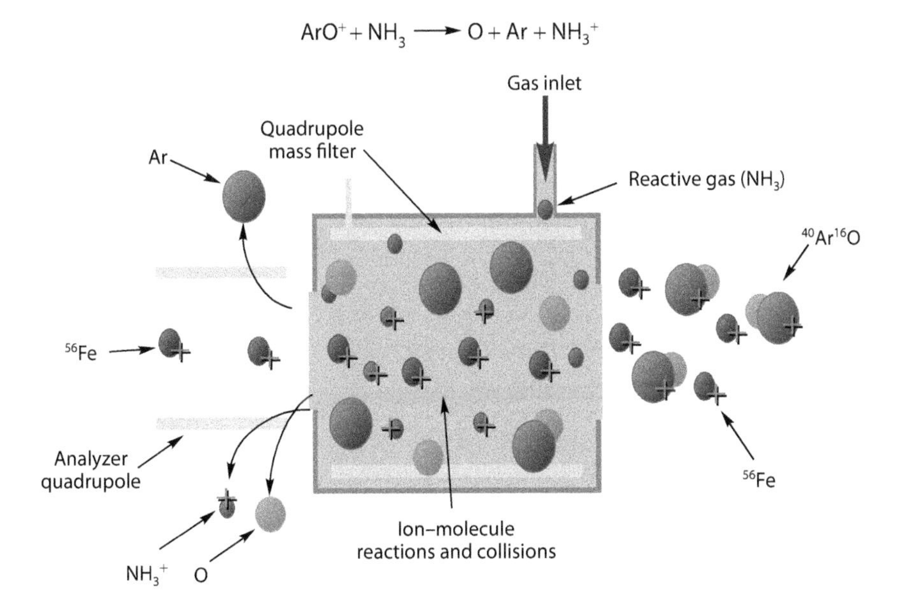

FIGURE 12.5 Elimination of the $^{40}Ar^{16}O^+$ interference with a DRC. (Copyright © 2013 PerkinElmer Inc., Waltham, MA. All rights reserved).

boundaries are efficiently and rapidly ejected from the cell. This means that additional reaction chemistries, which could potentially lead to new interferences, are successfully interrupted. In addition, the bandpass of the reaction cell quadrupole can be swept in concert with the bandpass of the quadrupole mass analyzer. This allows a dynamic bandpass to be defined for the reaction cell so that the analyte ion can be efficiently transferred to the analyzer quadrupole. The overall benefit is that within the reaction cell, the most efficient thermodynamic reaction chemistries can be used to minimize the formation of plasma- and matrix-based polyatomic interferences, in addition to simultaneously suppressing the formation of further reaction by-product ions.

The process described can be exemplified by the elimination of $^{40}Ar^+$ by NH_3 gas in the determination of $^{40}Ca^+$. The reaction between NH_3 gas and the $^{40}Ar^+$ interference, which is predominantly charge transfer or exchange, occurs because the ionization potential of NH_3 (10.2 eV) is low compared with that of Ar (15.8 eV). This makes the reaction extremely exothermic and fast. However, as the ionization potential of Ca (6.1 eV) is significantly less than that of NH_3, the reaction, which is endothermic, is not allowed to proceed.[15] This can be seen in greater detail in Figure 12.6.

Of course, other secondary reactions are probably taking place, which you would suspect with such a reactive gas as ammonia, but by careful selection of the cell quadrupole electrical fields, the optimum bandpass only allows the analyte ion to be transported to the analyzer quadrupole, free of the interfering species. This highly efficient reaction mechanism and selection process translates into a dramatic reduction of the spectral background at mass 40, which is shown graphically in Figure 12.7. It can be seen that at the optimum NH_3 flow, a reduction in the $^{40}Ar^+$ background signal of about eight orders of magnitude is achieved, resulting in a detection limit of approximately 0.1 ppt for $^{40}Ca^+$.

One final thing to point out is that when highly reactive gases are used, the purity of the gas is not so critical because the impurity is almost insignificant in determining the ion–molecule reaction mechanism. On the other hand, with collision and low-reactivity gases that contain impurities, such as carbon dioxide, hydrocarbons, or water vapor, the impurity could be the dominant reaction pathway, as opposed to the predicted collision/reaction with the bulk gas. In addition, the formation of unexpected by-product ions or other interfering species, which have the potential to interfere with other analyte ions in a KED-based collision cell, are not such a serious problem with the DRC system because of its ability to intercept and stop these side reactions using the bandpass mass filtering discrimination process.

These observations were, in fact, made by Hattendorf and Günter, who attempted to quantify the differences between KED and bandpass tuning with regard to suppression of interferences generated in a CRC for a group of mainly monoisotopic element (Sc, Y, La, and Th) oxides.[16] They observed that when the collision/reaction gas contains impurities, such as water vapor or ammonia, a broad range of additional interferences are produced in the cell. Depending on the relative mass of the precursor ions (ions that are formed in the plasma) compared with the by-product ions formed in the cell, there will be significant differences in the way these interferences are suppressed. They concluded that unless the mass (or energy) differences between the precursor and by-product ions are large, there will be significant overlap

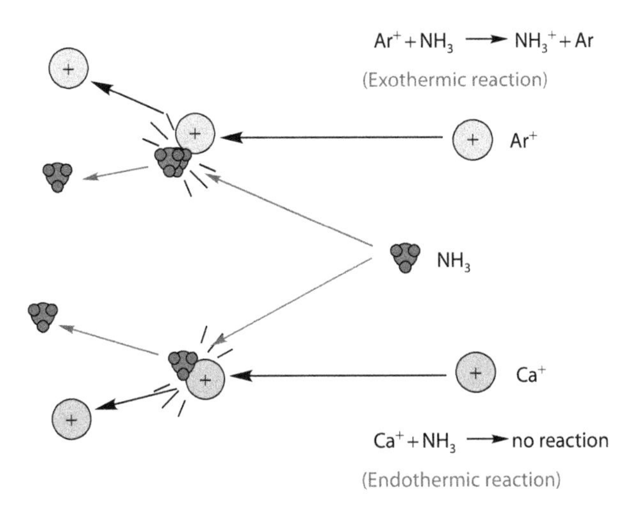

FIGURE 12.6 The reaction between NH_3 and Ar^+ is exothermic and fast, whereas there is no reaction between NH_3 and Ca^+ in the DRC. (Copyright © 2013 PerkinElmer Inc., Waltham, MA. All rights reserved.)

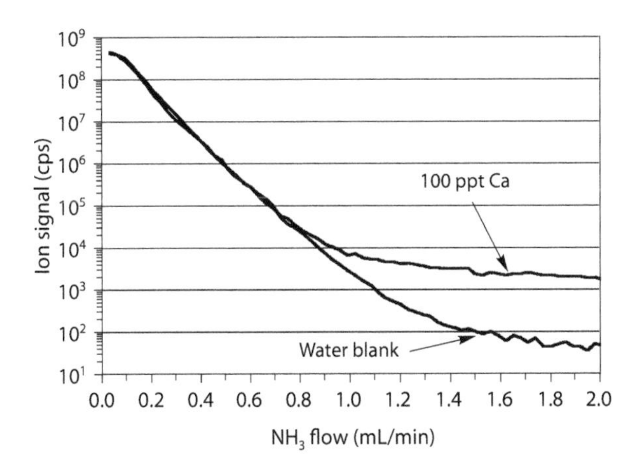

FIGURE 12.7 A reduction of eight orders of magnitude in the $^{40}Ar^+$ background signal is achievable with the DRC, resulting in a 0.1 ppt detection limit for $^{40}Ca^+$. (Copyright © 2013 PerkinElmer Inc., Waltham, MA. All rights reserved.)

of kinetic energy distribution, making it very difficult to separate them, which limits the effectiveness of KED to suppress the cell-generated ions. On the other hand, bandpass tuning can tolerate a much smaller difference in mass between the precursor and by-product interfering ions because of its ability to set the optimum mass or charge cutoff at the point where these interfering ions are rejected. In addition, they found that the bandpass tuning method can use a heavier or denser collision gas if desired, without suffering a loss of sensitivity due to scattering observed with the KED method. The overall conclusion of their study was that "under optimized conditions, the bandpass tuning approach provides superior analytical performance because it retains a significantly higher elemental sensitivity and provides more efficient suppression of cell-generated oxide ions, when compared to kinetic energy discrimination."

Low-Mass-Cutoff Collision/Reaction Cell

A variation on bandpass filtering which uses slightly different control of the filtering process. By operating the cell in the RF-only mode, the quadrupole's stability boundaries can be tuned to cut off low masses where the majority of the interferences occur.[17] This stops many of the problematic argon-, matrix-, and solvent-based ions from entering the CRC, therefore reducing the likelihood of creating new precursor ions in the cell, which have the potential to negatively impact the determination of the analytes. The basic principles of this technique are shown in Figure 12.8, which shows a typical Mathieu stability plot for a quadrupole. It can be seen with a fixed direct current (DC) electrical field (*a*) of zero (red horizontal line) and an RF electrical field (*q*) of 0.8 (blue vertical line); all masses above the red line and to the left of the blue line are stable and will pass through the quadrupole rods. While all masses that are to the right of the blue line will be unstable and ejected out of the electrical field. So, for example, a *q*-value of 0.8 might be equivalent to a 24 amu cutoff, which means that all masses above 24 amu are stable and will be transmitted,

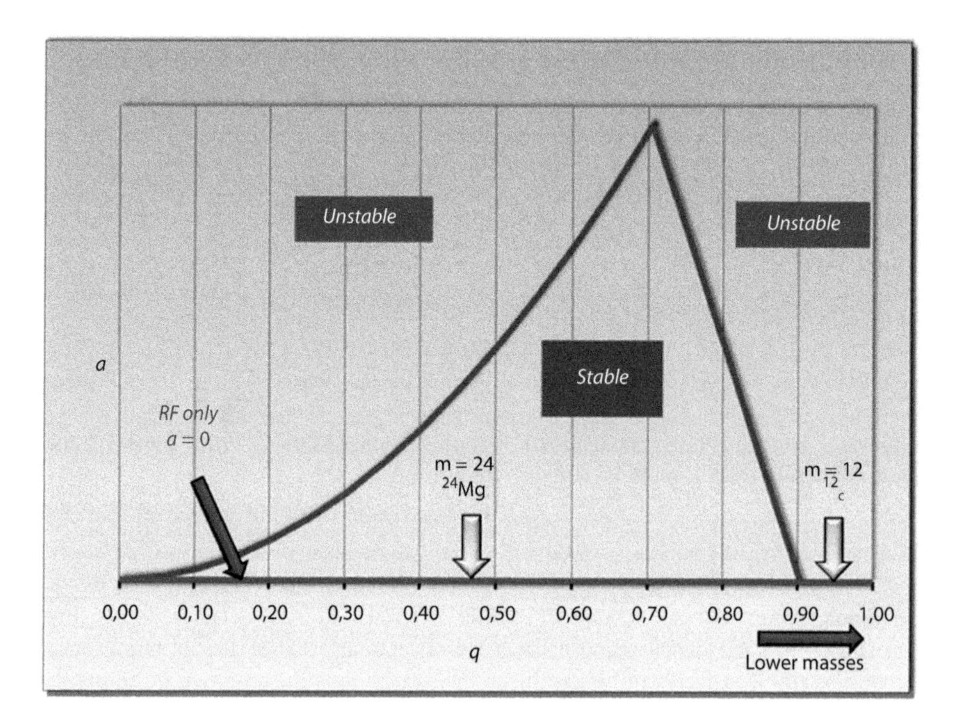

FIGURE 12.8 Basic principles of a low-mass-cutoff CRC.

Analyte	Cutoff mass	Potential interferent	Precursors
^{45}Sc	29	$^{13}C^{16}O_2$, $^{12}C^{16}O_2H$, ^{44}CaH, $^{32}S^{12}CH$, $^{32}S^{13}C$, $^{33}S^{12}C$	H, C, O, S, Ca
^{47}Ti	32	$^{31}P^{16}O$, ^{46}CaH, $^{35}Cl^{12}C$, $^{32}S^{14}NH$, $^{33}S^{14}N$	H, C, N, O, P, S, Cl, Ca
^{49}Ti	33	$^{31}P^{18}O$, ^{48}CaH, $^{35}Cl^{14}N$, $^{37}Cl^{12}C$, $^{32}S^{16}OH$, $^{33}S^{16}O$	H, C, N, O, P, S, Cl, Ca
^{50}Ti	34	$^{34}S^{16}O$, $^{32}S^{18}O$, $^{35}Cl^{14}NH$, $^{37}Cl^{12}CH$	H, C, N, O, S, Cl
^{51}V	35	$^{35}Cl^{16}O$, $^{37}Cl^{14}N$, $^{34}S^{16}OH$	H, O, N, S, Cl
^{52}Cr	36	$^{36}Ar^{16}O$, $^{40}Ar^{12}C$, $^{35}Cl^{16}OH$, $^{37}Cl^{14}NH$, $^{34}S^{18}O$	H, C, O, N, S, Cl, Ar
^{55}Mn	39	$^{37}Cl^{18}O$, $^{23}Na^{32}S$, $^{23}Na^{31}PH$,	H, O, Na, P, S, Cl, Ar
^{56}Fe	39	$^{40}Ar^{16}O$, $^{40}Ca^{16}O$	O, Ar, Ca
^{57}Fe	40	$^{40}Ar^{16}OH$, $^{40}Ca^{16}OH$	H, O, Ar, Ca
^{58}Ni	41	$^{40}Ar^{18}O$, $^{40}Ca^{18}O$, $^{23}Na^{35}Cl$	O, Na, Cl, Ar, Ca
^{59}Co	42	$^{40}Ar^{18}OH$, $^{43}Ca^{16}O$, $^{23}Na^{35}ClH$	H, O, Na, Cl, Ar, Ca
^{60}Ni	43	$^{40}Ca^{16}O$, $^{23}Na^{32}Cl$	O, Na, Cl, Ca
^{61}Ni	44	$^{44}Ca^{16}OH$, $^{38}Ar^{23}Na$, $^{23}Na^{37}ClH$	H, O, Na, Cl, Ca
^{63}Cu	45	$^{44}Ar^{23}Na$, $^{12}C^{15}O^{35}Cl$, $^{12}C^{14}N^{37}Cl$, $^{31}P^{32}S$, $^{31}P^{16}O_2$	C, N, O, Na, P, S, Cl
^{64}Zn	46	$^{32}S^{16}O_2$, $^{32}S_2$, $^{36}Ar^{12}C^{16}O$, $^{38}Ar^{12}C^{14}N$, $^{48}Ca^{16}O$	C, N, O, S, Ar, Ca
^{65}Cu	47	$^{32}S^{16}O_2H$, $^{32}S_2H$, $^{14}N^{15}O^{35}Cl$, $^{48}Ca^{16}OH$	H, N, O, S, Cl, Ca
^{66}Zn	47	$^{34}S^{16}O$, $^{32}S^{34}S$, ^{33}S,^{48}C,^{18}O	O, C, S
^{67}Zn	47	$^{32}S^{34}SH$, $^{33}S_2H$, $^{48}Ca^{18}OH$, $^{14}N^{16}O^{37}Cl$,$^{35}Cl^{16}O_2$	H, N, O, S, Cl, Ca
^{68}Zn	47	$^{32}S^{18}O_2$, $^{34}S_2$	O, S
^{69}Ga	47	$^{32}S^{18}O_2H$, $^{34}S_2H$, $^{37}Cl^{16}O_2$	H, O, S, Cl
^{70}Zn	47	$^{34}S^{18}O_2$, $^{35}Cl_2$	O, S, Cl
^{75}As	47	$^{40}Ar^{34}SH$, $^{40}Ar^{35}Cl$, $^{40}Ca^{35}Cl$, $^{37}Cl_2H$	H, S, Cl, Ca, Ar
^{77}Se	47	$^{40}Ar^{37}Cl$, $^{40}Ca^{37}Cl$	Cl, Ca, Ar
^{78}Se	47	$^{40}Ar^{38}Ar$	Ar
^{80}Se	47	$^{40}Ar_2$, $^{40}Ca_2$, $^{40}Ar^{40}Ca$, $^{32}S_2$$^{16}O$, $^{32}S^{16}O_3$	O, S, Ar, Ca

FIGURE 12.9 A list of elements with potential interferences that would benefit from the low mass cut-off approach (Note: Elements in red are above the low-mass cut off and could still contribute to a polyatomic interference).

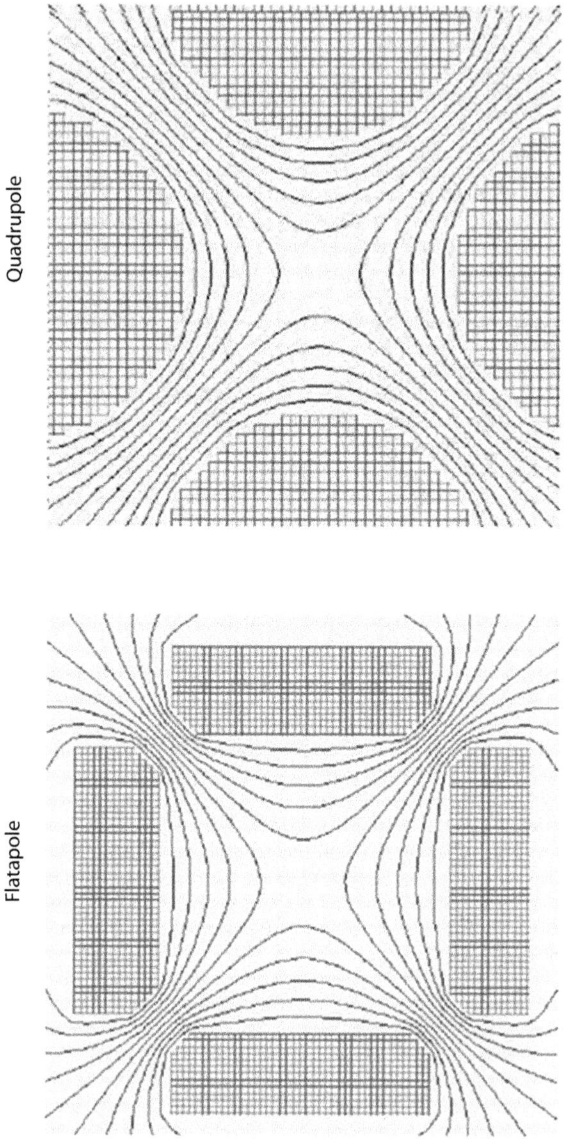

Quadrupole

Flatapole

FIGURE 12.10 Difference between the electric fields of a flatapole compared with a quadrupole.

while a q-value of >0.8 will be equivalent to all masses below 24 amu and will be rejected.

The benefit of this approach is seen in Figure 12.9, which shows a list of analytes in the far-left-hand column and the suggested cutoff mass in the next column. By using this cutoff mass, the potential polyatomic interferents in the third column won't be created because the majority of the precursor ions shown in the final column will be unstable and won't be present to react and combine with the other matrix- or solvent-based ions in the cell. However, it should be noted that the precursor elements in red in the final column are above the low-mass cut off and could still contribute to a polyatomic interference. For some elements, this can translate into a four- to fivefold improvement in detection capability over a traditional collision cell, particularly elements such as $^{31}P^{+}$ and $^{34}S^{+}$, which are prone to interferences from the $^{16}O^{+}$ ion.

The major difference of this technology is that it doesn't use a traditional quadrupole with spherical or elliptical-shaped rods. It uses a flatapole, which has beveled or straight edges, as opposed to round edges. The benefit is that the transmission efficiency tends to be similar to that of higher-order multipoles, thus allowing more ions through. The differences between the fields of a flatapole and quadrupole are shown in Figure 12.10.

Triple-Quadrupole Collision/Reaction Cell

It should be noted that up until now, only single multi-pole-based cells have realized commercial success, but a recent development has placed an additional quadrupole prior to the CRC multipole and the analyzer quadrupole. This first quadrupole acts as a simple mass filter to allow only the analyte masses to enter the cell, while rejecting all other masses. With all nonanalyte, plasma, and sample matrix ions excluded from the cell, sensitivity and interference removal efficiency are significantly improved compared with traditional CRC technology coupled with a single-quadrupole mass analyzer.

This very exciting collision/reaction technology is known as a "triple-quadrupole" CRC [18]—a name derived from the liquid chromatography tandem mass spectrometry (LC-MS-MS) technique, where three quadrupoles are used to separate, detect, and confirm the presence of organic molecules, such as proteins and peptides in biological samples. However, the term is not technically correct in this configuration used for ICP-MS. Even though the first quadrupole (Q1) is a mass filter and the second quadrupole (Q2) is the analyzer quadrupole, the middle multipole device is actually an octapole CRC. This means that the cell cannot be used as a conventional bandpass filter, like a single-quadrupole-based DRC or low-mass-cutoff device. Reactive gases can be used to initiate ion–molecule chemistry in the octopole cell, but they are then used just to pass all the product ions formed into the analyzer quadrupole (Q2), which is used to separate and select the mass or masses of interest. The principles of this technology are shown in Figure 12.11.

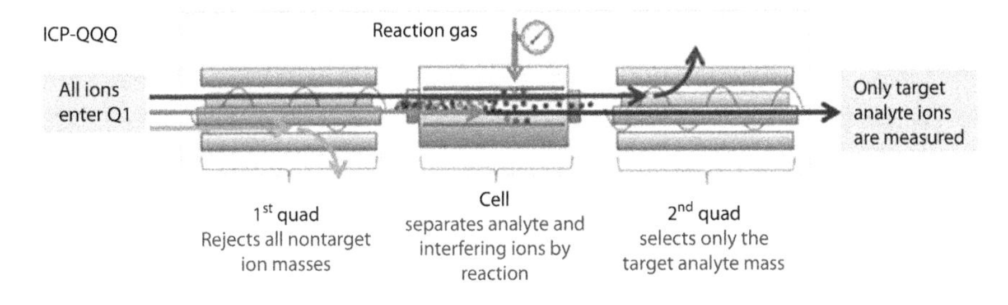

FIGURE 12.11 Fundamental principles of triple-quadrupole (QQQ) technology used in ICP-MS.

The capability and flexibility of the triple-quad CRC has enormous potential. There are a number of different ways it can be utilized depending on the severity of the analytical problem. The two most common modes of analysis are

- M/S mode
- MS/MS mode

M/S Mode

In its most basic configuration, the instrument can be utilized in the single M/S mode, where Q1 acts as a simple ion guide, allowing all ions through to the CRC, similar to a traditional single-quad ICP-MS system that uses an octopole-based collision cell. It can also be used in single M/S mode, where Q1 acts as a bandpass filter, allowing a "window" of masses through, above or below the Q2 mass range selected by the user. In this mode, it functions in a similar way as a single quad with a "scanning" bandpass filter-type CRC, except masses outside the bandpass window are rejected before they can enter the cell.

MS/MS Mode

The instrument can also be used in the MS/MS mode, where the first quadrupole is operated with a 1 amu fixed bandpass window, allowing only the target ions to enter the CRC. This process can be implemented in two different ways:

- On-mass mode
- Mass-shift mode

On-Mass MS/MS Mode

In this configuration, Q1 and Q2 are both set to the target mass. Q1 allows only the precursor ion mass to enter the cell (analyte and on-mass polyatomic interfering ions). The octopole CRC then separates the analyte ion from the interferences using the reaction chemistry of a reactive gas, while Q2 measures the analyte ion at the target mass after the on-mass interferences have been removed by reactions in the cell.

An example of this is in the removal of sulfur-based interferences using ammonia (NH_3) gas in the determination of vanadium in the presence of a sulfuric acid matrix. The major isotope of vanadium is $^{51}V^+$. However, in the presence of high concentrations of H_2SO_4, the interfering ions, $^{33}S^{18}O^+$ and $^{34}S^{16}OH^+$, overlap the $^{51}V^+$ ion. By using NH_3 as the reaction gas, which reacts very quickly with S-based polyatomic interferences, but is virtually unreactive with the vanadium ion, the $^{33}S^{18}O^+$ and $^{34}S^{16}OH^+$ ions are removed, thus allowing the $^{51}V^+$ ion to be detected, free of any interferences. So in this example, Q1 and Q2 would be set at mass 51 to take advantage of the reactive properties of ammonia to effectively remove the SO^+/SOH^+ interferences. No new analyte- or matrix-based NH_3 cluster ions can be created at mass 51, as no other ions are able to enter the cell. A detection limit in the order of 13 ppt is achievable for the determination of vanadium in percentage levels of sulfuric acid,

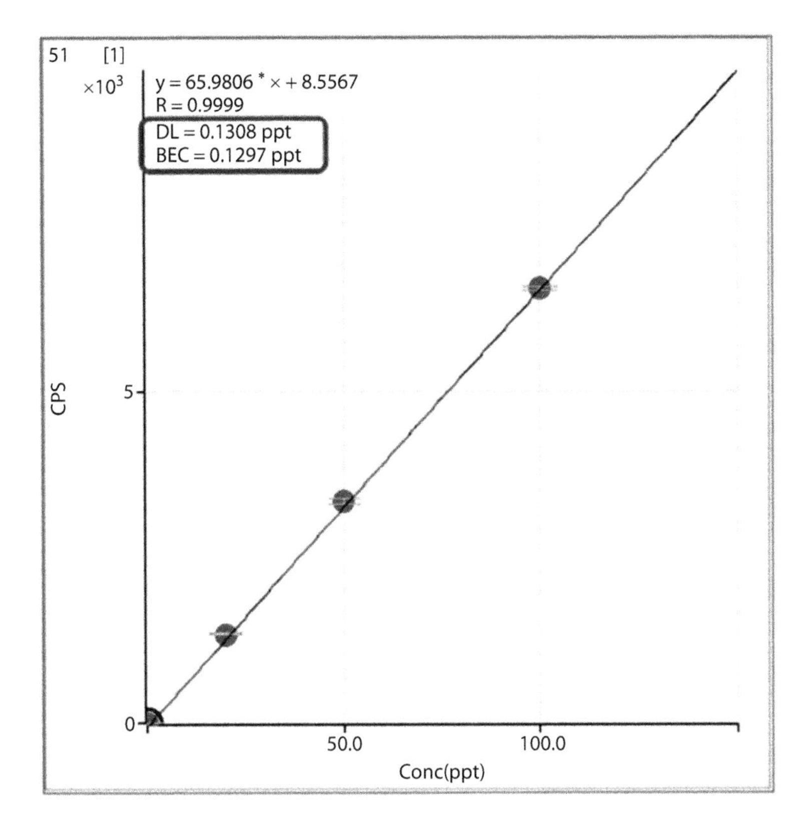

FIGURE 12.12 The detection capability of V in a sulfuric acid matrix using the on-mass MS/MS mode is in the order of 0.13 ppt, as shown in this calibration plot.

as demonstrated by the calibration seen in Figure 12.12. This detection capability is similar to what is achievable in an aqueous solution.

Mass-Shift MS/MS Mode

In this configuration, Q1 and Q2 are set to different masses. Similar to the on-mass mode, Q1 is set to the precursor ion mass (analyte and on-mass polyatomic interfering ions), controlling the ions that enter octapole CRC. However, in this mode Q2 is then set to the mass of a target reaction product ion containing the original analyte. The mass-shift mode is typically used when the analyte ion is reactive, while the interfering ions are unreactive with a particular CRC gas. The basic principles of this approach are shown in Figure 12.13.

An example of this is in the determination of arsenic in the presence of transition metal oxides. Because As is monoisotopic, it can only be determined at 75 amu, which is overlapped by the $^{59}Co^{16}O$ polyatomic interference. However, by reacting ^{75}As with oxygen gas in the cell, the $^{75}As^{16}O$ species at 91 amu can be used for quantitation. This is achieved by setting a bandpass to only allow ions of mass 75 amu to pass through Q1. All ions at 75 amu will then enter the octapole CRC, where they will interact with the oxygen cell gas. The ^{75}As will react with O_2 to form $^{75}As^{16}O$

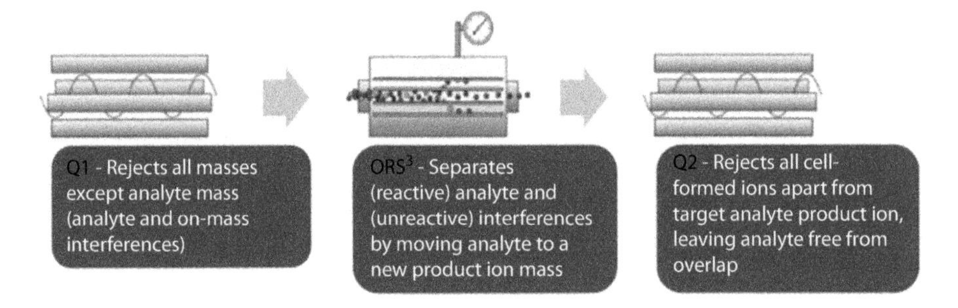

FIGURE 12.13 The basic principles of the mass-shift MS/MS mode.

species at 91 amu, while the $^{59}Co^{16}O$ at mass 75 amu will be unreactive. Q2 is then set at 91 amu to only allow the $^{75}As^{16}O$ ion through to be detected and quantified.

For more advanced, research-oriented applications, this technology offers other modes of interference reduction, including

- *Precursor ion scan*, where Q2 is set to the target ion mass, while Q1 scans over a user-defined mass range to select the precursor ions that enter the cell and react with the collision/reaction gas. An example of this is the monitoring of the by-product $^{14}NH_4$ ion in the determination of Hg using NH_3 as a reaction gas for a multielement method. In this example, NH_3 is the optimum reaction gas for the majority of the other elements, but by measuring the $^{14}NH_4$ ion, Hg can also be determined in the same suite.
- *Product ion scan*, where Q1 is set to allow only the target precursor ion mass to enter the cell, while Q2 scans to measure all the product ions formed in the cell, including controlled cluster ion analysis. An example of this is the use of NH_3 gas to create cluster ions of an analyte like titanium. By allowing only ^{48}Ti through Q1, only titanium cluster complex ions are formed in the cell and not other potentially interfering transition metal cluster ions, as with a traditional CRC.
- *Neutral gain scan*, where Q1 and Q2 are scanned together, with a user-defined mass difference. This mode allows monitoring of the product ions from a particular transition for all ions in the Q1 scan range. An example of this is in the determination of titanium using oxygen as the collision/reaction gas. By only scanning Q1 at the masses for titanium (^{46}Ti, ^{47}Ti, ^{48}Ti, ^{49}Ti, and ^{50}Ti) and Q2 at 16 amu higher ($^{46}Ti^{16}O$, $^{47}Ti^{16}O$, $^{48}Ti^{16}O$, $^{49}Ti^{16}O$, and $^{50}Ti^{16}O$), the isotopic abundances of the titanium oxide ions can be unambiguously measured, without the presence of other transition metals in the mass spectrum.

There is no question that the potential of this triple-quad approach to reducing interferences using a CRC analysis is a truly very exciting addition. However, it is probably more suited for research-type applications, or in an academic environment, where nonroutine investigations are being carried out. In my opinion, it will be competing with the double-focusing magnetic sector technology as a problem-solving tool and for the analysis of more complex sample matrices. The traditional, single-quadrupole ICP-MS instrumentation will still represent the vast majority of instruments sold in the marketplace, for carrying out high-throughput, routine applications. Although, for companies that are purchasing a second instrument, a triple quad probably represents a good investment.

Let us now discuss the collision reaction interface (CRI), otherwise known as the integrated collision reaction cell (iCRC) design.

INTEGRATED COLLISION/REACTION CELL

Unlike collision and reaction cells, the iCRC design does not use a pressurized multipole-based cell before the mass analyzer. Instead, it injects a reaction/collision gas (typically He, or H_2) at relatively high flow rates (100–150 mL/min) into the plasma through the aperture of the interface skimmer cone, where the plasma density is high.[19] This increases the rate of interactions between the introduced gas and interfering ions, giving improved attenuation of interfering ions. In addition, the reaction/collision gas is supplied directly to the plasma, which means that the plasma electrons are still available to assist in attenuating the interfering ions through electron–ion recombination. The presence of plasma electrons also significantly reduces the generation of secondary by-product ions produced from the interference attenuation process. The overall result is that most argon-based polyatomic interferences are destroyed or removed before they are extracted into the ion optics. Figure 12.14 shows the basic principles of the CRI or iCRC design.

The limitations of the earlier designs restricted their use for real-world samples because there appeared to be no way to effectively focus the ions, and therefore there was very little control over the collision process. So, even though the addition of a collision/reaction gas helped reduce plasma-based spectral interferences, it did virtually nothing for matrix-induced spectral interferences. In addition, there appeared to be no way to carry out KED in the interface region, and as a result, it was very difficult to take advantage of collisional mechanisms using an inert gas, such as helium.

However, in the most recent commercial design all the reaction/collision processes are actually taking place inside the tip of the skimmer and not between the sampler and skimmer cone, as with earlier designs. Because of this subtle difference, simple, loosely bonded polyatomic species can receive sufficient energy through collisional (vibrational and rotational) excitation mechanisms to bring about the dissociation of the interference, whereas the analyte ions simply lose energy as they collide with the gas molecules. And where the collisional impact is not suitable for interference reduction, as in the removal of the argon dimer ($^{40}Ar_2^+$) in the determination of $^{80}Se^+$, or the elimination of the $^{40}Ar^{12}C^+$ interference in the determination of $^{52}Cr^+$, a low-reactivity gas, such as hydrogen, can be used to initiate an ion–molecule reaction.

This can be seen in Figure 12.15, which shows the reduction of the $^{40}Ar_2^+$ and $^{40}Ar^{12}C^+$ interfering ions using hydrogen as the collision/reaction gas, in the determination of ^{80}Se and ^{52}Cr, respectively. The sensitivities of three internal standard elements, ^{45}Sc, ^{89}Y, and ^{115}In, were monitored at the same time as the two interfering ions. It should be noted that there is no Se or Cr in this solution, so the signals at mass 80 and 52 amu are contributions from the $^{40}Ar^{40}Ar$ and $^{40}Ar^{12}C$ polyatomic ions,

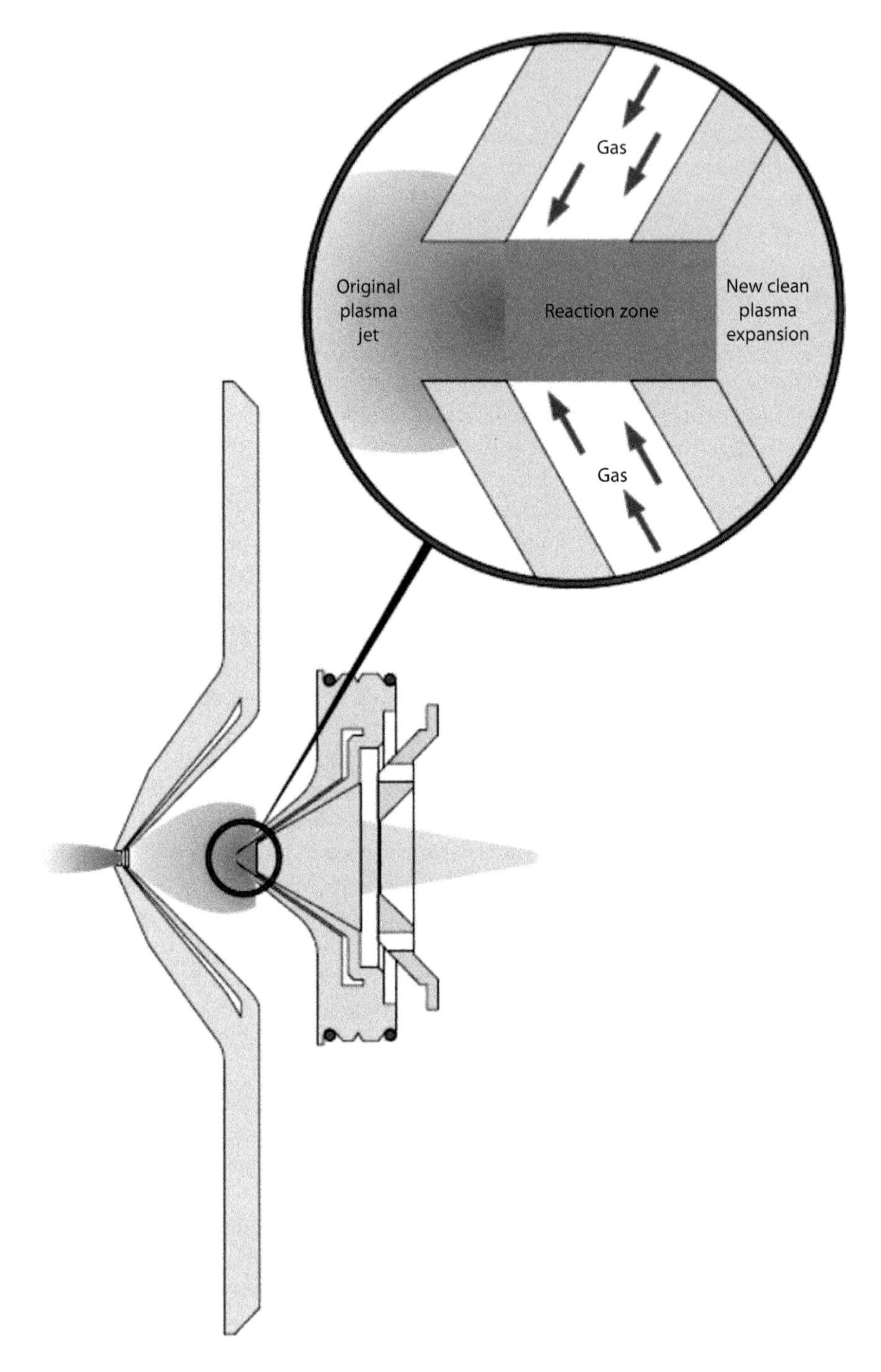

FIGURE 12.14 Principles of the CRI (also known as as the integrated collision reaction cell [iCRC] design). (Courtesy of Analytic Jena, Jena, Germany.)

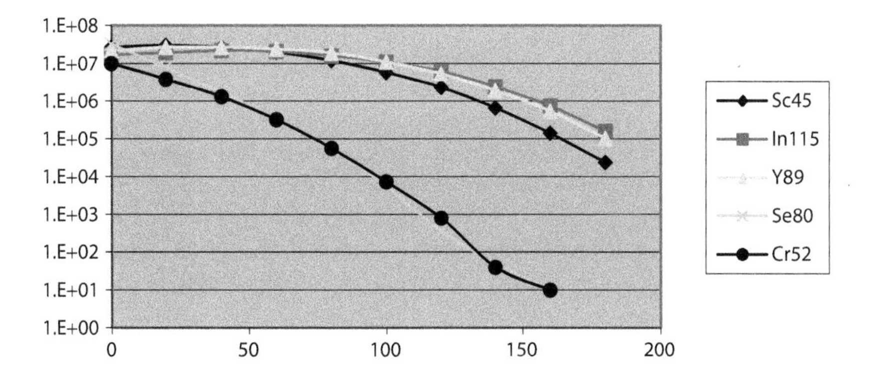

FIGURE 12.15 Optimization of hydrogen gas flow rate, showing three internal standard (Sc, In, and Y) signals while monitoring the interferences $_{40}Ar_{40}Ar$ ($_{80}Se$) and $_{40}Ar_{12}C$ ($_{52}Cr$). (Courtesy of Analytic Jena, Jena, Germany.)

respectively. It can be seen very clearly that there is sharper decrease of the interferent signals than those of the internal standards, showing evidence of the removal of the $ArAr^+$ and ArC^+ polyatomic interferences with increasing H_2 flow rate. The optimization plot shows that at a flow rate of 140 mL/min, signals of the interfering ions decrease by six orders of magnitude, while those of the Sc, Y, and In are only reduced by two orders of magnitude.[20]

The iCRC design looks to be a very interesting concept, which appears to offer a relatively straightforward, non-cell-based solution to minimizing plasma- and matrix-based spectral interferences in ICP-MS. Each year, more and more challenging applications appear in the public domain showing the capabilities of CRI systems. If there is an interest in this approach, it is worth checking out the vendor application notes, which show the capabilities of the iCRC design in many different, real-world matrices.[21]

USING REACTION MECHANISMS IN A COLLISION CELL

After almost 20 years of solving real-world application problems, the practical capabilities of both hexapole and octapole collision cells using KED are fairly well understood. It is clear that the majority of applications are being driven by the demand for routine multielement analysis of well-characterized matrices, where a rapid sample turnaround is required. This technique has also been promoted by the vendors as a fast, semiquantitative tool for unknown samples. The fact that it requires very little method development, just one collision gas and one set of tuning conditions, makes it very attractive for these kinds of applications.[22]

However, it is well recognized that this approach will not work well for many of the more complex interfering species, especially if the analyte is at ultratrace levels. For example, the collision mode using helium is not the best choice for quantifying selenium using its two major isotopes ($^{80}Se^+$ and $^{78}Se^+$) because it requires a large number of collisions to separate the argon dimers ($^{40}Ar_2^+$ and $^{40}Ar^{38}Ar^+$) from the analyte ions using KED. As a result of the inefficient interference reduction process, the signal intensity for the analyte ion is also suppressed. For this reason, it is well accepted that the best approach is to use a reaction gas in order to initiate some kind of charge transfer mechanism. This can be seen in Figure 12.16, which shows the calibration for $^{78}Se^{+}$ using hydrogen as the reaction gas to minimize the impact of the argon dimer, $^{40}Ar^{38}Ar^{+}$. Displayed in (a) are the 0, 1.0, 2.5, 5.0, and 10.0 ppb calibrations using helium, and in (b) are the same calibrations using hydrogen as the reaction gas. It can be seen very clearly that the signal intensity for the calibration standards using hydrogen gas is approximately five times higher than the calibration standards using helium, producing a selenium detection limit that is 15 times lower (1 ppt compared with 15 ppt); this is also seen in the table (c) under the calibration graphs.[23]

Likewise, helium has very little effect on reducing the $^{40}Ar^{+}$ interference in the determination of $^{40}Ca^{+}$. So, when using a collision cell with helium, the quantification of Ca must be carried out using the $^{44}Ca^{+}$ isotope, which is about 50 times less sensitive than $^{40}Ca^{+}$. For this reason, in order to achieve the lowest detection limits for calcium, a low-reactivity gas, such as pure hydrogen, is the better option.[22] Initiating an ion–molecule reaction allows the most sensitive calcium isotope at 40 amu to be used for quantitation. In fact, even though the use of hydrogen significantly improves the detection limit for calcium, the best interference reduction is achieved using a mixture of ammonia and helium.

Another example of the benefits of using more reactive gases, such as an ammonia and helium mixture over pure helium or hydrogen gas, is in the determination of vanadium in a high-concentration chloride matrix. The collision mode using helium works reasonably well on the reduction of the $^{35}Cl^{16}O^{+}$ interference at 51 amu. However, when 1% NH_3 in helium is used, the interference is dramatically reduced by the process of charge or electron transfer. This allows the most abundant isotope, $^{51}V^{+}$, to be used for the quantitation of vanadium in matrices such as seawater or hydrochloric acid. Vanadium detection capability in a chloride matrix is improved

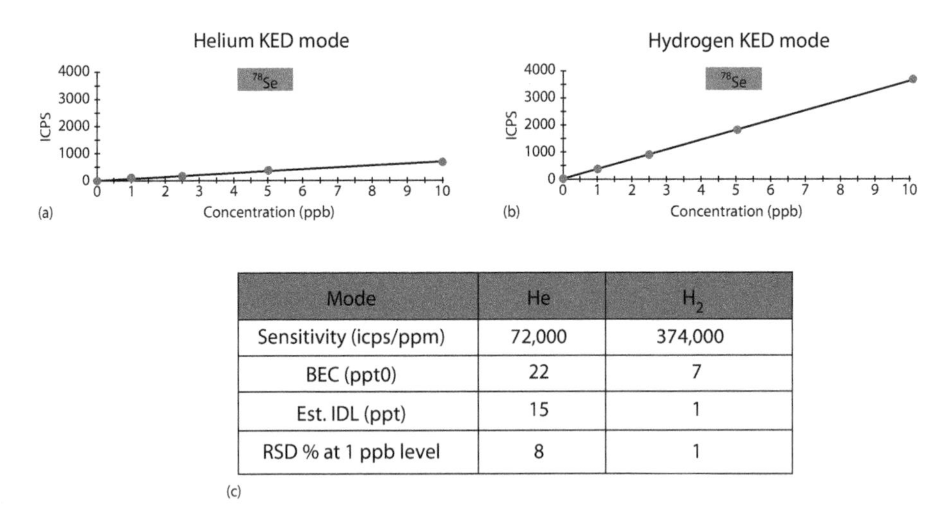

Mode	He	H_2
Sensitivity (icps/ppm)	72,000	374,000
BEC (ppt0)	22	7
Est. IDL (ppt)	15	1
RSD % at 1 ppb level	8	1

FIGURE 12.16 Comparison of $^{78}Se^{+}$ calibration plots and detection limits using the collision mode with helium (a) and the reaction mode with hydrogen (b). (Courtesy of Thermo Scientific, Waltham, MA.)

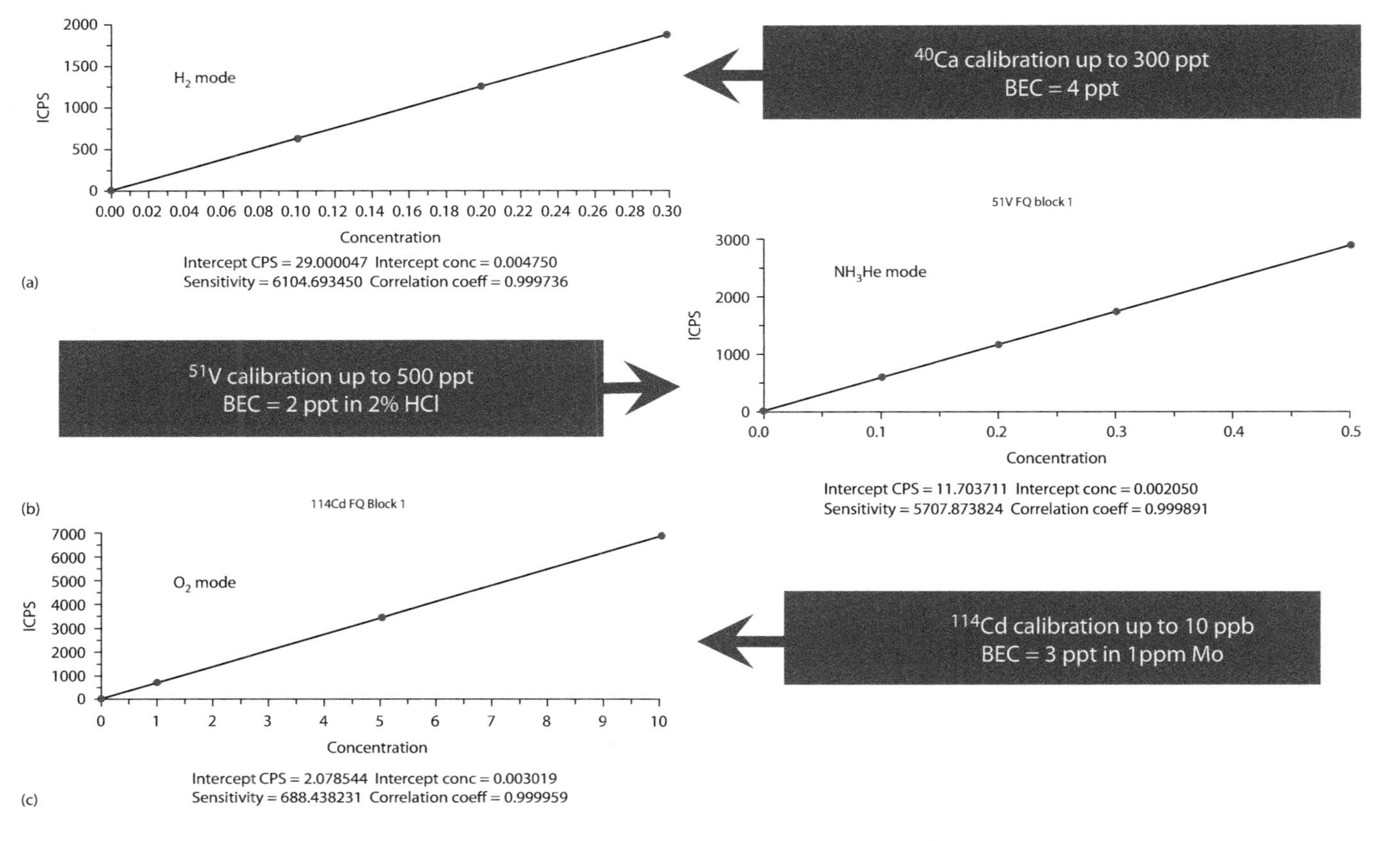

FIGURE 12.17 Calibration plots for (a) $40Ca^+$ using hydrogen, (b) $51V^+$ in 2% HCl using 1% NH_3 in helium, and (c) 114Cd+ in high concentrations f molybdenum using oxygen. (Courtesy of Thermo Scientific Waltham, MA.)

by a factor of 50–100 times using the reaction chemistry of NH_3 in helium compared with pure helium in the collision mode.[23]

In addition to ammonia–helium mixtures, oxygen is sometimes the best reaction gas to use because it offers the possibility of either moving the analyte ion to a region of the mass spectrum where the interfering ion does not pose a problem, or moving the interfering species away from the analyte ion by forming an oxygen-derived polyatomic ion 16 amu higher. An example of changing the mass of the interfering ion is in the determination of $^{114}Cd^+$ in the presence of high concentrations of molybdenum. In the plasma, the molybdenum forms a very stable oxide species, $^{98}Mo^{16}O^+$, at 114 amu, which interferes with the major isotope of cadmium, also at mass 114. By using pure oxygen as the reaction gas, the $^{98}Mo^{16}O^+$ interference is converted to the $^{98}Mo^{16}O^{16}O^+$ complex at 16 amu higher than the analyte ion, allowing the $^{114}Cd^+$ isotope to be used for quantitation.[23]

The benefits of using reaction mechanisms for the determination of calcium, vanadium, and cadmium are seen in Figure 12.17, which shows (a) a 0–300 ppt calibration plot for $^{40}Ca^+$ using hydrogen gas, (b) a 0–500 ppt calibration plot for $^{51}V^+$ in 2% hydrochloric acid using 1% NH_3 in helium, and (c) a 0–10 ppb calibration plot for $^{114}Cd^+$ in a molybdenum matrix using pure oxygen. Background equivalent concentration (BEC) values of 4, 2, and 3 ppt were achieved for calcium, vanadium, and cadmium, respectively.[23]

However, it is important to point out that even though a higher-order multipole collision cell with KED can use low-reactivity gases, it usually requires significantly more interactions than a reaction cell that uses a highly reactive gas.[23] Take, for example, the reduction of the $^{40}Ar^+$ interferences in the determination of $^{40}Ca^+$. In a collision cell, even though the kinetic energy of the $^{40}Ar^+$ ion will be reduced by reactive collisions with molecules of hydrogen gas, the $^{40}Ca^+$ will also lose kinetic energy because it, too, will collide with the reaction gas. Now, even though these interactions are basically nonreactive with respect to $^{40}Ca^+$, it will experience the same number of collisions because it has a similar cross-sectional area as the $^{40}Ar^+$ ion. So, in order to achieve many orders of interference reduction with a low-reactivity gas, such as hydrogen, a high number of collisions are required. This means that in addition to the interference being suppressed, the analyte will also be affected to a similar extent. As a result, the energy distribution of both the interfering species and the analyte ion at the cell exit will be very close, if not overlapping, resulting in compromised interference reduction capabilities compared with a reaction cell that uses highly reactive gases and discrimination by mass filtering. This compromise translates into a detection limit for $^{40}Ca^+$, with a KED-based collision cell using hydrogen being approximately 50–100 times worse than a DRC using pure ammonia.[24,25]

UNIVERSAL CELL

A recent development is the commercialization of an instrument that can be used in both the collision mode and the reaction mode. More commonly known as the "universal" cell, this innovative design can be utilized as both a simple collision cell using KED and a dynamic reaction using a pure reaction gas.[26] When operating in the collision cell mode, the universal cell works on the principal that the interfering ion is physically larger than the analyte ion. If both ions are allowed to pass through the cell containing the inert gas molecules, the interfering ion will collide more frequently with the inert gas atoms than the analyte ion due to its larger size. This results in a greater loss of kinetic energy by the interferent compared with the analyte ion. An energy barrier is then placed at the exit of the cell, so that the higher-energy analyte ions are allowed to pass through it, while the lower-energy interferences are not. As mentioned earlier in this chapter, this process is commonly referred to as KED.

When the universal cell is operating in the reaction mode, it takes advantage of whether the interaction of the analyte and interfering species with the reaction gas is either exothermic (fast) or endothermic (slow). The reactivity will be based on the relative ionization potential of the analyte and interfering species compared with the reaction gas. Typically, interfering ions tend to react exothermally with a reactive gas like ammonia, while analyte ions react endothermally. This means that if both interferent ions and analyte ions enter the cell, the interferent ions will react with the reactive gas and be converted to a new species with a different mass, while the analyte ions will be unaffected and pass through the reaction cell into the filtering quadrupole, free of the interference.

But it should be emphasized that while a reactive gas efficiently removes interferences, it is also capable of creating new interferences if not properly controlled. For that reason, an optimized reaction cell (coupled with a single mass analyzer) requires the use of a scanning quadrupole to prevent these new interferences from forming through the creation of a unit-resolution bandpass filter, to allow only a single mass to pass through.

The real benefit of the universal CRC approach is that it can be used in both the collision cell and the reaction cell modes. This means that the operator has the flexibility to operate the system in three different modes, all in the same multielement method—in the standard mode for elements where interferences are not present, in the collision mode for removal of a minor interference, and in the dynamic reaction mode for the most severe polyatomic spectral interferences.

DETECTION LIMIT COMPARISON

In general, highly reactive gases are recognized as being more efficient at reducing the interference and generating better signal-to-noise ratio than either an inert gas, such as helium, or low-reactivity gases, such as hydrogen or mixtures of ammonia and helium. However, it's very difficult to make a detection limit comparison with this technology because detection capability tends to be application specific. In other words, depending on the interference reduction capability of the CRC and iCRC device, it might offer better detection capability than another instrument in one particular sample matrix, but offer inferior detection limits in another. In addition, vendors' instrument detection limits are typically carried out in simple aqueous solutions, which is unlikely to show a significant difference in the performance of the different CRC and iCRC technology. The only way the interference reduction capability of a particular device can be truly evaluated is in a real sample containing matrix and solvent components. For that reason, refer to the cited references, which are a selection of technical and data sheets and application literature showing performance characteristics of the different CRC and iCRC approaches.[17,19,22,23,26]

FINAL THOUGHTS

There is no question that CRCs and interfaces have given a new lease on life to quadrupole mass analyzers used in ICP-MS. They have enhanced its performance and flexibility, and most definitely opened up the technique to more demanding applications that were previously beyond its capabilities. This is most definitely the case with the new triple-quad system, which will be competing with the magnetic sector, high-resolution systems for the more difficult, research-type applications. It should also be noted that when I wrote my previous ICP-MS textbook in 2014, there was only one vendor of triple-quad ICP-MS instrumentation. Today there is a second vendor, with a near certainty that other vendors will be offering the technology in the near future.[27]

However, it must be emphasized that when assessing CRC technology, it is critical that you fully understand the capabilities of the different approaches, especially how they match up to your application objectives. The KED-based collision cell using an inert gas such as helium is probably better suited to doing multielement analysis in a routine environment. However, you have to be aware that its detection capability is compromised for some elements, depending on the type of samples being analyzed. By using ion–molecule reactions as opposed to collisions, detection limits for many of these elements can be improved quite significantly. Of course, if two or even three different gases have to be used, the convenience of using one gas goes away.

On the other hand, using highly reactive gases with discrimination by mass filtering appears to offer the best performance and the most flexibility of all the different commercial approaches. By careful matching of the reaction gas with the analyte ion and polyatomic interference, extremely low detection limits can be achieved by ICP-MS, even for many of the notoriously difficult elements. It should be emphasized that selection of the optimum reaction gas and selection of the best quadrupole bandpass parameters can sometimes translate into quite lengthy method development, especially if there is very little application data available. However, most vendors do a very good job of generating application studies for some of the more routine applications. If you analyze out-of-the-ordinary or complex samples, you might initially need to spend the time to develop an analytical method that is both robust and routine. If you are uncertain which is the best approach for your application, you should consider investing in a system that offers both collision cell and reaction cell capability in the same instrument. The simple collision cell approach using helium could be used for your routine samples, while the more powerful and DRC could be utilized for the more difficult sample matrices. Even if you don't need both, it will at least give you peace of mind that you have the flexibility to tackle the most demanding applications if needed. And finally, if you are investing in a second instrument, a triple-quad system might be the best option.

So when evaluating this technique, pay attention not only to what the technique can do for your application problem but also to what it cannot do, which is equally important. In other words, make sure you evaluate its capabilities on the basis of all your present and future analytical requirements, such as ease of use, method development, flexibility, sample throughput, and detection capability. When assessing vendor-generated data, make sure the performance is achievable in your laboratory and on all your sample matrices.

REFERENCES

1. K. Sakata and K. Kawabata. *Spectrochimica Acta*, 49B, 1027, 1994.
2. J. T. Rowan and R. S. Houk. *Applied Spectroscopy*, 43, 976–980, 1989.
3. D. W. Koppenaal, C. J. Barinaga, and M. R. Smith. *Journal of Applied Analytical Chemistry*, 9, 1053–1058, 1994.
4. G. C. Eiden, C. J. Barinaga, and D. W. Koppenaal. *Journal of Applied Analytical Chemistry*, 11, 317–322, 1996.
5. P. Turner, T. Merren, J. Speakman, and C. Haines. *Plasma Source Mass Spectrometry: Developments and Applications*. 28–34, Royal Society of Chemistry, Cambridge, UK. 1996.
6. D. J. Douglas and J. B. French. *Journal of American Society of Mass Spectrometry*, 3, 398, 1992.
7. B. A. Thomson, D. J. Douglas, J. J. Corr, J. W. Hager, and C. A. Joliffe. *Analytical Chemistry*, 67, 1696–1704, 1995.
8. Thermo Scientific. X Series ICP-MS: Enhanced collision cell technology CCT. Thermo Scientific Product Specifications. July 2004. http://www.thermo.com/eThermo/CMA/PDFs/Articles/articlesFile_24138.pdf.
9. E. R. Denoyer, S. D. Tanner, and U. Voellkopf. *Spectroscopy*, 14, 2, 1999.
10. H. H. Willard, L. L. Merritt, J. A. Dean, and F. A. Settle. *Instrumental Methods of Analysis*. Belmont, CA: Wadsworth Publishing Co., 1988, pp. 465–507.
11. E. De Hoffman, J. Charette, and V. Stroobant. *Mass Spectrometry, Principles and Applications*. Paris: John Wiley & Sons, 1996.
12. Thermo Scientific. Analysis of ultra-trace levels of elements in seawaters using 3rd-generation collision cell technology. Thermo Scientific Product Application Note 40718. April 2007. http://www.thermo.com/eThermo/CMA/PDFs/Articles/articlesFile_26161.pdf.
13. J. Takahashi. Determination of impurities in semiconductor grade hydrochloric acid using the Agilent 7500 cs ICP-MS. Agilent Technologies Application Note 5989-4348EN. January 2006.
14. N. Yamada, J. Takahashi, and K. Sakata. *Journal of Analytical Atomic Spectrometry*, 17, 1213–1222, 2002.
15. S. D. Tanner and V. I. Baranov. *Atomic Spectroscopy*, 20(2), 45–52, 1999.
16. B. Hattendorf and D. Günter. *Journal of Analytical Atomic Spectrometry*, 19, 600–606, 2004.
17. Thermo Scientific. Product specifications of the Thermo Scientific iCAP Q: Dramatically different ICP-MS. https://static.thermoscientific.com/images/D20721~.pdf.
18. Agilent Technologies. Product overview of the Agilent Technologies' 8800 Triple Quadrupole ICP-MS. https://www.agilent.com/en/products/icp-ms/icp-ms-systems/8800-triple-quadrupole-icp-ms
19. Analytic Jena. Principles and performance of the integrated Collision Reaction Cell (iCRC). https://www.analytik-jena.de/fileadmin/content/pdf_analytical_instrumentation/ICP/ICP-MS/TechNote_ICP_MS_iCRC_en.pdf

20. M. Hamester, R. Chemnitzer, P. E. Riss, A. Gaal, X. D. Wang, and R. Thomas. Efficient removal of polyatomic spectral interferences for the multielement analysis of complex human biological samples by ICP-MS. *Spectroscopy Magazine*, 27(7), 20–27, 2012.
21. Analytic Jena. ICP-MS application library. https://www.analytik-jena.de/en/analytical-instrumentation/products/mass-spectrometry/plasmaquantr-ms-elite.html
22. Agilent Technologies. 7800 ICP-MS product overview. https://www.agilent.com/en-us/products/icp-ms/icp-ms-systems/7800-icp-ms
23. Theory and applications of collision/reaction cells: How collision and reaction cells work for interference removal using the Thermo Scientific iCAP Q ICP-MS. http://www.thermofisher.com/order/catalog/product/IQLAAGGAAQFAQKMBIT
24. S. D. Tanner, V. I. Baranov, and D. R. Bandura. *Spectrochimica Acta*, 57B(9), 1361–1452, 2002.
25. K. Kawabata, Y. Kishi, and R. Thomas. *Spectroscopy*, 18(1), 16–31, 2003.
26. PerkinElmer. Overview of the NexION 2000 ICP-MS using the "universal" collision/reaction cell.
27. Thermo Fisher Scientific. iCAP TQ Triple Quad ICP-MS landing page. http://www.perkinelmer.com/product/nexion-2000b-icp-ms-configuration-n8150044.

13 Ion Detectors

Chapter 13 looks at the detection system—an important area of the mass spectrometer that detects and quantifies the number of ions emerging from the mass analyzer. The detector converts the ions into electrical pulses, which are then counted using its integrated measurement circuitry. The magnitude of the electrical pulses corresponds to the number of analyte ions present in the sample, which is then used for trace element quantitation by comparing the ion signal with known calibration or reference standards. In this chapter, we take a look at conventional dynode detection, which monitors discrete ions emerging from the mass separation device in a sequential manner, in addition to describing the new breed of array detectors, which can monitor the entire mass spectrum simultaneously.

Since inductively coupled plasma mass spectrometry (ICP-MS) was first introduced in the early 1980s, a number of different ion detection designs have been utilized, the most popular being electron multipliers for low ion count rates and Faraday collectors for high count rates. Today, the majority of ICP-MS systems that are used for ultratrace analysis use detectors that are based on the active film or discrete dynode electron multiplier. They are very sophisticated pieces of equipment and are very efficient at converting ion currents emerging from the mass analyzer into electrical signals. The location of the detector in relation to the mass analyzer is shown in Figure 13.1.

Before I go on to describe discrete dynode detectors in greater detail, it is worth looking at two of the earlier designs—the channel electron multiplier (Channeltron®)[1] and the Faraday cup—to get a basic understanding of how the ICP-MS ion detection process works.

CHANNEL ELECTRON MULTIPLIER

The operating principles of the channel electron multiplier are similar to those of a photomultiplier tube used in inductively coupled plasma optical emission spectrometry (ICP-OES). However, instead of using individual dynodes to convert photons to electrons, the Channeltron is an open glass cone (coated with a semiconductor-type material) that generates electrons from ions that impinge on its surface. For the detection of positive ions, the front of the cone is biased at a negative potential while the far end, near the collector, is kept at ground. When the ion emerges from the quadrupole mass analyzer, it is attracted to the high negative potential of the cone. When the ion hits this surface, one or more secondary electrons are formed. The potential gradient inside the tube varies based on position, so the secondary electrons move further down the tube. As these electrons strike new areas of the coating, more secondary electrons are emitted. This process is repeated many times. The result is a discrete pulse, which contains many millions of electrons generated

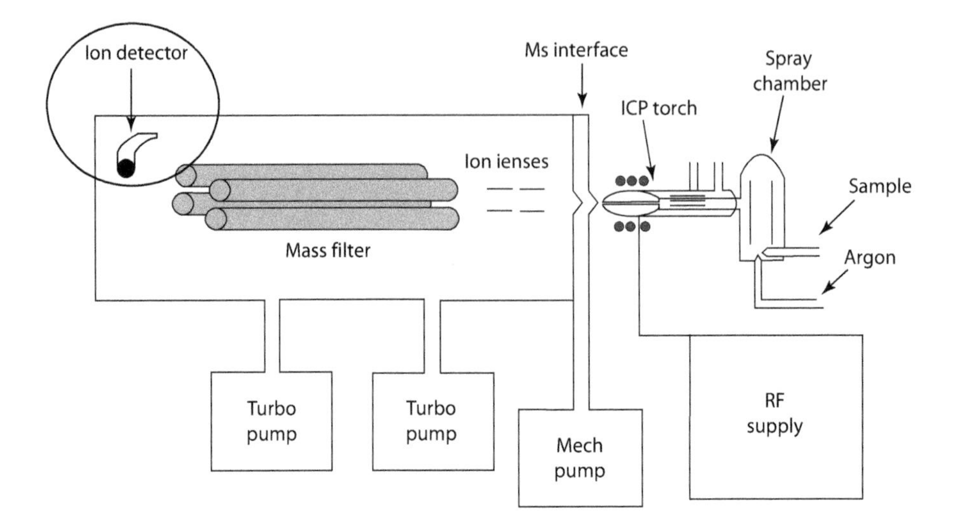

FIGURE 13.1 Location of the detector in relation to the mass analyzer. RF, radio frequency.

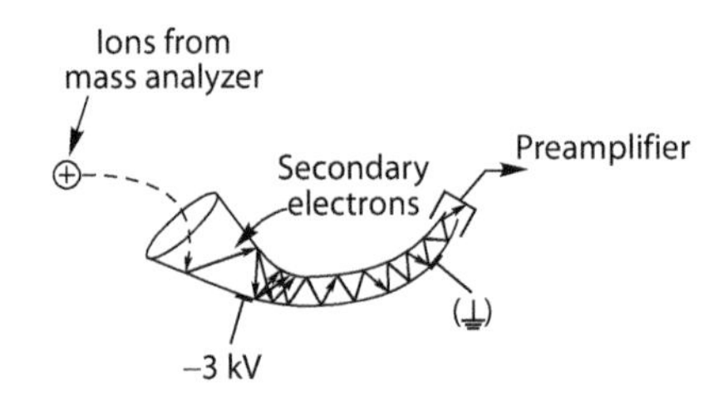

FIGURE 13.2 Basic principles of a channel electron multiplier. (From *Channeltron®: Electron Multiplier Handbook for Mass Spectrometry Applications*, Galileo Electro-Optic Corp., Sturbridge, MA, 1991. Channeltron is a registered trademark of Galileo Corp.)

from an ion that first hits the cone of the detector.[1] This process is shown simplistically in Figure 13.2.

This pulse is then sensed and detected by a very fast preamplifier. The output pulse from the preamplifier then goes to a digital discriminator and counting circuitry that only counts pulses above a certain threshold value. This threshold level needs to be high enough to discriminate against pulses caused by spurious emission inside the tube, from any stray photons from the plasma itself, or photons generated from fast-moving ions striking the quadrupole rods.

It is worth pointing out that the rate at which ions hit the detector is sometimes too high for the measurement circuitry to handle in an efficient manner. This is caused by ions arriving at the detector during the output pulse of the preceding ion and not being detected by the counting system. This "dead time," as it is known, is a fundamental limitation of the multiplier detector and is typically 30–50 s, depending on the detection system. Compensation in the measurement circuitry has to be made for this dead time to count the maximum number of ions hitting the detector.

FARADAY CUP

For some applications, where ultratrace detection limits are not required, the ion beam from the mass analyzer is directed into a simple metal electrode or Faraday cup. With this approach, there is no control over the applied voltage (gain), so they can only be used for high ion currents. Their lower working range is on the order of 10^4 cps, which means that if they are to be used as the only detector, the sensitivity of the ICP mass spectrometer will be severely compromised. For this reason, they are normally used in conjunction with a Channeltron or discrete dynode detector to extend the dynamic range of the instrument. An additional problem with the Faraday cup is that because of the time constant used in the DC amplification process to measure the ion current, they are limited to relatively low scan rates. This limitation makes them unsuitable for the fast scan rates required for traditional pulse counting used in ICP-MS and also limits their ability to handle fast transient peaks.

The Faraday cup was never sensitive enough for quadrupole ICP-MS technology, because it was not suitable for very low ion count rates. An attempt was made in the early 1990s to develop an ICP-MS system using a Faraday cup detector for the environmental market, but its sensitivity was compromised, and as a result it was considered more suitable for applications requiring ICP-OES trace-level detection capability. However, Faraday cup technology is still utilized in some magnetic sector instruments, particularly where high ion signals are encountered in the determination of high-precision isotope ratios, using a multicollector detection system.

DISCRETE DYNODE ELECTRON MULTIPLIER

These detectors, which are often called *active film multipliers*, work in a similar way as the Channeltron, but utilize discrete dynodes to carry out the electron multiplication.[2] Figure 13.3 illustrates the principles of operation of this device.

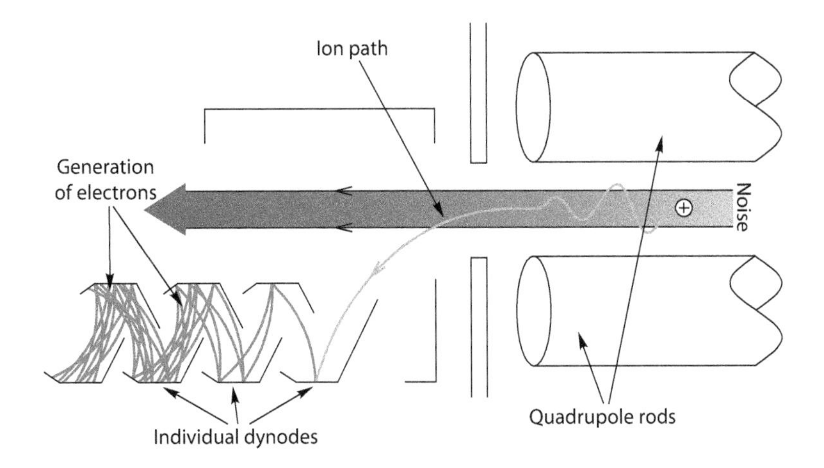

FIGURE 13.3 Schematic of a discrete dynode electron multiplier. (From Hunter, K., *Atomic Spectros.*, 15(1), 17–20, 1994.)

The detector is positioned off-axis to minimize the background noise from stray radiation and neutral species coming from the ion source. When an ion emerges from the quadrupole, it sweeps through a curved path before it strikes the first dynode. On striking the first dynode, it liberates secondary electrons. The electron optic design of the dynode produces acceleration of these secondary electrons to the next dynode, where they generate more electrons. This process is repeated at each dynode, generating a pulse of electrons that are finally captured by the multiplier collector or anode. Because of the materials used in the discrete dynode detector and the difference in the way electrons are generated, it is typically 50%–100% more sensitive than Channeltron technology.

Although most discrete dynode detectors are very similar in the way they work, there are subtle differences in the way the measurement circuitry handles low and high ion count rates. When ICP-MS was first commercialized, it could only handle up to five orders of dynamic range. However, when attempts were made to extend the dynamic range, certain problems were encountered. Before we discuss how modern detectors deal with this issue, let us first look at how it was addressed in earlier instrumentation.

EXTENDING THE DYNAMIC RANGE

Traditionally, ICP-MS using the pulse-counting measurement is capable of about five orders of linear dynamic range. This means that ICP-MS calibration curves, generally speaking, are linear from detection limits up to a few hundred parts per billion. However, there are a number of ways to extend the dynamic range of ICP-MS another four to five orders of magnitude, and working from sub-parts-per-trillion levels up to hundreds of parts per million. Here is a brief overview of some of the different approaches that have been used.

FILTERING THE ION BEAM

One of the very first approaches to extending the dynamic range in ICP-MS was to filter the ion beam. This was achieved by putting a nonoptimum voltage on one of the ion lens components or the quadrupole itself, to limit the number of ions reaching the detector. This voltage offset, which was set on an individual mass basis, acted as an energy filter to electronically screen the ion beam and reduce the subsequent ion signal to within a range covered by pulse-counting ion detection. The main disadvantage with this approach was that the operator had to have prior knowledge of the sample to know what voltage to apply to the high-concentration masses.

USING TWO DETECTORS

Another technique that was used on some of the early ICP-MS instrumentation was to utilize two detectors, such as a channel electron multiplier and a Faraday cup, to extend the dynamic range. With this technique, two scans would be made. In the first scan, it would measure the high concentration masses using the Faraday cup; in the second scan, it would skip over the high-concentration masses and carry out pulse counting of the low-concentration masses with a channel electron multiplier. This worked reasonably well, but struggled with applications that required rapid switching between the two detectors, because the ion beam had to be physically deflected to select the optimum detector. Not only did this degrade the measurement duty cycle, but detector switching and stabilization times of several seconds also precluded fast transient signal detection.

Using Two Scans with One Detector

A more recent approach is to use just one detector to extend the dynamic range. This has typically been done by using the detector in both pulse and analog modes, so high and low concentrations can be determined in the same sample. There are basically three approaches to using this type of detection system: two of them involve carrying out two scans of the sample, whereas the third only requires one scan.

The first approach uses an electron multiplier operated in both the digital and analog modes.[3] Digital counting provides the highest sensitivity, whereas operation in the analog mode (achieved by reducing the high voltage applied to the detector) is used to reduce the sensitivity of the detector, thus extending the concentration range for which ion signals can be measured. The system is implemented by scanning the spectrometer twice for each sample. The first scan, in which the detector is operated in the analog mode, provides signals for elements present at high concentrations. A second scan, in which the detector voltage is switched to the digital pulse-counting mode, provides high-sensitivity detection for elements present at low levels. A major advantage of this technology is that the user does not need to know in advance whether to use analog or digital detection, because the system automatically scans all elements in both modes. However, one of the drawbacks is that two independent mass scans are required to gather data across an extended signal range. This not only results in degraded measurement efficiency and slower analyses, but also means that the system is not ideally suited for fast transient signal analysis, because mode switching is generally too slow.

An alternative way of extending the dynamic range is similar to the first approach, except that the first scan is used as an investigative tool to examine the sample spectrum before analysis.[4] This first prescan establishes the mass positions at which the analog and pulse modes will be used for subsequently collecting the spectral signal. The second analytical scan is then used for data collection, switching the detector back and forth rapidly between pulse and analog mode at each analytical mass.

Even though these approaches worked very well, their main disadvantage was that two separate scans are required to measure high and low levels. With conventional nebulization, this is not such a major problem, except that it can impact sample throughput. However, it does become a concern when it comes to working with transient peaks found in laser sampling (LS), flow injection atomic spectrometry (FIAS), or electrothermal vaporization (ETV) ICP-MS. Because these transient peaks often only last a few seconds, all the available time must be spent measuring the masses of interest to get the best detection limits. When two scans have to be made, time is wasted collecting data, which is not contributing to the analytical signal.

Using One Scan with One Detector

The limitation of having to scan the sample twice led to the development of an improved design using a dual-stage discrete dynode detector.[5] This technology utilizes measurement circuitry that allows both high and low concentrations to be determined in one scan. This is achieved by measuring the ion signal as an analog signal at the midpoint dynode. When more than a threshold number of ions are detected, the signal is processed through the analog circuitry. When fewer than the threshold number of ions is detected, the signal cascades through the rest of the dynodes and is measured as a pulse signal in the conventional way. This process, which is shown in Figure 13.4, is completely automatic and means that both the analog and the pulse signals are collected simultaneously in one scan.[6]

The pulse-counting mode is typically linear from zero to about 10^6 cps, whereas the analog circuitry is suitable from 10^4 to 10^{10} cps. To normalize both ranges, a cross-calibration is carried out to cover concentration levels, which produces a pulse and an analog signal. This is possible because the analog and pulse outputs can be defined in identical terms of incoming pulse counts per second, based on knowing the voltage at the first analog stage, the output current, and a conversion factor defined by the detection circuitry electronics. By carrying out a cross-calibration across the mass range, a dual-mode detector of this type is capable of achieving approximately 9–10 orders of dynamic range in one simultaneous scan. This can be seen in Figures 13.5 and 13.6. Figure 13.5 shows that the pulse-counting calibration curve (left-hand line) is linear up to 10^6 cps, whereas the analog calibration curve (right-hand line) is linear from 10^4 to 10^9 cps. Figure 13.6 shows that after cross-calibration, the two curves are normalized, which means that the detector is suitable

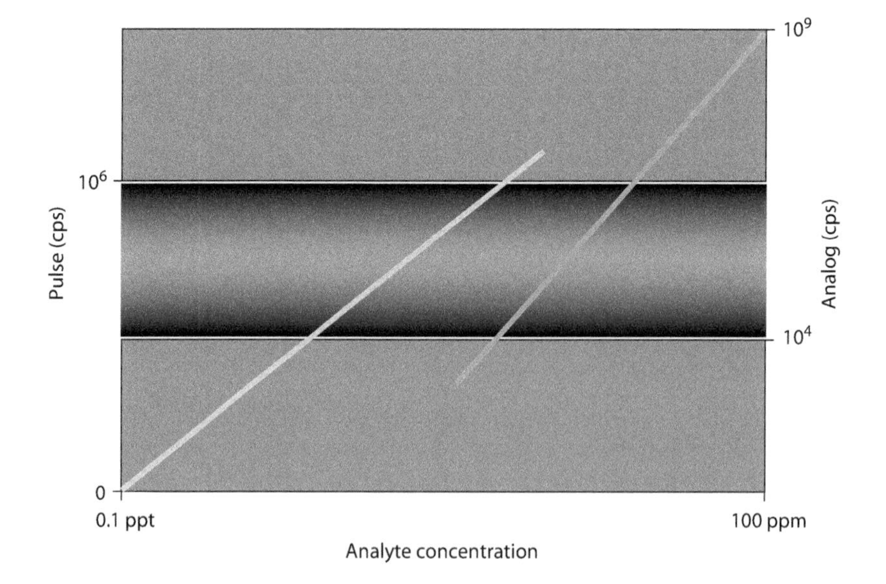

FIGURE 13.4 Dual-stage discrete dynode detector measurement circuitry. (From Denoyer, E. R., et al., *Spectroscopy*, 12(2), 56–61, 1997. Covered by U.S. Patent 5,463,219.)

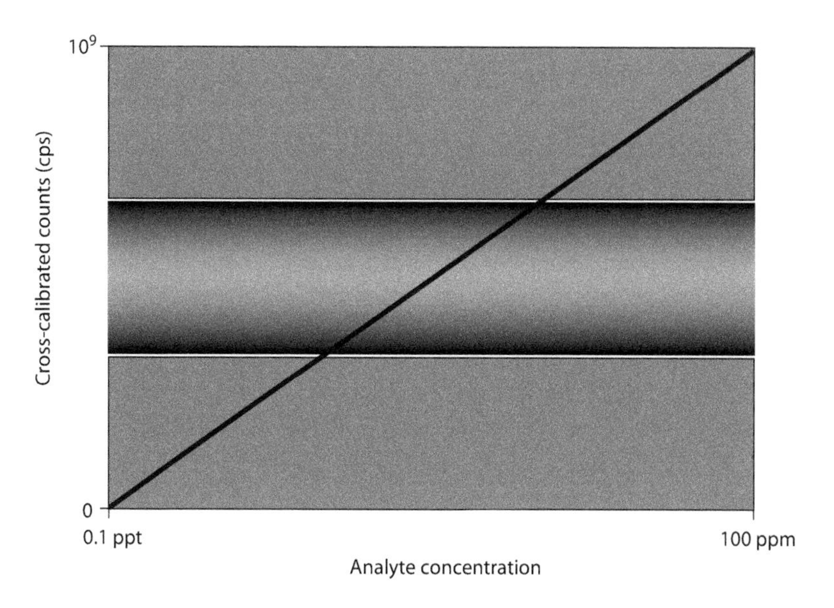

FIGURE 13.5 The pulse-counting mode covers up to 10^6 cps, while the analog circuitry is suitable from 10^4 to 10^9 cps, with a dual-mode discrete dynode detector. (From Denoyer, E. R., et al., *Spectroscopy*, 12(2), 56–61, 1997. Covered by U.S. Patent 5,463,219.)

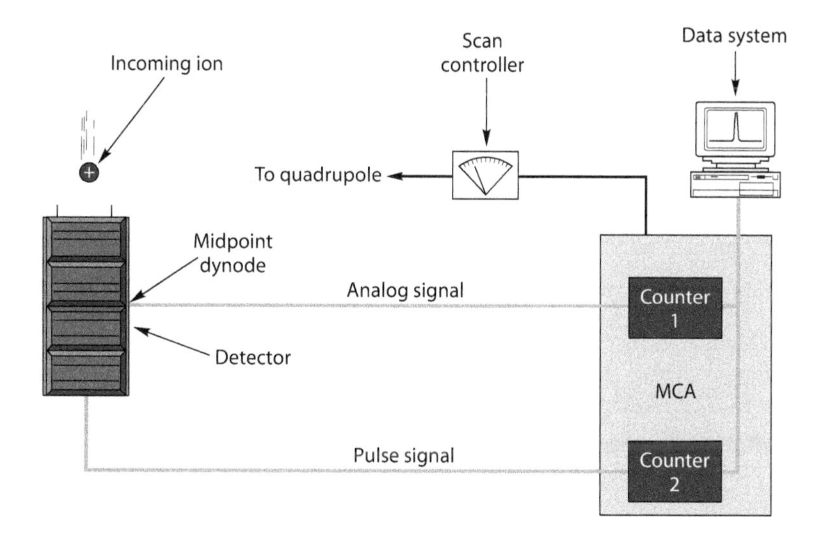

FIGURE 13.6 Using cross-calibration of the pulse and analog modes, quantitation from sub-parts-per-trillion to high-parts-per-million levels is possible. (From Denoyer, E. R., et al., *Spectroscopy*, 12(2), 56–61, 1997. Covered by U.S. Patent 5,463,219.)

for concentration levels between 0.1 ppt and 100 ppm and above—typically 9–10 orders of magnitude for most elements.[5]

There are subtle variations of this type of detection system, but its major benefit is that it requires only one scan to determine both high and low concentrations. It therefore not only offers the potential to improve sample throughput, but also means that the maximum data can be collected on a transient signal that only lasts a few seconds. This is described in greater detail in Chapter 14, where we discuss different measurement protocols and peak integration routines.

EXTENDING THE DYNAMIC RANGE USING PULSE-ONLY MODE

Another recent development in extending the dynamic range is to use the pulse-only signal. This is achieved by monitoring the ion flux at one of the first few dynodes of the detector (before extensive electron multiplication has taken place) and then attenuating the signal up to 10,000:1 by applying a control voltage. Electron pulses passed by the attenuation section are then amplified to yield pulse heights that are typical in normal pulse-counting applications.

There are basically three ways of implementing this technology based on the types of samples being analyzed. It can be run in conventional pulse-only mode for normal low-level work. It can also be run using an operator-selected attenuation factor if the higher-level elements being determined are known and similar in concentration. If the samples are complete unknowns and have not been well characterized beforehand, a dynamic attenuation mode of operation is available. In this mode, an additional premeasurement time is built into the quadrupole settling time to determine the optimum detector attenuation for the selected dwell times used.

This novel, pulse-only approach to extending the dynamic range looks to be a very interesting development that does not have the limitation of having to calibrate where pulse and analog signals cross over. However, it does require a preanalysis attenuation calibration to be carried out on a fairly frequent basis to determine the extent of signal attenuation required. The frequency of this calibration will vary depending on sample workload, but is expected to be on the order of once every 4 weeks.

However, it should be strongly emphasized that irrespective of which extended range technology is used, if low and high concentrations of the same analyte are expected in a suite of samples, it is unrealistic to think you can accurately quantitate down at the low end and at the top end of the linear range with the same calibration graph. If you want to achieve accurate and precise data at or near the limit of quantitation, you must run a set of appropriate calibration standards to cover your low-level samples. In addition, if you are expecting high and low concentrations in the same suite of samples, you have to be absolutely sure that a high-concentration sample has been thoroughly washed out from the spray chamber or nebulizer system, before a low-level sample in introduced. For this reason, caution must be taken when setting up the method with an autosampler, because if the read delay or integration times are not optimized for a suite of samples, erroneous results can be generated, which might necessitate a rerun under the manual supervision of the instrument operator.

SIMULTANEOUS ARRAY DETECTORS

Discrete dynode detectors are designed to handle a sequential stream of separated ions emerging from the mass spectrometer. Similar to a photomultiplier tube that converts photons from an optical emission signal into a pulse of electrons, these detectors cannot capture the entire mass spectrum at the same time. However, a new breed of ion detectors have recently been developed, that are based on solid-state, direct charge arrays, similar to the charge injection detector and charge-coupled detector technology used in ICP optical emission. By projecting all the separated ions from a mass separation device onto a two-dimensional array, these detectors can view the entire mass spectrum simultaneously.[7] Designed specifically for the Mattauch–Herzog double-focusing magnetic sector technology described in Chapter 8, this complementary metal oxide semiconductor (CMOS) ion-sensitive device is a 12 cm long array that covers the entire mass range simultaneously in 4800 separate channels.[8] The basic principle of the detector is similar to that of a Faraday cup: when a charged ion arrives at a detector array, it is discharged by receiving an electron, which generates a signal. The detector is referred to as a direct charge detector (DCD) because every ion arriving at the detector contributes to the signal. Furthermore, with 4800 detector arrays covering the mass range (from 5 to 240 amu), every mass unit is covered on average by 20 separate channels, resulting in a true mass spectrum rather than a single point for each atomic mass unit. A photograph of this DCD is shown in Figure 13.7.

FIGURE 13.7 DCD used for simultaneous measurement of ions separated by a Mattauch–Herzog double-focusing magnetic sector mass spectrometer.

To cover the linearity of up to eight orders of magnitude required in ICP-MS measurements, each detector channel incorporates separate high- and low-gain detector elements, allowing it to independently handle a wide range of signal levels in the basic integration cycle. The dynamic range can be further extended by optimizing the readout process. In the basic integration cycle, each channel is monitored by the electronics every 20 ms. If the signal integrated in this time interval nears the threshold of the channel, the integrated signal of that channel is automatically logged and the channel is reset and its measurement cycle repeated. This is repeated until the end of the defined measurement time, when all the collected data is integrated to produce the final signal. This means that the detector is always working within its linear response range and that longer integration times can be used without fear of detector saturation. For the benefits of simultaneous detection in ICP-MS, refer to Chapter 10 on magnetic sector technology and Chapter 14 on peak measurement protocol.

REFERENCES

1. *Channeltron®: Electron Multiplier Handbook for Mass Spectrometry Applications.* Sturbridge, MA: Galileo Electro-Optic Corp., 1991. (Channeltron is a registered trademark of Galileo Corp.)
2. K. Hunter. *Atomic Spectroscopy*, 15(1), 17–20, 1994.
3. R. C. Hutton, A. N. Eaton, and R. M. Gosland. *Applied Spectroscopy*, 44(2), 238–242, 1990.
4. Y. Kishi. *Agilent Technologies Application Journal*, August 1997.
5. E. R. Denoyer, R. J. Thomas, and L. Cousins. *Spectroscopy*, 12(2), 56–61, 1997. Covered by U.S. Patent 5,463,219.
6. J. Gray, R. Stresau, and K. Hunter. Ion counting beyond 10 GHz. Presented at Pittsburgh Conference and Exposition, Orlando, FL, 2003, poster 890-6P.
7. G. D. Schilling, S. J. Ray, A. A. Rubinshtein, J. A. Felton, R. P. Sperline, M. B. Denton, C. J. Barinaga, D. W. Koppenaal, and G. M. Hieftje. *Analytical Chemistry*, 81(13), 5467–5473, 2009.
8. Spectro Analytical Instruments. A new era in mass spectrometry. http://www.spectro.com/pages/e/p010402tab_overview.htm

14 Peak Measurement Protocol

With its multielement capability, superb detection limits, wide dynamic range, and high sample throughput, inductively coupled plasma mass spectrometry (ICP-MS) is proving to be a compelling technique for more and more diverse application areas. However, it is very unlikely that two different application areas have identical analytical requirements. For example, environmental and clinical contract laboratories, although wanting reasonably low detection limits, are not really pushing the technique to its extreme detection capability. Their main requirement usually is high sample throughput, because the number of samples these laboratories can analyze in a day directly impacts their revenue. On the other hand, a semiconductor fabrication plant or a supplier of high-purity chemicals to the electronics industry is interested in the lowest detection limits the technique can offer, because of the contamination problems associated with manufacturing high-performance electronic devices. This chapter looks at the many different measurement protocols associated with identifying and quantifying the analyte peak in ICP-MS and how they impact sample throughput and the quality of the data generated.

To meet such diverse application needs, modern ICP-MS instrumentation has to be very flexible if it is to keep up with the increasing demands of its users. Nowhere is this more important than in the area of peak integration and measurement protocol. The way the analytical signal is managed in ICP-MS has a direct impact on its multielement characteristics, isotopic capability, detection limits, dynamic range, and sample throughput—the five major strengths that attracted the trace element community to the technique almost 30 years ago. To understand signal management in greater detail and its implications on data quality, we discuss how measurement protocol is optimized based on the application's analytical requirements, and its impact on both continuous signals generated by traditional nebulization devices and transient signals produced by alternative sample introduction techniques, such as laser ablation, chromatographic separation, particle size studies, and flow injection.

MEASUREMENT VARIABLES

There are many variables that affect the quality of the analytical signal in ICP-MS. The analytical requirements of the application will often dictate this, but there is no question that instrumental detection and measurement parameters can have a significant impact on the quality of data in ICP-MS. Some of the variables that can potentially impact the quality of the data, particularly when carrying out multielement analysis, are as follows:

- Continuous or transient signal
- Temporal length of the sampling event
- Volume of sample available
- Number of samples being analyzed
- Number of replicates per sample
- Number of elements being determined
- Detection limits required
- Precision or accuracy expected
- Dynamic range needed
- Integration time used
- Peak quantitation routines

Before we go on to discuss these in greater detail and how they affect the data, it is important to remind ourselves how a scanning device like a quadrupole mass analyzer works. Although we will focus on quadrupole technology, the fundamental principles of measurement protocol will be very similar for all types of mass spectrometers that use a sequential approach for multielement peak quantitation.

MEASUREMENT PROTOCOL

The principles of scanning with a quadrupole mass analyzer are shown in Figure 14.1. In this simplified example, the analyte ion in front (black) and four other ions have arrived at the entrance to the four rods of the quadrupole. When a particular radio-frequency (RF) or direct current (DC) voltage is applied to each pair of rods, the positive or negative bias on the rods will electrostatically steer the analyte ion of interest down the middle of the four rods to the end, where it will emerge and be converted to an electrical pulse by the detector. The other ions of different mass-to-charge ratios will pass through the spaces between the rods and be ejected from the quadrupole. This scanning process is then repeated for another analyte at a completely different mass-to-charge ratio until all the analytes in a multielement analysis have been measured.

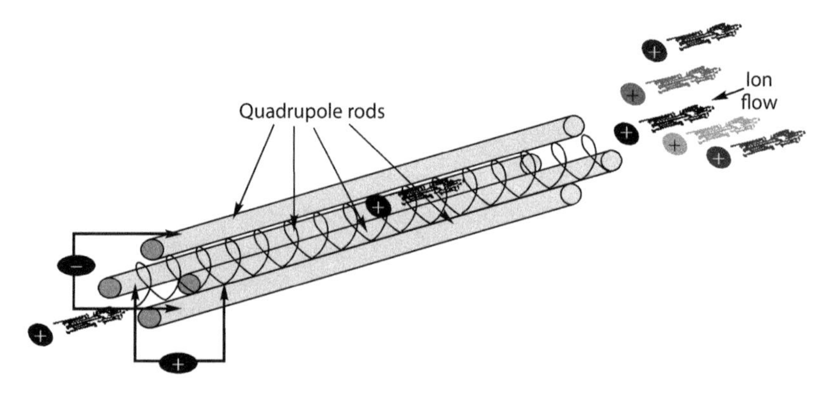

FIGURE 14.1 Principles of mass selection with a quadrupole mass filter.

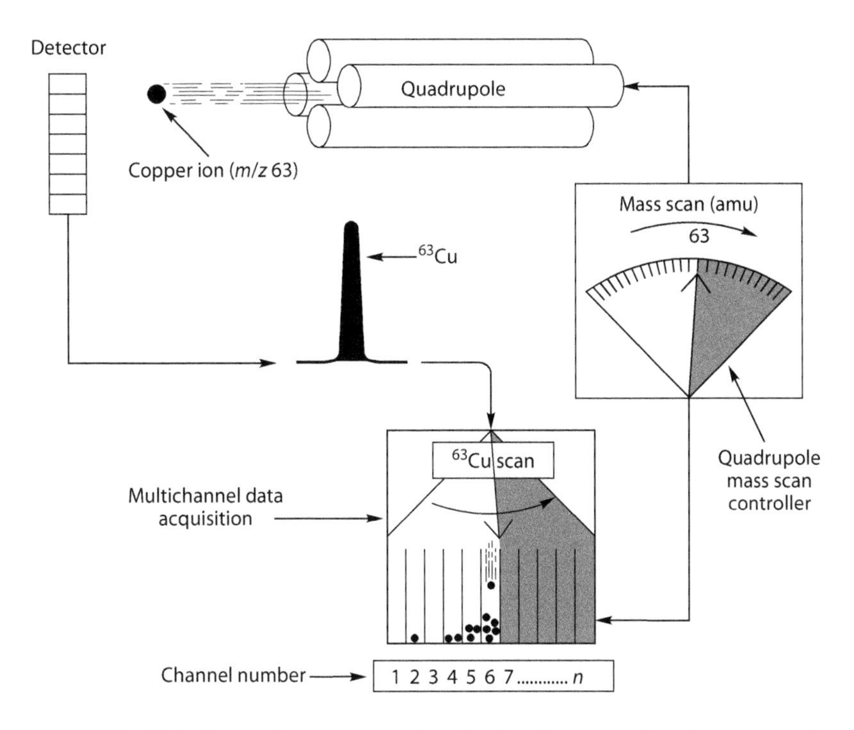

FIGURE 14.2 Detection and measurement protocol using a quadrupole mass analyzer. (From PerkinElmer, Integrated MCA technology in the ELAN ICP-mass spectrometer, Application Note TSMS-25, 1993.)

The process for the detection of one particular mass in a multielement run is represented in Figure 14.2. It shows a $^{63}Cu^+$ ion emerging from the quadrupole and being converted to an electrical pulse by the detector. As the optimum RF/DC ratio is applied for $^{63}Cu^+$ and repeatedly scanned, the ions as electrical pulses are stored and counted by a multichannel analyzer. This multichannel data acquisition system typically has 20 channels per mass, and as the electrical pulses are counted in each channel, a profile of the mass is built up over the 20 channels, corresponding to the spectral peak of $^{63}Cu^+$. In a multielement run, repeated scans are made over the entire suite of analyte masses, as opposed to just one mass, as represented in this example.

The principles of multielement peak acquisition are shown in Figure 14.3. In this example, signal pulses for two masses are continually collected as the quadrupole is swept across the mass spectrum, shown by sweeps 1–3. After a fixed number of sweeps (determined by the user), the total number of signal pulses in each channel is obtained, resulting in the final spectral peak.[1]

When it comes to quantifying an isotopic signal in ICP-MS, there are basically two approaches to consider. One is the multichannel ramp scanning approach, which uses a continuous smooth ramp of $1 - n$ channels (where n is typically 20) per mass across the peak profile. This is shown in Figure 14.4.

Also, there is the peak-hopping approach, in which the quadrupole power supply is driven to a discrete position on the peak (normally the maximum point) and allowed to settle, and a measurement is taken for a fixed amount of time. This is represented in Figure 14.5.

The multipoint scanning approach is best for accumulating spectral and peak shape information when doing mass scans. It is normally used for doing mass calibration and resolution checks, and as a classical qualitative method development tool to find out what elements are present in the sample and to assess their spectral implications on the masses of interest. Full peak profiling is not normally used for doing rapid quantitative analysis, because valuable analytical time is wasted taking data on the wings and valleys of the peak, where the signal-to-noise ratio is poorest.

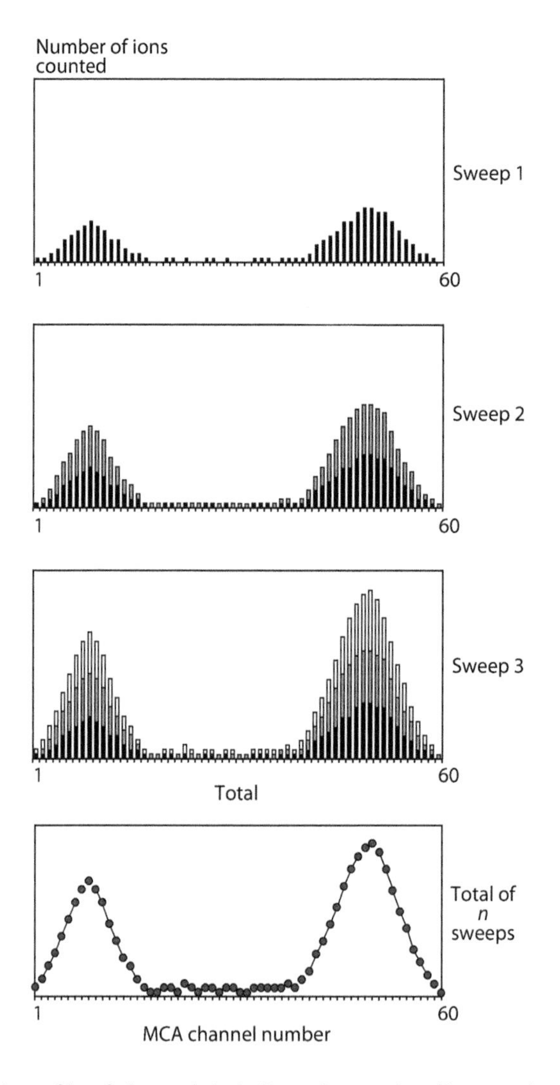

FIGURE 14.3 A profile of the peak is built up by continually sweeping the quadrupole across the mass spectrum. (From PerkinElmer, Integrated MCA technology in the ELAN ICP-mass spectrometer, Application Note TSMS-25, 1993.)

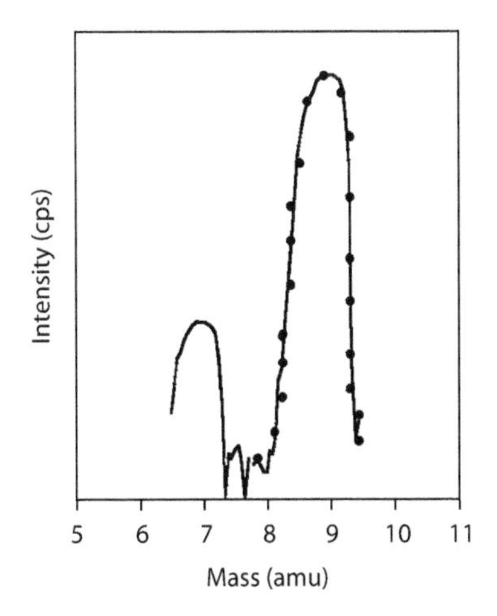

FIGURE 14.4 Multichannel ramp scanning approach using 20 channels per atomic mass unit.

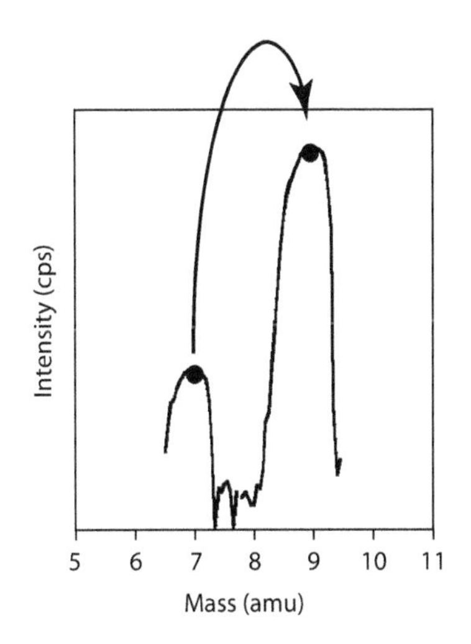

FIGURE 14.5 Peak-hopping approach.

When the best possible detection limits are required, the peak-hopping approach is best. It is important to understand that to get the full benefit of peak hopping, the best detection limits are achieved when single-point peak hopping at the peak maximum is chosen. However, to carry out single-point peak hopping, it is essential that the mass stability be good enough to reproducibly go to the same mass point every time. If good mass stability can be guaranteed (usually by thermostating the quadrupole power supply), measuring the signal at the peak maximum will always give the best detection limits for a given integration time. It is well documented that there is no benefit in spreading the chosen integration time over more than one measurement

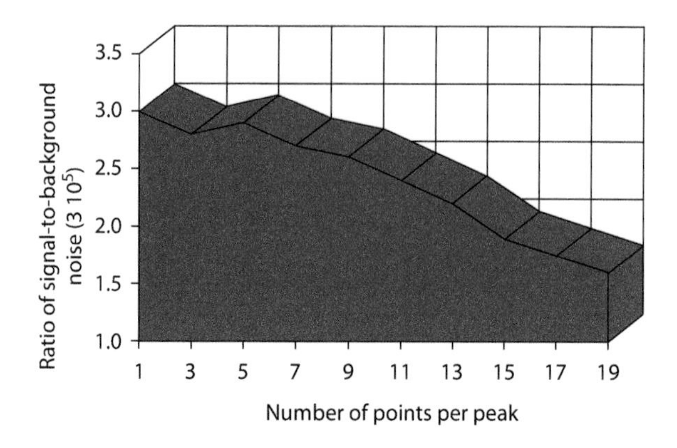

FIGURE 14.6 Signal-to-background noise degrades when more than one point, spread over the same integration time, is used for peak quantitation.

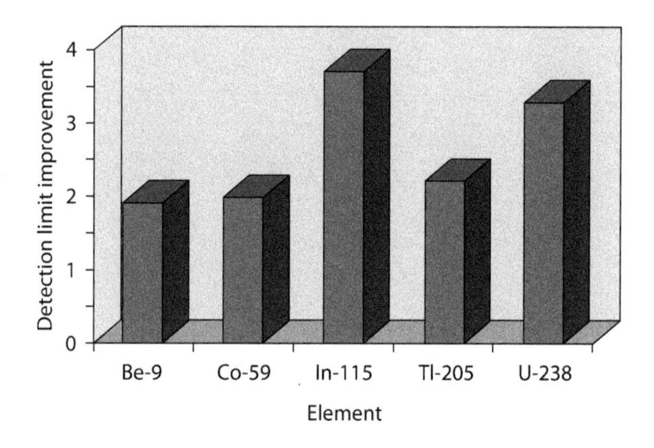

FIGURE 14.7 Detection limit improvement using one point per peak compared with 20 points per peak over the mass range. (From Denoyer, E. R., *Atomic Spectros.*, 13(3), 93–98, 1992.)

point per mass. If time is a major consideration in the analysis, then using multiple points is wasting valuable time on the wings and valleys of the peak, which contribute less to the analytical signal and more to the background noise. This is shown in Figure 14.6, which demonstrates the degradation in signal-to-background noise of 10 ppb Rh with an increase in the number of points per peak, spread over the same total integration time. Detection limit improvement for a selected group of elements using 1 point/peak compared with 20 points/peak is shown in Figure 14.7.

OPTIMIZATION OF MEASUREMENT PROTOCOL

Now that the fundamentals of the quadrupole measuring electronics have been described, let us now go into more detail on the impact of optimizing the measurement protocol based on the requirements of the application. When multielement analysis is being carried out by ICP-MS, there are a number of decisions that need to be made. First, we need to know if we are dealing with a continuous signal from a nebulizer or a transient signal from a sampling accessory, such as the laser ablation system of a chromatographic separation device. If it is a transient event, how long will the signal last? Another question that needs to be addressed is, how many elements are going to be determined? With a continuous signal, this is not such a major problem, but could be an issue if we are dealing with a transient signal that lasts only a few seconds. We also need to be aware of the level of detection capability required. This is a major consideration with a short laser pulse of a few seconds' duration, and even more critical when characterizing nanoparticles that only last for a few milliseconds. But it is also an issue with a continuous signal produced by a concentric nebulizer, where we might have to accept a compromise of detection limit based on the speed of analysis requirements or amount of sample available. What analytical precision is expected? If isotope ratio or dilution work is being done, how many ions do we have to count to guarantee good precision? Does increasing the integration time of the measurement help the precision? Finally, is there a time constraint on the analysis? A high-throughput laboratory might not be able to afford to use the optimum sampling time to get the ultimate detection limit. In other words, what compromises need to be made between detection limit, precision, and sample throughput? It is clear that before the measurement protocol can be optimized, the major analytical requirements of the application need to be defined. Let us look at this in greater detail.

MULTIELEMENT DATA QUALITY OBJECTIVES

Because multielement detection capability is probably the major reason why most laboratories invest in ICP-MS, it is important to understand the impact of measurement criteria on detection limits. We know that in a multielement analysis, the quadrupole's RF/DC ratio is "driven" or scanned to mass regions, which represent the elements of interest. The electronics are allowed to settle and then "sit" or dwell on the peak and take measurements for a fixed period of time. This is usually performed a number of times until the total integration time is fulfilled. For example, if a dwell time of 50 ms is selected for all masses and the total integration time is 1 s, then the quadrupole will carry out 20 complete sweeps per mass, per replicate. It will then repeat the same routine for as many replicates that have been built into the method. This is shown in a simplified manner in Figure 14.8, which displays the scanning protocol of a multielement scan of three different masses.

In this example, the quadrupole is scanned to mass A. The electronics are allowed to settle (settling time) and are left to dwell for a fixed period of time at one or multiple points on the peak (dwell time) and intensity measurements taken (based on the dwell time). The quadrupole is then scanned to masses B and C, and the measurement protocol is repeated. The complete multielement measurement cycle (sweep) is repeated as many times as needed to make up the total integration per peak. It should be emphasized that this is a generalization of the measurement routine—management of peak integration by the software will vary slightly based on different instrumentation.

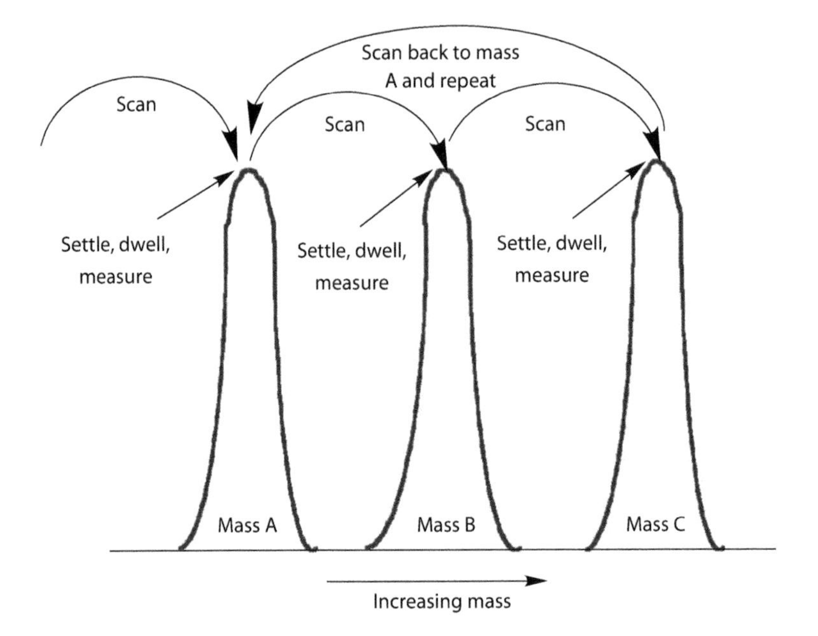

FIGURE 14.8 Multielement scanning and peak measurement protocol used in a quadrupole.

It is clear from this that during a multielement analysis, there is a significant amount of time spent scanning and settling the quadrupole, which does not contribute to the quality of the analytical signal. Therefore, if the measurement routine is not optimized carefully, it can have a negative impact on data quality. The dwell time can usually be selected on an individual mass basis, but the scanning and settling times are normally fixed because they are a function of the quadrupole and detector electronics. For this reason, it is essential that the dwell time, which ultimately affects detection limit and precision, dominate the total measurement time, compared with the scanning and settling times. It therefore follows that the measurement duty cycle (percentage of actual measuring time compared with total integration time) is maximized when the quadrupole and detector electronics settling times are kept to an absolute minimum. This can be seen in Figure 14.9, which shows a plot of percent measurement duty cycle against dwell time for four different quadrupole settling times—0.2, 1.0, 3.0, and 5.0 ms—for one replicate of a multielement scan of five masses, using one point per peak. In this example, the total integration time for each mass was 1 s, with the number of sweeps varying depending on the dwell time used. For this exercise, the percent duty cycle is defined by the following equation:

$$\frac{\textit{Dwell time} \times \# \textit{sweeps} \times \# \textit{elements} \times \# \textit{replicates}}{\{(\textit{Dwell time} \times \# \textit{sweeps} \times \# \textit{elements} \times \# \textit{replicates}) + (\text{Scanning/settingtime} \times \# \text{sweeps} \times \# \text{elements} \times \# \text{reps})\}} \times 100$$

To achieve the highest duty cycle, the nonanalytical time must be kept to an absolute minimum. This leads to more time being spent counting ions and less time scanning and settling, which do not contribute to the quality of the analytical signal.

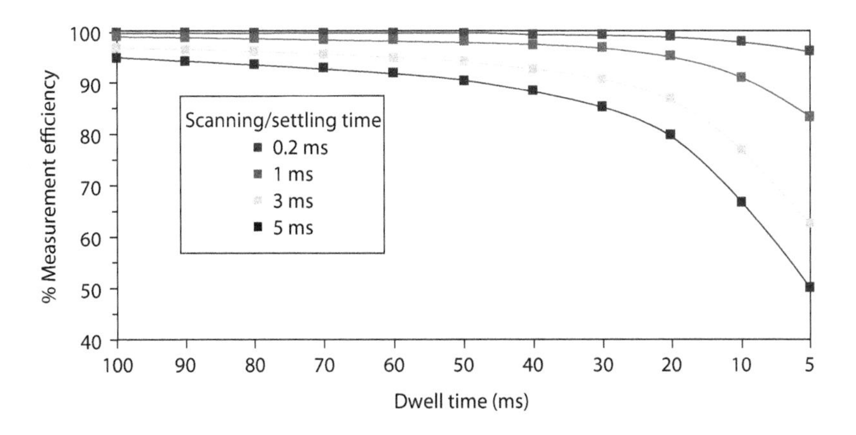

FIGURE 14.9 Measurement duty cycle as a function of dwell time with varying scanning and settling times.

This becomes critically important when a rapid transient peak is being quantified, because the available measuring time is that much shorter.[3] It is therefore a good rule of thumb, when setting up your measurement protocol in ICP-MS, to avoid using multiple points per peak and long settling times, because they ultimately degrade the quality of the data for a given integration time.

It can also be seen in Figure 14.9 that shorter dwell times translate into a lower duty cycle. For this reason, for normal quantitative analysis work, it is probably desirable to carry out multiple sweeps with longer dwell times (typically 50 ms) to get the best detection limits. So, if an integration time of 1 s is used for each element, this would translate into 20 sweeps of 50 ms dwell time per mass. Although 1 s is long enough to achieve reasonably good detection limits, longer integration times generally have to be used to reach the lowest possible detection limits. This is shown in Figure 14.10, which shows detection limit improvement as a function of integration time for $^{238}U^{+}$.

As would be expected, there is a fairly predictable improvement in the detection limit as the integration time is increased because more ions are counted without an increase in the background noise. However, this only holds true up to the point where the pulse-counting detection system becomes saturated and no more ions can be counted. In the case of $^{238}U^{+}$, it can be seen that this happens at around 25 s, because there is no obvious improvement in the detection limit (D/L) at a higher integration time. So from this data, we can say that there appears to be no real benefit in using longer than a 7 s integration time. When deciding the length of the integration time in ICP-MS, you have to weigh the detection limit improvement against the time taken to achieve that improvement. Is it worth spending 25 s measuring each mass to get a 0.02 ppt detection limit if 0.03 ppt can be achieved using a 7 s integration time? Alternatively, is it worth measuring for 7 s when 1 s will only degrade the performance by a factor of 3? It really depends on your data quality objectives.

For some applications, like isotope dilution and ratio studies, high precision is also a very important data quality objective.[4] However, to understand what is realistically achievable, we have to be aware of the practical limitations of measuring

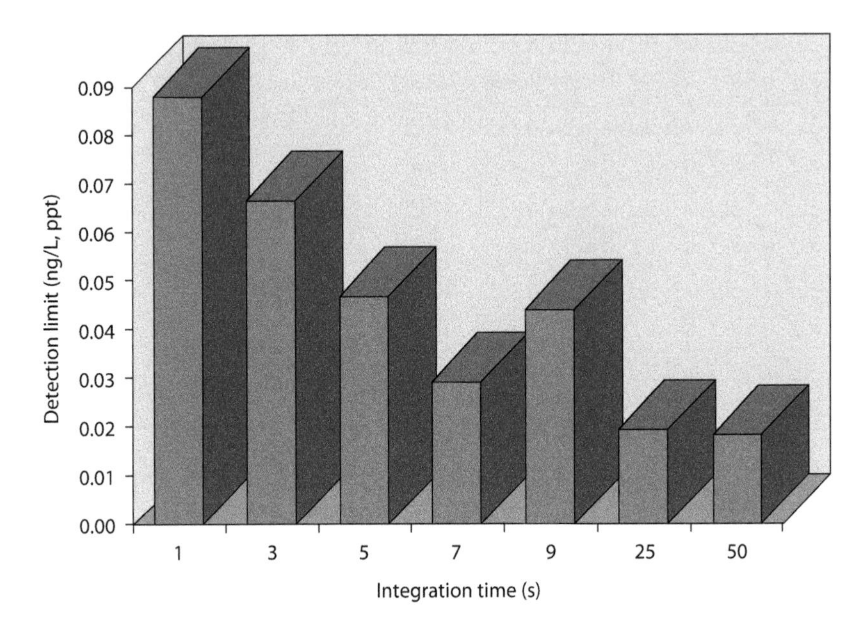

FIGURE 14.10 Plot of detection limit against integration time for $^{238}U^{+}$. (Copyright © 2013 PerkinElmer Inc., Waltham, MA. All rights reserved.)

a signal and counting ions in ICP-MS. Counting statistics tells us that the standard deviation of the ion signal is proportional to the square root of the signal. It follows therefore that the relative standard deviation (RSD) or precision should improve with an increase in the number (*N*) of ions counted, as shown by the following equation:

$$\%RDS = \sqrt{\frac{N}{N}} \times 100$$

In practice, this holds up very well as can be seen in Figure 14.11. In this plot of standard deviation as a function of signal intensity for $^{208}Pb^{+}$, the black dots represent the theoretical relationship as predicted by counting statistics. It can be seen that the measured standard deviation (black bars) follows theory very well up to about 100,000 cps. At that point, additional sources of noise (e.g., sample introduction pulsations and plasma fluctuations) dominate the signal, which lead to poorer standard deviation values.

So, based on counting statistics, it is logical to assume that the more ions that are counted, the better the precision will be. To put this in perspective, at least 1 million ions need to be counted to achieve an RSD of 0.1%. In practice, of course, these kinds of precision values are very difficult to achieve with a scanning quadrupole system because of the additional sources of noise. If this information is combined with our knowledge of how the quadrupole is scanned, we begin to understand what is required to get the best precision. This is confirmed by the spectral scan in Figure 14.12, which shows the predicted precision at all 20 channels of a 5 ppb $^{208}Pb^{+}$ peak.[2]

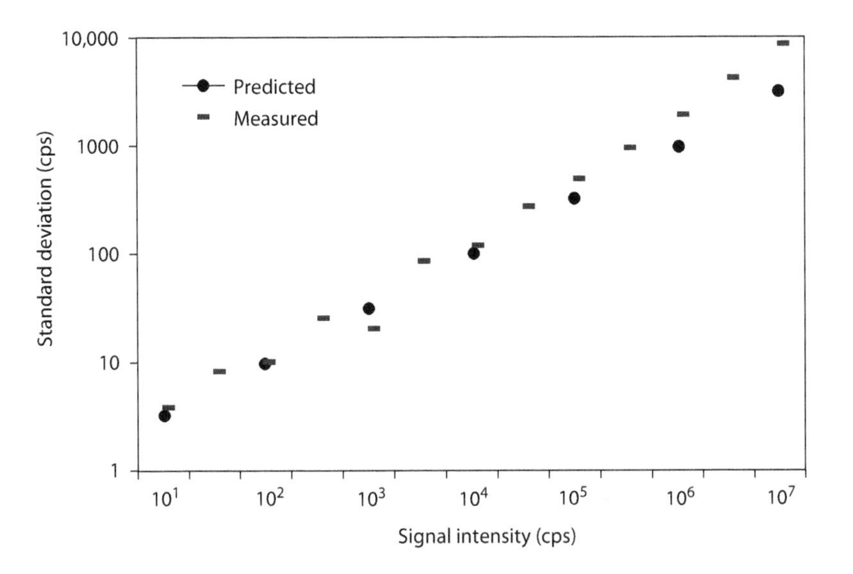

FIGURE 14.11 Comparison of measured standard deviation of a $^{208}Pb^+$ signal against that predicted by counting statistics. (From Denoyer, E. R., *Atomic Spectros.*, 13(3), 93–98, 1992.)

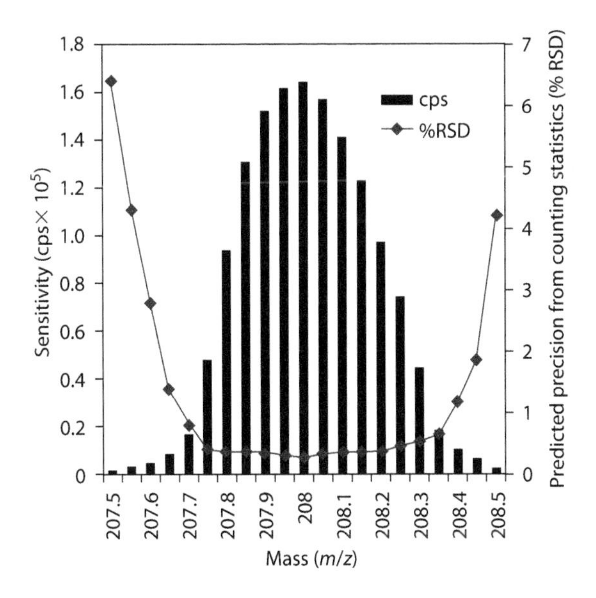

FIGURE 14.12 Comparison of percent RSD with signal intensity across the mass profile of a $^{208}Pb^+$ peak. (From Denoyer, E. R., *Atomic Spectros.*, 13(3), 93–98, 1992.)

This tells us that the best precision is obtained at the channels where the signal is highest, which, as we can see, are the ones at or near the center of the peak. For this reason, if good precision is a fundamental requirement of your data quality objectives, it is best to use single-point peak hopping with integration times on the order of 5–10 s. On the other hand, if high-precision isotope ratio or isotope dilution work is being done, where analysts would like to achieve precision values approaching counting statistics, then much longer measuring times are required. That is why integration times on the order of 5–10 min are commonly used for determining isotope ratios involving environmental pollutants[5] or clinical metabolism studies.[6] For this type of analysis, when two or more isotopes are being measured and ratioed to each other, it follows that the more simultaneous the measurement, the better the precision becomes. Therefore, the ability to make the measurement as simultaneous as possible is considered more desirable than any other aspect of the measurement. This is supported by the fact that the best isotope ratio precision data is achieved with multicollector, magnetic sector ICP-MS technology that carries out many isotopic measurements at the same time using multiple detectors.[7] Also, time-of-flight technology, which simultaneously samples all the analyte ions in a slice of the ion beam, offers excellent precision, particularly when internal standardization measurement is also carried out in a simultaneous manner.[8] So, the best way to approximate simultaneous measurement with a rapid scanning device, such as a quadrupole, is to use shorter dwell times (but not too short that insufficient ions are counted) and keep the scanning and settling times to an absolute minimum, which results in more sweeps for a given measurement time. This can be seen in Table 14.1, which shows the precision of Pb isotope ratios at different dwell times carried out by researchers at the Geological Survey of Israel.[9] The data is based on nine replicates of a NIST SRM-981 (75 ppb Pb) solution, using a 5.5 s integration time per isotope.

From these data, the researchers concluded that a dwell time of 10 or 25 ms offered the best isotope ratio precision measurement (quadrupole settling time was fixed at 0.2 ms). They also found that they could achieve slightly better precision by using a 17.5 s integration time (700 sweeps at 25 ms dwell time), but felt the marginal improvement in precision for nine replicates was not worth spending the approximately 3½ times longer analysis time. This can be seen in Table 14.2.

This work shows the benefit of being able to optimize the dwell time, settling time, and number of sweeps to get the best isotope ratio precision data. It also helps to be working with relatively healthy ion signals for the three Pb isotopes, ^{206}Pb, ^{207}Pb, and ^{208}Pb (24.1%, 22.1%, and 52.4% abundance, respectively). If the isotopic signals were dramatically different, as in the measurement of two of the uranium isotopes, ^{235}U to ^{238}U, which are 0.72% and 99.2745% abundant, respectively, then the ability to optimize the measurement protocol for individual isotopes becomes of even greater importance to guarantee good-precision data[10].

DATA QUALITY OBJECTIVES FOR SINGLE-PARTICLE ICP-MS STUDIES

The ability to optimize the measurement protocol is even more critical when using ICP-MS to characterize nanomaterials using the single-particle mode of analysis. Engineered nanomaterials, as they are known, refer to the process of producing materials that contain particles of <100 nm in size. They often possess different properties compared with bulk materials of the same composition, making them of great interest to a broad spectrum of industrial and commercial applications.

TABLE 14.1
Precision of Pb Isotope Ratio Measurement as a Function of Dwell Time Using a Total Integration Time of 5.5 s

Dwell Time (ms)	% RSD $^{207}Pb^+/^{206}Pb^+$	% RSD $^{208}Pb^+/^{206}Pb^+$
2	0.40	0.36
5	0.38	0.36
10	0.23	0.22
25	0.24	0.25
50	0.38	0.33
100	0.41	0.38

Source: Halicz, L., et al, *Atomic Spectros.*, 17(5), 186–189, 1996.

TABLE 14.2
Impact of Integration Time on the Overall Analysis Time for Pb Isotope Ratios

Dwell Time (ms)	No. of Sweeps	Integration Time (s)/Mass	% RSD $^{207}Pb^+/^{206}Pb^+$	% RSD $^{207}Pb^+/^{206}Pb^+$	Time for 9 Reps
25	220	5.5	0.24	0.25	2 min 29 s
25	500	12.5	0.21	0.19	6 min 12 s
25	700	17.5	0.20	0.17	8 min 29 s

Source: Halicz, L., et al, *Atomic Spectros.*, 17(5), 186–189, 1996.

Unfortunately, many of these nanomaterials, once they get into the environment, have proved to be harmful to humans. So in order to better understand the impact of nanoparticles on human health, several key properties need to be assessed, such as concentration, composition, particle size, and shape. Recent studies have shown that ICP-MS is proving to be a critical tool to characterize nanoparticles using the single-particle technique[11] However, these nanoparticles only exist for a few milliseconds, so the ability to measure them with good accuracy and precision is dependent on optimizing the dwell time, settling time, and speed of the data acquisition electronics. If the measurement protocol is not handled correctly, it can significantly impact the quality of data collected.

FINAL THOUGHTS

It is clear that the analytical demands put on ICP-MS are probably higher than those on any other trace element technique, because it is continually being asked to solve a wide variety of application problems at increasingly lower levels. However, by optimizing the measurement protocol to fit the analytical requirement, ICP-MS has shown that it has the unique capability to carry out rapid trace element analysis, with superb detection limits and good precision on both continuous and transient signals, and still meet the most stringent data quality objectives.

REFERENCES

1. PerkinElmer. Integrated MCA technology in the ELAN ICP-mass spectrometer. Application Note TSMS-25. 1993.
2. E. R. Denoyer. *Atomic Spectroscopy*, 13(3), 93–98, 1992.
3. E. R. Denoyer and Q. H. Lu. *Atomic Spectroscopy*, 14(6), 162–169, 1993.
4. T. Catterick, H. Handley, and S. Merson. *Atomic Spectroscopy*, 16(10), 229–234, 1995.
5. T. A. Hinners, E. M. Heithmar, T. M. Spittler, and J. M. Henshaw. *Analytical Chemistry*, 59, 2658–2662, 1987.
6. M. Janghorbani, B. T. G. Ting, and N. E. Lynch. *Microchemica Acta*, 3, 315–328, 1989.
7. J. Walder and P. A. Freeman. *Journal of Analytical Atomic Spectrometry*, 7, 571, 1992.
8. F. Vanhaecke, L. Moens, R. Dams, L. Allen, and S. Georgitis. *Analytical Chemistry*, 71, 3297, 1999.
9. L. Halicz, Y. Erel, and A. Veron. *Atomic Spectroscopy*, 17(5), 186–189, 1996.
10. D. R. Bandura and S. D. Tanner. *Atomic Spectroscopy*, 20(2), 69–72, 1999.
11. D. M. Mitrano, E. K. Leshner, A. Bednar, J. Monserud, C. P. Higgins, and J. F. Ranville. Detection of nanoparticulate silver using single particle inductively coupled plasma mass spectrometry. *Environmental Toxicology and Chemistry*, 31(1), 115–141, 2014.

15 Methods of Quantitation

There are many different ways to carry out trace element analysis by inductively coupled plasma mass spectrometry (ICP-MS), depending on your data quality objectives. Such is the flexibility of the technique that it allows detection from sub-parts-per-trillion up to high-parts-per-million levels using a wide variety of calibration methods, from full quantitative and semiquantitative analysis to one of the very powerful isotope ratioing techniques. This chapter looks at the most important quantitation methods available in ICP-MS.

This ability of ICP-MS to carry out isotopic measurements allows the technique to carry out quantitation methods that are not available to any other trace element technique. They include the following:

- Quantitative analysis
- Semiquantitative routines
- Isotope dilution
- Isotope ratio
- Internal standardization

Each of these techniques offers varying degrees of accuracy and precision; so, it is important to understand their strengths and weaknesses to know which one will best meet the data quality objectives of the analysis. In this chapter, we focus on using methods of quantitation for carrying out the analysis of liquids using continuous nebulization. However, even though the principles of calibration are similar, we covered the issues of quantitation of transient peaks in Chapter 14 and will deal with specific sampling accessories in Chapter 20.

Let's first take a look at each of the methods of quantitation in greater detail.

QUANTITATIVE ANALYSIS

As in other trace element techniquesm such as atomic absorption (AA) and inductively coupled plasma optical emission spectrometry (ICP-OES), quantitative analysis in ICP-MS is the fundamental tool used to determine analyte concentrations in unknown samples. In this mode of operation, the instrument is calibrated by measuring the intensity for all elements of interest in a number of known calibration standards that represent a range of concentrations likely to be encountered in your unknown samples. When the full range of calibration standards and blanks have been run, the software creates a calibration curve of the measured intensity versus concentration for each element in the standard solutions. Once calibration data is acquired, the unknown samples are analyzed by plotting the intensity of the elements of interest against the respective calibration curves. The software then calculates the concentrations for the analytes in the unknown samples.

This type of calibration is often called external standardization and is usually used when there is very little difference between the matrix components in the standards and the samples. However, when it is difficult to closely match the matrix of the standards with the samples, external standardization can produce erroneous results, because matrix-induced interferences will change analyte sensitivity based on the amount of matrix present in the standards and samples. When this occurs, better accuracy is achieved by using the method of standard addition or a similar approach, called addition calibration. Let us look at these three variations of quantitative analysis to see how they differ.

External Standardization

As explained earlier, this involves measuring a blank solution followed by a set of standard solutions to create a calibration curve over the anticipated concentration range. Typically, a blank and three standards containing different analyte concentrations are run. Increasing the number of points on the calibration curve by increasing the number of standards may improve accuracy in circumstances where the calibration range is very broad. However, it is seldom necessary to run a calibration with more than five standards. After the standards have been measured, the unknown samples are analyzed and their analyte intensities read against the calibration curve. Over extended analysis times, it is common practice to update the calibration curve, either by recalibrating the instrument with a full set of standards or by running one midpoint standard. The following protocol summarizes a typical calibration using external standardization:

1. Blank >
2. Standard 1 >
3. Standard 2 >
4. Standard 3 >
5. Sample 1 >
6. Sample 2 >
7. Sample … *n*
8. Recalibrate
9. Sample $n + 1$, etc.

This can be seen more clearly in Figure 15.1, which shows a typical calibration curve using a blank and three standards of 2, 5, and 10 ppb. This calibration curve shows a simple *linear regression*, but usually other modes of calibration are also available, like *weighted linear*, to emphasize measurements at the low-concentration region of the curve, and *linear through zero*, where the linear regression is forced through zero. Whatever approach is used, it is critically important to select your range of standards, based on the expected concentration levels in your samples, if you want to ensure to optimum accuracy.

It should be emphasized that this graph represents a single-element calibration. However, because ICP-MS is usually used for multielement analysis, multielement standards are typically used to generate calibration data. For that reason, it

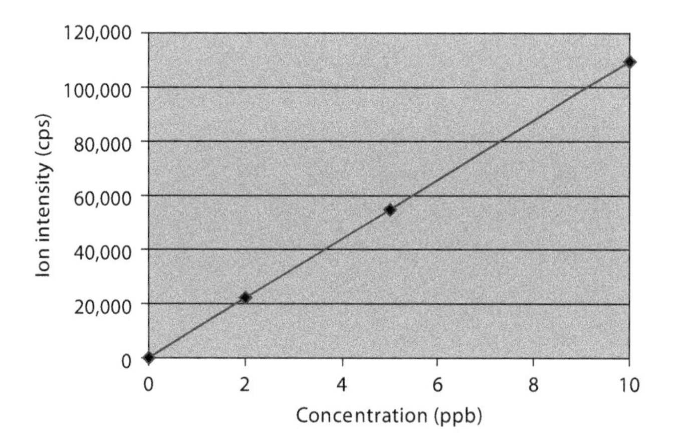

FIGURE 15.1 Simple linear regression calibration curve.

is absolutely essential to use multielement standards that have been manufactured specifically for ICP-MS. Single-element AA standards are not suitable, because they usually have only been certified for the analyte element and not for any others. The purity of the standard cannot be guaranteed for any other element and, as a result, cannot be used to make up multielement standards for use with ICP-MS. For the same reason, ICP-OES multielement standards are not advisable either, because they are only certified for a group of elements and could contain other elements at higher levels, which will affect the ICP-MS multielement calibration.

Standard Additions

This mode of calibration provides an effective way to minimize sample-specific matrix effects by spiking samples with known concentrations of analytes.[1,2] In standard addition calibration, the intensity of a blank solution is first measured. Next, the sample solution is "spiked" with known concentrations of each element to be determined. The instrument measures the response for the spiked samples and creates a calibration curve for each element for which a spike has been added. The calibration curve is a plot of the blank subtracted intensity of each spiked element against its concentration value. After creating the calibration curve, the unspiked sample solutions are then analyzed and compared with the calibration curve. Based on the slope of the calibration curve and where it intercepts the x-axis, the instrument software determines the unspiked concentration of the analytes in the unknown samples. This can be seen in Figure 15.2, which shows a calibration of the sample intensity and the sample spiked with 2 and 5 ppb of the analyte. The concentration of sample is where the calibration line intercepts the negative side of the x-axis.

The following protocol summarizes a typical calibration using the method of standard additions:

1. Blank >
2. Spiked sample 1 (spike concentration 1) >

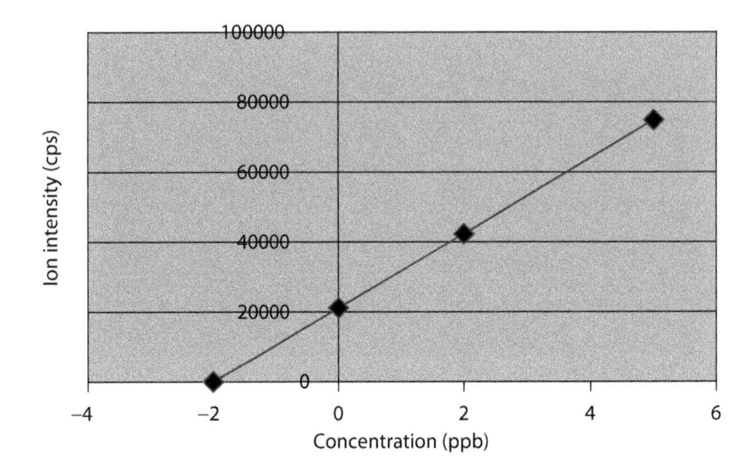

FIGURE 15.2 Typical "method of additions" calibration curve.

3. Spiked sample 1 (spike concentration 2) >
4. Unspiked sample 1 >
5. Blank >
6. Spiked sample 2 (spike concentration 1) >
7. Spiked sample 2 (spike concentration 2) >
8. Unspiked sample 2 >
9. Blank >
10. Etc.

Addition Calibration

Unfortunately, with the method of standard additions, each and every sample has to be spiked with all the analytes of interest, which becomes extremely labor-intensive when many samples have to be analyzed. For this reason, a variation of standard additions called *addition calibration* is more widely used in ICP-MS. However, this method can only be used when all the samples have a similar matrix. It uses the same principle as standard additions, but only the first (or representative) sample is spiked with known concentrations of analytes, and then analyzes the rest of the sample batch against the calibration, assuming all samples have a matrix similar to the first one. The following protocol summarizes a typical calibration using the method of addition calibration:

1. Blank >
2. Spiked sample 1 (spike concentration 1) >
3. Spiked sample 1 (spike concentration 2) >
4. Unspiked sample 1 >
5. Unspiked sample 2 >
6. Unspiked sample 3 >
7. Etc.

SEMIQUANTITATIVE ANALYSIS

If your data quality objectives for accuracy and precision are less stringent, ICP-MS offers a very rapid semiquantitative mode of analysis. This technique enables you to automatically determine the concentrations of up to 75 elements in an unknown sample, without the need for calibration standards.[3,4] This is an approach that could be extremely useful for initially screening samples, before quantitative analysis is carried out.

There are slight variations in the way different instruments approach semiquantitative analysis, but the general principle is to measure the entire mass spectrum, without specifying individual elements or masses. It relies on the principle that each element's natural isotopic abundance is fixed. By measuring the intensity of all their isotopes; correcting for common spectral interferences, including molecular, polyatomic, and isobaric species; and applying heuristic, knowledge-driven routines in combination with numerical calculations, a positive or negative confirmation can be made for each element present in the sample. Then, by comparing the corrected intensities against a stored isotopic response table, a good semiquantitative approximation of the sample components can be made.

Semiquant, as it is often called, is an excellent approach to rapidly characterize unknown samples. Once the sample has been characterized, you can choose to either update the response table with your own standard solutions to improve analytical accuracy, or switch to the quantitative analysis mode to focus on specific elements and determine their concentrations with even greater accuracy and precision. Whereas a semiquantitative determination can be performed without using a series of standards, the use of a small number of standards is highly recommended for improved accuracy across the full mass range. Unlike traditional quantitative analysis, in which you analyze standards for all the elements you want to determine, semiquant calibration is achieved using just a few elements distributed across the mass range. This calibration process, shown more clearly in Figure 15.3, is used to update the reference response curve data that correlates measured ion intensities to the concentrations of elements in a solution. During calibration, this response data is adjusted to account for changes in the instrument's sensitivity due to variations in the sample matrix.

This process is often called semiquantitative analysis using external calibration, and like traditional quantitative analysis using external standardization, it works

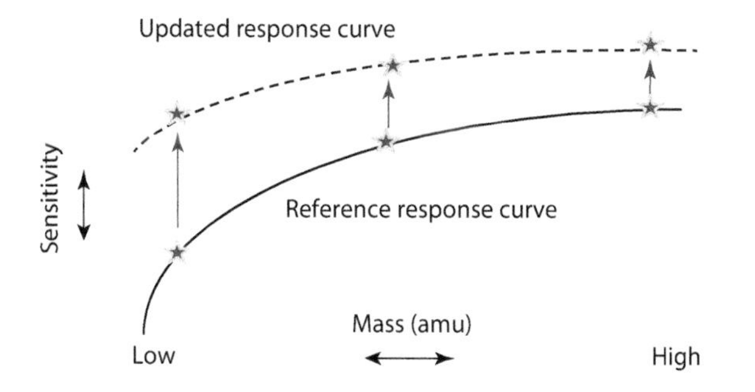

FIGURE 15.3 In semiquantitative analysis, a small group of elements are used to update the reference response curve to improve the accuracy as the sample matrix changes.

extremely well for samples that have a similar matrix. However, if you are analyzing samples containing widely different concentrations of matrix components, external calibration does not work very well because of the matrix-induced suppression effects on the analyte signal. If this is the case, semiquant using a variation of standard addition calibration should be used. Similar to standard addition calibration used in quantitative analysis, this procedure involves adding known quantities of specific elements to every unknown sample before measurement. The major difference with semiquant is that the elements you add must not already be present in significant quantities in the unknown samples, because they are being used to update the stored reference response curve. As with external calibration, the semiquant software then adjusts the stored response data for all remaining analytes relative to the calibration elements. This procedure works very well, but tends to be very labor-intensive because the calibration standards have to be added to every unknown sample.

The other thing to be wary of with semiquantitative analysis is the spectral complexity of unknown samples. If you have a spectrally rich sample and are not making any compensation for spectral overlaps close to the analyte peaks, this could possibly give you a false positive for that element. Therefore, you have to be very cautious when reporting semiquantitative results on completely unknown samples. They should be characterized first, especially with respect to the types of spectral interferences generated by the plasma gas, the matrix, and the solvents, acids, or chemicals used for sample preparation. Collision/reaction cells and interfaces can help in the reduction of some of these interferences, but extreme care should be taken, as these devices are known to have no effect on some polyatomic interferences, and in some cases can increase the spectral complexity by generating other interfering complexes.

ISOTOPE DILUTION

Although quantitative and semiquantitative analysis methods are suitable for the majority of applications, there are other calibration methods available, depending on your analytical requirements. For example, if your application requires even greater accuracy and precision, the isotope dilution technique may offer some benefits. Isotope dilution is an absolute means of quantitation based on altering the natural abundance of two isotopes of an element by adding a known amount of one of the isotopes, and is considered one of the most accurate and precise approaches to elemental analysis.[5–8]

For this reason, a prerequisite of isotope dilution is that the element must have at least two stable isotopes. The principle works by spiking a known weight of an enriched stable isotope into your sample solution. By knowing the natural abundance of the two isotopes being measured, the abundance of the spiked enriched isotopes, the weight of the spike, and the weight of the sample, the original trace element concentration can be determined by using the following equation:

$$C = \frac{[\text{Aspike} - (R \times \text{Bspike})] \times \text{Wspike}}{[R \times (\text{Bsample} - \text{Asample})] \times \text{Wsample}}$$

where

C	=	Concentration of trace element
Aspike	=	percent of higher-abundance isotope in spiked enriched isotope
Bspike	=	percent of lower-abundance isotope in spiked enriched isotope
Wspike	=	Weight of spiked enriched isotope
R	=	Ratio of the percent of higher-abundance isotope to lower-abundance isotope in the spiked sample
Bsample	=	Percent of higher-natural-abundance isotope in sample
Asample	=	Percent of lower-natural-abundance isotope in sample
Wsample	=	Weight of sample

This might sound complicated, but in practice, it is relatively straightforward. This is illustrated in Figure 15.4, which shows an isotope dilution method for the determination of copper in a 250 mg sample of orchard leaves, using the two copper isotopes ^{63}Cu and ^{65}Cu.

In the bar graph on top, it can be seen that the natural abundances of the two isotopes are 69.09% and 30.91%, respectively, for ^{63}Cu and ^{65}Cu. The middle graph shows that 4 µg of an enriched isotope of 100% ^{65}Cu (and 0% ^{63}Cu) is spiked into the sample, which now produces a spiked sample containing 71.4% of ^{65}Cu and 28.6% of ^{63}Cu, as seen in the bottom plot.[9] If we plug this data into the preceding equation, we get

$$C = \frac{[100 - (71.4/28.6 \times 0)] \times 4\mu g}{[(71.4/28.6 \times 69.09) - 30.91] \times 0.25\,g}$$

$$C = 400/35.45 = 11.3\mu g/g$$

The major benefit of the isotope dilution technique is that it provides measurements that are extremely accurate because you are measuring the concentration of the isotopes in the same solution as your unknown sample, and not in a separate external calibration solution. In addition, because it is a ratioing technique, loss of solution during the sample preparation stage has no influence on the accuracy of the result. The technique is also extremely precise, because using a simultaneous detection system, such as a magnetic sector multicollector, or a simultaneous ion sampling device, such as a time-of-flight ICP-MS, the results are based on measuring the two-isotope solution at the same instant in time, which compensates for imprecision of the signal due to sources of sample introduction–related noise, such as plasma instability, peristaltic pump pulsations, and nebulization fluctuations. Even when using a scanning mass analyzer such as a quadrupole, the measurement protocol can be optimized to scan very rapidly between the two isotopes and achieve very good precision. However, isotope dilution has some limitations, which makes it suitable only for certain applications. These limitations include the following:

- The element you are determining must have more than one isotope, because calculations are based on the ratio of one isotope to another isotope of the same element; this makes it unsuitable for approximately 15 elements that can be determined by ICP-MS.

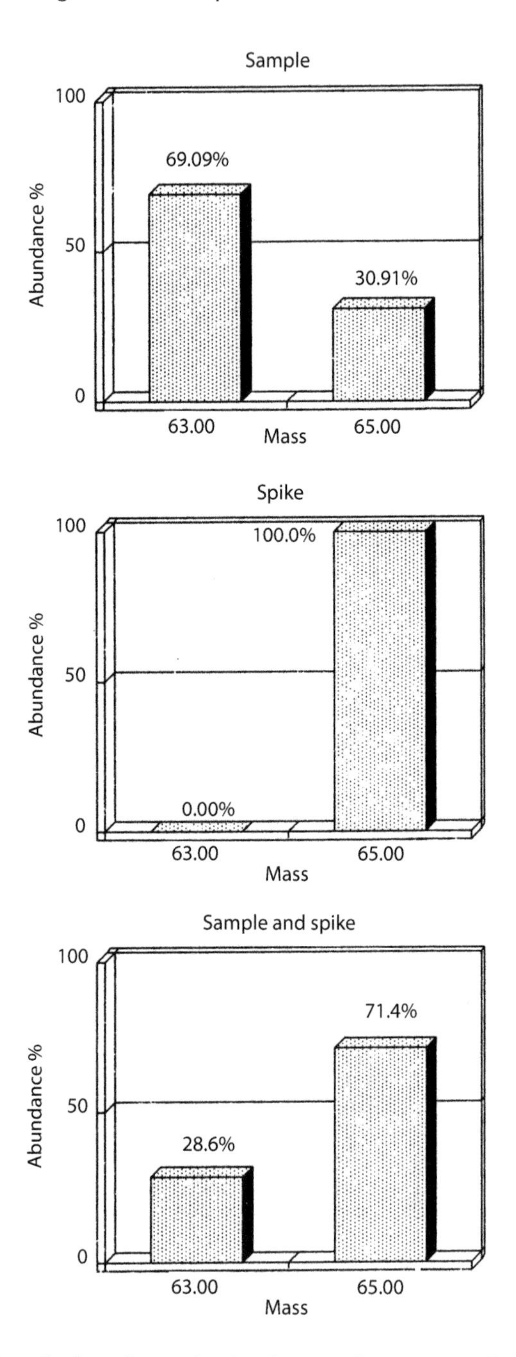

FIGURE 15.4 Quantitation of trace levels of copper in a sample of SRM orchard leaves using isotope dilution methodology. (From PerkinElmer, Multielemental isotope dilution using the Elan ICP-MS elemental analyzer, ICP-MS Technical Summary TSMS-1, 1985.)

- It requires certified enriched isotopic standards, which can be very expensive, especially those that are significantly different from the normal isotopic abundance of the element.
- It compensates for interferences due to signal enhancement or suppression, but does not compensate for spectral interferences. For this reason, an external blank solution must always be run.

ISOTOPE RATIOS

The ability of ICP-MS to determine individual isotopes also makes it suitable for another isotopic measurement technique called "isotope ratio" analysis. The ratio of two or more isotopes in a sample can be used to generate very useful information, including an indication of the age of a geological formation and a better understanding of animal metabolism, and it can also help to identify sources of environmental contamination.[10–14] Similar to isotope dilution, isotope ratio analysis uses the principle of measuring the exact ratio of two isotopes of an element in the sample. With this approach, the isotope of interest is typically compared with a reference isotope of the same element. For example, you might want to compare the concentration of ^{204}Pb with that of ^{206}Pb. Alternatively, the requirement might be to compare one isotope with all remaining reference isotopes of an element, for example, the ratio of ^{204}Pb with ^{206}Pb, ^{207}Pb, and ^{208}Pb. The ratio is then expressed in the following manner:

Isotope ratio = Intensity of isotope interest/Intensity of reference isotope

As this ratio can be calculated from within a single sample measurement, classic external calibration is not normally required. However, if there is a large difference between the concentrations of the two isotopes, it is recommended to run a standard of known isotopic composition. This is done to verify that the higher-concentration isotope is not suppressing the signal of the lower-concentration isotope and biasing the results. This effect, called mass discrimination, is less of a problem if the isotopes are relatively close in concentration, for example, ^{107}Ag to ^{109}Ag, which are 51.839% and 48.161% abundant, respectively. However, it can be an issue if there is a significant difference in their concentration values, for example, ^{235}U to ^{238}U, which are 0.72% and 99.275% abundant, respectively. Mass discrimination effects can be reduced by running an external reference standard of known isotopic concentration, comparing the isotope ratio with the theoretical value, and then mathematically compensating for the difference. The principles of isotope ratio analysis and how to achieve optimum precision values are explained in greater detail in Chapter 14.

INTERNAL STANDARDIZATION

Another method of standardization commonly employed in ICP-MS is called *internal standardization*. It is not considered an absolute calibration technique, but instead is used to correct for changes in analyte sensitivity caused by variations in the concentration and type of matrix components found in the sample. An internal standard is a nonanalyte isotope that is added to the blank solution, standards, and samples before analysis. It should also be noted that the chosen internal standard element must not be present in the sample matrix. It is typical to add three or four internal standard elements to the samples to cover the analyte elements of interest. The software adjusts the analyte concentration in the unknown samples by comparing the intensity values of the internal standard intensities in the unknown sample with those in the calibration standards.

The implementation of internal standardization varies according to the analytical technique that is being used. For quantitative analysis, the internal standard elements are selected on the basis of the similarity of their ionization characteristics to the analyte elements. Each internal standard is bracketed with a group of analytes. The software then assumes that the intensities of all elements within a group are affected in a similar manner by the matrix. Changes in the ratios of the internal standard intensities are then used to correct the analyte concentrations in the unknown samples.

For semiquantitative analysis that uses a stored response table, the purpose of the internal standard is similar, but a little different in implementation from quantitative analysis. A semiquant internal standard is used to continuously compensate for instrument drift or matrix-induced suppression over a defined mass range. If a single internal standard is used, all the masses selected for the determination are updated by the same amount based on the intensity of the internal standard. If more than one internal standard is used, which is recommended for measurements over a wide mass range, the software interpolates the intensity values based on the distance in mass between the analyte and the nearest internal standard element.

It is worth emphasizing that if you do not want to compare your intensity values with a calibration graph, most instruments allow you to report raw data. This enables you to analyze your data using external data processing routines, to selectively apply a minimum set of ICP-MS data processing methods, or to just view the raw data file before reprocessing it. The availability of raw data is primarily intended for use in nonroutine applications, such as chromatography separation techniques and laser sampling devices that produce a time-resolved transient peak, or by users whose sample set requires data processing using algorithms other than those supplied by the instrument software.

REFERENCES

1. D. Beauchemin, J. W. McLaren, A. P. Mykytiuk, and S. S. Berman. *Analytical Chemistry*, 59, 778, 1987.
2. E. Pruszkowski, K. Neubauer, and R. Thomas. *Atomic Spectroscopy*, 19(4), 111–115, 1998.
3. M. Broadhead, R. Broadhead, and J. W. Hager. *Atomic Spectroscopy*, 11(6), 205–209, 1990.
4. E. Denoyer. *Journal of Analytical Atomic Spectrometry*, 7, 1187, 1992.
5. J. W. McLaren, D. Beauchemin, and S. S. Berman. *Analytical Chemistry*, 59, 610, 1987.
6. H. Longerich. *Atomic Spectroscopy*, 10(4), 112–115, 1989.
7. A. Stroh. *Atomic Spectroscopy*, 14(5), 141–143, 1993.
8. T. Catterick, H. Handley, and S. Merson. *Atomic Spectroscopy*, 16(10), 229–234, 1995.
9. PerkinElmer, Multi-elemental isotope dilution using the Elan ICP-MS elemental analyzer, ICP-MS Technical Summary TSMS-1, 1985.

10. B. T. G. Ting and M. Janghorbani. *Analytical Chemistry*, 58, 1334, 1986.
11. M. Janghorbani, B. T. G. Ting, and N. E. Lynch. *Microchemica Acta*, 3, 315–328, 1989.
12. T. A. Hinners, E. M. Heithmar, T. M. Spittler, and J. M. Henshaw. *Analytical Chemistry*, 59, 2658–2662, 1987.
13. L. Halicz, Y. Erel, and A. Veron. *Atomic Spectroscopy*, 17(5), 186–189, 1996.
14. M. Chaudhary-Webb, D. C. Paschal, W. C. Elliott, H. P. Hopkins, A. M. Ghazi, B. C. Ting, and I. Romieu. *Atomic Spectroscopy*, 19(5), 156, 1998.

16 Review of ICP-MS Interferences

Now that we have covered the fundamental principles of inductively coupled plasma mass spectrometry (ICP-MS) and its measurement and calibration routines, let us turn our attention to the technique's most common interferences and the methods that are used to compensate for them. Although interferences are reasonably well understood in ICP-MS, it can often be difficult and time-consuming to compensate for them, particularly in complex sample matrices. Prior knowledge of the interferences associated with a particular set of samples will often dictate the sample preparation steps and the instrumental methodology used to analyze them.

Interferences in ICP-MS are generally classified into three major groups: spectral, matrix, and physical. Each of them has the potential to be problematic in its own right, but modern instrumentation and good software, combined with optimized analytical methodologies, have minimized their negative impact on trace element determinations by ICP-MS. Let us look at these interferences in greater detail and describe the different approaches used to compensate for them.

SPECTRAL INTERFERENCES

Spectral overlaps are probably the most serious types of interferences seen in ICP-MS. The most common are known as a polyatomic or molecular spectral interference and are produced by the combination of two or more atomic ions. They are caused by a variety of factors, but are usually associated with the plasma or nebulizer gas used, matrix components in the solvent or sample, other elements in the sample, or entrained oxygen or nitrogen from the surrounding air. For example, in the argon plasma, spectral overlaps caused by argon ions and combinations of argon ions with other species are very common. The most abundant isotope of argon is at mass 40, which dramatically interferes with the most abundant isotope of calcium at mass 40, whereas the combination of argon and oxygen in an aqueous sample generates the $^{40}Ar^{16}O^+$ interference, which has a significant impact on the major isotope of Fe at mass 56. The complexity of these kinds of spectral problems can be seen in Figure 16.1, which shows a mass spectrum of deionized water from mass 40 to mass 90.

In addition, argon can form polyatomic interferences with elements found in the acids used to dissolve the sample. For example, in a hydrochloric acid medium, $^{40}Ar^+$ combines with the most abundant chlorine isotope at 35 amu to form $^{40}Ar^{35}Cl^+$, which interferes with the only isotope of arsenic at mass 75, whereas in an organic solvent matrix, argon and carbon combine to form $^{40}Ar^{12}C^+$, which interferes with $^{52}Cr^+$, the most abundant isotope of chromium. Sometimes, matrix or solvent ions combine to form spectral interferences of their own. A good example is in a sample that contains

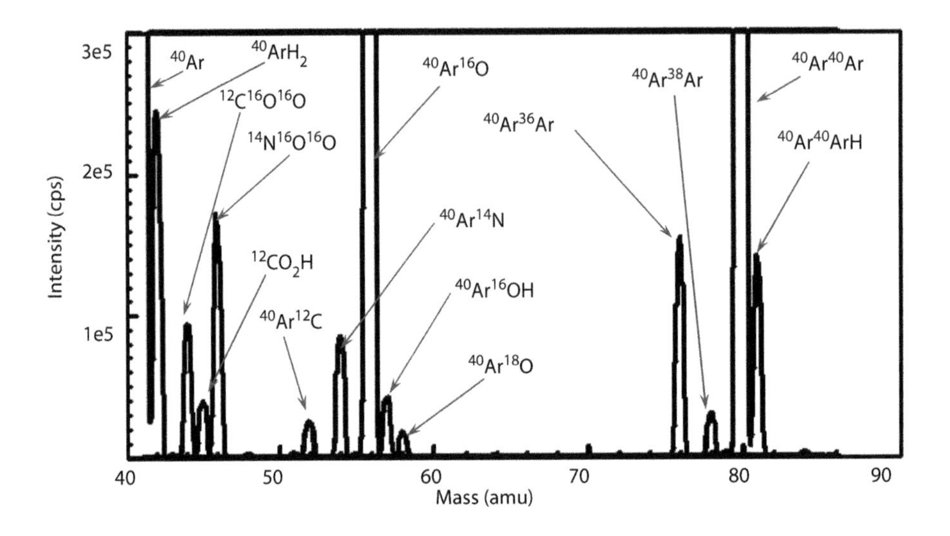

FIGURE 16.1 ICP mass spectrum of deionized water from mass 40 to mass 90.

sulfuric acid. The dominant sulfur isotope, $^{32}S^+$, combines with two oxygen ions to form a $^{32}S^{16}O^{16}O^+$ molecular ion, which interferes with the major isotope of Zn at mass 64. In the analysis of samples containing high concentrations of sodium, such as seawater, the most abundant isotope of Cu at mass 63 cannot be used because of interference from the $^{40}Ar^{23}Na^+$ molecular ion. There are many more examples of these kinds of polyatomic and molecular interferences, which have been comprehensively reviewed.[1] Table 16.1 represents some of the most common matrix–solvent spectral interferences seen in ICP-MS.

Oxides, Hydroxides, Hydrides, and Doubly Charged Species

Another type of spectral interference is produced by elements in the sample combining with H^+, $^{16}O^+$, or $^{16}OH^+$ (either from water or air) to form molecular hydrides (+ H^+), oxides (+ $^{16}O^+$), and hydroxides (+ $^{16}OH^+$), which occur at 1, 16, and 17 mass units, respectively, higher than the element's mass.[2] These interferences are typically produced in the cooler zones of the plasma, immediately before the interface region. They are usually more serious when rare earth or refractory-type elements are present in the sample, because many of them readily form molecular species (particularly oxides), which create spectral overlap problems on other elements in the same group. If the oxide species is mainly derived from entrained air around the plasma, it can be reduced by using either an elongated outer tube to the torch or a metal shield between the plasma and the radio-frequency (RF) coil.

Associated with oxide-based spectral overlaps are doubly charged spectral interferences. These are species that are formed when an ion is generated with a double-positive charge, as opposed to a normal single charge, and produces an isotopic peak at half its mass. Similar to the formation of oxides, the level of doubly charged species is related to the ionization conditions in the plasma and can

TABLE 16.1
Some Common Plasma-, Matrix-, and Solvent-Related Polyatomic Spectral Interferences Seen in ICP-MS

Element/Isotope	Matrix/Solvent	Interference
$^{39}K^+$	H_2O	$^{38}ArH^+$
$^{40}Ca^+$	H_2O	$^{40}Ar^+$
$^{56}Fe^+$	H_2O	$^{40}Ar^{16}O^+$
$^{80}Se^+$	H_2O	$^{40}Ar^{40}Ar^+$
$^{51}V^+$	HCl	$^{35}Cl^{16}O^+$
$^{75}As^+$	HCl	$^{40}Ar^{35}Cl^+$
$^{28}Si^+$	HNO_3	$^{14}N^{14}N^+$
$^{44}Ca^+$	HNO_3	$^{14}N^{14}N\ ^{16}O^+$
$^{55}Mn^+$	HNO_3	$^{40}Ar^{15}N^+$
$^{48}Ti^+$	H_2SO_4	$^{32}S^{16}O^+$
$^{52}C^+r$	H_2SO_4	$^{34}S^{18}O^+$
$^{64}Zn^+$	H_2SO_4	$^{32}S^{16}O^{16}O^+$
$^{63}Cu^+$	H_3PO_4	$^{31}P^{16}O^{16}O^+$
$^{24}Mg^+$	Organics	$^{12}C^{12}C^+$
$^{52}Cr^+$	Organics	$^{40}Ar^{12}C^+$
$^{65}Cu^+$	Minerals	$^{48}Ca^{16}OH^+$
$^{64}Zn^+$	Minerals	$^{48}Ca^{16}O^+$
$^{63}Cu^+$	Seawater	$^{40}Ar^{23}Na^+$

usually be minimized by careful optimization of the nebulizer gas flow, RF power, and sampling position within the plasma. It can also be impacted by the severity of the secondary discharge present at the interface,[3] which was described in greater detail in Chapter 7. Table 16.2 shows a selected group of elements that readily form oxides, hydroxides, hydrides, and doubly charged species, together with the analytes affected by them.

Isobaric Interferences

The final classification of spectral interferences is called isobaric overlaps, produced mainly by different isotopes of other elements in the sample creating spectral interferences at the same mass as the analyte. For example, vanadium has two isotopes at 50 and 51 amu. However, mass 50 is the only practical isotope to use in the presence of a chloride matrix because of the large contribution from the $^{16}O^{35}Cl^+$ interference at mass 51. Unfortunately, mass 50 amu, which is only 0.25% abundant, also coincides with isotopes of titanium and chromium, which are 5.4% and 4.3% abundant, respectively. This makes the determination of vanadium in the presence of titanium and chromium very difficult unless mathematical corrections are made. Figure 16.2 shows all the possible naturally occurring isobaric spectral overlaps in ICP-MS.[4]

TABLE 16.2
Some Elements That Readily Form Oxides, Hydroxides, Hydrides, and Doubly Charged Species in the Plasma, Together with the Analytes Affected by the Interference

Oxide, Hydroxide, Hydride, Doubly Charged Species	Analyte Affected by Interference
$^{40}Ca^{16}O^+$	$^{56}Fe^+$
$^{48}Ti^{16}O^+$	$^{64}Zn^+$
$^{98}Mo^{16}O^+$	$^{114}Cd^+$
$^{138}Ba^{16}O^+$	$^{154}Sm^+$, $^{154}Gd^+$
$^{139}La^{16}O^+$	$^{155}Gd^+$
$^{140}Ce^{16}O^+$	$^{156}Gd^+$, $^{156}Dy^+$
$^{40}Ca^{16}OH^+$	$^{57}Fe^+$
$^{31}P^{18}O^{16}OH^+$	$^{66}Zn^+$
$^{79}BrH^+$	$^{80}Se^+$
$^{31}P^{16}O^2H^+$	$^{64}Zn^+$
$^{138}Ba^{2+}$	$^{69}Ga^+$
$^{139}La^{2+}$	$^{69}Ga^+$
$^{140}Ce^{2+}$	$^{70}Ge^+$, $^{70}Zn^+$

Ways to Compensate for Spectral Interferences

Let us now look at the different approaches used to compensate for spectral interferences. One of the very first ways used to get around severe matrix-derived spectral interferences was to remove the matrix somehow. In the early days, this involved precipitating the matrix with a complexing agent and then filtering off the precipitate. However, more recently, this has been carried out by automated matrix removal and analyte preconcentration techniques using chromatography-type equipment. In fact, this is the preferred method for carrying out trace metal determinations in seawater, because of the matrix and spectral problems associated with such high concentrations of sodium and chloride ions.[5]

Mathematical Correction Equations

Another method that has been successfully used to compensate for isobaric interferences and some less severe polyatomic overlaps (when no alternative isotopes are available for quantitation) is to use mathematical interference correction equations. Similar to interelement corrections (IECs) in inductively coupled plasma optical emission spectrometry (ICP-OES), this method works on the principle of measuring the intensity of the interfering isotope or interfering species at another mass, which is ideally free of any interferences. A correction is then applied by knowing the ratio of the intensity of the interfering species at the analyte mass to its intensity at the alternate mass. Let us look at a "real-world" example to exemplify this type of

Isotope	%	%	%
1	H 99.985		
2	H 0.015		
3		He 0.000137	
4		He 99.999863	
5			
6			Li 7.5
7			Li 92.5
8			
9	Be 100		
10		B 19.9	
11		B 80.1	
12			C 98.90
13			C 1.10
14	N 99.643		
15	N 0.366		
16		O 99.762	
17		O 0.038	
18		O 0.200	
19			F 100
20	Ne 90.48		
21	Ne 0.27		
22	Ne 9.25		
23		Na 100	
24			Mg 78.99
25			Mg 10.00
26			Mg 11.01
27	Al 100		
28		Si 92.23	
29		Si 4.67	
30		Si 3.10	
31			P 100
32	S 95.02		
33	S 0.75		
34	S 4.21		
35		Cl 75.77	
36	S 0.02		Ar 0.337
37		Cl 24.23	
38			Ar 0.063
39	K 93.2581		
40	K 0.0117	Ca 96.941	Ar 99.600
41	K 6.7302		
42		Ca 0.647	
43		Ca 0.135	
44		Ca 2.086	
45			Sc 100
46	Ti 8.0	Ca 0.004	
47	Ti 7.3		
48	Ti 73.8	Ca 0.187	
49	Ti 5.5		
50	Ti 5.4	V 0.250	Cr 4.345
51		V 99.750	
52			Cr 83.789
53			Cr 9.501
54	Fe 5.8		Cr 2.365
55		Mn 100	
56	Fe 91.72		
57	Fe 2.2		
58	Fe 0.28		Ni 68.077
59		Co 100	
60			Ni 26.223

Isotope	%	%	%
61			Ni 1.140
62			Ni 3.634
63	Cu 69.17		
64		Zn 48.6	Ni 0.926
65	Cu 30.83		
66		Zn 27.9	
67		Zn 4.1	
68		Zn 18.8	
69			Ga 60.108
70	Ge 21.23	Zn 0.6	
71			Ga 39.892
72	Ge 27.66		
73	Ge 7.73		
74	Ge 35.94	Se 0.89	
75			As 100
76	Ge 7.44	Se 9.36	
77		Se 7.63	
78	Kr 0.35	Se 23.78	
79			Br 50.69
80	Kr 2.25	Se 49.61	
81			Br 49.31
82	Kr 11.6	Se 8.73	
83	Kr 11.5		
84	Kr 57.0	Sr 0.56	
85			Rb 72.165
86	Kr 17.3	Sr 9.86	
87		Sr 7.00	Rb 27.835
88		Sr 82.58	
89			Y 100
90	Zr 51.45		
91	Zr 11.22		
92	Zr 17.15	Mo 14.84	
93			Nb 100
94	Zr 17.38	Mo 9.25	
95		Mo 15.92	
96	Zr 2.80	Mo 16.68	Ru 5.52
97		Mo 9.55	
98		Mo 24.13	Ru 1.88
99			Ru 12.7
100		Mo 9.63	Ru 12.6
101			Ru 17.0
102	Pd 1.02		Ru 31.6
103		Rh 100	
104	Pd 11.14		Ru 18.7
105	Pd 22.33		
106	Pd 27.33	Cd 1.25	
107			Ag 51.839
108	Pd 26.46	Cd 0.89	
109			Ag 48.161
110	Pd 11.72	Cd 12.49	
111		Cd 12.80	
112	Sn 0.97	Cd 24.13	
113		Cd 12.22	In 4.3
114	Sn 0.65	Cd 28.73	
115	Sn 0.34		In 95.7
116	Sn 14.53	Cd 7.49	
117	Sn 7.68		
118	Sn 24.23		
119	Sn 8.59		
120	Sn 32.59	Te 0.096	

Isotope	%	%	%
121			Sb 57.36
122	Sn 4.63	Te 2.603	
123		Te 0.908	Sb 42.64
124	Sn 5.79	Te 4.816	Xe 0.10
125		Te 7.139	
126		Te 18.95	Xe 0.09
127	I 100		
128		Te 31.69	Xe 1.91
129			Xe 26.4
130	Ba 0.106	Te 33.80	Xe 4.1
131			Xe 21.2
132	Ba 0.101		Xe 26.9
133		Cs 100	
134	Ba 2.417		Xe 10.4
135	Ba 6.592		
136	Ba 7.854	Ce 0.19	Xe 8.9
137	Ba 11.23		
138	Ba 71.70	Ce 0.25	La 0.0902
139			La 99.9098
140		Ce 88.48	
141			Pr 100
142	Nd 27.13	Ce 11.08	
143	Nd 12.18		
144	Nd 23.80	Sm 3.1	
145	Nd 8.30		
146	Nd 17.19		
147		Sm 15.0	
148	Nd 5.76	Sm 11.3	
149		Sm 13.8	
150	Nd 5.64	Sm 7.4	
151			Eu 47.8
152	Gd 0.20	Sm 26.7	
153			Eu 52.2
154	Gd 2.18	Sm 22.7	
155	Gd 14.80		
156	Gd 20.47	Dy 0.06	
157	Gd 15.65		
158	Gd 24.84	Dy 0.10	
159			Tb 100
160	Gd 21.86	Dy 2.34	
161		Dy 18.9	
162	Er 0.14	Dy 25.5	
163		Dy 24.9	
164	Er 1.61	Dy 28.2	
165			Ho 100
166	Er 33.6		
167	Er 22.95		
168	Er 26.8	Yb 0.13	
169			Tm 100
170	Er 14.9	Yb 3.05	
171		Yb 14.3	
172		Yb 21.9	
173		Yb 16.12	
174		Yb 31.8	Hf 0.162
175	Lu 97.41		
176	Lu 2.59	Yb 12.7	Hf 5.206
177			Hf 18.606
178			Hf 27.297
179			Hf 13.629
180	Ta 0.012	W 0.13	Hf 35.100

Isotope	%	%	%
181	Ta 99.988		
182		W 26.3	
183		W 14.3	
184	Os 0.02	W 30.67	
185			Re 37.40
186	Os 1.58	W 28.6	
187	Os 1.6		Re 62.60
188	Os 13.3		
189	Os 16.1		
190	Os 26.4		Pt 0.01
191		Ir 37.3	
192	Os 41.0		Pt 0.79
193		Ir 62.7	
194			Pt 32.9
195			Pt 33.8
196	Hg 0.15		Pt 25.3
197		Au 100	
198	Hg 9.97		Pt 7.2
199	Hg 16.87		
200	Hg 23.10		
201	Hg 13.18		
202	Hg 29.86		
203			Tl 29.524
204	Hg 6.87	Pb 1.4	
205			Tl 70.476
206		Pb 24.1	
207		Pb 22.1	
208		Pb 52.4	
209	Bi 100		
210			
211			
212			
213			
214			
215			
216			
217			
218			
219			
220			
221			
222			
223			
224			
225			
226			
227			
228			
229			
230			
231	Pa 100		
232	Th 100		
233			
234	U 0.0055		
235	U 0.7200		
236			
237			
238	U 99.2745		

FIGURE 16.2 Relative isotopic abundances of the naturally occurring elements, showing all the potential isobaric interferences. (From IUPAC, *Pure Appl. Chem.*, 75(6), 683–799, 2003.)

correction. The most sensitive isotope for cadmium is at mass 114. However, there is also a minor isotope of tin at mass 114. This means that if there is any tin in the sample, quantitation using $^{114}Cd^+$ can only be carried out if a correction is made for $^{114}Sn^+$. Fortunately, Sn has a total of 10 isotopes, which means that probably at least one of them is going to be free of a spectral interference. Therefore, by measuring the intensity of Sn at one of its most abundant isotopes (typically, $^{118}Sn^+$) and ratioing it to $^{114}Sn^+$, a correction is made in the method software in the following manner:

Total counts at mass 114 = $^{114}Cd^+$ + $^{114}Sn^+$
Therefore, $^{114}Cd^+$ = Total counts at mass 114 – $^{114}Sn^+$

To find out the contribution from $^{114}Sn^+$, it is measured at the interference-free isotope of $^{118}Sn^+$ and a correction of the ratio of $^{114}Sn^+/^{118}Sn^+$ is applied, which means $^{114}Cd^+$ = Counts at mass 114 – ($^{114}Sn^+/^{118}Sn^+$) × ($^{118}Sn^+$).

Now, the ratio ($^{114}Sn^+/^{118}Sn^+$) is the ratio of the natural abundances of these two isotopes (065%/24.23%) and is always constant.
Therefore, $^{114}Cd^+$ = Mass 114 – (0.65%/24.23%) × ($^{118}Sn^+$)
or $^{114}Cd^+$ = Mass 114 – (0.0268) × ($^{118}Sn^+$)

An interference correction for $^{114}Cd^+$ would then be entered in the software as

– (0.0268) × ($^{118}Sn^+$)

This is a relatively simple example, but it explains the basic principles of the process. In practice, especially in spectrally complex samples, corrections often have to be made to the isotope being used for the correction, in addition to the analyte mass, which makes the mathematical equation far more complex.

This approach can also be used for some less severe polyatomic-type spectral interferences. For example, in the determination of V at mass 51 in diluted brine (typically 1000 ppm NaCl), there is a substantial spectral interference from $^{35}C^{16}1O^+$ at mass 51. By measuring the intensity of the $^{37}C^{16}1O^+$ at mass 53, which is free of any interference, a correction can be applied in a manner similar to that of the previous example.

Cool and Cold Plasma Technology

If the intensity of the interference is large, and the analyte intensity is extremely low, mathematical equations are not ideally suited as a correction method. For that reason, alternative approaches have to be considered to compensate for the interference. One such approach, which has helped to reduce some of the severe polyatomic overlaps, is to use cold or cool plasma conditions. This technology, which was reported in the literature in the late 1980s, uses a low-temperature plasma to minimize the formation of certain argon-based polyatomic species.[6] Under normal plasma conditions (typically 1000–1400 W RF power and 0.8–1.0 L/min of nebulizer gas flow), argon ions combine with matrix and solvent components to generate problematic spectral interferences, such as $^{38}ArH^+$, $^{40}Ar^+$, and $^{40}Ar^{16}O^+$, which impact

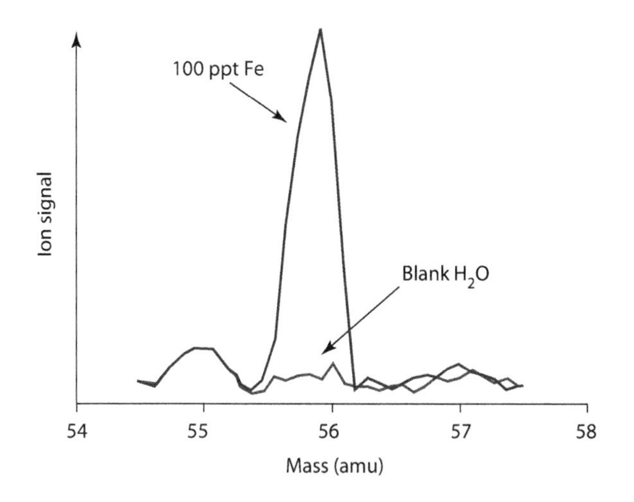

FIGURE 16.3 Spectral scan of 100 ppt ^{56}Fe and deionized water using cool plasma conditions. (From Tanner, S. D., et al., *Atomic Spectrosc.*, 16(1), 16, 1995.)

the detection limits of a small number of elements, including K, Ca, and Fe. By using cool plasma conditions (500–800 W RF power and 1.5–1.8 L/min nebulizer gas flow), the ionization conditions in the plasma are changed so that many of these interferences are dramatically reduced. The result is that detection limits for this group of elements are significantly enhanced.[7] An example of this improvement is shown in Figure 16.3. It shows a spectral scan of 100 ppt $^{56}Fe^+$ (its most sensitive isotope) using cool plasma conditions. It can be clearly seen that there is virtually no contribution from $^{40}Ar^{16}O^+$, as indicated by the extremely low background for deionized water, resulting in single-figure parts per trillion detection limits for iron. Under normal plasma conditions, the $^{40}Ar^{16}O^+$ intensity is so large that it would completely overlap the $^{56}Fe^+$ peak[8].

Unfortunately, even though the use of cool plasma conditions is recognized as being a very useful tool for the determination of a small group of elements, its limitations are well documented.[9] A summary of the limitations of cool plasma technology includes the following:

- As a result of less energy being available in a cool plasma, elements that form a strong bond with one of the matrix or solvent ions cannot be easily decomposed, and as a result, their detection limits are compromised.
- Elements with high ionization potentials cannot be ionized, because there is much less energy compared with a normal, high-temperature plasma.
- Elemental sensitivity is severely affected by the sample matrix; so cool plasma often requires the use of standard additions or matrix matching to achieve satisfactory results.
- When carrying out multielement analysis, normal plasma conditions must also be used. This necessitates the need for stabilization times on the order of 3 min to change from a normal to a cool plasma, which degrades productivity and results in higher sample consumption.

For this reason, it is not ideally suited for the analysis of complex samples, but it does offer real detection limit improvement for elements with low ionization potential, such as sodium and lithium, which benefit from the ionization conditions of the cooler plasma. However, recent advances in solid-state, free-running RF generators appear to have enhanced the capability of cold plasma technology. Researchers have reported that by using the combination of highly reactive gases in a collision/reaction cell with cold plasma conditions, efficient reduction of some polyatomic interferences was achieved. This new approach allowed ultratrace levels of a suite of elements to be accurately measured in high-purity hydrochloric acid using both cold and normal plasma operating conditions.[10]

Collision/Reaction Cells

The earlier limitations of cool plasma technology have led to the development of collision/reaction cells and interfaces, which utilize ion–molecule collisions and reactions to cleanse the ion beam of harmful polyatomic and molecular interferences before they enter the mass analyzer. Quadrupole mass analyzers fitted with these devices are showing enormous potential to eliminate many spectral interferences, allowing the use of the most sensitive elemental isotopes that were previously unavailable for quantitation.

A full description and review of collision/reaction cell and interface technology and how they handle spectral interferences is given in Chapter 12. However, it's worth noting that one of the unique features of using reaction chemistry in a collision/reaction cell is that it can be used in the mass-shift mode, which is particularly beneficial if a spectral interference is encountered with an analyte that is monoisotopic (only one mass is available for quantitation). An example of this is that when high levels of chloride are present in the sample (whether from the matrix itself or from hydrochloric acid used in the sample digestion procedure), it can be problematic to quantitate arsenic and vanadium at low levels, because of polyatomic spectral interferences $^{40}Ar^{75}Cl$ and $^{35}Cl^{16}O$, which interfere with the major isotopes of ^{75}As and ^{51}V, respectively. A collision cell with kinetic energy discrimination can reduce these interferences to a certain level, but if the requirement is for ultratrace determinations of these analytes, reaction chemistry is the better option. Pure ammonia works extremely well to reduce the $^{35}Cl^{16}O$ interference when determining trace levels of vanadium, while oxygen gas is best suited for the determination of low levels of arsenic because only one mass (^{75}As) is available for quantitation. When using oxygen to determine arsenic, reaction chemistry works by implementing the mass-shift mode. In this approach, the arsenic reacts rapidly with oxygen to form $^{75}As^{16}O$ at mass 91, 16 amu away from the $^{40}Ar^{35}Cl$ interference at mass 75. Because $^{40}Ar^{35}Cl$ does not react with oxygen, $^{75}As^{16}O$ is measured free of interferences at 91 amu. Figure 16.4 exemplifies this, showing the conversion of ^{75}As to $^{75}As^{16}O$ as a function of oxygen flow. As the O_2 flow increases, the signal for ^{75}As decreases (blue plot), while the signal for $^{75}As^{16}O$ increases (red plot), demonstrating complete conversion.[10]

However, it should be noted that An and co-workers improved the detection capability of ^{75}As in a high chloride matrix by the addition of small amounts methanol to the samples, using a conventional helium-based collision cell (CC). By optimizing the methanol concentration (~3%) and the He flow rate (~ 3 mL/min), the $^{40}Ar^{35}Cl$ interference could be significantly reduced to improve the As sensitivity by up to 3-fold.[11]

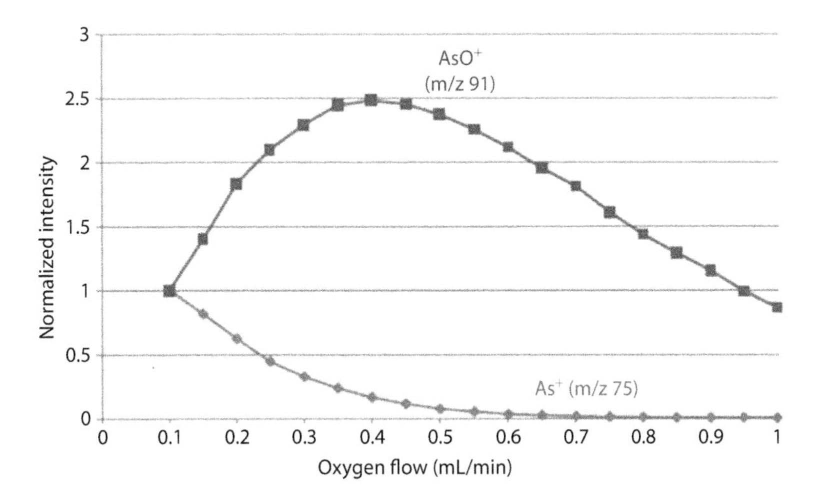

FIGURE 16.4 Conversion of ^{75}As to $^{75}As^{16}O$ as a function of oxygen flow.

High-Resolution Mass Analyzers

The best and probably most efficient way to remove spectral overlaps is to resolve them using a high-resolution mass spectrometer.[12] Over the past 10 years, this approach, particularly double-focusing magnetic sector mass analyzers, has proved to be invaluable for separating many of the problematic polyatomic and molecular interferences seen in ICP-MS, without the need to use cool plasma conditions or collision/reaction cells. This can be seen in Figure 16.5, which shows a spectral peak for 10 ppb $^{75}As^+$ resolved from the $^{40}Ar^{35}Cl^+$ interference in a 1% hydrochloric acid matrix, using a resolution setting of 5000.[13]

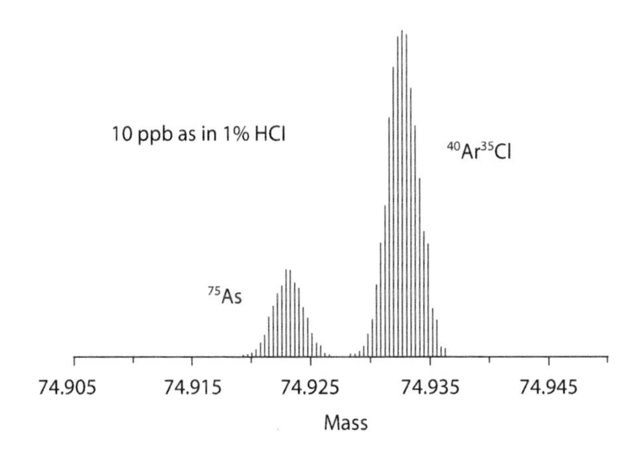

FIGURE 16.5 Separation of $^{75}As^+$ from $^{40}Ar^{35}Cl^+$ using high resolving power (5000) of a double-focusing magnetic sector instrument. (From Tittes, W., et al., The capabilities of a new sector-based ICP-MS for trace element analysis, presented at Winter Conference on Plasma Spectrochemistry, San Diego, 1994.)

Although their resolving capability is far more powerful than that of quadrupole-based instruments, there is a sacrifice in sensitivity if an extremely high resolution is used, which can often translate into a degradation in detection capability for some elements, compared with other spectral interference correction approaches. A full review of magnetic sector technology for ICP-MS is given in Chapter 10.

MATRIX INTERFERENCES

Let us now look at the other class of interference in ICP-MS—suppression of the signal by the matrix itself. There are basically three types of matrix-induced interferences. The first and simplest to overcome is often called a *sample transport effect* and is a physical suppression of the analyte signal, brought on by the level of dissolved solids or acid concentration in the sample. It is caused by the sample's impact on droplet formation in the nebulizer or droplet size selection in the spray chamber. In the case of organic matrices, it is usually caused by variations in the pumping rate of solvents with different viscosities. The second type of matrix suppression is caused when the sample affects the ionization conditions of the plasma discharge. This results in the signal being suppressed by varying amounts, depending on the concentration of the matrix components. This type of interference is exemplified when different concentrations of acids are aspirated into cool plasma. The ionization conditions in the plasma are so fragile that higher concentrations of acid result in severe suppression of the analyte signal. This can be seen very clearly in Figure 16.6, which shows sensitivity for a selected group of elements in varying concentrations of nitric acid in a cool plasma.[9]

COMPENSATION USING INTERNAL STANDARDIZATION

The classic way to compensate for a physical interference is to use internal standardization (IS). With this method of correction, a small group of elements (usually at the parts per billion level) are spiked into the samples, calibration standards, and blank

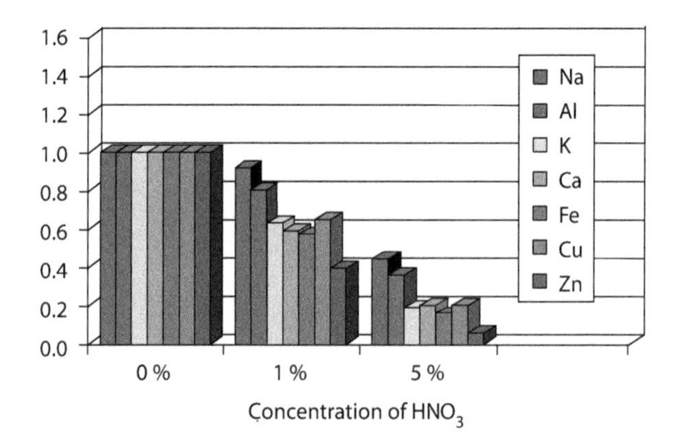

FIGURE 16.6 Matrix suppression caused by increasing concentrations of HNO_3, using cool plasma conditions (RF power, 800 W; nebulizer gas, 1.5 L/min). (From Collard, J. M., et al., *Micro*, January 2002.)

to correct for any variations in the response of the elements caused by the matrix. As the intensities of the internal standards change, the element responses are updated every time a sample is analyzed. The following criteria are typically used for selecting an internal standard:

- It is not present in the sample.
- The sample matrix or analyte elements do not spectrally interfere with it.
- It does not spectrally interfere with the analyte masses.
- It should not be an element that is considered an environmental contaminant.
- It is usually grouped with analyte elements of a similar mass range. For example, a low-mass internal standard is grouped with the low-mass analyte elements, and so on, up the mass range.
- It should be of a similar ionization potential to the group of analyte elements, so it behaves in a similar manner in the plasma.

Some of the most common elements and masses reported to be good candidates for internal standards include ^{9}Be, ^{45}Sc, ^{59}Co, ^{74}Ge, ^{89}Y, ^{103}Rh, ^{115}In, ^{169}Tm, ^{175}Lu, ^{187}Re, and ^{232}Th. An internal standard is also used to compensate for long-term signal drift as a result of matrix components slowly blocking the sampler and skimmer cone orifices. Even though total dissolved solids are usually kept below 0.2% in ICP-MS, this can still produce instability of the analyte signal over time with some sample matrices. It should also be emphasized that the difference in intensities of the internal standard elements across the mass range will indicate the flatness of the mass response curve. The flatter the mass response curve (i.e., less mass discrimination), the easier it is to compensate for matrix-based suppression effects using IS.

Space Charge–Induced Matrix Interferences

Many of the early researchers reported that the magnitude of signal suppression in ICP-MS increased with decreasing atomic mass of the analyte ion.[14] More recently, it has been suggested that the major cause of this kind of suppression is the result of poor transmission of ions through the ion optics due to matrix-induced space charge effects.[15] This has the effect of defocusing the ion beam, which leads to poor sensitivity and detection limits, especially when trace levels of low-mass elements are being determined in the presence of large concentrations of high-mass matrices. Unless any compensation is made, the high-mass matrix element will dominate the ion beam, pushing the lighter elements out of the way.[16] This can be seen in Figure 16.7, which shows the classic space charge effects of a uranium (major isotope $^{238}U^+$) matrix on the determination of $^{7}Li^+$, $^{9}Be^+$, $^{24}Mg^+$, $^{55}Mn^+$, $^{85}Rb^+$, $^{115}In^+$, $^{133}Cs^+$, $^{205}Tl^+$, and $^{208}Pb^+$. It can clearly be seen that the suppression of the low-mass elements, such as Li and Be, is significantly higher than that with the high-mass elements, such as Tl and Pb, in the presence of 1000 ppm uranium.

There are a number of ways to compensate for space charge matrix suppression in ICP-MS. IS has been used, but unfortunately it does not address the fundamental cause of the problem. The most common approach used to alleviate or at least reduce space charge effects is to apply voltages to individual lens components of the ion

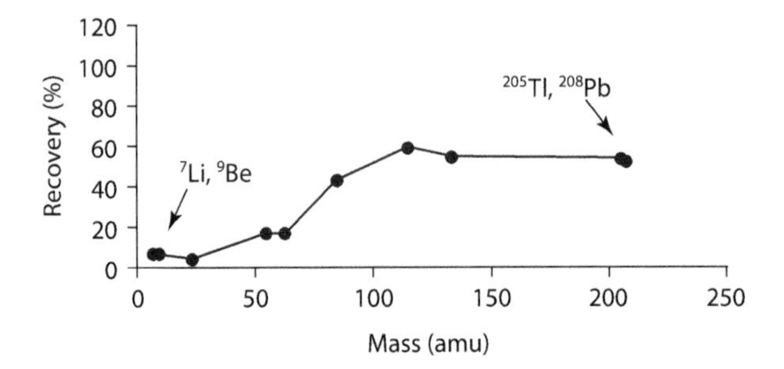

FIGURE 16.7 Space charge matrix suppression caused by 1000 ppm uranium is significantly higher on low-mass elements, such as Li and Be, than it is with the high-mass elements, such as Tl and Pb. (From Tanner, S. D., *J. Anal. At. Spectrom.*, 10, 905, 1995.)

optics. This is achieved in a number of different ways, but irrespective of the design of the ion-focusing system, its main function is to reduce matrix-based suppression effects by steering as many of the analyte ions through to the mass analyzer while rejecting the maximum number of matrix ions. For more details on space charge effects and different designs of ion optics, refer to Chapter 8 on ion optics.

REFERENCES

1. M. A. Vaughan and G. Horlick. *Applied Spectroscopy*, 41(4), 523, 1987.
2. S. N. Tan and G. Horlick. *Applied Spectroscopy*, 40(4), 445, 1986.
3. D. J. Douglas and J. B. French. *Spectrochimica Acta*, 41B(3), 197, 1986.
4. IUPAC. Isotopic composition of the elements. *Pure and Applied Chemistry*, 75(6), 683–799, 2003.
5. S. N. Willie, Y. Iida, and J. W. McLaren. *Atomic Spectroscopy*, 19(3), 67, 1998.
6. S. J. Jiang, R. S. Houk, and M. A. Stevens. *Analytical Chemistry*, 60, 1217, 1988.
7. K. Sakata and K. Kawabata. *Spectrochimica Acta*, 49B, 1027, 1994.
8. S. D. Tanner, M. Paul, S. A. Beres, and E. R. Denoyer. *Atomic Spectroscopy*, 16(1), 16, 1995.
9. J. M. Collard, K. Kawabata, Y. Kishi, and R. Thomas. *Micro*, January 2002.
10. K. Neubauer. The use of ion-molecule chemistry combined with optimized plasma conditions to meet SEMI tier C guidelines for elemental impurities in semiconductor grade hydrochloric acid by ICP-MS. *Spectroscopy*, October 2017.
11. Jinsung An, Junseok Lee, Gyuri Lee, Kyoungphile Nam, Hye-On Noon, Combined use of collision cell technique and methanol addition for the analysis of arsenic in a high-chloride-containing sample by ICP-MS. http://www.sciencedirect.com/science/journal/0026265X Microchemical Journal Vol. 120, 77–81, May 2015.
12. R. Hutton, A. Walsh, D. Milton, and J. Cantle. *ChemSA*, 17, 213–215, 1992.
13. N. Jakubowski, and D. Stuewer. Plasma Source MS: Enhancement of Versatility by Consequent Optimization. Presented at Winter Conference on Plasma Spectrochemistry, San Diego, 1994.
14. J. A. Olivares and R. S. Houk. *Analytical Chemistry*, 58, 20, 1986.
15. S. D. Tanner, D. J. Douglas, and J. B. French. *Applied Spectroscopy*, 48, 1373, 1994.
16. S. D. Tanner. *Journal of Analytical Atomic Spectrometry*, 10, 905, 1995.

17 Routine Maintenance

The components of an inductively coupled plasma (ICP) mass spectrometer are generally more complex than those of other atomic spectroscopic techniques, and as a result, more time is required to carry out routine maintenance to ensure that the instrument is performing to the best of its ability. Some tasks involve a simple visual inspection of a part, whereas others involve cleaning or changing components on a regular basis. However, routine maintenance is such a critical part of owning an inductively coupled plasma mass spectrometry (ICP-MS) system that it can impact both the performance and lifetime of the instrument. This chapter covers this topic in greater detail.

The fundamental principle of ICP-MS, which gives the technique its unequaled isotopic selectivity and sensitivity, also unfortunately contributes to some of its weaknesses. The fact that the sample "flows into" the spectrometer and is not "passed by it" at right angles, such as flame atomic absorption (AA) and radial inductively coupled plasma optical emission spectroscopy (ICP-OES), means that the potential for thermal problems, corrosion, chemical attack, blockage, matrix deposits, and drift is much higher than with the other atomic spectrometry (AS) techniques. However, being fully aware of this fact and carrying out regular inspection of instrumental components can reduce and sometimes eliminate many of these potential problem areas. There is no question that a laboratory that initiates a routine maintenance schedule stands a much better chance of having an instrument ready and available for analysis whenever it is needed, compared with a laboratory that basically ignores these issues and assumes the instrument will look after itself.

Let us now look at the areas of the instrument that a user needs to pay attention to. I will not go into great detail but just give a brief overview of what is important, so you can compare it with maintenance procedures of trace element techniques you are more familiar with. These areas should be very similar with all commercial ICP-MS systems, but depending on the design of the instrument and the types of samples being analyzed, the regularity of changing or cleaning components might be slightly different (particularly if the instrument is being used for laser ablation work). The main areas that require inspection and maintenance on a routine or semiroutine basis include the following:

- Sample introduction system
- Plasma torch
- Interface region
- Ion optics
- Roughing pumps
- Air and water filters

Other areas of the instrument require less attention, but nevertheless the user should also be aware of maintenance procedures required to maximize their lifetime. They will be discussed at the end of this section.

SAMPLE INTRODUCTION SYSTEM

The sample introduction system, comprising the peristaltic or pneumatic pump, nebulizer, spray chamber, and drain system, takes the initial abuse from the sample matrix and, as a result, is an area of the ICP mass spectrometer that requires a great deal of attention. The principles of the sample introduction area have been described in great detail in Chapter 5, so let us now examine what kind of routine maintenance it requires.

Peristaltic Pump Tubing

If the instrument uses a peristaltic pump, the sample is pumped at about 1 mL/min into the nebulizer. The constant motion and pressure of the pump rollers on the pump tubing, which is typically made from a polymer-based material, ensure a continuous flow of liquid to the nebulizer. However, over time, this constant pressure of the rollers on the pump tubing has the tendency to stretch it, which changes its internal diameter, and therefore the amount of sample being delivered to the nebulizer. The impact could be a change in the analyte intensity, and therefore a degradation in short-term stability.

Therefore, the condition of the pump tubing should be examined every few days, particularly if your laboratory has a high sample workload or if extremely corrosive solutions are being analyzed. The peristaltic pump tubing is probably one of the most neglected areas, so it is absolutely essential that it be a part of your routine maintenance schedule. Here are some suggested tips to reduce pump tubing–based problems:

- Manually stretch the new tubing before use.
- Maintain the proper tension on the tubing.
- Ensure the tubing is placed correctly in the channel of the peristaltic pump.
- Periodically check the flow of the sample delivery, and throw away tubing if in doubt.
- Replace tubing if there is any sign of wear; do not wait until it breaks.
- With a high sample workload, change the tubing every day or every other day.
- Release pressure on the pump tubing when the instrument is not in use.
- Pump and capillary tubing can be a source of contamination.
- Pump tubing is a consumable item—keep a large supply of it on hand.

A very useful tool to diagnose any problems associated with the peristaltic pump tubing (or the nebulizer) is a digital thermoelectric flowmeter. By inserting this device in the sample line, you always know the actual rate of sample uptake to your nebulizer. This enhances the day-to-day reproducibility of your results and reduces

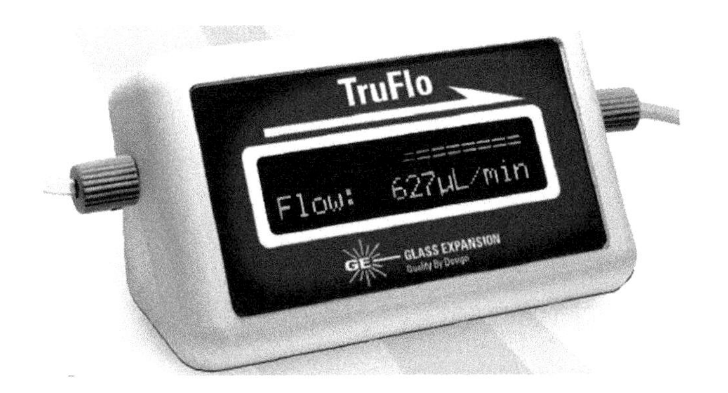

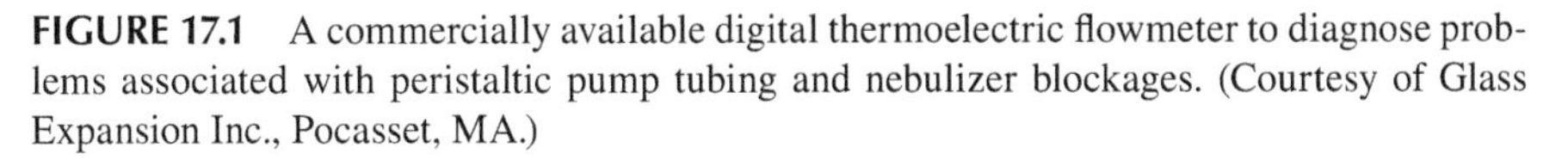

FIGURE 17.1 A commercially available digital thermoelectric flowmeter to diagnose problems associated with peristaltic pump tubing and nebulizer blockages. (Courtesy of Glass Expansion Inc., Pocasset, MA.)

the need to repeat measurements due to a blocked nebulizer, worn pump tubing, or incorrect clamping of the pump tube. In addition, the borosilicate glass sample path ensures that there is no memory effect or sample contamination. A commercially available digital thermoelectric flowmeter is shown in Figure 17.1.

Nebulizers

The frequency of nebulizer maintenance will primarily depend on the types of samples being analyzed and the design of the nebulizer being used. For example, in a cross-flow nebulizer, the argon gas is directed at right angles to the sample capillary tip, in contrast to the concentric nebulizer, where the gas flow is parallel to the capillary. This can be seen in Figures 17.2 and 17.3, which show schematics of a concentric and cross-flow nebulizer, respectively.

The larger diameter of the liquid capillary and longer distance between the liquid and gas tips of the cross-flow design make it far more tolerant to dissolved solids and suspended particles in the sample than the concentric design. On the other hand, aerosol generation of a cross-flow nebulizer is far less efficient than that of a

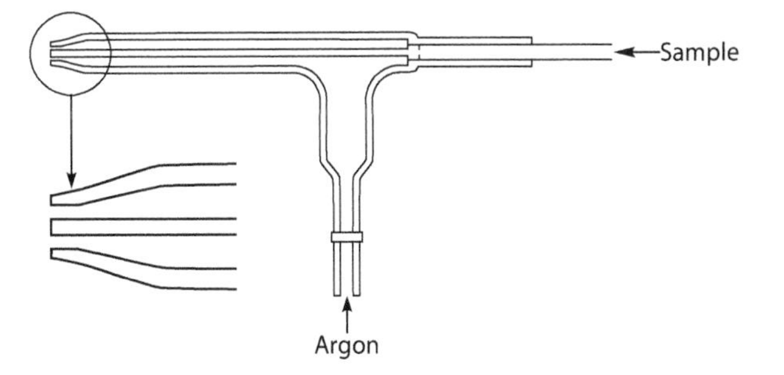

FIGURE 17.2 Schematic of a concentric nebulizer. (Courtesy of Meinhard Glass Products, Golden, CO.)

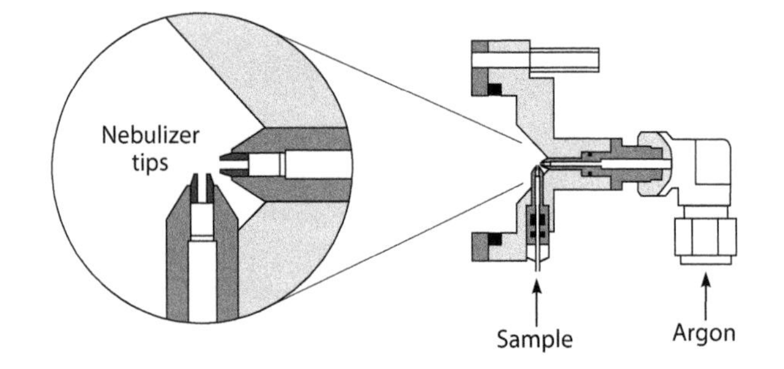

FIGURE 17.3 Schematic of a cross-flow nebulizer. (Copyright © 2013 PerkinElmer Inc., Waltham, MA. All rights reserved.)

concentric nebulizer, and therefore it produces droplets of less optimum size than that required for the ionization process. As a result, concentric nebulizers generally produce higher sensitivity and slightly better precision than the cross-flow design, but are more prone to clogging.

So, the choice of which nebulizer to use is usually based on the types of samples being aspirated and the data quality objectives of the analysis. However, whichever type is being used, attention should be paid to the tip of the nebulizer to ensure it is not getting blocked. Sometimes, microscopic particles can build up on the tip of the nebulizer without the operator noticing, which, over time, can cause a loss of sensitivity, imprecision, and poor long-term stability. In addition, O-rings and the sample capillary can be affected by the corrosive solutions being aspirated, which can also degrade performance. For these reasons, the nebulizer should always be a part of the regular maintenance schedule. Some of the most common things to do include the following:

- Visually check the nebulizer aerosol by aspirating water—a blocked nebulizer will usually result in an erratic spray pattern with lots of large droplets.
- Remove blockage by either using backpressure from the argon line or dissolving the material by immersing the nebulizer in an appropriate acid or solvent—an ultrasonic bath can sometimes be used to aid dissolution, but check with the manufacturer first in case it is not recommended. (Note: Never stick any wires down the end of the nebulizer, because it could do permanent damage.)
- Ensure that the nebulizer is securely seated in the spray chamber end cap.
- Check all O-rings for damage or wear.
- Ensure that the sample capillary is inserted correctly into the sample line of the nebulizer.
- Inspect the nebulizer every 1–2 weeks, depending on the workload.

The digital thermoelectric flowmeter described earlier is also very useful to diagnose problems with the nebulizer, even if you are using a self-aspirating nebulizer, because you are concerned about imprecision from the pulsing of a peristaltic pump. By placing the device in-line, you always know what your sample uptake is and can take immediate corrective action if there is any change. You can also record your sample flow in order to check that you are using the same flow from day to day.

If the flowmeter indicates a blocked nebulizer tip, there are also nebulizer-cleaning devices offered by most of the third-party consumables and accessories companies. Traditionally, if particulate matter from the sample lodged itself in the end of the nebulizer, cleaning wires or ultrasonic baths were the only way to remove the obstruction, which often resulted in permanent damage. These new cleaning devices are designed to efficiently deliver a pressurized cleanser through the nebulizer capillary to safely dislodge particle buildup and thoroughly clean the nebulizer, without fear of damage.

Spray Chamber

By far the most common design of spray chamber used in commercial ICP-MS instrumentation is the double-pass design, which selects the small droplets by directing the aerosol into a central tube. The larger droplets emerge from the tube and, by gravity, exit the spray chamber via a drain tube. The liquid in the drain tube is kept at positive pressure (usually by way of a loop), which forces the small droplets back between the outer wall and the central tube; they emerge from the spray chamber into the sample injector of the plasma torch. Scott double-pass spray chambers come in a variety of shapes, sizes, and materials, but are generally considered the most rugged design for routine use. Figure 17.4 shows a double-pass spray chamber (made of a polymer material) coupled to a cross-flow nebulizer.

The most important maintenance with regard to the spray chamber is to make sure that the drain is functioning properly. A malfunctioning or leaking drain can produce a change in the spray chamber backpressure, producing fluctuations in the analyte signal,

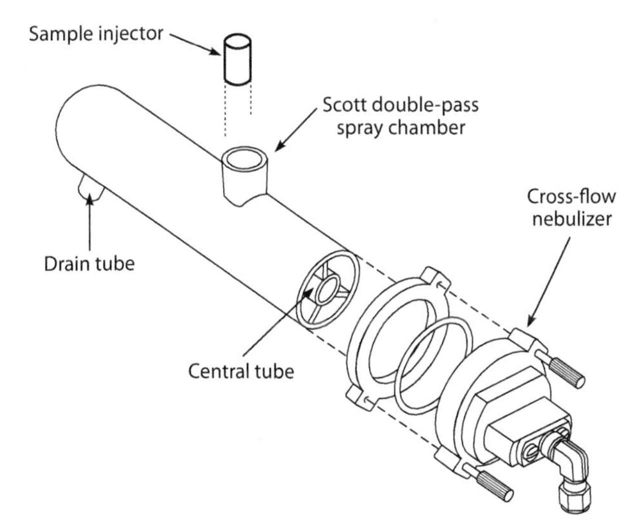

FIGURE 17.4 A double-pass spray chamber coupled to a cross-flow nebulizer. (Copyright © 2013 PerkinElmer Inc., Waltham, MA. All rights reserved.)

resulting in erratic and imprecise data. Less frequent problems can result from degradation of O-rings between the spray chamber and sample injector of the plasma torch. Typical maintenance procedures regarding the spray chamber include the following:

- Make sure the drain tube fits tightly and there are no leaks.
- Ensure that the waste solution is being pumped from the spray chamber into the drain properly.
- If a drain loop is being used, make sure the level of liquid in the drain tube is constant.
- Check the O-ring or ball joint between the spray chamber exit tube and torch sample injector—ensure the connection is snug.
- The spray chamber can be a source of contamination with some matrices or analytes, so flush thoroughly between samples.
- Empty the spray chamber of liquid when the instrument is not in use.
- The spray chamber and drain should be inspected every 1–2 weeks, depending on the workload.

Plasma Torch

Not only are the plasma torch and sample injector exposed to the sample matrix and solvent, but they also have to sustain the analytical plasma at approximately 10,000 K. This combination makes for a very hostile environment, and therefore is an area of the system that requires regular inspection and maintenance. A plasma torch positioned in the radio-frequency (RF) coil is shown in Figure 17.5.

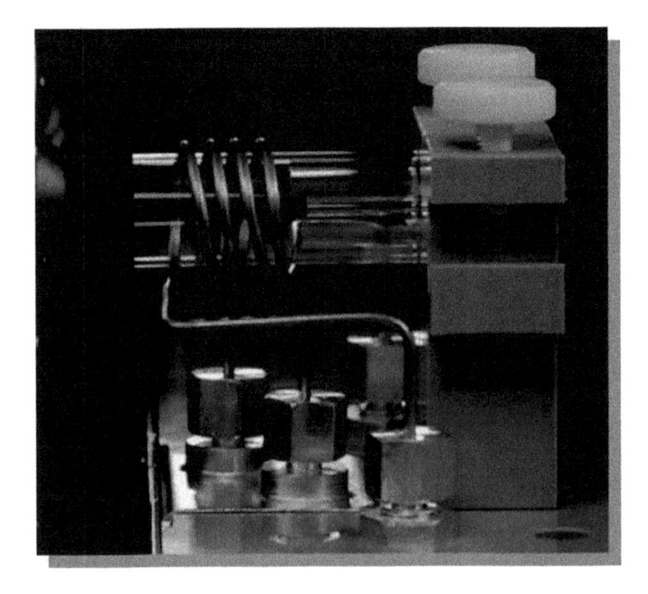

FIGURE 17.5 A plasma torch mounted in the torch box. (Courtesy of Analytic Jena, Jena, Germany.)

As a result, one of the main problems is staining and discoloration of the outer tube of the quartz torch because of heat and the corrosiveness of the liquid sample. If the problem is serious enough, it has the potential to cause electrical arcing. Another potential problem area is blockage of the sample injector due to matrix components in the sample. As the aerosol exits the sample injector, desolvation takes place, and the sample changes from small liquid droplets to minute solid particles prior to entering the base of the plasma. Unfortunately, with some sample matrices, these particles can deposit themselves on the tip of the sample injector over time, leading to possible clogging and drift. In fact, this can be a potentially serious problem when aspirating organic solvents, because carbon deposits can rapidly build up on the sample injector and cones unless a small amount of oxygen is added to the nebulizer gas flow. Some torches also use metal plates or shields to reduce the secondary discharge between the plasma and the interface. These are consumable items, because of the intense heat and the effect of the RF field on the shield. A shield in poor condition can affect instrument performance, so the user should always be aware of this and replace it when necessary.

Some useful maintenance tips with regard to the torch area include the following:

- Look for discoloration or deposits on the outer tube of the quartz torch. Remove material by soaking the torch in appropriate acid or solvent if required.
- Check the torch for thermal deformation. A nonconcentric torch can cause loss of signal.
- Check the sample injector for blockages. If the injector is demountable, remove the material by immersing it in an appropriate acid or solvent if required (if the torch is one piece, soak the entire torch in the acid).
- Ensure that the torch is positioned in the center of the load coil and at the correct distance from the interface cone when replacing the torch assembly.
- If the coil has been removed for any reason, make sure the gap between the turns is correct as per recommendations in the operator's manual.
- Inspect any O-rings or ball joints for wear or corrosion. Replace if necessary.
- If a shield or plate is used to ground the coil, ensure that it is always in good condition; otherwise, replace when necessary.
- The torch should be inspected every 1–2 weeks, depending on the workload.

INTERFACE REGION

As the name suggests, the interface is the region of the ICP mass spectrometer where the plasma discharge at atmospheric pressure is "coupled" to the mass spectrometer at 10^{-6} torr by way of two interface cones—a sampler and skimmer. This coupling of a high-temperature ionization source, such as an ICP, to the metallic interface of the mass spectrometer imposes demands on this region of the instrument that are unique to this AS technique. When this is combined with matrix, solvent, and analyte ions, together with particulates and neutral species being directed at high velocity at the interface cones, an extremely harsh environment is the result. The most common types of problems associated with the interface are blocking or corrosion of the

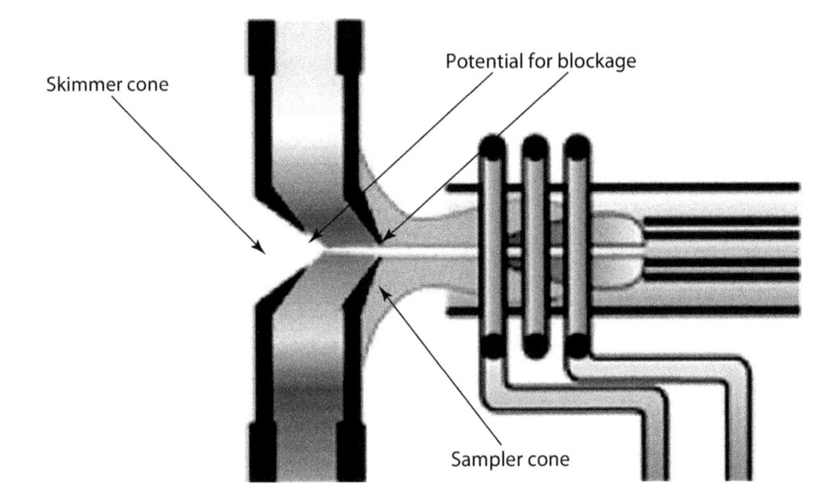

FIGURE 17.6 A schematic of the interface cones showing potential areas of blockage.

sampler cone and, to a lesser extent, the skimmer cone. A schematic of the interface cones showing potential areas of blockage is shown in Figure 17.6.

A blockage is not always obvious, because often the buildup of material on the cone or corrosion around the orifice can take a long time to reveal itself. For that reason, the sampler and skimmer interface cones have to be inspected and cleaned on a regular basis. The frequency will often depend on the types of samples being analyzed and also the design of the ICP mass spectrometer. For example, it is well documented that a secondary discharge at the interface can prematurely discolor and degrade the sampler cone, especially when complex matrices are being analyzed or if the instrument is being used for high sample throughput.

Besides the cones, the metal interface housing itself is also exposed to the high-temperature plasma. Therefore, it needs to be cooled by a recirculating water system, usually containing some kind of antifreeze or corrosion inhibitor, or by a continuous supply of mains water. Recirculating systems are probably more widely used because the temperature of the interface can be controlled much better. There is no real routine maintenance involved with the interface housing, except maybe to check the quality of the coolant from time to time, to make sure there is no corrosion of the interface cooling system. If for any reason the interface gets too hot, there are usually built-in safety interlocks that will turn the plasma off. Some useful hints to prolong the lifetime of the interface and cones include the following:

- Check that both the sampler and skimmer cone are clean and free of sample deposits. The typical frequency is weekly, but will depend on sample type and workload.
- If necessary, remove and clean cones using the manufacturer's recommendations. Typical approaches include immersion in a beaker of weak acid or detergent placed in a hot water or ultrasonic bath. Abrasion with fine wire wool or a coarse polishing compound has also been used.

- Never stick any wire into the orifice; it could do permanent damage.
- Nickel cones will degrade rapidly with harsh sample matrices. Use platinum cones for highly corrosive solutions and organic solvents.
- Periodically check the cone orifice diameter and shape with a magnifying glass (10× to 20× magnification). An irregular-shaped orifice will affect instrument performance.
- Thoroughly dry cones before installing them back into the instrument because water or solvent could be pulled back into the mass spectrometer.
- Check coolant in the recirculating system for signs of interface corrosion, such as copper or aluminum salts (or predominant metal of interface).

ION OPTICS

The ion optic system is usually positioned just behind or close to the skimmer cone to take advantage of the maximum number of ions entering the mass spectrometer. There are many different commercial designs and layouts, but they all have one attribute in common, and that is to transport the maximum number of analyte ions while allowing the minimum number of matrix ions through to the mass analyzer.

The ion-focusing system is not traditionally thought of as a component that needs frequent inspection, but because of its proximity to the interface region, it can accumulate minute particulates and neutral species that over time can dislodge, find their way into the mass analyzer, and affect instrument performance. Signs of a dirty or contaminated ion optic system are poor stability or a need to gradually increase lens voltages over time. For that reason, no matter what design of ion optics is used, inspection and cleaning every 3–6 months (depending on workload and sample type) should be an integral part of a preventative maintenance plan. Some useful maintenance tips for the ion optics to ensure maximum ion transmission and good stability include the following:

- Look for sensitivity loss over time, especially in complex matrices.
- If sensitivity is still low after cleaning the sample introduction system, torch, and interface cones, it could indicate that the ion lens system is becoming dirty.
- Try retuning or reoptimizing the lens voltages.
- If voltages are significantly different (usually higher than previous settings), it probably means lens components are getting dirty.
- When the lens voltages become unacceptably high, the ion lens system will probably need replacing or cleaning. Use recommended procedures outlined in the operator's manual.
- Depending on the design of the ion optics, some single-lens systems are considered consumables and are discarded after a period of time, whereas multicomponent lens systems are usually cleaned using abrasive papers or polishing compounds, and rinsed with water and an organic solvent.
- If cleaning ion optics, make sure that they are thoroughly dry because water or solvent could be sucked back into the mass spectrometer.

- Gloves are usually recommended when reinstalling an ion optic system because of the possibility of contamination.
- Do not forget to inspect or replace O-rings or seals when replacing ion optics.
- Depending on instrument workload, you should expect to see some deterioration in the performance of the ion lens system after 3–4 months of use. This is a good approximation of when it should be inspected and cleaned or replaced if necessary.
- With some instruments, you will need to break the vacuum to get to the ion optic region. Even though vacuum can be reestablished very quickly, this should be a consideration when carrying out your own ion lens cleaning procedures.

ROUGHING PUMPS

Typically, two roughing pumps are used in commercial instruments. One pump is used on the interface region, and the other is used as a backup to the turbomolecular pumps on the main vacuum chamber. They are usually oil-based rotary or diffusion pumps, where the oil needs to be changed on a regular basis, depending on the instrument usage. The oil in the interface pump will need changing more often than the oil in the pump on the main vacuum chamber because it is pumping for a longer period. A good indication of when the oil needs to be changed is the color in the "viewing glass." If it appears dark brown, there is a good chance that heat has degraded its lubricating properties, and it needs to be changed. With the roughing pump on the interface, the oil should be changed every 1–2 months, and with the main vacuum chamber pump, it should be changed every 3–6 months. These times are only approximations and will vary depending on the sample workload and the time the instrument is actually running. Some important tips when changing the roughing pump oil include

- Do not forget to turn the instrument and the vacuum off. If the oil is being changed from "cold," it might be useful to run the instrument for 10–15 min beforehand to get the oil to flow better.
- Drain the oil into a suitable vessel; be cautious, as the oil might be very hot if the instrument has been running all day.
- Fill the oil to the required level in the viewing glass.
- Check for any loose hose connections.
- Replace the oil filter if necessary.
- Turn the instrument back on. Check for any oil leaks around the filling cap, and tighten if necessary.

AIR FILTERS

Most of the electronic components, especially the ones in the RF generator, are air-cooled. Therefore, the air filters should be checked, cleaned, or replaced on a fairly regular basis. Although this is not carried out as routinely as the sample introduction system, a typical time frame to inspect the air filters is every 3–6 months, depending on the workload and instrument usage.

OTHER COMPONENTS TO BE PERIODICALLY CHECKED

It is also important to emphasize that other components of the ICP mass spectrometer have a finite lifetime, and will need to be replaced or at least inspected from time to time. These components are not considered a part of the routine maintenance schedule, and usually require a service engineer (or at least an experienced user) to clean or change them. The areas to be cleaned are described in the following text.

DETECTOR

Depending on the usage and levels of ion signals measured on a routine basis, the electron multiplier should last about 12 months. A sign of a failing detector is a rapid decrease in the "gain" setting despite attempts to increase the detector voltage. The lifetime of a detector can be increased by avoiding measurements at masses that produce extremely high ion signals, such as those associated with the argon gas, solvent or acid used to dissolve the sample (e.g., hydrogen, oxygen, and nitrogen), or any mass associated with the matrix itself. It is important to emphasize that the detector should be replaced by an experienced person wearing gloves, to reduce the possibility of contamination from grease or organic or water vapor from the operator's hands. It is advisable that a spare detector be purchased with the instrument.

TURBOMOLECULAR PUMPS

Most of the instruments running today use two turbomolecular pumps to create the operating vacuum for the main mass analyzer or detector chamber and the ion optic region. However, some of the newer instruments use a single, twin-throated turbo pump. The lifetime of turbo pumps, in general, is dependent on a number of factors, including the pumping capacity of the pump (usually expressed as L/s), the size (or volume) of the vacuum chamber to be pumped, the orifice diameter of the interface cones (in mm), and the time the instrument is running. Although some instruments still use the same turbo pumps after 5–10 years of operation, the normal lifetime of a pump in an instrument that has a reasonably high sample workload is on the order of 3–4 years. This is an approximation and will obviously vary depending on the make and design of the pump (especially the type of bearings used). As the turbomolecular pump is one of the most expensive components of an ICP-MS system, this should be factored into the overall running costs of the instrument over its operating lifetime.

It is worth pointing out that although the turbo pump is not generally included in routine maintenance, most instruments use a Penning (or similar) gauge to monitor the vacuum in the main chamber. Unfortunately, this gauge can become dirty over time and lose its ability to measure the correct pressure. The frequency of this is almost impossible to predict but is closely related to the types and numbers of samples analyzed. A sudden drop in pressure or fluctuations in the signal are two of the most common indications of a dirty Penning gauge. When this happens, the gauge must be removed and cleaned. This should be performed by an experienced operator or service engineer because removing the gauge, cleaning it, maintaining the correct electrode geometry, and reinstalling it correctly into the instrument is a fairly complicated procedure. It is further complicated by the fact that a Penning gauge is operated at high voltage.

Mass Analyzer and Collision/Reaction Cell

Under normal circumstances, there is no need for the operator to be concerned about the routine maintenance of the mass analyzer or collision/reaction cell. With modern turbomolecular pumping systems, it is highly unlikely that there will be any pump contamination problems associated with the quadrupole, magnetic sector, or time-of-flight mass analyzer. And very few sample matrix components ever make it into the mass spectrometer region, which dramatically reduces the frequency of routine maintenance tasks. This certainly was not the case with some of the early instruments that used oil-based diffusion pumps, because many researchers found that the quadrupole and prefilters were contaminated by oil vapors from the pumps. Today, it is fairly common for turbomolecular-based mass analyzers to require no maintenance of the analyzer or the collision/reaction cell quadrupole rods over the lifetime of the instrument, other than an inspection carried out by a service engineer on an annual basis. However, in extreme cases, particularly with older instruments, removal and cleaning of the quadrupole assembly might be required to get acceptable peak resolution and abundance sensitivity performance.

FINAL THOUGHTS

The overriding message I would like to leave you with on this subject is that routine maintenance cannot be overemphasized in ICP-MS. Even though it might be considered a mundane and time-consuming chore, it can have a significant impact on the uptime of your instrument. Read the routine maintenance section of the operator's manual and understand what is required. It is essential that time be scheduled on a weekly, monthly, and quarterly basis for preventative maintenance on your instrument. In addition, you should budget for an annual preventative maintenance contract under which the service engineer checks out all the important instrumental components and systems on a regular basis to make sure they are all working correctly. This might not be as critical if you work in an academic environment, where the instrument might be down for extended periods, but in my opinion, it is absolutely critical if you work in a commercial laboratory, which is using the instrument to generate revenue. There is no question that spending the time to keep your ICP mass spectrometer in good working order can mean the difference between owning an instrument whose performance could be slowly degrading without your knowledge or one that is always working in "peak" condition.

However, I must give credit to the instrument designers and accessory suppliers in making today's instrumentation extremely easy to maintain and keep clean. The technique will be 35 years old in 2018, and its commercial success has been built on acceptance by the analytical community for applying the technique to truly routine, real-world analysis. The only areas that need cleaning on a regular basis are the sample introduction system and interface cones, depending on usage and the sample matrices being aspirated. And many of today's instruments have alarms that can be set to remind the operator when it's time for the few preventative maintenance tasks that are required, such as oil changes and tubing replacement. Some systems will even display how many hours various components have been used and when they might need attention. So even though routine maintenance is very important, instrument vendors have put a great deal of time and effort into making this as straightforward and seamless as possible.[1] Refer to the websites of all the instrument vendors and sample introduction and accessories companies in Chapter 29 for troubleshooting hints and tips.

REFERENCE

1. R, Brennan, J. Dulude, R. Thomas (2015). Approaches to Maximize Performance and Reduce the Frequency of Routine Maintenance in ICP-MS. Spectroscopy, 30,(10),12-25.

18 Collecting and Preparing the Sample for Analysis

Collecting and preparing a sample for analysis is extremely important in inductively coupled plasma mass spectrometry (ICP-MS), particularly with regard to possible contamination issues. If you have been using flame atomic absorption (AA) or inductively coupled plasma optical emission spectrometry (ICP-OES), you will probably have to rethink your sample preparation procedures for ICP-MS. Although sample digestion and dissolution procedures for drug materials are covered later in the book, Chapter 18 gives a general overview of sample collection and preparation and the major causes of contamination and analyte loss in ICP-MS, and how they affect both the analysis and the method development process. Chapter 19 goes into greater detail about sample digestion techniques for pharmaceutical-type samples.

There are many factors that influence the ability to get the correct result with any trace element technique. Unfortunately, with ICP-MS, the problem is magnified even more because of its extremely high sensitivity. So, in order to ensure that the data reported is an accurate reflection of the sample in its natural state, the analyst must be aware of not only all the potential sources of contamination, but also the many reasons why analyte loss is a problem in ICP-MS. Figure 18.1 shows the major factors than can impact the analytical result in ICP-MS.

COLLECTING THE SAMPLE

Collecting the sample and maintaining its integrity is a science of its own and beyond the scope of this book. However, it is worth discussing briefly, in order to understand its importance in the overall scheme of collecting, preparing, and analyzing a pharmaceutical-type sample. The object of sampling is to collect a portion of the material that is small enough in size to be conveniently transported and handled, and at the same time accurately represents the bulk material being sampled. Depending on the sampling requirements and the type of sample, there are basically three main types of sampling procedures:

- *Random sampling* is the most basic type of sampling and only represents the composition of the bulk material at the time and place it was sampled. If the composition of the material is known to vary with time, individual samples collected at suitable intervals and analyzed separately can reflect the extent, frequency, and duration of these variations.
- *Composite sampling* is when a number of samples are collected at the same point, but at different times and mixed together before being analyzed.
- *Integrated sampling* is achieved by mixing together a number of samples that have been collected simultaneously from different points.

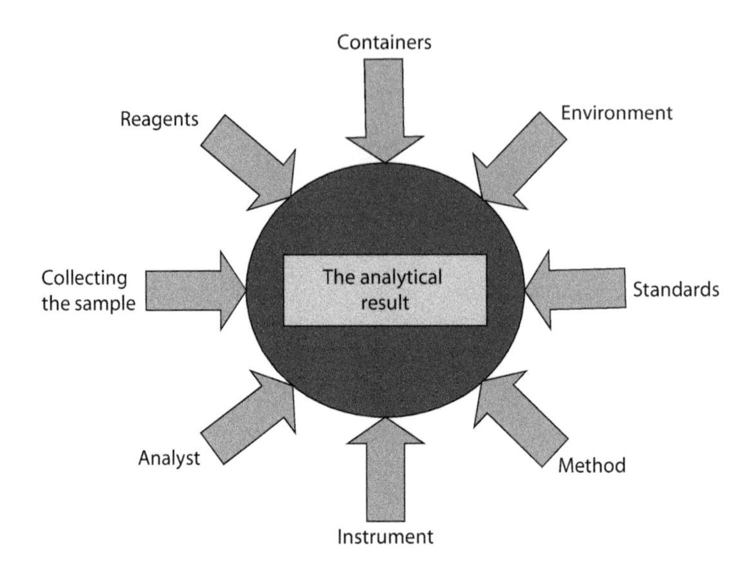

FIGURE 18.1 Major factors that can influence the analytical result in ICP-MS.

We will not go into which type of sampling is the most effective, but it must be emphasized that unless the correct sampling or subsampling procedure is used, the analytical data generated by the ICP-MS instrumentation may be seriously flawed because it may not represent the original bulk material. If the sample is a liquid, it is also important to collect the sample in clean containers (as shown later) that have been thoroughly washed. In addition, if the sample is to be kept for a long period of time before analysis, it is essential that the analytes stay in solution in a preservative such as a dilute acid (this will also help stop the analytes from being absorbed into the walls of the container). It is also important to keep the samples as cool as possible to avoid losses through evaporation. Kratochvil and Taylor give an excellent review of the importance of sampling for chemical analysis.[1]

PREPARING THE SAMPLE

As mentioned previously, ICP-MS was originally developed for the analysis of liquid samples, even though it can be either modified or coupled to another piece of equipment to analyze solid materials. Thus, if the sample is not in a liquid form, some kind of sample preparation has to be carried out in order to make it so. There is no question that collecting a solid sample, preparing it, and getting it into solution probably represent the most crucial steps in the overall ICP-MS analytical methodology because of the potential sources of contamination from grinding, sieving, weighing, dissolving, and diluting the sample. Let us take a look at these steps in greater detail and, in particular, focus on their importance when being used for ICP-MS. This will be expanded upon in the next chapter on sample digestion procedures for pharmaceutical materials.

GRINDING THE SAMPLE

Some fine-powder solid samples are ready to be dissolved without grinding; for them, mere passage through a fine-mesh sieve (mesh is typically 0.1–0.2 mm^2) is enough. Other types of coarser solid samples, such as soils, need to be passed through a coarse-mesh sieve (typically 2 mm^2 mesh) to make it ready for dissolution.[2] However, if the solid sample is not in a convenient form to be dissolved, it has to be ground to a smaller particle size, mainly to improve the homogeneity of the original sample taken and make it more representative when taking a subsample. The ideal particle size will vary depending on the sample, but the sample is typically ground to pass through a fine-mesh sieve (0.1 mm^2 mesh). This uniform particle size ensures that the particles in the test portion are the same size as the particles in the rest of the ground sample. Another reason for grinding the sample into small uniform particles is that they are easier to dissolve.

The process of grinding a sample with a mortar and pestle or ball mill and passing it through a metallic sieve can be a major cause of contamination. This can occur from the remains of a previous sample that had been prepared earlier or from materials used in the manufacture of the grinding or sieving equipment. For example, if tungsten carbide equipment is used to grind the sample, major elements, like tungsten and carbon, as well as additive elements, like cobalt and titanium, can also be a problem. Additionally, sieves, which are made from stainless steel, bronze, or nickel, can also introduce metallic contamination into the sample. In order to minimize some of these problems, plastic sieves are often used. However, still remaining is the problem of contamination from the grinding equipment. For this reason, it is usual to discard the first portion of the sample or even to use different grinding and sieving equipment for different kinds of samples.

CRYOGENIC GRINDING

It should be noted that over the past few years, a new breed of sample grinding techniques have become available. These grinders are based on the principle of cryogenically freezing the sample in liquid nitrogen. At this kind of temperature (–196°C), many materials, including pharmaceutical materials and products, become very brittle, and a result can be ground to a fine powder ready for a dissolution procedure. The technology incorporates an insulated tub into which liquid nitrogen is poured. The grinding mechanism is a magnetic coil assembly suspended in the liquid nitrogen bath. Cooling materials to temperatures approaching –200°C makes samples extremely brittle, so they can be pulverized quickly by impact milling. This ability allows difficult-to-process samples (bone, rocks, polymers, metals, food, and pharmaceutical capsules) to be more efficiently processed before analysis. The sample is placed in a closed grinding vial and thoroughly cooled before grinding by the magnetic coil shuttling the impactor rapidly back and forth, pulverizing the sample against the end plugs of the vial, as shown in Figure 18.2. The type of vial and impactor for grinding samples is selected to reduce any potential cross-contamination of metals. For example, for pharmaceutical samples, it might be better to use polymer vials and polymer-encased impactors instead of metal ones to reduce the possibility of metal contamination.

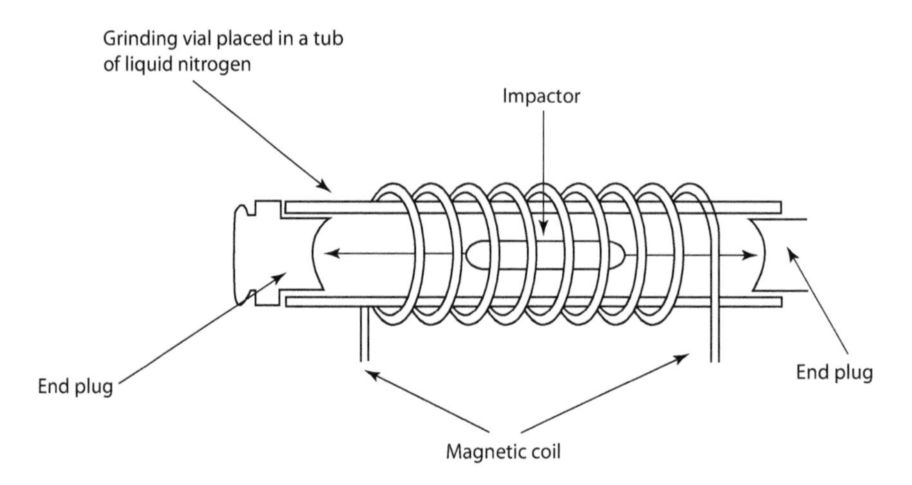

FIGURE 18.2 Principles of a cryogenic freezer mill. (Courtesy of SPEX SamplePrep, Metuchen, NJ.)

SAMPLE DISSOLUTION METHODS

Unfortunately, there is no single dissolution procedure that can be used for all types of solid samples. There are many different approaches to getting solid samples into solution. For some samples, this is fairly straightforward and fast, whereas for others it can be very complex and time-consuming. However, all the successful sample dissolution procedures used in ICP-MS typically have a number of things in common:

- Complete dissolution is a usual requirement.
- Ultrapure reagents should be used.
- The reagents should not contaminate or interfere with the analysis.
- Equipment should show no chemical attack or corrosion.
- There should be no loss of analyte.
- In speciation studies, the integrity of elemental form, valency state, and species should be maintained.
- Ideally, dissolution should be fast.
- Safety is paramount.

Even though the contamination issues are exaggerated with ICP-MS, the most common approaches to getting samples into solution are very similar to the ones used for other trace element techniques. The most common dissolution techniques include the following:

- Hot plates, pressure bombs,[3] or microwave digestion[4] using concentrated acids and oxidizing agents—such as nitric acid, perchloric acid, hydrofluoric acid, aqua regia, hydrogen peroxide, or various mixtures of these—are among the most common approaches to dissolution and are typically used for metals, soils or sediments,[5] minerals,[6] and biological samples.[7]

- Dissolution with strong bases such as caustic or trimethyl ammonium hydroxide (TMAH)—typically used for biological samples.[8]
- Heating with fusion mixtures or fluxes such as lithium metaborate, sodium carbonate, or sodium peroxide in a metal crucible (e.g., platinum, silver, or nickel) and redissolving in a dilute mineral acid—typically used for ceramics, stubborn minerals, ores, rocks, and slags.[9,10]
- Dry ashing using a flame, heat lamp, or a heated muffle furnace and redissolving the residue in a dilute mineral acid—typically used for organic or biological matrices.[11]
- Wet ashing using concentrated acids (usually with some kind of heat)—typically used for organic, petrochemical, or biomedical samples.[12]
- Dissolution with organic solvents—typically used for organic or oil-type samples.[13]

The choice of which dissolution technique to use is often very complicated and depends on criteria such as the size of the sample, the matrix components in the sample, the elements to be analyzed, the concentration of elements being determined, the types of interferences anticipated, the type of ICP-MS equipment being used, the time available for analysis, safety concerns, and the expertise of the analyst. However, with ICP-MS, contamination issues are probably the greatest concern. For that reason, the most common approach to sample preparation is to keep the process as simple as possible, because the more steps that are involved, the more chance there is of contaminating the sample. This means that, ideally, if the sample is already a liquid, a simple acidification might be all that is needed. If the sample is a solid, a straightforward acid dissolution is preferred over the more complex and time-consuming fusion and ashing procedures. An excellent handbook of decomposition methods used for analytical chemistry was written by Bock in 1979.[14]

It is also important to emphasize that many acids that are used for AA and ICP-OES are not ideal for ICP-MS because of the polyatomic spectral interferences they produce. Although this is not strictly a contamination problem, it can significantly impact your data if not taken into consideration. For example, if vanadium or arsenic is being determined, it is advisable not to use hydrochloric acid (HCl) or perchloric acid ($HClO_4$) because they generate polyatomic ions such as $^{35}Cl^{16}O^+$ and $^{40}Ar^{35}Cl^+$, which interfere with the isotopes $^{51}V^+$ and $^{75}As^+$, respectively. Sulfuric acid (H_2SO_4) and phosphoric acid (H_3PO_4) are also acids that should be avoided if possible because they generate sulfur- and phosphorus-based polyatomic ions. Therefore, if there is a choice of which acid to use for dissolution, nitric acid (HNO_3) is the preferred one to use. Even though it can generate interferences of its own, they are generally less severe than those of the other acids.[15] Table 18.1 shows the kinds of polyatomic spectral interferences generated by the most common mineral acids and dissolution chemicals.

In addition, fusion mixtures present unique problems for ICP-MS—not only because the major elements form polyatomic spectral interferences with the argon gas, but also because the high levels of dissolved solids in the sample can cause blockage of the interface cones, which over time can lead to signal drift. An additional problem with a fusion procedure is the risk of losing volatile analytes due to the high temperature of the muffle furnace or flame used to heat the crucible.

TABLE 18.1
Typical Polyatomic Spectral Interferences Generated by Common Mineral Acids and Dissolution Chemicals

Acid–Solvent–Fusion Mixture	Interference	Element/Isotope
HCl	$^{35}Cl^{16}O^+$	$^{51}V^+$
HCl	$^{40}Ar^{35}Cl^+$	$^{75}As^+$
HNO_3	$^{14}N^{14}N^+$	$^{28}Si^+$
HNO_3	$^{14}N^{14}N^{16}O^+$	$^{44}Ca^+$
HNO_3	$^{40}Ar^{15}N^+$	$^{55}Mn^+$
H_2SO_4	$^{32}S^{16}O^+$	$^{48}Ti^+$
H_2SO_4	$^{34}S^{18}O^+$	$^{52}Cr^+$
H_2SO_4	$^{32}S^{16}O^{16}O^+$	$^{64}Zn^+$
H_3PO_4	$^{31}P^{16}O^{16}O^+$	$^{63}Cu^+$
Any organic solvent	$^{12}C^{12}C^+$	$^{24}Mg^+$
Any organic solvent	$^{40}Ar^{12}C^+$	$^{52}Cr^+$
Lithium-based fusion mixtures	$^{40}Ar^{7}Li^+$	$^{47}Ti^+$
Boron-based fusion mixtures	$^{40}Ar^{11}B^+$	$^{51}V^+$
Sodium-based fusion mixtures	$^{40}Ar^{23}Na^+$	$^{63}Cu^+$

CHOICE OF REAGENTS AND STANDARDS

Careful consideration must be given to the choice and purity of reagents, especially if sub-parts-per-trillion concentration levels are expected. General laboratory- or reagent-grade chemicals used for AA or ICP-OES sample preparation are not usually pure enough. For that reason, most manufacturers of laboratory chemicals now offer ultra-high-purity grades of chemicals, acids, and fusion mixtures specifically for use with ICP-MS. It is therefore absolutely essential that the highest-grade chemicals and water be used in the preparation and dilution of the sample. In fact, the grade of deionized water used for dilution and the cleaning of vessels and containers are very important in ICP-MS.

Less pure water, such as single-distilled or deionized water, is fine for flame AA or ICP-OES, but is probably not suitable for use with ICP-MS because it could possibly contain contaminants, such as dissolved inorganic or organic matter, suspended dust or scale particles, and microorganisms. All these contaminants can affect reagent blank levels and negatively impact instrument and method detection limits. This necessitates the use of the most chemically pure water for ICP-MS work. There are a number of water purification systems on the market, which use combinations of filters, ion exchange cartridges, and/or reverse osmosis systems to remove the particulates, organic matter, and trace metal contaminants. These ultra-high-purity water systems (similar to the ones used for semiconductor processing) typically produce water with a resistance of better than 18 mΩ.[16]

Another area of concern with regard to contamination is in the selection of calibration standards. Because ICP-MS is a technique capable of quantifying up to 75 different elements, it will be detrimental to the analysis to use calibration standards that are developed for a single-element technique, such as AA. These single-element standards are usually certified only for the analyte element and not for any others, although they are often quoted on the certificate. It is therefore absolutely critical to use calibration standards that have been specifically made for a multielement technique such as ICP-MS. It does not matter whether they are single or multielement standards, as long as the certificate contains information on the suite of analyte elements you are interested in, as well as any other potential interferents.

It is also desirable that the certified values be confirmed by both a classical wet technique and an instrumental technique, all of which are traceable to National Institute of Standards and Technology (NIST) reference materials. It is also important to fully understand the uncertainty or error associated with a certified value, so that you know how it impacts the data you report.[17] Figure 18.3 is a certificate for a 1000 mg/L erbium certified reference standard used in ICP-MS, showing values for more than 30 trace metal contaminants. Make sure that all your calibration standards come with similar certification, so you have the confidence that your reported data can be scrutinized to the highest standards.

The same case applies if a calibration standard is being made from a high-purity salt of the metal. The salt has to be certified for not only the element of interest, but also for the full suite of analyte elements, as well as other elements that could be potential interferents. It is also important to understand the shelf life of these standards and chemicals and how long-term storage affects the concentration of the analyte elements, especially at low levels.

VESSELS, CONTAINERS, AND SAMPLE PREPARATION EQUIPMENT

The containers used for preparation, dilution, storage, and introduction of the sample can have a huge impact on your data in ICP-MS. Traditional glassware, such as beakers, volumetric flasks, and autosampler tubes, which are fine for AA and ICP-OES work, are not ideally suited to ICP-MS. The major problem is potential contamination from the major elemental components of the glassware. For example, glass made from soda lime contains percentage concentrations of silicon, sodium, calcium, magnesium, and aluminum; also, borosilicate glass contains high levels of boron. Besides these major elements, glass might also contain minor concentrations of Zr, Li, Ba, Fe, K, and Mn. Unfortunately, if the sample solution is highly acidic, there is a strong possibility that these elements can be leached out of the glassware.

In addition to the contamination issues, analytes can be absorbed into the walls of volumetric flasks and beakers made of glass. This can be a serious problem if the sample or standard is to be stored for extended periods of time, especially if the analyte concentrations are extremely low. If using glassware is unavoidable, it is a good idea to clean the glassware on a regular basis using chromic acid or some kind of commercial glass detergent, such as Decon™ or Citranox™. If long-term storage is a necessity, either avoid using glassware or minimize the analyte loss by keeping the solutions acidified (~pH 2), so there is very little chance of absorption into the walls of the glassware.[18]

Glassware is such a universal material used for sample preparation that it is very difficult to completely avoid it. However, serious consideration should be given to looking for alternative materials in as many of the ICP-MS sample preparation steps

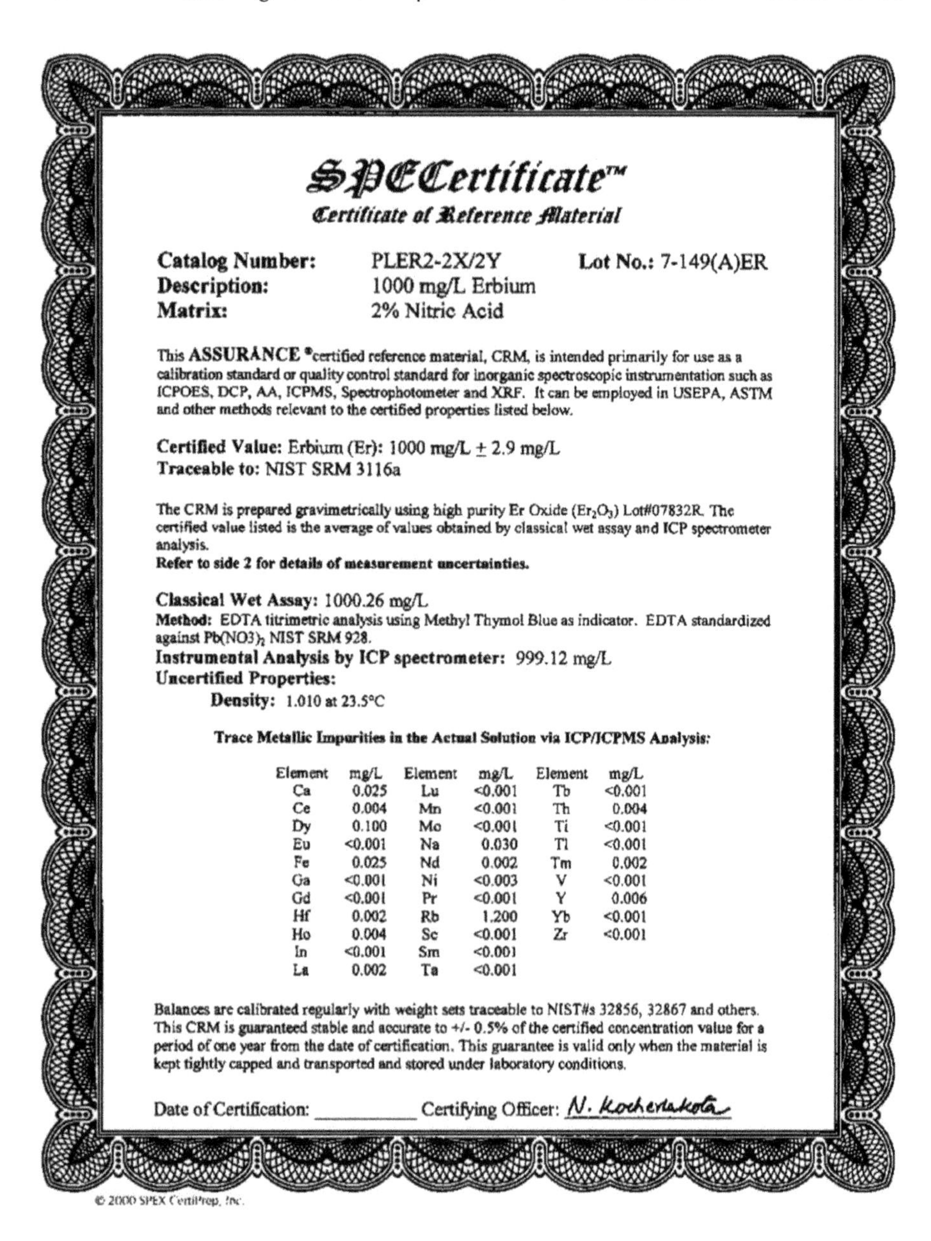

SPECertificate™

Certificate of Reference Material

Catalog Number: PLER2-2X/2Y **Lot No.:** 7-149(A)ER
Description: 1000 mg/L Erbium
Matrix: 2% Nitric Acid

This **ASSURANCE** ®certified reference material, CRM, is intended primarily for use as a calibration standard or quality control standard for inorganic spectroscopic instrumentation such as ICPOES, DCP, AA, ICPMS, Spectrophotometer and XRF. It can be employed in USEPA, ASTM and other methods relevant to the certified properties listed below.

Certified Value: Erbium (Er): 1000 mg/L ± 2.9 mg/L
Traceable to: NIST SRM 3116a

The CRM is prepared gravimetrically using high purity Er Oxide (Er_2O_3) Lot#07832R. The certified value listed is the average of values obtained by classical wet assay and ICP spectrometer analysis.
Refer to side 2 for details of measurement uncertainties.

Classical Wet Assay: 1000.26 mg/L
Method: EDTA titrimetric analysis using Methyl Thymol Blue as indicator. EDTA standardized against $Pb(NO3)_2$ NIST SRM 928.
Instrumental Analysis by ICP spectrometer: 999.12 mg/L
Uncertified Properties:
Density: 1.010 at 23.5°C

Trace Metallic Impurities in the Actual Solution via ICP/ICPMS Analysis:

Element	mg/L	Element	mg/L	Element	mg/L
Ca	0.025	Lu	<0.001	Tb	<0.001
Ce	0.004	Mn	<0.001	Th	0.004
Dy	0.100	Mo	<0.001	Ti	<0.001
Eu	<0.001	Na	0.030	Tl	<0.001
Fe	0.025	Nd	0.002	Tm	0.002
Ga	<0.001	Ni	<0.003	V	<0.001
Gd	<0.001	Pr	<0.001	Y	0.006
Hf	0.002	Rb	1.200	Yb	<0.001
Ho	0.004	Sc	<0.001	Zr	<0.001
In	<0.001	Sm	<0.001		
La	0.002	Ta	<0.001		

Balances are calibrated regularly with weight sets traceable to NIST#s 32856, 32867 and others. This CRM is guaranteed stable and accurate to +/- 0.5% of the certified concentration value for a period of one year from the date of certification. This guarantee is valid only when the material is kept tightly capped and transported and stored under laboratory conditions.

Date of Certification: __________ Certifying Officer: N. Kocherlakota

FIGURE 18.3: Certificate for a 1000 mg/L erbium certified reference standard used in ICP-MS, showing values for more than 30 trace metal contaminants. (Courtesy of SPEX CertiPrep, Metuchen, NJ.)

as possible. Today, the most common materials used to manufacture beakers, volumetric containers, and autosampler tubes for ultratrace element techniques, such as GFAA and ICP-MS, are mainly plastic based. Over the past 15–20 years, the demand for these kinds of materials has increased significantly because of the contamination issues associated with glassware.

Some plastics are more inert and more pure than others, so thought should be given as to which one is optimal for your samples. Selection should be made based on the suite of elements being analyzed, analyte concentration levels, matrix components, or whether it is an aqueous-, acid-, or organic-based solution. Some of the most common plastic materials used in the manufacture of sample preparation vessels and sample introduction components include polypropylene (PP), polyethylene (PE), polysulfide (PS), polycarbonate (PC), polyvinylchloride (PVC), polyvinylfluoride (PVF), perfluoroalkoxy (PFA), and polytetrafluoroethylene (PTFE).

It is generally felt that PTFE and PFA probably represent the cleanest materials, and even though they are the most expensive, they are considered the most suitable for ultratrace ICP-MS work. However, even though these types of plastics are generally much cleaner than glass, they still contain some trace elements. For example, certain plastics might contain phosphorus from the mold-releasing agent, and some plastic tube caps and covers are manufactured with barium compounds to enhance their color. These are all potential sources of contamination that can cause serious problems in ICP-MS, especially if heat is involved in sample preparation. This is particularly true if microwave dissolution is used to prepare the sample because of the potential for high-temperature breakdown of the polymer material over time. Table 18.2, which was taken from a publication from more than 30 years ago, gives trace element contamination levels of some common plastics used in the manufacture of laboratory beakers, volumetric ware, and autosampler tubes.[19] It should be strongly emphasized that this data might not be representative of current-day products, but is an approximation for comparison purposes.

Even though microwave dissolution is rapidly becoming the sample dissolution method of choice over conventional hot-plate digestion methods, it will not be discussed in great detail in this chapter. There are a multitude of books and reference papers in the public domain covering just about every type of sample being analyzed by ICP-MS, including geological materials,[20] soils,[21] sediments,[22] waters,[23] biological materials,[24] and foodstuffs.[25] In addition, Chapter 19 goes into great depth about the use of microwave digestion procedures for pharmaceutical matrices.

TABLE 18.2
Typical Trace Element Contamination Levels of Some Common Plastic Materials Used in the Manufacture of Laboratory Beakers, Volumetric Ware, and Autosampler Tubes

Material	Na (ppm)	Al (ppm)	K (ppm)	Sb (ppm)	Zn (ppm)
Chlorinated Polyethylene (CPE)	1.3	0.5	5	0.005	—
Linear Polyethylene (LPE)	15	30	0.6	0.2	520
Polypropylene (PP)	4.8	55	—	0.6	—
Polysulfide (PS)	2.2	0.5	—	—	—
Polycarbonate (PC)	2.7	3.0	—	—	—
Polyvinylchloride (PVC)	20	—	—	—	—
Polytetrafluoroethylene (PTFE)	0.16	0.23	90	—	—

Source: Moody, J. R., and Lindstrom, R. M., *Anal. Chem.*, 49(14), 2264–2267, 1977.

Consideration should also be given to the selection of other equipment and materials used in sample preparation, as they can impact the analysis; some areas of concern include the following:

- The filtering materials, if the sample needs to be filtered—whether to use conventional filter papers or ones made from cellulose or acetate glass, or the method of vacuum filtration using sintered disks.
- If biological fluids are to be analyzed, the cleanliness of the syringe or sampling technique, and in particular the material, can contribute to contamination of the sample.
- Paper towels used for many different reasons in a laboratory. These are generally high in zinc and also contain trace levels of transition metals, such as Fe, Cr, and Co, so avoid using them in and around your sample preparation areas.
- Pipettes, pipette tips, and suction bulbs can all contribute to trace metal contamination levels, so the disposable variety is recommended.

It is important to emphasize that whatever containers, vessels, beakers, volumetric ware, or equipment is used to prepare the sample for ICP-MS analysis, it is absolutely critical that when not in use, they be soaked and washed in a dilute acid (1%–2% HNO_3 is typical). In addition, if they are not being used for extended periods, they should be stored with dilute acid in them. Wherever possible, disposable equipment such as autosampler tubes and pipette tips should be used and then thrown away after use, to cut down on contamination.

THE ENVIRONMENT

The environment in this case refers to the cleanliness of the surrounding area where the instrumentation is installed and where sample preparation is carried out, and any other area the sample comes in contact with. It is advisable that the sample preparation area be as close to the instrument as possible, without actually being in the same room, so that the sample is not exposed to any additional sources of contamination. It is recommended that dissolution be carried out in clean, metal-free fume extraction hoods and, if possible, in an area separate from where samples are to be prepared for less sensitive techniques, such as flame AA or ICP-OES. In addition to having a clean area for dissolution, it is also important to carry out other sample preparation tasks, such as weighing, filtering, pipetting, and diluting, in a clean environment.

These kinds of environmental contamination problems are everyday occurrences in the semiconductor industry because of the strict cleanliness demands required for the fabrication of silicon wafers and production of semiconductor devices. The purity of silicon wafers has a direct effect on the yield of devices, so it is crucial that trace element contamination levels are kept to a minimum in order to reduce defects. This means that any analytical methodology used to determine purity levels on the surface of silicon wafers, or in the high-purity chemicals used to manufacture the devices, must use spotlessly clean instruments. These unique demands of the semiconductor industry have led to the development of special air filtration systems that continually pump air through ultraclean high-efficiency particulate air (HEPA) filters to remove the majority of airborne particulates.

The efficiency of particulate removal will depend on the analytical requirements, but for the semiconductor industry, it is typical to work in environments that contain 1 or 10 particles (<0.2 μ) per cubic foot of air (class 1 and 10 clean rooms, respectively). These kinds of precautions are absolutely necessary to maintain low instrument background levels for the analysis of semiconductor-related samples, but might not be required for other types of applications, such as in a pharmaceutical manufacturing plant. So, even though contamination-free analysis is important, it might be sufficient to work in a class 100, 1000, or 10,000 clean room and still meet your cleanliness objectives.[26]

These clean rooms tend to be very expensive to build, so if your budget does not stretch to a "full-blown" clean room, it might be worth investing in special HEPA filter enclosures just for your instrument and sample preparation area. These are typically either mobile units that can be wheeled around the laboratory and placed around different equipment or hood-based enclosures that are placed over a particular instrument. Whatever system is used, their objective is to ensure that the area around the equipment is free of airborne contamination and the instrument background levels are as low as possible.

THE ANALYST

The expertise of the analyst who actually prepares the samples and carries out the analysis can be a major factor in getting the right result by ICP-MS. Even if all precautions have been taken to cut down on contamination, if the analyst is not experienced in working with ICP-MS and does not understand all the potential pitfalls, the analysis could be doomed. For example, analysts have to be aware of all the potential contaminants that are generated by their own bodies, body parts, or the clothes and jewelry they are wearing. Table 18.3 shows some common trace elements found on the human body. It is by no means an exhaustive list, but at least it gives you an idea of the problem.

TABLE 18.3
Some Common Trace Element Contaminants Found on and around the Human Body

Source of Contamination	Trace Metal Contaminant
Hair	Zn, Cu, Fe, Pb, Mn
Skin	Zn, Cu
Nails	Ca, Si
Jewelry	Au, Ag, Cu, Fe, Ni, Cr
Cigarette smoke	Cd, As, K, Fe, B
Cosmetics	Zn, Bi
Deodorants	Al

These kinds of contamination problems are the reason you often see operators of equipment used in the semiconductor industry wearing "bunny suits." These are white suits that cover the entire body of the operator, including head, hands, and feet, to stop any human-based contamination from getting into the equipment or instrumentation. They are not so important for higher levels of quantitation, but are absolutely necessary for the kind of ultratrace contamination levels found in the electronics industry.

INSTRUMENT AND METHODOLOGY

The instrument and the methodology itself can also be potential sources of error. It is therefore important to be aware of this and to understand what is required when developing a method to carry out the determination of ultratrace levels by ICP-MS. As mentioned previously, the choice of sample preparation methodology can impact the analysis by either causing corrosion problems for some of the instrument components, producing spectral interferences on the analyte, or creating matrix-induced signal drift problems. However, in addition to optimizing sample preparation, a great deal of thought must also go into the choice of instrumental components and to understanding how they impact the method development process. Some of the criteria that should be under consideration when deciding on the analytical methodology include the following:

1. The acid concentration in the final solution being presented to the instrument should ideally be 2%–3% maximum because of the sample transport interferences associated with high concentrations of mineral acids.
2. If highly corrosive acids, such as hydrofluoric acid, are being used, appropriate corrosion-resistant sample introduction components, such as plastic spray chamber and nebulizer, sapphire sample injector, and platinum interface cones, should be used.
3. Hydrochloric, sulfuric, and phosphoric acids should be avoided if possible because of the spectral problems created by the high concentration of chloride, sulfur, and phosphorus ions in the matrix.
4. The choice of fusion mixture should be given serious consideration because of the potential for the lithium-, sodium-, or potassium-based salts to deposit themselves around the sampler or skimmer cone orifice, which over time can lead to serious drift problems.
5. The sample weight might have to be compromised if a fusion mixture is required, because 0.2% is the maximum level of dissolved solids that can be aspirated into the ICP mass spectrometer.
6. There are many grades of argon gas available for spectrochemical analysis. For ultratrace determinations by ICP-MS, the highest grade should always be used (usually ultra-high-purity-grade argon is 99.99999% pure).
7. The use of high-purity collision gases is absolutely critical when collision cells are being used, because of the potential to create additional interfering species. In kinetic energy discrimination–based collision cell technology,

the cell relies on interactions of the interfering ion with an inert or low-reactivity gas, such that it can be separated from the analyte based on their differences in kinetic energy. If the gas contains impurities such as water vapor or hydrocarbons, the impurity could be the dominant reaction pathway, as opposed to the predicted collision/reaction with bulk gas, resulting in the formation of additional and unexpected spectral interferences. For this reason, it is strongly advised that the highest-purity collision/reaction gases be used or a gas purifier system be placed in the gas line to cleanse the collision/reaction gas of any impurities.

8. Petrochemical-type samples usually require the addition of oxygen to the nebulizer gas flow in order to burn off the organic matrix, so the highest quality of oxygen should be used.
9. The choice of pneumatic tubing should be compatible with the sample solution. For example, when analyzing organic samples, suitable pump tubing and sample capillary should be used that are resistant to the organic solvent.
10. There are many different kinds of pump tubing and capillary. If a PVC-based tubing is used, chlorine could potentially be leached out and may cause spectral interferences.
11. Peristaltic pump speed, washout times, read delays, and stability times should be optimized based on the sample matrix and suite of elements, because of memory effects in the sample introduction and interface areas—which may facilitate contamination from the previously analyzed sample.
12. What are the expected analyte concentrations and matrix levels? This will decide whether the sample can be diluted or whether the analytes need to be preconcentrated or the matrix components removed.
13. If the samples are completely unknown, it is a good strategy to dilute the sample 1:100 and get an approximation of the analyte concentrations using the instrument's semiquant routine. This can also give you an insight into understanding the potential interferences from the other elements in the sample.

These are generally considered some of the most important criteria for deciding on an analytical methodology to analyze a set of samples by conventional solution nebulization. However, it should be emphasized that the strategy might also include the use of sampling accessories, such as laser ablation or flow injection. For example, the ability to analyze a solid directly by laser ablation eliminates most of the contamination issues with the preparation, dilution, and aspiration of liquid samples. Even though this might sound attractive, solid sampling has unique problems of its own. So, before this approach is chosen, it is important to also understand all its limitations, especially for a particular set of samples. On the other hand, if solution nebulization is the preferred approach, it should be determined whether there will be any benefit to using segmented flow analysis to reduce the amount of matrix entering the mass spectrometer. Clearly, for some matrices it is advantageous, but for others it might not be worth the effort. It is therefore important to understand these issues before a decision is made.

FINAL THOUGHTS

Whatever analytical methodology approach is used, the issue of contamination must always be at the forefront of the decision.[27] ICP-MS is such a sensitive technique that to take advantage of its unparalleled detection capability and sample throughput capabilities, analytical cleanliness and optimized method development are of utmost importance. If attention is paid to these areas, there is no question that data of the highest quality can be obtained, even at the ultratrace level. This chapter is not intended to be an exhaustive look at sample collection, preparation, or contamination issues, but just to make the reader aware that in order to get the right result in ICP-MS, it is important to examine all aspects of analysis, from first collection of the sample all the way through to quantitation by the instrumental technique.

REFERENCES

1. B. Kratochvil and J. K. Taylor. *Analytical Chemistry*, 53(8), 925A–938A, 1981.
2. Environmental Protection Agency. ICP-MS method for soils and sediments. 2008.
3. B. Bernas. *Analytical Chemistry*, 40(11), 1682–1586, 1986.
4. H. M. Kingston and L. B. Jassie, eds. *Introduction to Microwave Sample Preparation—Theory and Practice*. Washington, DC: American Chemical Society, 1988.
5. A. Hewitt and C. M. Reynolds. *Atomic Spectroscopy*, 11(5), 187–192, 1990.
6. R. A. Nadkarni. *Analytical Chemistry*, 56, 2233–2237, 1984.
7. A. Abu-Samra. *Analytical Chemistry*, 47(8), 1475–1477, 1975.
8. E. Pruszkowski, K. Neubauer, and R. Thomas. *Atomic Spectroscopy*, 19(4), 111–115, 1998.
9. C. O. Ingamells. *Analytica Chimica Acta*, 52, 323–334, 1970.
10. C. B. Belcher. *Talanta*, 10, 75–81, 1963.
11. T. Christensen, L. Pederson, J. Tjell. *Comparison of Sample Preparation Methods for the Determination of Metals in Sewage Sludges*. International Journal of Environmental Analytical Chemistry. Vol 9(3), 1981, 41–50,
12. S. Bajo and U. Suter. *Analytical Chemistry*, 54(1), 49–51, 1982.
13. F. McElroy, A. Mennito, E. Debrah, and R. Thomas. *Spectroscopy*, 13(2), 42–53, 1998.
14. R. Bock. *A Handbook for Decomposition Methods in Analytical Chemistry*. International Textbook Company Ltd., John Wiley and Sons, New York, NY. 1979.
15. S. Tan, G. Horlick. *Applied Spectroscopy*, 40, 445, 1986.
16. Semiconductor Equipment and Materials International. Suggested guidelines for pure water. In *Book of SEMI Standards (BOSS)*. San Jose, CA: Semiconductor Equipment and Materials International, 201–202, 2002.
17. N. Kocherlakota, R. Obernauf, and R. Thomas. *Spectroscopy*, 17(7), 20–28, 2002.
18. D. E. Robertson. *Analytical Chemistry*, 40(7), 1067–1072, 1968.
19. J. R. Moody and R. M. Lindstrom. *Analytical Chemistry*, 49(14), 2264–2267, 1977.
20. M. Totland, I. Jarvis, and K. E. Jarvis. *Chemical Geology*, 95, 35–62, 1992.
21. V. L. Verma and T. M. McKee. *Comparison of procedures for TCLP extract digestion: Conventional vs Microwave*. Presented at the Seventh Annual Waste Testing and Quality Assurance Symposium (EnvirACS), Washington, DC, July 10, 1991.
22. American Society for Testing and Materials. Standard practice for acid extraction of elements from sediments using closed vessel microwave heating. ASTM Method Number D5258-92. 1992.
23. American Society for Testing and Materials. Standard practice for sample digestion using closed vessel microwave heating technique for the determination of total recoverable metals in water. ASTM Method Number D4309-91. 1991.

24. H. T. McCarthy and P. C. Ellis. *Journal of Analytical Chemistry*, 74(3), 566–569, 1991.
25. D. Sears Jr. and Z. Grosser. *Food Testing and Analysis*, June/July 1997.
26. T. Talasek. *Solid State Technology*, December 1993, 44–46.
27. M. Zief and J. W. Mitchel. *Contamination Control in Trace Metal Analysis*. New York: John Wiley & Sons, 1976.

19 Sample Digestion Techniques for Pharmaceutical Samples

This chapter specifically focuses on sample digestion techniques for pharmaceutical-type matrices. This is a very important aspect of characterizing drug compounds for elemental impurities by inductively coupled plasma optical emission spectrometry (ICP-OES) or inductively coupled plasma mass spectrometry (ICP-MS). Unless the material under test is completely in solution when presented to the instrument, an accurate measurement cannot be carried out. For that reason, the final solution has to be clear and colorless with no obvious particulate matter, indicating that all the organic and inorganic components of the sample are completely dissolved. Only then can the analyst be assured that the data generated is indicative of the elemental impurities in the sample matrix[1].

SAMPLE PREPARATION PROCEDURES AS DESCRIBED IN USP CHAPTER <233>

Let's first remind ourselves what Chapter <233> says about sample preparation. The selection of the appropriate sample preparation procedure is dependent on the material being analyzed and is the responsibility of the analyst. The procedures described below have all been shown to be appropriate. It should also be pointed out that all liquid samples should be weighed.

- Neat: This approach is applicable for liquids that can be analyzed with no sample dilution.
- Direct aqueous solution: This procedure is used when the sample is soluble in an aqueous solvent.
- Direct organic solution: This procedure is appropriate where the sample is soluble in an organic solvent.
- Indirect solution: This is used when a material is not directly soluble in aqueous or organic solvents. It is preferred that a total metal extraction sample preparation be carried out in order to obtain an indirect solution, such as open-vessel acid dissolution, or a closed-vessel approach, such as microwave digestion, similar to the one described below. The sample preparation scheme should yield sufficient sample to allow quantification of each element at the elemental impurity limits specified in Chapter <232>.
- Closed-vessel digestion: The benefit of closed-vessel digestion is that it minimizes the loss of volatile impurities. The choice of what concentrated

mineral acid to use depends on the sample matrix and its impact of any potential interferences on the analytical technique being used. An example procedure is given in the chapter, which is described below.

> Weigh accurately 0.5 g of the dried sample in an appropriate flask and add 5 mL of the concentrated acid. Allow the flask to sit loosely covered for 30 min in a fume hood then add an additional 10 mL of the acid, and digest using a closed vessel technique, until digestion is complete (please follow the manufacturer's recommended procedures to ensure safe use). Make up to an appropriate volume and analyze using the technique of choice. Alternatively, a leaching extraction may be appropriate with justification following scientifically validated metal dissolution studies of the specific metal in the drug product under test.

Note 1: As discussed in Chapter 18, the use of a cryogenic freezer mill might be beneficial for getting certain types of pharmaceutical materials or capsules into a convenient powder form ready for the digestion procedure.[2]

Note 2: It should also be noted that if any trace element spiked additions need to be made to comply with the validation protocols, they are carried out before the sample preparation step to ensure that there is no loss of analyte or contamination from the microwave digestion procedure. This could also indicate whether any matrix suppression or enhancement effects are occurring from the dissolution acids or chemicals.

SAMPLE PREPARATION GUIDANCE

Let's now take a closer look at how the digestion technique can be optimized for different sample types. I think it's fair to say that the ideal scenario is that the sample under investigation is in a liquid form, so that it can be analyzed by direct aspiration or perhaps by simply diluting in an aqueous or organic solvent. However, if the sample is a solid or powdered material, the chances are that it will have to be brought into solution via either a hot-plate dissolution technique using concentrated mineral acids, or a closed-vessel microwave digestion procedure.

WHY DISSOLVE SAMPLES?

Sample dissolution using acid digestion techniques can add a significant amount of time to the overall analytical procedure. For that reason, it's important to fully understand the benefits of working with a solution, which are outlined below:

- Solid sampling techniques are notoriously prone to sampling inhomogeneity. Taking multiple portions of the solid material under test and dissolving them represents the best option for working with a homogeneous sample.
- Solution-based analytical techniques need a homogenous sample, which is representative of the sample matrix under test. Taking one-off solid samples, such as a tablet, can produce erroneous results, because it may not be truly representative of the batch of samples.

- Measurements take a finite amount of time where the signal must stay constant. Dissolving the sample and obtaining a clear solution is the best way to achieve signal stability.

It's also important to understand that the sample weight and final volume will be dictated by the expected impurity levels and total dissolved solid (TDS) limitations of the instrumental technique being used. However, it's fair to say that if the dissolution technique requires a microwave digestion system, it introduces a level of complexity that needs to be addressed before the analysis is carried out. In addition, the dilution factor used in the sample preparation step will ultimately have an impact on the ability of the technique to detect the impurity levels. So it is inevitable that there will have to be a certain level of compromise with the dissolution of the sample, based on the level of acceptable TDSs for the analytical technique, particularly if ICP-MS is being used, to ensure that the analyte is measurable above the limit of quantitation (LOQ) of the instrument.

Microwave Digestion Considerations

Chapter <233> actually recommends the use of closed-vessel microwave digestion in the case of insoluble pharmaceutical and nutraceutical final products and excipients in order to completely destroy and solubilize the sample matrix. Microwave digestion systems are commonly used for trace elemental analysis studies in a multitude of application areas, including pharmaceutical materials, to get the samples into solution because they are easy to use and can rapidly process many samples at a time, which makes them ideally suited for high-sample-throughput pharmaceutical production environments.[1]

Why Microwave Digestion?

So let's remind ourselves why closed or pressurized microwave digestion offers the best way to get samples into solution:

- Dissolution temperatures above the boiling point of the solvent can be achieved.
- The oxidation potential of reagents is higher at elevated temperatures, which means digestion is faster and more complete.
- Under these conditions, concentrated nitric acid and/or hydrochloric can be used for the majority of pharmaceutical materials.
- Microwave dissolution conditions and parameters can be reproduced from one sample to the next.
- It is safer for laboratory personnel, as there is less need to handle hot acids.
- Samples can be dissolved very rapidly.
- The digestion process can be fully automated.
- High sample throughput can be achieved.
- There are less hazardous fumes in the laboratory.

Typically, 0.1–0.5 g of sample is weighed and placed into a plastic vessel, along with the appropriate acids, which are then sealed with a tight-fitting cap to create a pressurized environment. Once samples are digested, which takes 10–30 min, depending on the matrix, the resulting liquid is then transferred to a volumetric flask and made up to the required volume using high-purity water.

Choice of Acids

The choice of acids used for the preparation of digested samples is also important. Typically, concentrated nitric and/or hydrochloric acids are used in various concentrations, depending on the sample type. The presence of hydrochloric acid is useful for stabilization of the platinum group elements, but can sometimes produce insoluble chlorides, particularly if there is any silver in the sample. The presence of chloride can also be detrimental when ICP-MS is the chosen technique, as the chloride ions combine with other ions in the sample matrix and the argon plasma to generate polyatomic spectral interferences. An example of this is the formation of the $^{40}Ar^{35}Cl$ polyatomic ion in the determination of ^{75}As and $^{35}Cl^{16}O$ in the determination of ^{51}V. These polyatomic interferences can usually be removed by the use of collision/reaction cell (CRC) technology if the ICP-MS system offers that capability. However, CRC technology can slow the analysis down because stabilization times have to be built into a multielement method to determine analytes that require both cell and no-cell conditions.

Nitric acid and hydrogen peroxide are often used for the dissolution of organic matrices, as they are both strong oxidizing agents that effectively destroy the organic matter. However, care must be taken when testing for osmium, as this can form volatile osmium oxides, which are easily lost from the sample. In some cases, hydrofluoric acid (HF) may need to be used to dissolve certain silicate-based excipients and fillers that have been used in the final product. In cases where HF is required, specialized plastic (polytetrafluoroethylene) sample introduction components need to be used, including the use of buffering agents like boric acid to dissolve insoluble fluorides and neutralize excess HF. It should be emphasized that HF is a highly corrosive acid, and extreme caution should be taken whenever it is used.[3]

However, it's important to understand that the more complex the sample preparation, the longer the analytical procedure will become, which will have a negative impact on the overall analysis time, particularly in a lab with a high sample workload. In addition, the sample preparation steps could potentially have an effect on the overall TDS levels, so it is important to consider this when looking at the preparation of these samples.

MICROWAVE TECHNOLOGY

There are a number of different microwave digestion technologies on the market, including batch and single-reaction chamber systems, which can be run in either a sequential or a simultaneous manner. Depending on the types and variety of samples being digested and the degree of automation, they both have their own strengths and weaknesses. However, if a suitable digestion procedure has been optimized for a particular type of pharmaceutical material, the digestion chemistry should be applicable to both types of systems. For example, many pharmaceutical matrices can be digested by taking 0.5 g of sample, adding 10 mL of acid mixture (9 mL of HNO_3 and 1 mL of HCl), and placing it in a microwave digestion vessel. Depending on the sample and program, it can be fully dissolved to a clear or colorless liquid in approximately 10 min.

It is not the intent to favor one microwave digestion approach over the other. There is much application information in the public domain to help users make that decision themselves based on sample variety and workload.[4,5] However, the benefit of the single-reaction chamber is that many different pharmaceutical matrices can be loaded into the system and all run together irrespective of sample weight, acid medium, or matrix components. This could be attractive to a pharmaceutical or dietary supplement manufacturer that has lots of different drug products or raw material to analyze.

On the other hand, the batch system appears to be advantageous when there are large numbers of similar materials being analyzed. This approach has proven to be a good choice for some labs, because with the use of an autosampler, it allows the user to load a full batch of similar samples and then run them unattended. If other matrices are to be analyzed, additional autosampler racks can then be set up and run as separate batches. Note: For more detailed application material, refer to the contact information for microwave digestion companies in Chapter 29, where you will find a multitude of application reports for different types of pharmaceutical materials.

Sampling Procedures for Mercury

It's also worth pointing out that element-specific sample preparation techniques might also be necessary. For example, when preparing samples for the determination of mercury, care must be taken not to lose the analyte because of its volatility. This is especially relevant when carrying out microwave digestion. Some of the steps taken to minimize losses of mercury include the use of hydrochloric acid in the dissolution step to produce an excess of chloride ions or the addition of gold (typically a few parts per million) to stabilize the mercury in solution.[6]

Under the right chemistry conditions, mercury can also be determined by the cold vapor generation technique. This technique is used in commercially available mercury analyzers, where the mercury in solution is reduced to its atomic state and the elemental mercury vapor is detected using either atomic absorption or atomic fluorescence. These instruments are extremely sensitive and are capable of carrying out both the chemistry and detection steps online, in an automated manner.[7] In conventional mercury analyzers, the samples must be either liquids or brought into solution using a dissolution step, which limits their applicability for the analysis of solid or powdered pharmaceutical materials. However, there is a variation of this method, which can handle solids directly. In this approach, the sample is first heated to 900°C in a combustion furnace to volatilize the sample, and then is swept into a catalyst to release the mercury vapor, and next concentrated onto the surfaces of a gold amalgamation trap, where it is eventually heated and swept into an atomic absorption for detection and quantitation[8].

REFERENCES

1. N. Lewen. Preparation of pharmaceutical samples for elemental impurities analysis: Some potential approaches. *Spectroscopy Magazine*, 31(4), 36–43, 2016.
2. Spex SamplePrep. Sample preparation techniques for pharmaceutical labs: The benefits of cryogenic freezing. Metuchen, NJ: Spex SamplePrep. https://www.spexsampleprep.com/knowledge-base/resources/application_notes/0118-105546-Pharma-prep_flyer_2017_secured.pdf.
3. National Institute for Occupational Safety and Health. Safe use of hydrogen fluoride and hydrofluoric acid. http://www.cdc.gov/niosh/ershdb/emergencyresponsecard_29750030.html.
4. S. Hussein and T. Michel. The application of single-reaction-chamber microwave digestion to the preparation of pharmaceutical samples in accordance with USP <232> and <233>. *Spectroscopy Magazine*, Special Issue, 27(10), 2012.
5. CEM Corp. Microwave digestion of pharmaceutical samples followed by ICP-MS analysis for USP Chapters <232> and <233>. Application note. https://cem.sharefile.com/download.aspx?id=sc6d12fed0084ab4b#.
6. U.S. Environmental Protection Agency. Mercury preservation techniques. https://www.inorganicventures.com/sites/default/files/mercury_preservation_techniques.pdf.
7. J. Forsberg. Ultra-Trace Mercury Determination in Water, Using EPA Method 1631, Spectroscopy Magazine, June, 2015, http://www.spectroscopyonline.com/ultra-trace-mercury-determination-bottled-water-epa-method-1631-using-teledyne-leeman-labs-quicktrac.
8. S. Phatak, J. Gerbino. *Determination of Mercury in Soil Using Direct Mercury Analysis*. Spectroscopy Magazine, January, 2015, http://www.spectroscopyonline.com/determination-mercury-soil-using-direct-mercury-analysis.

20 Performance and Productivity Enhancement Techniques

Conventional sample introduction systems using a spray chamber and nebulizer account for the majority of inductively coupled plasma mass spectrometry (ICP-MS) applications being carried out today. However, nonstandard sampling accessories, such as laser ablation systems, flow injection (FI) analyzers, electrothermal vaporizers, cooled spray chambers, desolvation equipment, direct injection nebulizers, automated sample delivery systems, autodilutors, and on-line chemistry procedures, are considered critical to enhancing the practical capabilities of the technique. Initially regarded as novel sampling devices, they have since proved themselves to be invaluable for solving real-world application problems by enhancing the flexibility, performance, and productivity of the technique. Although they are not all suitable for pharmaceutical-type samples, these tools can be very beneficial for labs that are looking to analyze a suite of samples in the most efficient and cost-effective manner. This chapter describes the basic principles of these accessories and gives an overview of their practical capabilities.

It is recognized that standard ICP-MS instrumentation using a traditional sample introduction system comprising a spray chamber and nebulizer has certain limitations, particularly when it comes to the analysis of complex samples. Some of these known limitations include the following:

- Inability to analyze solids directly.
- Contamination issues with samples requiring multiple sample preparation steps.
- Liquid aerosol can impact the ionization process.
- Total dissolved solids must be kept below 0.2%.
- If matrix has to be removed, it has to be done offline.
- Long washout times required for samples with a heavy matrix.
- Dilutions and the addition of internal standards can be labor-intensive and time-consuming.
- Matrix components can generate severe spectral overlaps on many analytes.
- The analysis of slurries is very difficult.
- Matrix suppression can be quite severe with some samples.
- Spectral interferences generated by solvent-induced species can limit detection capability.
- Organic solvents can present unique problems.

- Sample throughput is limited by the sample introduction process.
- Not suitable for the determination of elemental species or oxidation states.

Such were the demands of real-world users to overcome these kinds of problem areas that instrument manufacturers developed different strategies based on the type of samples being analyzed. Some of these strategies involved parameter optimization or modification of instrument components, but it was clear that this approach alone was not going to solve every conceivable problem. For this reason, they turned their attention to the development of sampling accessories, which were optimized for a particular application problem or sample type. Over the past 10–15 years, this demand has led to the commercialization of specialized performance and productivity enhancement tools, manufactured not only by the instrument manufacturers themselves, but also by third-party vendors specializing in these kinds of sampling techniques. The most common ones used today include the following:

- Laser ablation systems
- FI analyzers
- Electrothermal vaporizors (ETVs)
- Chilled spray chambers and desolvation systems
- Direct injection nebulizers
- Fast automated sampling procedures
- Autodilution and autocalibration systems
- Automated sample identification and tracking

Although many of them are not well suited for the analysis of pharmaceutical materials according to USP Chapter <233>, many of them are applicable to contract labs that may be analyzing different materials in addition to pharmaceutical samples. Let us take a closer look at some of these enhanced techniques to understand their basic principles and what benefits they bring to ICP-MS and inductively coupled plasma optical emission spectrometry (ICP-OES).

PERFORMANCE-ENHANCING TECHNIQUES

LASER ABLATION

The limitation of ICP-MS to analyze solids, without dissolving the material, led to the development of laser ablation. The principle behind this approach is the use of a high-powered laser to ablate the surface of a solid and sweep the sample aerosol into the ICP mass spectrometer for analysis in the conventional way.[1]

Before I go on to describe some typical applications suited to laser ablation ICP-MS, let us first take a brief look at the history of analytical lasers and how they eventually became such a useful sampling tool. The use of lasers as vaporization devices was first investigated in the early 1960s. When light energy with an extremely high power density interacts with a solid material, the photon-induced energy is converted into thermal energy, resulting in vaporization and removal of the material from the surface of the solid.[2] Some of the early researchers used ruby lasers to induce a plasma discharge on the surface of the sample and measure the emitted light with an atomic emission spectrometer.[3] Although this proved useful for certain applications, the technique suffered from low sensitivity, poor precision, and severe matrix effects caused by nonreproducible excitation characteristics. Over the years, various improvements were made to this basic design with very little success,[4] because the sampling process and the ionization and excitation process (both under vacuum) were still intimately connected and interacted strongly with each other.

This limitation led to the development of laser ablation as a sampling device for atomic spectroscopy instrumentation, where the sampling step was completely separated from the excitation or ionization step. The major benefit is that each step can be independently controlled and optimized. These early devices used a high-energy laser to ablate the surface of a solid sample, and the resulting aerosol was swept into some kind of atomic spectrometer for analysis. Although initially used with atomic absorption[5,6] and plasma-based emission techniques,[7,8] it was not until the mid-1980s, when lasers were coupled with ICP-MS, that the analytical community sat up and took notice.[9] For the first time, researchers were coming up with evidence that virtually any type of solid could be vaporized, irrespective of electrical characteristics, surface topography, size, or shape, and be transported into the ICP for analysis by atomic emission or mass spectrometry. This was an exciting breakthrough for ICP-MS, because it meant the technique could be used for the bulk sampling of solids or, if required, for the analysis of small spots or microinclusions, in addition to being used for the analysis of solutions.

Commercial Laser Ablation Systems for ICP-MS

The first laser ablation systems developed for ICP instrumentation were based on solid-state ruby lasers, operating at 694 nm. These were developed in the early 1980s, but did not prove to be successful for a number of reasons, including poor stability, low power density, low repetition rate, and large beam diameter, which made them limited in their scope and flexibility as a sample introduction device for trace element analysis. It was at least another 5 years before any commercial instrumentation became available. These early commercial laser ablation systems, which were specifically developed for ICP-MS, used the neodymium-doped yttrium aluminum garnet (Nd:YAG) design, operated at the primary wavelength of 1064 nm—in the infrared.[10] They initially showed a great deal of promise because analysts were finally able to determine trace levels directly in the solid without sample dissolution. However, it soon became apparent that they did not meet the expectations of the analytical community, for many reasons, including complex ablation characteristics, poor precision, nonoptimization for microanalysis, and poor laser coupling, making them unsuitable for many types of solids. By the early 1990s, most of the laser ablation systems purchased were viewed as novel and interesting, but not suited to solving real-world application problems.

These basic limitations in infrared laser technology led researchers to investigate the benefits of shorter wavelengths. Systems were developed that were based on Nd:YAG technology at the 1064 nm primary wavelength, but utilizing optical components to double (532 nm), quadruple (266 nm), and quintuple (213 nm) the frequency. Innovations in lasing materials and electronic design, together with better thermal characteristics, produced higher energy with higher pulse-to-pulse stability. These more advanced ultraviolet (UV) lasers showed significant improvements, particularly in the area of coupling efficiency, making them more suitable for a wider array of sample types. In addition, the use of higher-quality optics allowed for a more homogeneous laser beam profile, which provided the optimum energy density to couple with the sample matrix. This resulted in the ability to make spots much smaller and with more controlled ablations, irrespective of sample material, which were critical for the analysis of surface defects, spots, and microinclusions. Figure 20.1 shows the optical layout of a commercially available frequency-quintupled 213 nm Nd:YAG laser ablation system.

Excimer Lasers

The successful trend toward shorter wavelengths and the improvements in the quality of optical components also drove the development of UV gas-filled lasers, such as XeCl (308 nm), KrF (248 nm), and ArF (193 nm) excimer lasers. These showed great promise, especially the ones that operated at shorter wavelengths and were specifically designed for ICP-MS. Unfortunately, they necessitated a more sophisticated beam delivery system, which tended to make them more expensive. In addition, the complex nature of the optics and the fact that gases had to be changed on a routine basis made them a little more difficult to use and maintain, and as a result, they required a more skilled operator to run them. However, their complexity was far

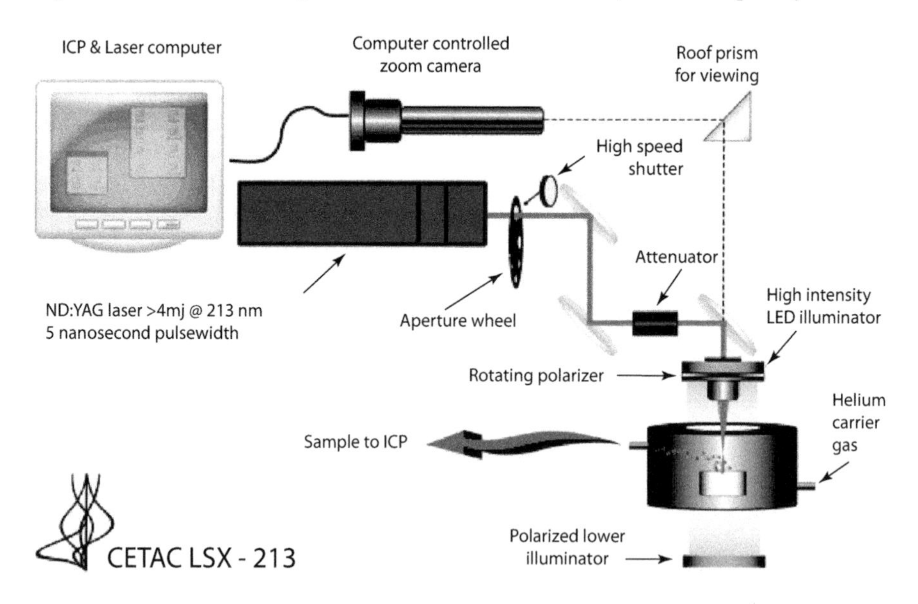

FIGURE 20.1 Schematic of a commercially available frequency-quintupled 213 nm Nd:YAG laser ablation system. (Courtesy of Teledyne Cetac Technologies.)

outweighed by their better absorption capabilities for UV-transparent materials, such as calcites, fluorites, and silicates; smaller particle size; and higher flow of ablated material. There was also evidence to suggest that the shorter-wavelength excimer lasers exhibit better elemental fractionation characteristics (typically defined as the intensity of certain elements varying with time, relative to the dry aerosol volume) than the longer-wavelength Nd:YAG design, because they produce smaller particles that are easier to volatilize.

Even though excimer lasers are optically more complex than other designs, it's worth mentioning that today's instruments are far more rugged and robust than the earlier-designed systems, and as a result are being used for more and more routine applications. And because the higher-grade optical components used in the excimer technology are available at a more realistic cost, their price has come down significantly in the past few years.

Benefits of Laser Ablation for ICP-MS

Today there are a number of commercial laser ablation systems on the market designed specifically for ICP-MS, including 266 and 213 nm Nd:YAG and 193 nm ArF excimer lasers. They all have varying output energy, power density, and beam profiles, and even though each one has different ablation characteristics, they all work extremely well, depending on the types of samples being analyzed and the data quality requirements. Laser ablation is now considered a very reliable sampling technique for ICP-MS, which is capable of producing data of the very highest quality directly on solid samples and powders. Some of the many benefits offered by this technique include the following:

- Direct analysis of solids without dissolution
- Ability to analyze virtually any kind of solid material, including rocks, minerals, metals, ceramics, polymers, plastics, plant material, and biological specimens
- Ability to analyze a wide variety of powders by pelletizing with a binding agent
- No requirement for sample to be electrically conductive
- Sensitivity in the parts per billion to parts per trillion range, directly in the solid
- Labor-intensive sample preparation steps are eliminated, especially for samples such as plastics and ceramics that are extremely difficult to get into solution
- Contamination is minimized because there are no digestion or dilution steps
- Reduced polyatomic spectral interferences compared with solution nebulization
- Examination of small spots, inclusions, defects, or microfeatures on the surface of the sample
- Elemental mapping across the surface of a mineral
- Depth profiling to characterize thin films or coatings

Let us now take a closer look at the strengths and weaknesses of the different laser designs with respect to application requirements.

Optimum Laser Design Based on the Application Requirements

The commercial success of laser ablation was initially driven by its ability to directly analyze solid materials, such as rocks, minerals, ceramics, plastics, and metals, without going through a sample dissolution stage. Table 20.1 represents some typical multielement detection limits in NIST 612 glass generated with a 266 nm Nd:YAG design coupled to an ICP-MS system.[11] It can be seen that sub-parts-per-billion detection limits in the solid material are achievable for most of the elements. This kind of performance is typically obtained using larger spot sizes on the order of 100–1000 μm in diameter, which is ideally suited to 266 nm laser technology.

However, the desire for ultratrace analysis of optically challenging materials, such as calcite, quartz, glass, and fluorite, combined with the capability to characterize small spots and microinclusions, proved very challenging for the 266 nm design. The major reason is that the ablation process is not very controlled and precise, and as a result, it is difficult to ablate a minute area without removing some of the surrounding material. In addition, erratic ablating of the sample initially generates larger particles (>1 μm size), which are not efficiently ionized in the plasma and therefore contribute to poor precision.[12] Even though modifications helped improve

TABLE 20.1
Typical Detection Limits Achievable in NIST 612 SRM Glass Using a 266 nm Nd:YAG Laser Ablation System Coupled to an ICP Mass Spectrometer

Element	3σ DLs (ppb)	Element	3σ DLs (ppb)
B	3.0	Ce	0.05
Sc	3.4	Pr	0.05
Ti	9.1	Nd	0.5
V	0.4	Sm	0.1
Fe	13.6	Eu	0.1
Co	0.05	Gd	1.5
Ni	0.7	Dy	0.5
Ga	0.2	Ho	0.01
Rb	0.1	Er	0.2
Sr	0.07	Yb	0.4
Y	0.04	Lu	0.04
Zr	0.2	Hf	0.4
Nb	0.5	Ta	0.1
Cs	0.2	Th	0.02
Ba	0.04	U	0.02
La	0.05		

Source: Courtesy of Teledyne Cetac Technologies Omaha, NE.
Note: DLs, detection limits.

ablation behavior, it was not totally successful because of the basic limitation of the 266 nm to couple efficiently to UV-transparent materials. The drawbacks in 266 nm technology eventually led to the development of 213 nm lasers[13] because of the recognized superiority of shorter wavelengths to exhibit a higher degree of absorbance in transparent materials.[14]

Analytical chemists, particularly in the geochemical community, welcomed 213 nm UV lasers with great enthusiasm, because they now had a sampling tool that offered much better control of the ablation process, even for easily fractured minerals. This is demonstrated in Figures 20.2 and 20.3, which show the ablation differences between 266 and 213 nm, respectively, for NIST 612 glass standard reference material (SRM). It can be seen that the 200 μm ablation crater produced with the 266 nm laser is irregular and shows redeposited ablated material around the edges of the crater, whereas the crater with the 213 nm system is very clean and symmetrical, with no ablated material around the edges. The absence of any redeposited material with the 213 nm laser means that a higher proportion of ablated material actually makes it to the plasma. Both craters are shown at 10× magnification.

This significant difference in crater geometry between the two systems is predominantly a result of the effective absorption of laser energy and the difference in the delivered power per unit area—also known as laser irradiance (or fluence per laser pulse width).[15] The result is a difference in depth penetration and the size and volume of particles reaching the plasma. With the 266 nm laser system, a high-volume burst of material is initially observed, whereas with the 213 nm laser, the signal gradually increases and levels off quickly, indicating a more consistent stream of small particles being delivered to the plasma and the mass spectrometer. Therefore, when analyzing this type of mineral with the 266 nm design, it is typical that the first 100–200 shots of the ablation process are filtered out to ensure that no data is taken during the initial burst of material. This can be somewhat problematic when analyzing small spots or inclusions, because of the limited amount of sample being ablated.

FIGURE 20.2 A 200 μm crater produced by the ablation of a NIST 612 glass SRM, using a 266 nm laser ablation system, showing excess ablated material around the edges of the crater. (Courtesy of Teledyne Cetac Technologies.)

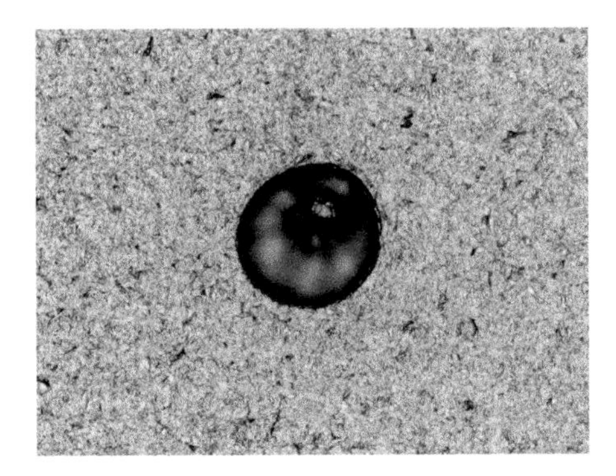

FIGURE 20.3 A 200 μm crater produced by the ablation of a NIST 612 glass SRM using a 213 nm laser system, showing a symmetrical, well-defined crater. (Courtesy of Teledyne Cetac Technologies.)

193 nm LASER TECHNOLOGY

The benefits of 213 nm lasers emphasize that matrix independence, high spatial resolution, and the ability to couple with UV-transparent materials without fracturing (particularly for small spots or depth analysis studies) were very important for geochemical-type applications. These findings led researchers to study even shorter wavelengths, in particular, 193 nm ArF excimer technology. Besides their accepted superiority in coupling efficiency, a major advantage of the 193 nm design is that it utilizes a fundamental wavelength, and therefore achieves much higher energy transfer than a Nd:YAG solid-state system that utilizes crystals to quadruple or quintuple the frequency. Additionally, the less coherent nature of the excimer beam enables better optical homogenization, resulting in an even flatter beam profile. The overall benefit is that cleaner, flatter craters are produced down to approximately 3–4 μm in diameter, at energy densities up to 45 J/cm^2. This provides far better control of the ablation process, which is especially important for depth profiling and fluid inclusion analysis. This is demonstrated in Figure 20.4, which shows a scanning electron microscope (SEM) image (1200 magnification) of a NIST 1612 glass SRM ablated with a 193 nm ArF excimer laser using a highly homogenized, flattop optical beam profile.[16] It can be seen that the 160 μm ablation crater produced by 50 laser pulses is extremely flat and smooth around the edges. The 213 nm Nd:YAG design just would not be capable of the kind of high precision required for depth analysis and small-spot and fluid inclusion studies carried out by the geological community.

It is very important to emphasize that the geochemical community initially drove the design of excimer lasers, because they were interested in characterizing extremely small spots and inclusions on the surface of minerals and geological specimens, which was very difficult using either of the other approaches. As a result, the performance and the analytical capabilities were focused on their needs, which included the requirement for finely controlled, "homogenizer-flat" ablations with high sensitivity and split-second response. Also, fire-on-the-fly lasing synchronized to the stage's motion, combined with fast-washout ablation cells, made high-precision depth profiling of spots, lines, and areas possible, allowing for high-spatial-resolution elemental mapping. Additionally, the combination of ultrashort pulse length and the 193 nm wavelength produced very high coupling efficiency. This meant higher absorbance in a wide range of materials, which produced smaller particles on average than those produced by the 213 nm YAG design. This resulted in greater ionization within the plasma, leading to better sensitivity and less deposition at the ionization source. Today's commercial excimer lasers can ablate all materials, from opaque to highly transparent, including delicate powders, hard quartz, and resilient carbonates, with a depth penetration in the tens of nanometers per shot. The beam energy profile is homogenized to ensure uniform ablations across the entire range of spot sizes and on a wide range of materials.

The benefits of laser ablation coupled with ICP-MS are now well documented by the large number of application references in the public domain, which describe the analysis of metals, ceramics, polymers, minerals, biological tissue, pharmaceutical tablets, and many other sample types.[17–22] These references should be investigated further to better understand the optimum configuration, design, and wavelength of laser ablation equipment for different types of sample matrices. It should also be emphasized that there are many overlapping areas when selecting the optimum laser system for the sample type. Roy and Neufeld published a very useful article that offered some guidelines on the importance of matching the laser hardware to the application.[23]

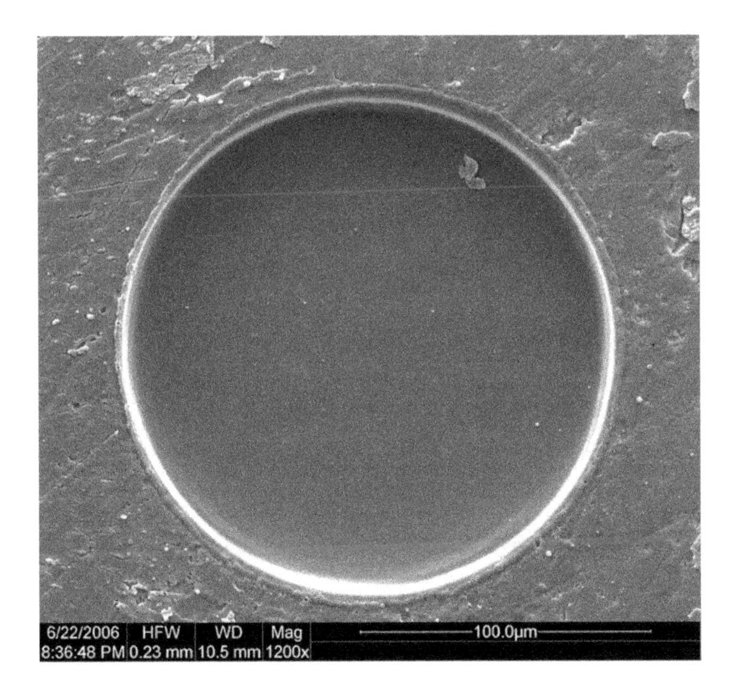

FIGURE 20.4 Scanning electron microscope (SEM) image (1200 magnification) of a 160 μm crater produced by the ablation (50 pulses) of NIST 1612 glass SRM using an optically homogenized flat beam, 193 nm ArF excimer laser system. (Courtesy of Teledyne Cetac Technologies.) (From SEM photo courtesy of Dr. Honglin Yuan, Northwest University, Xi'an, China.)

FLOW INJECTION ANALYSIS

FI is a powerful front-end sampling accessory for ICP-MS that can be used for preparation, pretreatment, and delivery of the sample. Originally described by Ruzicka and Hansen,[24] FI involves the introduction of a discrete sample aliquot into a flowing carrier stream. Using a series of automated pumps and valves, procedures can be carried out online to physically or chemically change the sample or analyte before introduction into the mass spectrometer for detection. There are many benefits of coupling FI procedures to ICP-MS, including the following:

- Automation of online sampling procedures, including dilution and additions of reagents
- Minimum sample handling translates into less chance of sample contamination
- Ability to introduce low sample or reagent volumes
- Improved stability with harsh matrices
- Extremely high sample throughput using multiple loops

In its simplest form, FI-ICP-MS consists of a series of pumps and an injection valve preceding the sample introduction system of the ICP mass spectrometer. A typical manifold used for microsampling is shown in Figure 20.5.

In the fill position, the valve is filled with the sample. In the inject position, the sample is swept from the valve and carried to the ICP by means of a carrier stream. The measurement is usually a transient profile of signal versus time, as shown by the signal profile in Figure 20.5.

The area of the signal profile measured is greater for larger injection volumes, but for volumes of 500 μL or greater, the signal peak height reaches a maximum equal to that observed using continuous solution aspiration. The length of a transient peak in FI is typically 20–60 s, depending on the size of the loop. This means that if multielement determinations are a requirement, all the data quality objectives for the analysis, including detection limits, precision, dynamic range, and number of elements, must be achieved in this time frame. Similar to laser ablation, if a sequential mass analyzer such as a quadrupole or single-collector magnetic sector system is used, the electronic scanning, dwelling, and settling times must be optimized in order to capture the maximum amount of multielement data in the duration of the transient event.[25] This can be seen in greater detail in Figure 20.6, which shows a three-dimensional transient plot of intensity versus mass in the time domain for the determination of a group of elements.

Some of the many online procedures that are applicable to FI-ICP-MS include the following:

- Microsampling for improved stability with heavy matrices[26]
- Automatic dilution of samples and standards[27]
- Standards addition[28]
- Cold vapor and hydride generation for enhanced detection capability for elements such as Hg, As, Sb, Bi, Te, and Se[29]

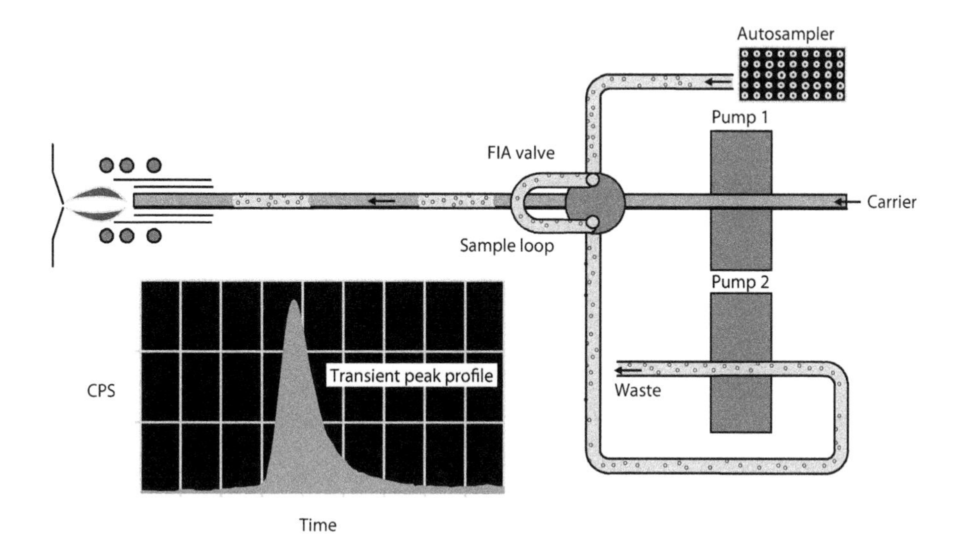

FIGURE 20.5 Schematic of a flow injection system used for the process of microsampling.

- Matrix separation and analyte preconcentration using ion exchange procedures[30]
- Elemental speciation[31]
- Maximize sample throughput

FI coupled to ICP-MS has shown itself to be very diverse and flexible in meeting the demands presented by complex samples, as indicated in the foregoing references. However, one of the most interesting areas of research is in the direct analysis of seawater by FI-ICP-MS. Traditionally, seawater is very difficult to analyze by ICP-MS because of two major problems. First, the high NaCl content will block the sampler cone orifice over time, unless a 10- to 20-fold dilution is made of the sample. This is not such a major problem with coastal waters, because the levels are high enough. However, if the sample is open-ocean seawater, this is not an option, because the trace metals are at a much lower level. The other difficulty associated with the analysis of seawater is that ions from the water, the chloride matrix, and the plasma gas can combine to generate polyatomic spectral interferences, which are a problem, particularly for the first-row transition metals.

Attempts have been made over the years to remove the NaCl matrix and preconcentrate the analytes using various types of chromatography and ion exchange column technology. One such early approach was to use an HPLC system coupled to an ICP mass spectrometer utilizing a column packed with silica-immobilized 8-hydroxyquinoline.[32] This worked reasonably well, but was not considered a routine method, because silica-immobilized 8-hydroxyquinoline was not commercially available, and also, spectral interferences produced by HCl and HNO_3 (used to elute the analytes) precluded determination of a number of the elements, such as Cu, As, and V. More recently, chelating agents based on the iminodiacetate acid functionality

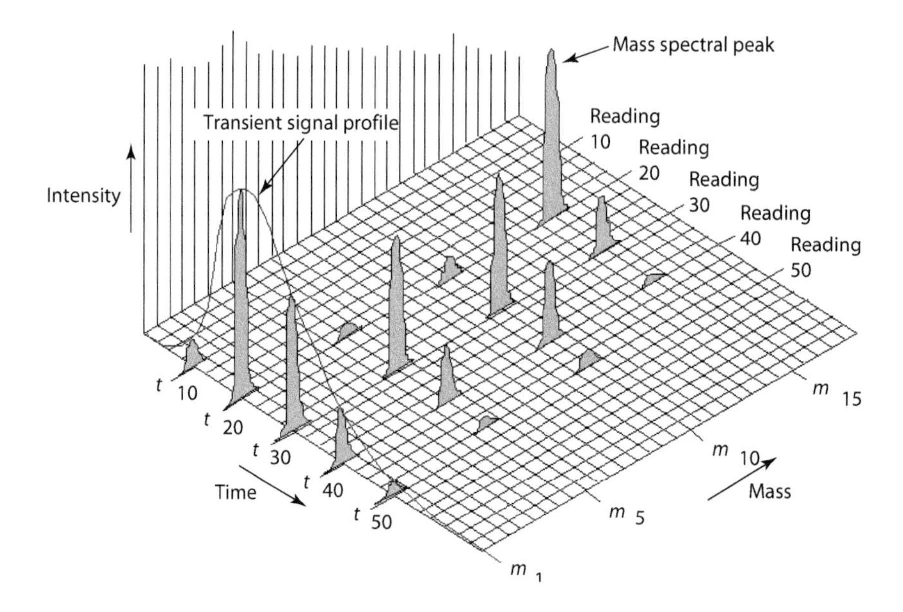

FIGURE 20.6 3D plot of intensity versus mass in the time domain for the determination of a group of elements in a transient peak. (Copyright © 2013, all rights reserved, PerkinElmer Inc.)

group have gained wider success, but are still not considered truly routine for a number of reasons, including the necessity for calibration using standard additions, the requirement of large volumes of buffer to wash the column after loading the sample, and the need for conditioning between samples because some ion exchange resins swell with changes in pH.[33–35]

However, a research group at the National Research Council (NRC) in Canada has developed a very practical online approach, using an FI sampling system coupled to an ICP mass spectrometer.[30] Using a special formulation of a commercially available, iminodiacetate ion exchange resin (with a macroporous methacrylate backbone), trace elements can be separated from the high concentrations of matrix components in the seawater, with a pH 5.2 buffered solution. The trace metals are subsequently eluted into the plasma with 1 M HNO_3, after the column has been washed out with deionized water. The column material has sufficient selectivity and capacity to allow accurate determinations at parts per trillion levels using simple aqueous standards, even for elements such as V and Cu, which are notoriously difficult in a chloride matrix. This can be seen in Figure 20.7, which shows spectral scans for a selected group of elements in a certified reference material open-ocean seawater sample (NASS-4), and Table 20.2, which compares the results for this methodology with the certified values, together with the limits of detection (LODs). Using this online method, the turnaround time is less than 4 min per sample, which is considerably faster than that of other high-pressure chelation techniques reported in the literature.

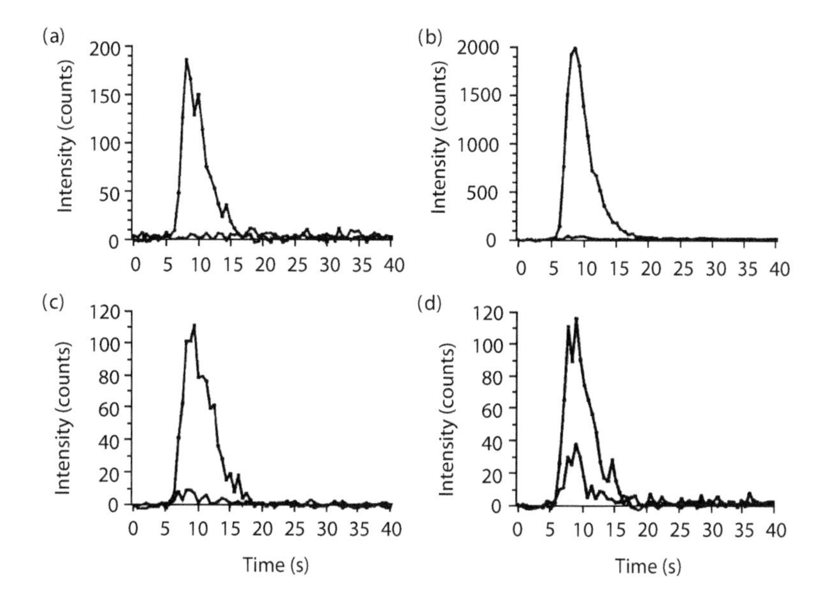

FIGURE 20.7 Analyte and blank spectral scans of (a) Co, (b) Cu, (c) Cd, and (d) Pb in NASS-4 open-ocean seawater certified reference material, using flow injection coupled to ICP-MS. (From S. N. Willie, Y. Iida and J. W. McLaren, *Atomic Spectroscopy*, 19[3], 67, 1998.)

ELECTROTHERMAL VAPORIZATION

Electrothermal atomization (ETA) for use with AA has proved to be a very sensitive technique for trace element analysis over the last three decades. However, the possibility of using the atomization or heating device for ETV sample introduction into an ICP mass spectrometer was identified in the late 1980s.[36] The ETV sampling process relies on the basic principle that a carbon furnace or metal filament can be used to thermally separate the analytes from the matrix components and then sweep them into the ICP mass spectrometer for analysis. This is achieved by injecting a small amount of the sample (usually 20–50 μL via an autosampler) into a graphite tube or onto a metal filament. After the sample is introduced, drying, charring, and vaporization are achieved by slowly heating the graphite tube or metal filament. The sample material is vaporized into a flowing stream of carrier gas, which passes through the furnace or over the filament during the heating cycle. The analyte vapor recondenses in the carrier gas and is then swept into the plasma for ionization.

One of the attractive characteristics of ETV for ICP-MS is that the vaporization and ionization steps are carried out separately, which allows for the optimization of each process. This is particularly true when a heated graphite tube is used as the vaporization device, because the analyst typically has more control of the heating process and, as a result, can modify the sample by means of a very precise thermal program before it is introduced to the ICP for ionization. By boiling off and sweeping the solvent and volatile matrix components out of the graphite tube,

TABLE 20.2
Analytical Results for NASS-4 Open-Ocean Seawater Certified Reference Material, Using FI-ICP-MS Methodology

		NASS-4 (ppb	
Isotope	**LOD (ppt)**	**Determined**	**Certified**
51V+	4.3	1.20 ± 0.04	Not certified
63Cu+	1.2	0.210 ± 0.008	0.228 ± 0.011
60Ni+	5	0.227 ± 0.027	0.228 ± 0.009
66Zn+	9	0.139 ± 0.017	0.115 ± 0.018
55Mn+	Not reported	0.338 ± 0.023	0.380 ± 0.023
59Co+	0.5	0.0086 ± 0.0011	0.009 ± 0.001
208Pb+	1.2	0.0090 ± 0.0014	0.013 ± 0.005
114Cd+	0.7	0.0149 ± 0.0014	0.016 ± 0.003

Source: Willie, S. N., et al., *Atomic Spectros.*, 19(3) 67, 1998.

spectral interferences arising from the sample matrix can be reduced or eliminated. The ETV sampling process consists of six discrete stages: sample introduction, drying, charring (matrix removal), vaporization, condensation, and transport. Once the sample has been introduced, the graphite tube is slowly heated to drive off the solvent. Opposed gas flows, entering from each end of the graphite tube, purge the sample cell by forcing the evolving vapors out the dosing hole. As the temperature increases, volatile matrix components are vented during the charring steps. Just prior to vaporization, the gas flows within the sample cell are changed. The central channel (nebulizer) gas then enters from one end of the furnace, passes through the tube, and exits out of the other end. The sample-dosing hole is then automatically closed, usually by means of a graphite tip, to ensure that no analyte vapors escape. After this gas flow pattern has been established, the temperature of the graphite tube is ramped up very quickly, vaporizing the residual components of the sample. The vaporized analytes either recondense in the rapidly moving gas stream or remain in the vapor phase. These particulates and vapors are then transported to the ICP in the carrier gas, where they are ionized by the ICP for analysis in the mass spectrometer.

Another benefit of decoupling the sampling and ionization processes is the opportunity for chemical modification of the sample. The graphite furnace itself can serve as a high-temperature reaction vessel where the chemical nature of compounds within it can be altered. In a manner similar to that used in atomic absorption, chemical modifiers can change the volatility of species to enhance matrix removal and increase elemental sensitivity.[37] An alternative gas, such as oxygen, may also be introduced into the sample cell to aid in the charring of the carbon in organic matrices, such as biological or petrochemical samples. Here, the organically bound carbon reacts with the oxygen gas to produce CO_2, which is then vented from the system. A typical ETV sampling device, showing the two major steps of sample pretreatment (drying and ashing) and vaporization into the plasma, is seen schematically in Figure 20.8.

Over the past 20 years, ETV sampling for ICP-MS has mainly been used for the analysis of complex matrices, including geological materials,[38] biological fluids,[39] seawater,[40] and coal slurries,[41] which has proved difficult or impossible by conventional nebulization. By removal of the matrix components, the potential for severe spectral and matrix-induced interferences is dramatically reduced. Even though ETV-ICP-MS was initially applied to the analysis of very small sample volumes, the advent of low-flow nebulizers has limited its use for this type of work.

An example of the benefit of ETV sampling is in the analysis of samples containing high concentrations of mineral acids, such as HCl, HNO_3, and H_2SO_4. Besides physically suppressing analyte signals, these acids generate massive polyatomic spectral overlaps, which interfere with many analytes, including As, V, Fe, K, Si, Zn, and Ti. By carefully removing the matrix components with the ETV device, the determination of these elements becomes relatively straightforward. This is illustrated in Figure 20.9, which shows a spectral display in the time domain for 50

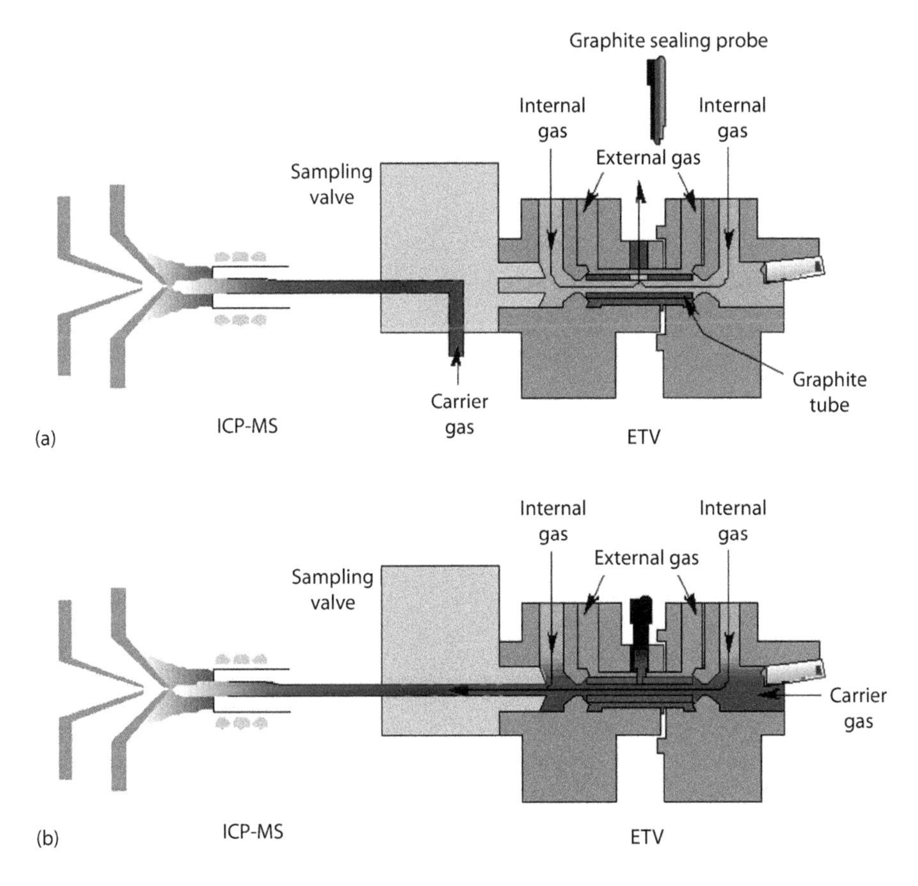

FIGURE 20.8 A graphite furnace ETV sampling device for ICP-MS, showing the two distinct steps of sample pretreatment (a) and vaporization (b) into the plasma. (Copyright © 2013, all rights reserved, PerkinElmer Inc.)

pg spikes of a selected group of elements in concentrated hydrochloric acid (37% w/w), using a graphite furnace–based ETV-ICP-MS.[42] It can be seen, in particular, that good sensitivity is obtained for $^{51}V^+$, $^{56}Fe^+$, and $^{75}As^+$, which would have been virtually impossible by direct aspiration because of spectral overlaps from $^{39}ArH^+$, $^{35}Cl16O^+$, $^{40}Ar16O^+$, and $^{40}Ar^{35}Cl^+$, respectively. The removal of the chloride and water from the matrix translates into parts per trillion detection limits directly in 37% HCl, as shown in Table 20.3.

It can also be seen in Figure 20.9 that the elements are vaporized off the graphite tube in order of their boiling points. In other words, magnesium, which is the most volatile, is driven off first, whereas V and Mo, which are the most refractory, come off last. However, even though they emerge at different times, the complete transient event lasts less than 3 s. This physical time limitation, imposed by the duration of the transient signal, makes it imperative that all isotopes of interest be measured under the highest signal-to-noise conditions throughout the entire event.

The rapid nature of the transient has also limited the usefulness of ETV sampling for routine multielement analysis, because realistically only a small number of elements can be quantified with good accuracy and precision in less than 3 s. In addition, the development of low-flow nebulizers, desolvation devices, automated online chemistry, cool plasma technology, and collision/reaction cells and interfaces has meant that multielement analysis can now be carried out on difficult matrices without the need for ETV sample introduction. This has limited its use to more research-oriented applications.

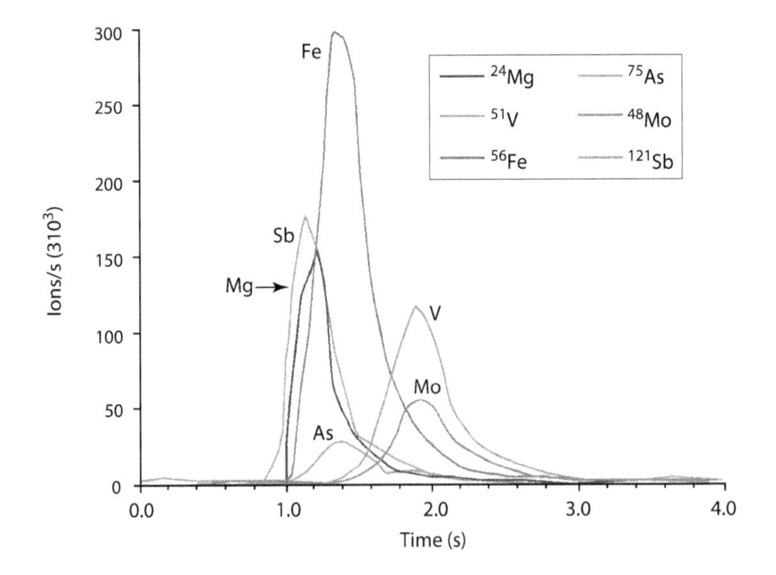

FIGURE 20.9 A temporal display of 50 pg of Mg, Sb, As, Fe, V, and Mo in 37% hydrochloric acid by ETV-ICP-MS. (From S. A. Beres, E. R. Denoyer, R. Thomas, P. Bruckner, *Spectroscopy*, 9(1), 20–26, 1994.)

TABLE 20.3
Detection Limits for V, Fe, and As in 37% Hydrochloric Acid by ETV-ICP-MS

Element	DL (ppt)
$^{51}V+$	50
$^{56}Fe+$	20
$^{75}As+$	40

Source: Beres, S. A., et al., *Spectroscopy*, 9(1), 20–26, 1994.
Note: DL, detection limit.

CHILLED SPRAY CHAMBERS AND DESOLVATION DEVICES

Chilled or cooled spray chambers and desolvation devices are becoming more and more common in ICP-MS, primarily to cut down the amount of liquid entering the plasma in order to reduce the severity of the solvent-induced spectral interferences, such as oxides, hydrides, hydroxides, and argon- or solvent-based polyatomic interferences. They are very useful for aqueous-type samples, but probably more important for volatile organic solvents, because there is a strong possibility that the sample aerosol would extinguish the plasma unless modifications are made to the sampling procedure. The most common chilled spray chambers and desolvation systems being used today include the following:

- Water-cooled spray chambers
- Peltier-cooled spray chambers
- Ultrasonic nebulizers (USNs)
- USNs coupled with membrane desolvation
- Specialized microflow nebulizers coupled with desolvation techniques

Let us take a closer look at these devices.

Water-Cooled and Peltier-Cooled Spray Chambers

Water-cooled spray chambers have been used in ICP-MS for many years and are standard on a number of today's commercial instrumentation to reduce the amount of water or solvent entering the plasma. However, the trend today is to cool the sample using a thermoelectric device called a Peltier cooler. Thermoelectric cooling (or heating) uses the principle of generating a hot or cold environment by creating a temperature gradient between two different materials. It uses electrical energy via a solid-state heat pump to transfer heat from a material on one side of the device to a different material on the other side, thus producing a temperature gradient across the device (similar to a household air conditioning system). Peltier cooling devices, which are typically air-cooled (but water cooling is an option), can be used with any kind of spray chamber and nebulizer, but commercial products for use with ICP-MS are normally equipped with a cyclonic spray chamber and a low-flow pneumatic nebulizer.

The main purpose of cooling the sample aerosol is to reduce the amount of water or solvent entering the plasma by lowering the temperature of the spray chamber. This can be a few degrees below ambient or as low –20°C, depending on the type of samples being analyzed. This can have a threefold effect: First, it helps to minimize solvent-based spectral interferences, such as oxides and hydroxides formed in the plasma, and second, because very little plasma energy is needed to vaporize the solvent, it allows more energy to be available to excite and ionize the analyte ions. Third, there is evidence to suggest that cooling the spray chamber will help minimize signal drift due to external environmental temperature changes in the laboratory.

Cooling the spray chamber to as low as –20°C by either Peltier cooling or a recirculating system using ethylene glycol as the coolant is particularly useful when it comes to analyzing some volatile organic samples. It has the effect of reducing the amount of organic solvent entering the interface, and when combined with the addition of a small amount of oxygen into the nebulizer gas flow, it is beneficial in reducing the buildup of carbon deposits on the sampler cone orifice and also minimizing the problematic carbon-based spectral interferences.[43]

Some systems also have the ability to increase the temperature of the spray chamber above ambient. Studies have shown that the sensitivity for many analytes is enhanced by a factor of two- to threefold by running the spray chamber as high as 60°C, a feature that is particularly important for samples with limited volume, or viscous matrices, such as engine or edible oils.

Ultrasonic Nebulizers

Ultrasonic nebulization was first developed in the late 1980s for use with ICP optical emission.[44] Its major benefit was that it offered an approximately 10× improvement in detection limits, because of its more efficient aerosol generation. However, this was not such an obvious benefit for ICP-MS, because more matrix entered the system than with a conventional nebulizer, increasing the potential for signal drift, matrix suppression, and spectral interferences. This was not such a major problem for simple aqueous-type samples, but was problematic for real-world matrices. The elements that showed the most improvement were the ones that benefited from lower solvent-based spectral interferences. Unfortunately, many of the other elements exhibited higher background levels and, as a result, showed no significant improvement in detection limit. In addition, the increased amount of matrix entering the mass spectrometer usually necessitated the need for larger dilutions of the sample, which again negated the benefit of using a USN with ICP-MS for samples with a heavier matrix. This limitation led to the development of a USN fitted with an additional membrane desolvator. This design virtually removed all the solvent from the sample, which dramatically improved detection limits for a large number of the problematic elements and also lowered metal oxide levels by at least an order of magnitude.[45]

The principle of aerosol generation using a USN is based on a sample being pumped onto a quartz plate of a piezoelectric transducer. An electrical energy of 1–2 MHz frequency is coupled to the transducer, which causes it to vibrate at high frequency. These vibrations disperse the sample into a fine-droplet aerosol, which is carried in a stream of argon. With a conventional USN, the aerosol is passed through a heating tube and a cooling chamber, where most of the sample solvent is removed as a condensate before it enters the plasma. If a membrane desolvation system is fitted to the USN, it is positioned after the cooling unit. The sample aerosol enters the membrane desolvator, where the remaining solvent vapor passes through the walls of a tubular microporous membrane. A flow of argon gas removes the volatile vapor from the exterior of the membrane, while the analyte aerosol remains inside the tube and is carried into the plasma for ionization. Membrane desolvation systems also have the capability to add a secondary gas, such as nitrogen, which has shown to be very beneficial in changing the ionization conditions to reduce levels of oxides in the plasma. The combination of membrane desolvation with a USN can be seen more clearly in Figures 20.10 and 20.11 shows the principles of membrane desolvation with water vapor as the solvent.

For ICP-MS, the system is best operated with both desolvation stages working, although for less demanding ICP-OES analysis, the membrane desolvation stage can be bypassed if required. The power of the system when coupled to an ICP mass spectrometer can be seen in Table 20.4, which compares the sensitivity (counts per second) and signal-to-background ratio of a membrane desolvation USN with those of a conventional cross-flow nebulizer for two classic solvent-based polyatomic interferences, $^{12}C^{16}O_2^+$ on $^{44}Ca^+$ and $^{40}Ar^{16}O^+$ on $^{56}Fe^+$, using a quadrupole ICP-MS system.

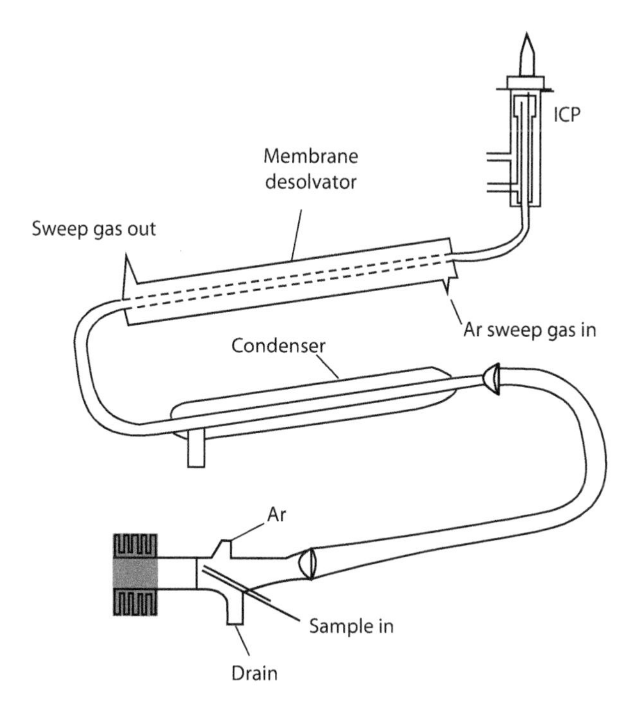

FIGURE 20.10 Schematic of an ultrasonic nebulizer fitted with a membrane desolvation system. (Courtesy of Teledyne Cetac Technologies.)

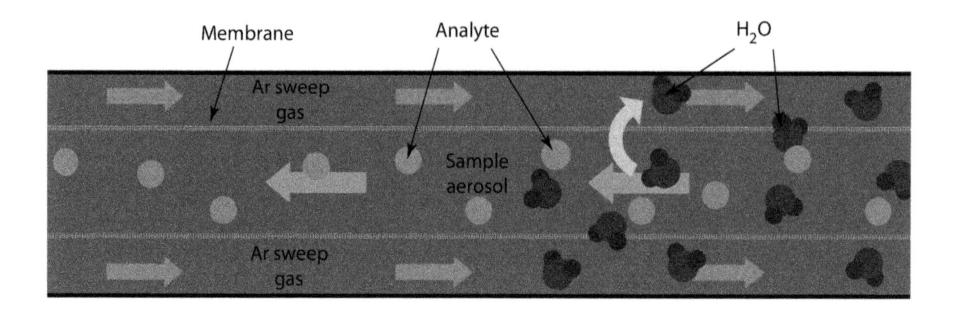

FIGURE 20.11 Principles of membrane desolvation showing the water molecules passing through a microporous membrane and being swept away by the argon gas, while the analyte is transported through the tube to the plasma. (Courtesy of Elemental Scientific Inc.)

TABLE 20.4
Comparison of Sensitivity and S/B Ratios—^{44}Ca+ and ^{56}Fe+—Using a Cross-Flow Nebulizer and a USN Membrane Desolvation System

Analytical Mass	Cross-Flow Nebulizer (cps)	Signal/BG	USN with Membrane Desolvation (cps)	Signal/BG
25 ppb 44Ca+ (BG subtracted)	2,300	2,300/7,640 = 0.30	20,800	20,800/1730 = 12.0
12C16O2+ (BG)	7,640		1,730	
10 ppb 56Fe+ (BG subtracted)	95,400	95,400/868,00 = 0.11	262,000	262,000/8,200 = 32.0
40Ar16O+ (BG)	868,000		8,200	

Source: Courtesy of Cetac Technologies, Omaha, NE.
Note: Signal/BG is calculated as the background subtracted signal divided by the background.

It can be seen that for the two analyte masses, the signal-to-background ratio is significantly better with the membrane-desolvated USN than with the cross-flow design, which is a direct result of the reduction of the solvent-related spectral background levels. This approach is even more beneficial for the analysis of organic solvents because when they are analyzed by conventional nebulization, modifications have to be made to the sampling process, such as the addition of oxygen to the nebulizer gas flow, the use of a low-flow nebulizer, and probably external cooling of the spray chamber. With a membrane desolvation USN system, volatile solvents such as isopropanol can be directly aspirated into the plasma with relative ease. However, it should be mentioned that depending on the sample type, this approach does not work for analytes that are bound to organic molecules. For example, the high volatility of certain mercury and boron organometallic species means that they could pass through the microporous polytetrafluoroethylene (PTFE) membrane and never make it into the ICP-MS. In addition, samples with high dissolved solids, especially ones that are biological in nature, could possibly result in clogging the microporous membrane unless substantial dilutions are made. For these reasons, caution must be used when using a membrane desolvation system for the analysis of certain types of sample matrices.

Specialized Microflow Nebulizers with Desolvation Techniques

Microflow or low-flow nebulizers, which were described in greater detail in Chapter 3, are being used more and more for routine applications. The most common ones used in ICP-MS are based on the microconcentric design, which operates at sample flows of 20–500 μL/min. Besides being ideal for small sample volumes, the major benefit of microconcentric nebulizers is that they are more efficient and produce smaller droplets than a conventional nebulizer. In addition, many microflow nebulizers use chemically inert plastic capillaries, which makes them well suited for the analysis of highly corrosive chemicals. This kind of flexibility has made low-flow nebulizers very popular, particularly in the semiconductor industry, where it is essential to analyze high-purity materials using a sample introduction system that is free of contamination.[46]

Such is the added capability and widespread use of these nebulizers across all application areas that manufacturers are developing application-specific integrated systems that include the spray chamber and a choice of different desolvation techniques to reduce the amount of solvent aerosol entering the plasma. Depending on the types of samples being analyzed, some of these systems include a low-flow nebulizer, Peltier-cooled spray chambers, heated spray chambers, Peltier-cooled condensers, and membrane desolvation technology. Some of the commercially available equipment includes the following:

- *Microflow nebulizer coupled with a Peltier-cooled spray chamber:* An example of this is the PC3 from Elemental Scientific Inc. (ESI). This system is offered with or without the nebulizer and utilizes a Peltier-cooled cyclonic spray chamber made from either quartz, borosilicate glass, or a fluoropolymer. Options include the ability to reduce the temperature to −20°C for analyzing organics and a dual-spray chamber for improved stability.
- *Microflow nebulizer with heated spray chamber and Peltier-cooled condenser:* An example of this design is the Apex inlet system from ESI. This unit includes a microflow nebulizer, heated cyclonic spray chamber (up to 140°C), and Peltier multipass condenser or cooler (down to −5°C). A number of different spray chamber and nebulizer options and materials are available, depending on the application requirements. Also, the system is available with Teflon or Nafion microporous membrane desolvation, depending on the types of samples being analyzed. Figure 20.12 shows a schematic of the Apex sample inlet system with the cross-flow nebulizer.

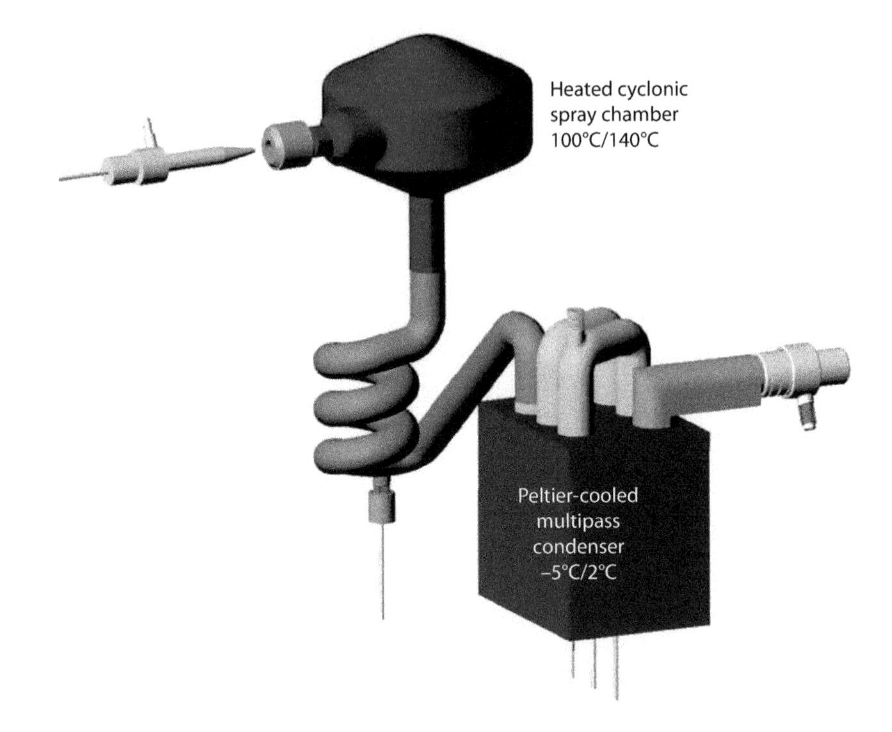

FIGURE 20.12 A schematic of the Apex sample inlet system. (Courtesy of Elemental Scientific Inc.)

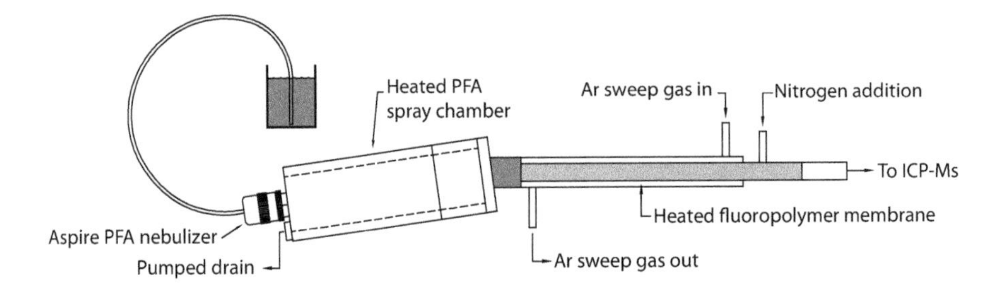

FIGURE 20.13 A schematic of the Aridus II microflow nebulizer with membrane desolvation. (Courtesy of Teledyne Cetac Technologies.)

- *Microflow nebulizer coupled with membrane desolvation:* An example of this is the Aridus system from Teledyne Cetac Technologies. The aerosol from the nebulizer is either self-aspirated or pumped into a heated perfluoroalkoxy spray chamber (up to 110°C) to maintain the sample in a vapor phase. The sample vapor then enters a heated PTFE membrane desolvation unit, where a counterflow of argon sweep gas is added to remove solvent vapors that permeate the microporous walls of the membrane. Nonvolatile sample components do not pass through the membrane walls, but are transported to the ICP-MS for analysis. Figure 20.13 shows a schematic of the Aridus II microflow nebulizer with membrane desolvation.

There is an extremely large selection of these specialized sample introduction techniques, so it is critical that you talk to the vendors so they can suggest the best solution for your application problem. They may not be required for the majority of your application work, but there is no question that they can be very beneficial for analyzing certain types of sample matrices and for elements that might be prone to solvent-based spectral overlaps.

DIRECT INJECTION NEBULIZERS

Direct injection nebulization is based on the principle of injecting a liquid sample under high pressure directly into the base of the plasma torch.[47] The benefit of this approach is that no spray chamber is required, which means that an extremely small volume of sample can be introduced directly into the ICP-MS with virtually no carryover or memory effects from the previous sample. Because they are capable of injecting less than 5 μL of liquid, they have found use in applications where sample volume is limited or where the material is highly toxic or expensive.

They were initially developed more than 15 years ago and found some success in certain niche applications that could not be adequately addressed by other nebulization systems, such as introducing samples from a chromatography separation device into an ICP-MS or the determination of mercury by ICP-MS, which is prone to severe memory effects. Unfortunately, they were not considered particularly user-friendly and, as a result, became less popular when other sample introduction devices were developed to handle microliter sample volumes. More recently, a refinement of the direct injection nebulizer has been developed, called the direct injection high-efficiency nebulizer (DIHEN), which appears to have overcome many of the limitations of the original design.[48] The advantage of the DIHEN is its ability to introduce microliter volumes into the plasma at extremely low sample flow rates (1–100 μL/min), with an aerosol droplet size similar to that of a concentric nebulizer fitted with a spray chamber. The added benefit is that it is almost 100% efficient and has extremely low memory characteristics. A schematic of a commercially available DIHEN system is shown in Figure 20.14.

PRODUCTIVITY-ENHANCING TECHNIQUES

With the increasing demand to analyze more and more samples, carry out automated dilutions and additions of internal standards, and perform online chemistry procedures, manufacturers of autosamplers and sample introduction accessories are designing automated sampling systems to maximize sample throughput, minimize sample preparation times, and increase productivity.[49–51] Depending on the application requirements, this is being achieved in a number

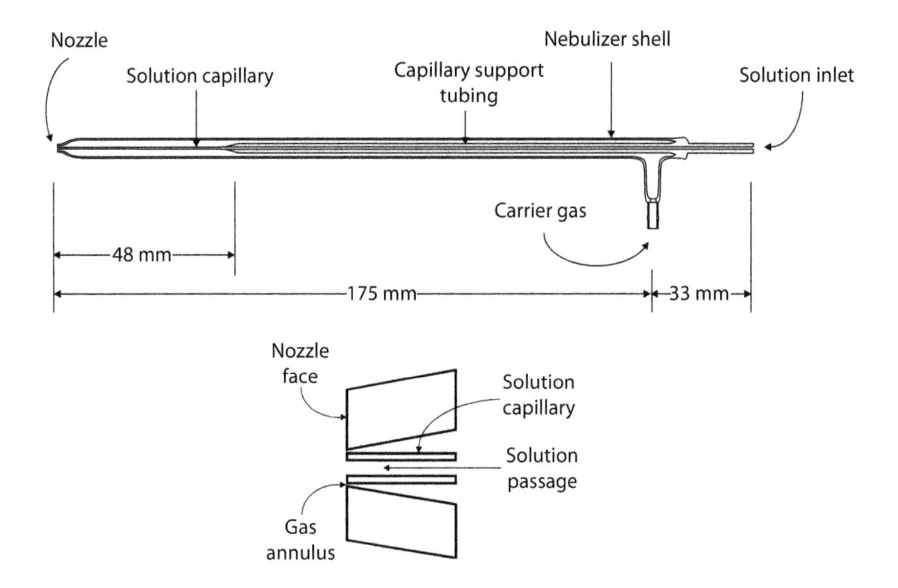

FIGURE 20.14 A schematic of a commercially available DIHEN system. (Courtesy of Meinhard Glass Products, a part of Elemental Scientific.)

of different ways using a variety of components, including multiport and switching valves; loops; vacuum, piston, or syringe pumps; mixing chambers; and ion exchange and preconcentration columns. There are basically three approaches to enhancing productivity in ICP-MS, depending on the application requirements:

1. Achieve faster analysis times by optimizing sample delivery to the instrument.
2. Perform online dilutions, internal standard additions. and calibrations to save manual operations.
3. Carry out automated chemistry online to remove sample matrices and/or preconcentrate the samples to reduce interferences and minimize labor-intensive, manual sample preparation steps.

Let's take a more detailed look at each of these approaches.

Faster Analysis Times

This is basically a rapid sampling approach integrated into an intelligent autosampler, which significantly reduces analysis times by optimizing the sample delivery process to reduce the pre- and postmeasurement time. There are a number of these systems on the market, which work slightly differently, but basically they all use piston, syringe, or vacuum pumps and switching valves and loops to control the delivery of the sample and standards to and from the ICP-MS. Besides significantly faster analysis times, other benefits include improved precision and accuracy, reduced carryover, and longer lifetime of sample introduction consumables. Depending on the design of the system, some typical areas of optimization include the following:

Autosampler response is the time it takes for the instrument to send a signal to the autosampler to move the sample probe to the next sample. By moving the autosampler probe over to the next sample while the previous sample is being analyzed, a significant amount of time will be saved over the entire automated run.

Sample uptake is the time taken for a sample to be drawn into the autosampler probe and pass through the capillary and pump tubing into the nebulizer. By using a small vacuum pump to rapidly fill the sample loop, which is positioned in close proximity of the sample loop to the nebulizer, sample uptake time is minimized.

Signal stabilization is the time required to allow the plasma to stabilize after air has entered the line from the autosampler probe dipping in and out of the sample tubes (this can also be exaggerated if the pump speed is increased to help in sample delivery). However, if the pump delivering the sample to the plasma remains at a constant flow rate, and the injection valve ensures no air is introduced into the sample line, very little stabilization time is required.

Rinse-out is the time required to remove the previous sample from the sample tubing and sample introduction system. So, if the probe is being rinsed during the sample analysis, minimal rinse time is needed.

Overhead time is the time spent by the ICP performing calculations and printing results, so if this time is used to ensure that the previous sample has reached baseline, minimal rinse time is required for the next sample.

Another slightly different approach is to use a rapid-rinse accessory, based on an FI loop and a positive displacement piston pump coupled to an autosampler. With this system, the multiport valve switches between two positions. In the first position, a loop of capillary tubing is filled with the sample, while in the second position the sample is delivered to the nebulizer. This is seen in greater detail in Figure 20.15a, which shows the piston pump rapidly filling the sample loop. At the same time, rinse and internal standard solutions are delivered to the nebulizer, washing out the nebulizer and spray chamber, and ensuring that plasma stability is maintained. Figure 20.15b shows the actual aspiration process where the valve switches position so that the rinse solution pushes the sample into the nebulizer. The internal standard is then mixed with the sample within the valve. At the same time, the autosampler probe and sample uptake tubing are rinsed by the piston pump.

There are a number of these enhanced productivity sampling systems on the market, which all work in a slightly different way. However, they all have one thing in common, and that is they offer at least a twofold reduction in analysis time compared with traditional autosamplers. Some of the other benefits that are realized with this time saving include

- Improved precision because of no pulsing from the peristaltic pump
- Better accuracy due to online dilution and addition of internal standards

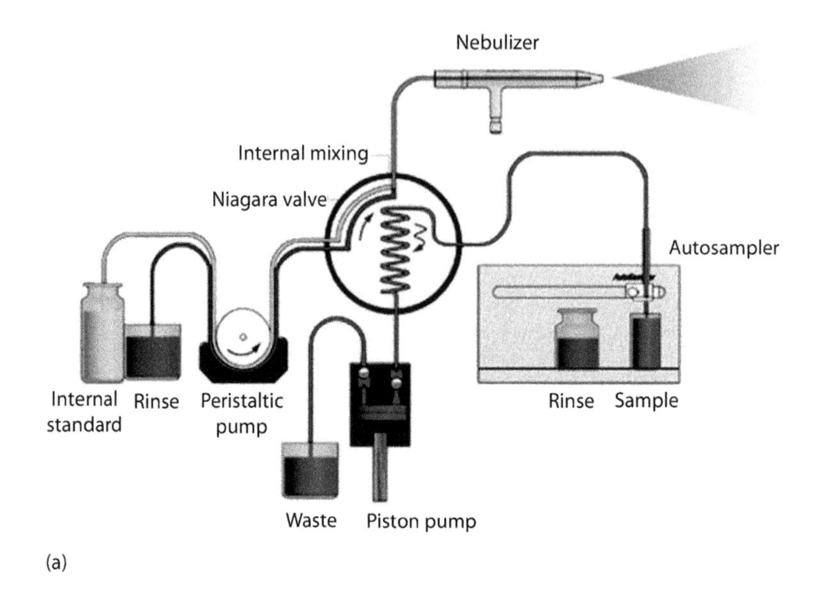

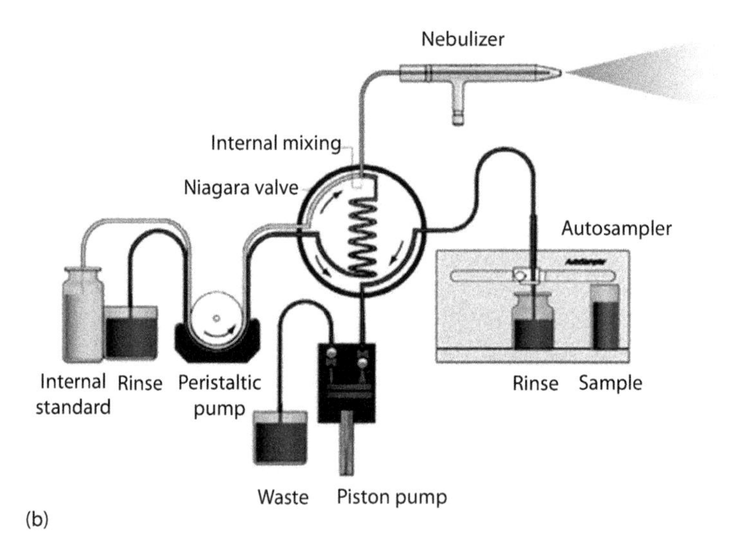

FIGURE 20.15 (a): The piston pump rapidly fills the sample loop, while at the same time, rinse and internal standard solutions are delivered to the nebulizer (courtesy of Glass Expansion Inc). (b): The valve switches position so that the rinse solution pushes the sample into the nebulizer, while the internal standard is mixed with the sample and the autosampler probe and sample uptake tubing are rinsed by the piston pump (courtesy of Glass Expansion Inc).

- Constant flow of solutions to plasma reduces stabilization times
- Less sample volume used
- Lower argon consumption

- Reduced cost of consumables
- Less routine maintenance
- Much lower chemical waste

There is no question that all these benefits can make a significant improvement in the overall cost of analysis, especially in high-workload routine environmental laboratories, where high sample throughput is an absolute requirement.[52,53]

Automated In-Line Autodilution and Autocalibration

A new range of automated sampling accessories has recently been developed, which performs very precise and accurate online autodilutions and autocalibration procedures using syringe and piston pumps.[54] Samples are rapidly and reproducibly loaded from each autosampler location into a sample loop. From there, the sample is injected into a diluent liquid stream and transported to a tee located between the valve and nebulizer. The internal standard is added in the tee to obtain final dilution factors defined by the operator. At the heart of the system is a syringe pump, which delivers the sample over a wide range of flow rates, ensuring rapid and reliable in-line dilutions. The benefits of fully automated in-line autodilution and autocalibration include

- Real-time dilutions
- Dilution in valve head and tee
- No additional tube or reagents required
- Eliminates manual dilutions
- Rapid uptake and washout
- Lowers risk of contamination
- Sample analysis time constant, independent of dilution factor

This approach has shown itself to be ideally-suited to the demands of analyzing samples according to the new USP chapters on elemental impurities in pharmaceutical products. In a recent study, this automated sampling technique coupled with ICP-MS was applied to the analysis of a group of over-the-counter medications according to Chapters <232> and <233>. It demonstrated that ICP-MS can comfortably achieve the permissable daily exposure (PDE) limits defined in Chapter <232> for the four major drug delivery categories. In addition, it also showed that an in-line auto-dilution/auto-calibration sample delivery system was well-suited to meet the strict quality control and validation protocols outlined in Chapter <233>, including J-value based drift requirements, detectability, repeatability, specificity, accuracy, ruggedness, and spike recovery specifications. The study also showed that the labor-intensive steps of preparation of calibration standards, dilution of samples/standards, adding spiked additions, adding internal standards, as well as dilutions and reruns of over range samples and so on, can all be fully-automated if needed. Furthermore, it significantly lowered the risk human error, as well as contamination of the sample, standards or blanks, because all these functions are being carried out in-line, with no manual intervention by the analyst.[55]

AUTOMATED SAMPLE IDENTIFICATION AND TRACKING

A recent development in automation is in the area of advanced, automated sample identification and tracking systems that can accurately associate stored information with a sample throughout the sample collection, preparation, and introduction process. Through a series of four distinct stages, this technology uses barcodes to enter, store, and reference data associated with a sample from initial collection to taring and final dilution.[56]

For example, from the point of sample collection, the technology associates collection time and global positioning system (GPS) location with sample container barcode. This then relates the information from its location and tracks it through the dilution process, inputting weight measurements before the sample preparation process. The software then has the ability to automatically apply sample digestion procedures according to sample ID code. Finally, the system presents the samples for ICP-MS and ICP-OES analysis while confirming sample identity and providing legally defensible data for regulatory inspection.

This technology is ideally suited to any application area, such as the pharmaceutical industry, where it is critical to keep track of the sample, as it moves from collection to preparation, dilution, and so forth, to finally being analyzed by the instrumentation of choice that is generating the data to make a decision about the levels of elemental impurities in a drug compound or pharmaceutical raw material.

REFERENCES

1. E. R. Denoyer, K. J. Fredeen, and J. W. Hager. *Analytical Chemistry*, 63(8), 445–457A, 1991.
2. J. F. Ready. *Effects of High Power Laser Radiation*. New York: Academic Press, 1972, chaps. 3–4.
3. L. Moenke-Blankenburg. *Laser Microanalysis*. New York: Wiley, 1989.
4. E. R. Denoyer, R. Van Grieken, F. Adams, and D. F. S. Natusch. *Analytical Chemistry*, 54, 26A, 1982.
5. J. W. Carr and G. Horlick. *Spectrochimica Acta*, 37B, 1, 1982.
6. T. Kantor et al. *Talanta*, 23, 585, 1979.
7. H. C. G. Human et al. *Analyst*, 106, 265, 1976.
8. M. Thompson, J. E. Goulter, and F. Seiper. *Analyst*, 106, 32, 1981.
9. A. L. Gray. *Analyst*, 110, 551, 1985.
10. P. A Arrowsmith and S. K. Hughes. *Applied Spectroscopy*, 42, 1231–1239, 1988.
11. T. Howe, J. Shkolnik, and R. Thomas. *Spectroscopy*, 16(2), 54–66, 2001.
12. D. Günther and B. Hattendorf. *Mineralogical Association of Canada—Short Course Series*, 29, 83–91, 2001.
13. T. E. Jeffries, S. E. Jackson, and H. P. Longerich. *Journal of Analytical Atomic Spectrometry*, 13, 935–940, 1998.
14. R. E. Russo, X. L. Mao, O. V. Borisov, and L. Haichen. *Journal of Analytical Atomic Spectrometry*, 15, 1115–1120, 2000.
15. H. Liu, O. V. Borisov, X. Mao, S. Shuttleworth, and R. Russo. *Applied Spectroscopy*, 54(10), 1435, 2000.
16. SEM photo courtesy of Dr. Honglin Yuan, Northwest University, Xi'an, China.
17. S. E. Jackson, H. P. Longerich, G. R. Dunning, and B. J. Fryer. *Canadian Mineralogist*, 30, 1049–1064, 1992.

18. D. Günther and C. A. Heinrich. *Journal of Analytical Atomic Spectrometry*, 14, 1369, 1999.
19. D. Günther, I. Horn, and B. Hattendorf. *Fresenius' Journal of Analytical Chemistry*, 368, 4–14, 2000.
20. R. E. Wolf, C. Thomas, and A. Bohlke. *Applied Surface Science*, 127–129, 299–303, 1998.
21. J. Gonzalez, X. L. Mao, J. Roy, S. S. Mao, and R. E. Russo. *Journal of Analytical Atomic Spectrometry*, 17, 1108–1113, 2002.
22. R. Lam and E.D. Salin. *Journal of Analytical Atomic Spectrometry*, 19, 938–940, 2004.
23. J. Roy and L. Neufeld. *Spectroscopy*, 19(1), 16–28, 2004.
24. J. Ruzicka and E. H. Hansen. *Analytic Chimica Acta*, 78, 145, 1975.
25. R. Thomas. *Spectroscopy*, 17(5), 54–66, 2002.
26. A. Stroh, U. Voellkopf, and E. Denoyer. *Journal of Analytical Atomic Spectrometry*, 7, 1201, 1992.
27. Y. Israel, A. Lasztity, and R. M. Barnes. *Analyst*, 114, 1259, 1989.
28. Y. Israel and R. M. Barnes. *Analyst*, 114, 843, 1989.
29. M. J. Powell, D. W. Boomer, and R. J. McVicars. *Analytical Chemistry*, 58, 2864, 1986.
30. S. N. Willie, Y. Iida, and J. W. McLaren. *Atomic Spectroscopy*, 19(3), 67, 1998.
31. R. Roehl and M. M. Alforque. *Atomic Spectroscopy*, 11(6), 210, 1990.
32. J. W. McLaren, J. W. H. Lam, S. S. Berman, K. Akatsuka, and M. A. Azeredo. *Journal of Analytical Atomic Spectrometry*, 8, 279–286, 1993.
33. L. Ebdon, A. Fisher, H. Handley, and P. Jones. *Journal of Analytical Atomic Spectrometry*, 8, 979–981, 1993.
34. D. B. Taylor, H. M. Kingston, D. J. Nogay, D. Koller, and R. Hutton. *Journal of Analytical Atomic Spectrometry*, 11, 187–191, 1996.
35. S. M. Nelms, G. M. Greenway, and D. Koller. *Journal of Analytical Atomic Spectrometry*, 11, 907–912, 1996.
36. C. J. Park, J. C. Van Loon, P. Arrowsmith, and J. B. French. *Analytical Chemistry*, 59, 2191–2196, 1987.
37. R. D. Ediger and S. A. Beres. *Spectrochimica Acta*, 47B, 907, 1992.
38. C. J. Park and M. Hall. *Journal of Analytical Atomic Spectrometry*, 2, 473–480, 1987.
39. C. J. Park and J. C. Van Loon. *Trace Elements in Medicine*, 7, 103, 1990.
40. G. Chapple and J. P. Byrne. *Journal of Analytical Atomic Spectrometry*, 11, 549–553, 1996.
41. U. Voellkopf, M. Paul, and E. R. Denoyer. *Fresenius' Journal of Analytical Chemistry*, 342, 917–923, 1992.
42. S. A. Beres, E. R. Denoyer, R. Thomas, and P. Bruckner. *Spectroscopy*, 9(1), 20–26, 1994.
43. F. McElroy, A. Mennito, E. Debrah, and R. Thomas. *Spectroscopy*, 13(2), 42–53, 1998.
44. K. W. Olson, W. J. Haas Jr., and V. A. Fassel. *Analytical Chemistry*, 49(4), 632–637, 1977.
45. J. Kunze, S. Koelling, M. Reich, and M. A. Wimmer. Atomic Spectroscopy, 19, 5, 1998.
46. G. Settembre and E. Debrah. *Micro*, June 1998.
47. D. R. Wiederin and R. S. Houk. *Applied Spectroscopy*, 45(9), 1408–1411, 1991.
48. J. A. McLean, H. Zhang, and A. Montaser. *Analytical Chemistry*, 70, 1012–1020, 1998.
49. Glass Expansion. Characterization of a customized valve for enhanced productivity in ICP/ICP-MS. Application note. http://www.geicp.com/site/images/application_notes/NiagaraApplicationsWhitePaper_Feb2010.pdf.
50. Elemental Scientific Inc. Automated sample introduction with the SC fast. Application note. http://www.icpms.com/pdf/SC-FAST.pdf.
51. Teledyne Cetac Technologies. ASXpress Plus rapid sample introduction system. http://www.teledynecetac.com/products/automation/asxpress-plus.

52. Elemental Scientific Inc. Improving throughput of environmental samples by ICP-MS following EPA Method 200.8. Application note. http://www.icpms.com/products/sc-fast-enviro.php.
53. M. P. Field, M. LaVigne, K. R. Murphy, G. M. Ruiz, and R. M. Sherrell. *Journal of Analytical Atomic Spectrometry*, 22, 1145, 2007.
54. Elemental Scientific Inc. The evolution of automation. Prepfast data sheet. http://www.icpms.com/products/prepfast.php.
55. L. Davidowski, A. Shultz, K. Uhlmeyer, E. Pruszkowski, and R. Thomas. ICP-MS with AutoDilution and Auto Calibration for Implementing the New USP Chapters on Elemental Impurities, Applications of ICP & ICP-MS Techniques for Today's Spectroscopist, Supplement to Spectroscopy Magazine, November, 40-47, 2012.
56. Elemental Scientific Inc. PlasmaTrax: Automated sample identification and tracking system. Application note. http://www.icpms.com/pdfv1/15054-1%20PlasmaTRAX-2D%20Barcode%20Automation.pdf.

21 Coupling ICP-MS with Chromatographic Separation Techniques for Speciation Studies

The specialized sample introduction techniques described in Chapter 20 were mainly developed as a result of a basic limitation of inductively coupled plasma mass spectrometry (ICP-MS) in carrying out elemental determinations on certain types of complex sample matrices. However, even though all these sampling accessories significantly improved the flexibility, performance, and productivity of the technique, they were still being used to measure the total metal content of the samples being analyzed. If the requirement was to learn more about the oxidation state or speciated form of the element, the trace metal analytical community had to look elsewhere for answers. Then, in the early 1990s, researchers started investigating the use of ICP-MS as a detector for chromatography systems, which triggered an explosion of interest in this exciting new hyphenated technique, especially for environmental and biomedical applications. In this chapter, we look at what drove this research and discuss the use of chromatographic separation techniques, including high-performance liquid chromatography (HPLC), with ICP-MS to carry out trace element speciation determinations. This chapter will be of particular interest if there is a requirement to measure different species of arsenic and/or mercury in a drug compound, or pharmaceutical or dietary supplement raw material.

ICP-MS has gained popularity over the years, mainly because of its ability to rapidly quantitate ultratrace metal contamination levels. However, in its basic design, ICP-MS cannot reveal anything about the metal's oxidation state or alkylated form, how it is bound to a biomolecule, or how it interacts at the cellular level. The desire to understand in what form or species an element exists led researchers to investigate the combination of chromatographic separation devices with ICP-MS. The ICP mass spectrometer becomes a very sensitive detector for trace element speciation studies when coupled to a chromatographic separation device based on HPLC, low-pressure liquid chromatography (LC), ion chromatography (IC), gas chromatography (GC), size exclusion chromatography (SEC), capillary electrophoresis (CE), and so forth. In these hyphenated techniques, elemental species are separated based on their chromatographic retention, mobility, or molecular size, and then eluted or passed into the ICP mass spectrometer for detection.[1] The intensities of the eluted peaks are then displayed for each isotopic mass of interest, in the time domain, as shown in Figure 21.1. The figure shows a typical time-resolved chromatogram for a selected group of masses.

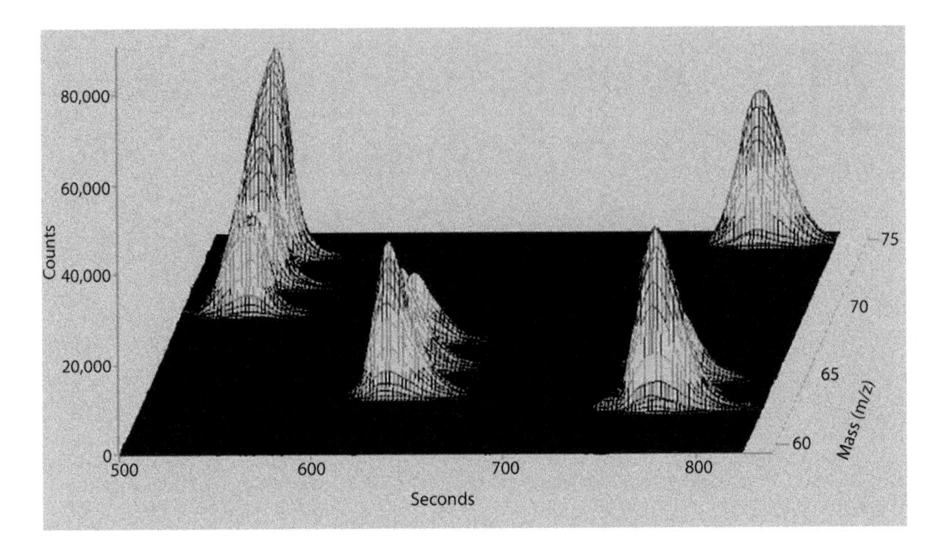

FIGURE 21.1 A typical time-resolved chromatogram generated using chromatography coupled with ICP-MS, showing a temporal display of intensity against mass (Copyright © 2013, all rights reserved, PerkinElmer Inc.).

There is no question that ICP-MS has allowed researchers in the environmental, biomedical, geochemical, and nutritional fields to gain a much better insight into the impact of different elemental species on humans and their environment. Even though elemental speciation studies were being carried out using other atomic spectrometry (AS) detection techniques, it was the commercialization of ICP-MS in the early 1980s, with its extremely low detection capability, that saw a dramatic increase in the number of trace element speciation studies being carried out. Today, the majority of these studies are being driven by environmental regulations. In fact, the U.S. Environmental Protection Agency (EPA) has published a number of speciation methods involving chromatographic separation with ICP-MS, including Method 321.8 for the speciation of bromine compounds in drinking water and wastewater, and Method 6800 for the measurement of various metal species in potable and wastewaters by isotope dilution mass spectrometry. However, other important areas of interest include nutritional and metabolic studies, toxicity testing, bioavailability measurements, and now with the new United States Pharmacopeia (USP) and International Conference on Harmonisation of Technical Requirements for Registration of Pharmaceuticals for Human Use (ICH) elemental impurity guidelines, the measurement of different arsenic and mercury species in pharmaceutical and nutraceutical materials. Speciation studies cross over many different application areas, but the majority of determinations being carried out can be classified into three major categories:

- *Measurement of different oxidation states*—For example, hexavalent chromium, Cr (VI), is a powerful oxidant and is extremely toxic, but in soil and water systems, it reacts with organic matter to form trivalent chromium,

Cr (III), which is the more common form of the element and an essential micronutrient for plants and animals.[2]

- *Measurement of alkylated forms*—Very often the natural form of an element can be toxic, although its alkylated form is relatively harmless, or vice versa. A good example of this is the element arsenic. Inorganic forms of the element, such as As (III) and As (V), are toxic, whereas many of its alkylated forms, such as monomethylarsonic acid (MMA) and dimethylarsonic acid (DMA), are relatively innocuous.[3]
- *Measurement of metallobiomolecules*—These molecules are formed by the interaction of trace metals with complex biological molecules. For example, in animal farming studies, the activity and mobility of an innocuous arsenic-based growth promoter are determined by studying its metabolic impact and excretion characteristics. So, measurement of the biochemical form of arsenic is crucial in order to know its growth potential.[4]

Table 21.1 represents a small cross section of both inorganic and organic species of interest classified under these three categories.

There are 400–500 speciation papers published every year, the majority of which are based on toxicologically significant elements, such as As, Cr, Hg, Se, and Sn.[5] The following is a small selection of some of the most recent research that can be found in the public domain:

- Determination of chromium (VI) in drinking water samples, using HPLC-ICP-MS[6]
- Determination of trivalent and hexavalent chromium in pharmaceutical and biological materials by IC-ICP-MS[7]
- The use of LC-ICP-MS in better understanding the role of inorganic and organic forms of selenium in biological processes[8]

TABLE 21.1
Some Typical Inorganic and Organic Species That Have Been Studied by Researchers Using Chromatographic Separation Techniques

Oxidation States	Alkylated Forms	Biomolecules
Se^{+4}	Methyl—Hg, Ge, Sn, Pb, As, Sb, Se, Te, Zn, Cd, Cr	Organometallic complexes—As, Se, Cd
Se^{+6}		Metalloporphyrines
As^{+3}	Ethyl—Pb, Hg	Metalloproteins
As^{+5}	Butyl, phenyl, cyclohexyl—Sn	Metallodrugs
Sn^{+2}		Metalloenzymes
Sn^{+4}		Metals at the cellular level
Cr^{+3}		
Cr^{+6}		
Fe^{+2}		
Fe^{+3}		

- Identification of selenium compounds in contaminated estuarine waters using IC-ICP-MS[9]
- Determination of organoarsenic species in marine samples, using cation exchange HPLC-ICP-MS[10]
- Determination of biomolecular forms of arsenic in chicken manure by CE-ICP-MS[11]
- Analysis of tributyltin (TBT) in marine samples using HPLC-ICP-MS[12]
- Measurement of anticancer platinum compounds in human serum by HPLC-ICP-MS[13]
- Bioavailability of cadmium and lead in beverages and foodstuffs using SEC-ICP-MS[14]
- Investigation of sulfur speciation in petroleum products by capillary GC with ICP-MS detection[15]
- Analysis of methyl mercury in water and soils by HPLC-ICP-MS[16]
- Analysis of arsenic and mercury species from botanicals and dietary supplements using LC-ICP-MS[17]

HPLC COUPLED WITH ICP-MS

It can be seen from this brief snapshot of speciation publications that by far the most common chromatographic separation techniques being used with ICP-MS are the many different types of HPLC, such as adsorption, ion exchange, size exclusion, gel permeation, and normal- or reverse-phase chromatography. To get a better understanding of how the technique works, particularly when attempting to develop a routine method to simultaneously measure multiple species in the same analytical run, let us take a more detailed look at how the HPLC system is coupled to the ICP mass spectrometer. Figure 21.2 shows a typical setup of the hardware components.

The coupling of the ICP-MS system to the liquid chromatograph hardware components is relatively straightforward, connecting a capillary tube from the end of the HPLC column, through a switching valve, to the sample introduction system of the ICP mass spectrometer. However, matching the column flow with the uptake of the ICP-MS sample introduction system is not a trivial task. Therefore, to develop a successful trace element speciation method, it is important to optimize not only the chromatographic separation, but also selection of the nebulization process to match the flow of the sample being eluted off the column, together with finding the best ICP-MS operating conditions for the analytes or species of interest. Let us take a closer look at this.

CHROMATOGRAPHIC SEPARATION REQUIREMENTS

Traditionally, the measurement of trace levels of elemental species by HPLC has been accomplished by separating the species using column separation technology and detecting them, one element at a time, as they elute. This approach works well for one element or species, but is extremely slow and time-consuming for the determination of multiple species or elements, because the chromatographic separation

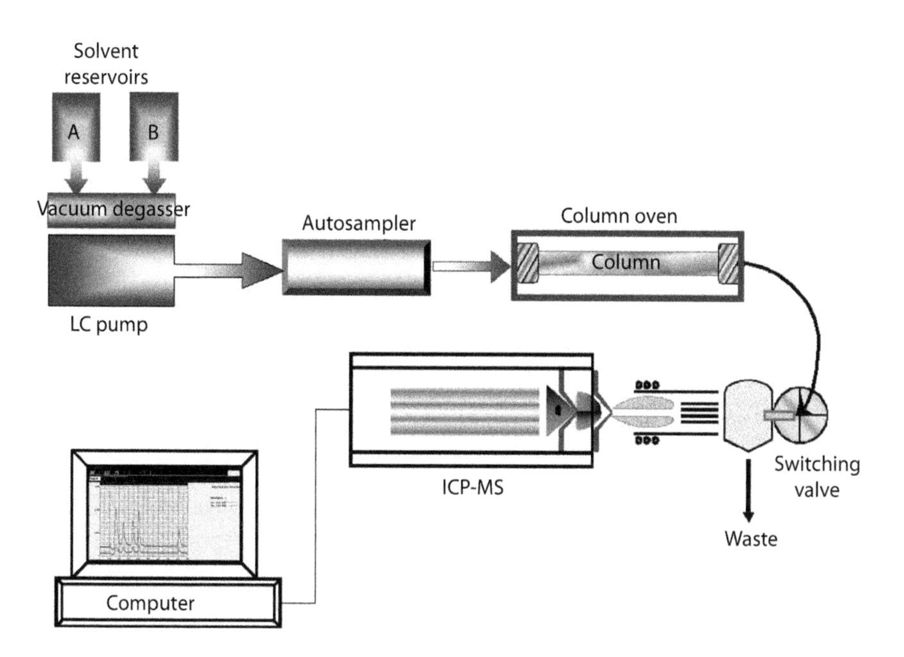

FIGURE 21.2 A typical configuration of an HPLC system interfaced with an ICP mass spectrometer. (Copyright © 2013, all rights reserved, PerkinElmer Inc.)

process has to be optimized for each species being measured. Therefore, the limitation to multielement speciation is rarely the detection of the elements, but is usually the separation of the species. The inherent problem is that LC works on the principle of equilibration between the species of interest, the mobile phase, and the column material. Because the chemistries of elements and their species differ, it is difficult to find common conditions capable of separating species of more than one element simultaneously. For example, toxic elements of environmental interest, such as arsenic (As), chromium (Cr), and selenium (Se), have different reaction chemistries, thus requiring different chromatographic conditions to separate them. So, to achieve the analytical goal of fast, simultaneous measurement of different species of these elements in a single sample injection, the chromatographic separation has to be optimized.

Let us first examine the chromatography, focusing on two common forms of separation (ion exchange and reverse-phase ion-pairing chromatography), together with two commonly employed HPLC elution techniques (isocratic and gradient). Each process is described, and the advantages and disadvantages of each are discussed in the context of choosing a separation scheme to achieve our analytical goals. One final note before we discuss the separation process: it is extremely important when preparing the sample that the speciated form of the element not be changed or altered in any way. This is not such a serious problem with a simple matrix, such as drinking water, which typically involves straightforward acidification. However, when more complex samplesm such as soils, biological samples, or pharmaceutical matrices, are being analyzed, it is quite common to use strong acids, oxidizing agents, or high temperatures to get the samples into solution. So, maintaining the integrity of the valency or oxidation state or species should always be an extremely important decision when preparing a sample for speciation analysis.

Ion Exchange Chromatography

In this technique, separation is based on the exchange of ions (anions or cations) between the mobile phase and the ionic sites on a stationary phase bound to a support material in the column. A charged species is covalently bound to the surface of the stationary phase. The mobile phase, typically a buffer solution, contains a large number of ions that have a charge opposite to that of the surface-bound ions. These mobile-phase ions, referred to as counterions, establish an equilibrium with the stationary phase. Sample ions passing through the column and having the same ionic charge as the counterion compete with the mobile-phase ions for sites on the column material, resulting in a disruption of the equilibrium and retention of that analyte. It is the differential competition of various analytes with the mobile-phase ions that ultimately produces species separation and chromatographic peaks, as detected by ICP-MS. Therefore, analyte retention is based on the affinity of different ions for the support material and on other solution parameters, including counterion type, ionic strength (buffer concentration), and pH. Varying these mobile-phase parameters changes the separation.

The principle of ion exchange chromatography (IEC) is schematically represented in Figure 21.3 for an anion exchange separation. In this figure, the anions

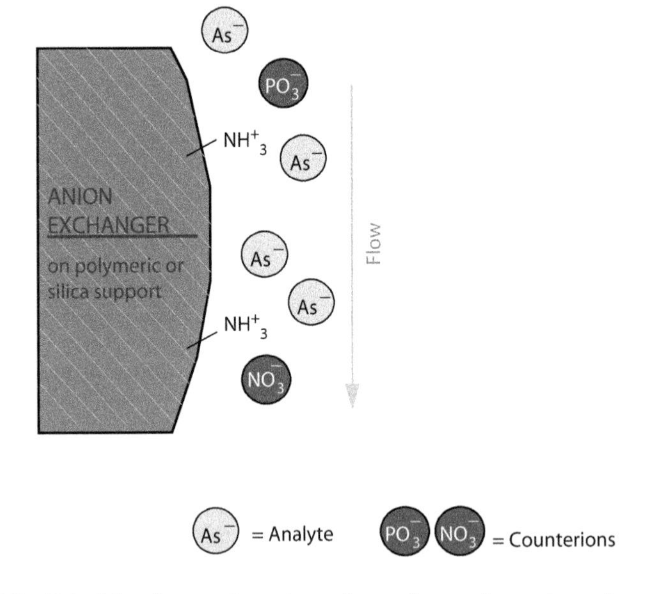

FIGURE 21.3 Principle of separation using anion exchange chromatography.

are denoted by A^- and represent the analyte species, whereas the counterions are symbolized by X^-. The main benefit of this approach is that it has a high tolerance to matrix components, and the main disadvantages are that these columns tend to be expensive and somewhat fragile.

Reverse-Phase Ion Pair Chromatography

Ion pair chromatography typically uses a reverse-phase column in conjunction with a special type of chemical in the mobile phase called an *ion-pairing reagent.* "Reverse phase" essentially means that the column's stationary phase is nonpolar (less polar, more organic) than the mobile-phase solvents. For reverse-phase ion pair chromatography (RP-IPC), the stationary phase is a carbon chain, most often consisting of 8 or 18 carbon atoms (C8, C18) bonded to a silica support. The mobile-phase solvents usually consist of water mixed with a water-miscible organic solvent, such as acetonitrile or methanol.

The ion-pairing reagent is a compound that has both an organic and an ionic end. To promote ionic interaction, the ionic end should have a charge opposite of that of the analytes of interest. For anionic IPC, a commonly used ion-pairing reagent is tetrabutylammonium hydroxide (TBAOH); for cationic IPC, hexanesulfonic acid (HSA) is often chosen. In principle, the ionized or ionizable analytes interact with the ionic end of the ion-pairing reagent, whereas the ion-pairing reagent's organic end interacts with the C18 stationary phase. Retention and separation selectivity are primarily affected by characteristics of the mobile phase, including pH, selection or concentration of ion-pairing reagent, and ionic strength, as well as the use of additional mobile-phase modifiers. With this scheme, a reverse-phase column acts like an ion exchange column.

The benefits of this type of separation are that the columns are generally less expensive and tend to be more rugged than ion exchange columns. The main disadvantages, compared with IEC, are that the separation may not be as good, and there may be less tolerance to high sample matrix concentrations. A simplified representation of this approach is shown in Figure 21.4.

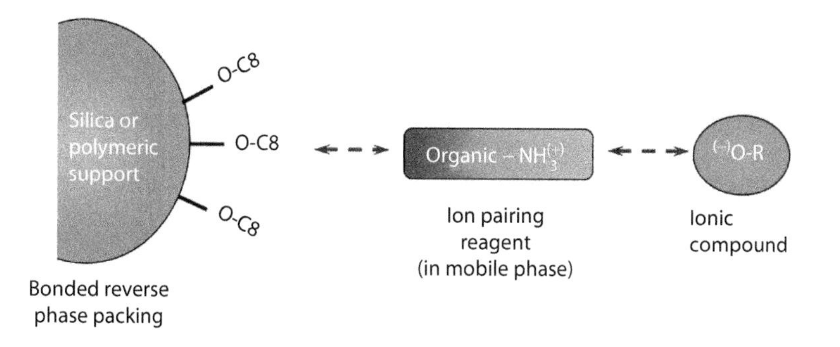

FIGURE 21.4 Principle of ion pair chromatography (anionic mode).

Column Material

As previous explained, several variables of the mobile phase can be modified to affect separations, by both IEC and RP-IPC. A very important consideration when choosing a column is the pH of the mobile phase used to achieve the separation. The pH is critical in the selection of the column for both IEC and RP-IPC because the support materials may be pH sensitive. One of the most common column materials is silica, because it is inexpensive and rugged, and has been used for many years. The main disadvantage of silica columns is that they are useful only over a pH range of 2–8. Outside this range, the silica dissolves. For applications requiring pH values outside of this range, polymer columns are commonly used. Polymer columns are useful over the entire pH range. The downside to these columns is that they are typically more expensive and fragile. Columns consisting of other support materials are also available.

Isocratic or Gradient Elution

Another consideration is how best to elute the species from the column. This can be accomplished using either an isocratic or gradient elution scheme. Isocratic refers to using the same solvent throughout the analysis. Because of this, samples can be injected immediately after the preceding one has finished eluting. This type of elution can be performed on all HPLC pumping systems. The advantages of isocratic elutions are simplicity and higher sample throughput.

Gradient elution involves variation of the mobile phase composition over time by a number of steps, such as changing the organic content, altering the pH, changing the concentration of the buffer, or using a completely different buffer. This is usually accomplished by having two or more bottles of mobile phases connected to the HPLC pump. The pump is then programmed to vary the amount of each mobile-phase component. The mobile-phase components are mixed online before reaching the column, and therefore a more complex pumping system is required to handle this task.

The main reason for using gradient elution is that the mobile-phase composition can be varied to fine-tune the separation. As a result, component separations are better, and chromatograms are generally shorter than with isocratic elutions. However, overall sample throughput is much lower than with isocratic separations because of the variation in mobile-phase composition. So, after an elution is complete, the mobile phase must be changed back to its original composition, as at the start of the chromatogram. This means that the equilibrium between the mobile phase and the column must be reestablished before the next sample can be injected. If the equilibrium is not established, the peaks will elute at different times compared with the previous sample. The equilibration time varies with column, but it is usually between 5 and 30 min.

Therefore, if a fast, automated, routine method for the measurement of multispecies or elements is the desired analytical goal, it is often best to attempt an isocratic separation method first, because of the complexity of method development and the low sample throughput of gradient elution methods. In fact, a simultaneous method

TABLE 21.2
HPLC Separation Parameters and Conditions for Measuring Inorganic As, Cr, and Se Species in Potable Waters

HPLC configuration	Quaternary pump, column oven, and autosampler
Column	C8, reduced activity, 3.3 × 0.46 cm (3 μm packing)
Column temperature	35°C
Mobile phase	mM TBAOH + 0.15 mM NH_4CH_2COOH + 0.15 mM EDTA (K salt) + 5% MeOH
pH	7.5
pH adjustment	Dilute HNO_3, NH_4OH
Injection volume	50 mL
Sample flow rate	1.5 mL/min
Samples	Various potable waters
Sample preparation	Dilute with mobile phase (2–10); heat at 50°C–55°C for 10 min

Source: K. R. Neubauer, P. A. Perrone, W. Reuter, R. Thomas, *Current Trends in Mass Spectroscopy*, May 2006.

for the separation of As, Cr, and Se species in drinking water samples was demonstrated by Neubauer and coworkers; they developed a method to determine inorganic forms of arsenic (As^{+3} and As^{+5}), chromium (Cr^{+3} and Cr^{+6}), and selenium (Se^{+4}, Se^{+6}, and $SeCN^-$) by reverse-phase ion-pairing chromatography with isocratic elution.[18] Details of the HPLC separation parameters and conditions they used are shown in Table 21.2.

It is important to emphasize that if these species were being measured on a single-element basis, the optimum chromatographic conditions would be different. However, the goal of this study was to determine all the species in a single multielement run so that the method could be applied to a routine, high-throughput environment. The 3.3 cm long column was packed with a C8 hydrocarbon material (3 μm particle size). The mobile phase was a mixture of TBAOH, ammonium acetate (NH_4CH_2COOH), the potassium salt of EDTA, and methanol. The pH was adjusted to 7.5 (prior to the addition of methanol) using 10% nitric acid and 10% ammonium hydroxide. The sample preparation consisted of dilution with the mobile phase and heating at 50°C–55°C to speed the formation of the Cr (III)–EDTA complex. For each analysis, 50 μL of sample was injected into the column, which resulted in a sample flow of 1.5 mL/min being eluted off the column into the ICP mass spectrometer.

SAMPLE INTRODUCTION REQUIREMENTS

When coupling an HPLC system to an ICP mass spectrometer, it is very important to match the flow of sample being eluted off the column with the ICP-MS nebulization system. With today's choice of sample introduction components, there are specialized nebulizers and spray chambers on the market that can handle sample flows from 20 up to 3000 μL/min. The most common type of nebulizer used for chromatography applications is the concentric design, because it is self-aspirating and generates an aerosol with extremely small droplets, which tends to produce better signal stability than a cross-flow design. The choice of which type of concentric nebulizer to use should therefore be based on the sample flow coming off the column. If the sample flow is on the order of 1 mL/min, a higher-flow concentric nebulizer should be used, and if the flow is much lower, such as in nano- or microflow LC work, a specialized low-flow nebulizer should be used.

It is also very important that the dead volume of the sample introduction process be kept to an absolute minimum to optimize peak integration of the separated species over the length of the transient signal. For this reason, the length of sample capillary from the end of the column to the nebulizer should be kept to a minimum; the internal volume of the nebulizer should be as small as possible; the connectors, fittings, and valves should all have low dead volume; and a self-aspirating nebulizer should be used to avoid the need for peristaltic pump tubing. In addition, a spray chamber with a short aerosol path should be selected, which will not add additional dead volume to the method. However, depending on the total flow of the sample and the type of nebulizer, a spray chamber may not even be required. Cutting down on the sample introduction dead volume and minimizing peak broadening by careful selection of column technology will ultimately dictate the number of species that can be separated in a given time—an important consideration when developing a routine method for high sample throughput. Figure 21.5 shows a high-efficiency concentric nebulizer (HEN) designed for HPLC-ICP-MS work, and Figure 21.6 shows the difference between the capillary of this nebulizer (a) and a standard concentric type A nebulizer (b).

Another important reason to match the nebulizer with the flow coming off the column is that concentric nebulizers are mainly self-aspirating. For this reason, the column flow must be high enough to ensure that the nebulizer can sustain a consistent and reproducible aerosol. On the other hand, if the column flow is too low, a makeup flow might need to be added to the column flow to meet the flow requirements of the nebulizer being used. This has the additional benefit of being able to add

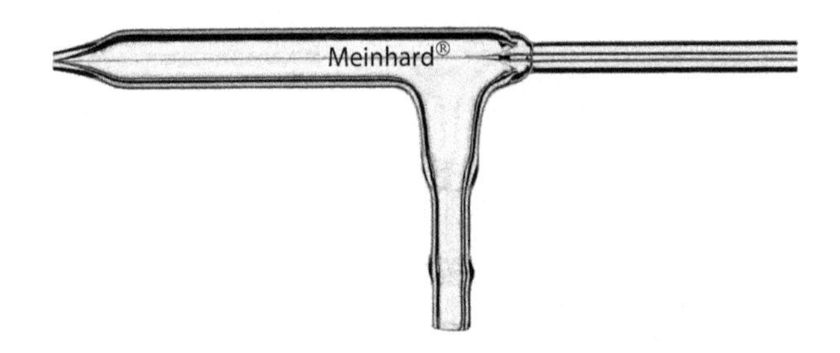

FIGURE 21.5 A high-efficiency concentric nebulizer (HEN) designed for HPLC-ICP-MS work. (Courtesy of Meinhard Glass Products.)

FIGURE 21.6 The difference between the capillary of a high-efficiency concentric nebulizer for HPLC work (a) and a standard concentric type A nebulizer (b). (Courtesy of Meinhard Glass Products.)

an internal standard after the column with another pump to correct for instrument drift or matrix effects in gradient elution work.

OPTIMIZATION OF ICP-MS PARAMETERS

In the early days of trace element speciation studies using chromatography coupled with ICP-MS, researchers had no choice but to interface their own LC pumps, columns, autosamplers, and so forth, to the ICP mass spectrometer, because off-the-shelf systems were not commercially available. However, the analytical objectives of a research project are a little different from the requirements for routine analysis. With a research project, there are fewer time constraints to optimize the chromatography and detection parameters, whereas in a commercial environment, there are often financial penalties if the laboratory cannot be up and running real samples and generating revenue as quickly as possible. This demand, especially from commercial laboratories in the environmental and biomedical communities, for routine trace element speciation methods convinced the instrument manufacturers and vendors to develop fully integrated HPLC-ICP-MS systems.

The availability of these off-the-shelf systems rapidly drove the growth of this hyphenated technique, so much so that vendors who did not offer it with full application and hardware and software support were at a disadvantage. As this technique is maturing and is being used more as a routine analytical tool, it is becoming clear that the requirements of the ICP-MS system doing speciation analysis are different from those of an instrument carrying out trace element determinations using conventional nebulization. With that in mind, let us take a closer look at the typical requirements of an ICP mass spectrometer that is being utilized as a multielement detector for trace element speciation studies.

COMPATIBILITY WITH ORGANIC SOLVENTS

The requirements of the sample introduction system, and in particular the nebulization process, have been described earlier in this chapter. However, some reverse-phase HPLC separations use gradient elution with mixtures of organic solvents, such as methanol or acetonitrile. If this is the case, consideration must be given to the fact that some volatile organic solvents will extinguish the plasma.[19] Therefore, modifications to the sample introduction might need to be made, such as adding small amounts of oxygen to the nebulizer gas flow, or perhaps using a cooled spray chamber or a desolvation device to stop the buildup of carbon deposits on the sampler cone. Other approaches, such as direct injection nebulization, have been used to introduce the sample eluent into the ICP-MS, but historically, they have not gained widespread acceptance because of usability issues.

Collision/Reaction Cell or Interface Capability

Another requirement of the ICP-MS system for speciation work is the collision/reaction cell or interface capability. It is becoming clear that as more and more speciation methods are being developed, the ability to minimize polyatomic spectral interferences generated by the solvent, buffer, mobile phase, pH-adjusting acids or bases, plasma gas, and so forth, is of crucial importance. Take, for example, the elements discussed earlier. As, Cr, and Se are notoriously difficult elements for ICP-MS analysis because their major isotopes suffer from argon- and sample-based polyatomic interferences. Arsenic has only one isotope (*m/z* 75), which is difficult to quantify in chloride-containing samples because of the presence of $^{40}Ar^{35}Cl^+$. Low-level chromium analysis is difficult because of the presence of the $^{40}Ar^{12}C^+$ and $^{40}Ar^{13}C^+$ interferences, which overlap the two major isotopes of chromium at masses 52 and 53. These interferences are nearly always present, but are especially strong in samples with organic content. The argon dimers ($^{40}Ar^{40}Ar^+$ and $^{40}Ar^{38}Ar^+$) at masses 80 and 78 interfere with the major isotopes of Se at masses 80 and 78, respectively, and bromine, which is usually present in natural waters, forms $^{79}BrH^+$ and $^{81}BrH^+$, which interferes with the Se masses at 80 and 82, respectively. The effects of these and other interferences have been reduced somewhat with conventional ICP-MS instrumentation, using alternate masses, interference correction equations, cool plasma technology, and desolvation techniques, but these approaches have not shown themselves to be particularly useful for these elements, especially at ultratrace levels.

For these reasons, it will be very beneficial if the ICP-MS instrumentation is fitted with a collision or reaction cell or interface and has the capability to minimize the formation of these undesired polyatomic interferences, using either collisional mechanisms with kinetic energy discrimination[20,21] or ion–molecule reaction kinetics with mass bandpass tuning,[22] or by introduction of a collision/reaction gas into the interface region.[23] The best approach will depend on the type of sample being analyzed, the number of species being determined, and the detection limit requirements, but there is no doubt that for multielement work, it is beneficial if one gas can be used for all the analyte species. This is exemplified by the research of Neubauer and coworkers described earlier, who used oxygen as the reaction gas to reduce the polyatomic spectral interferences described earlier to determine seven different species of As, Cr, and Se in potable water. In fact, even though the oxygen was used to remove the argon-carbide and argon dimer interferences to quantify the chromium and selenium species, respectively, it was used in a different way to quantify the arsenic species.

TABLE 21.3
ICP-MS Instrumental Conditions for the Speciation Analysis of As^{+3}, As^{+5}, Cr^{+3}, Cr^{+6}, Se^{+4}, Se^{+6}, and SeCN in Potable Water Samples

Nebulizer	Quartz concentric
Spray chamber	Quartz cyclonic
RF power	1500 W
Collision/reaction cell technology	Dynamic reaction cell
Reaction gas	O_2 = 0.7 mL/min
Analytical masses	AsO^+ (*m*/*z* 91), Se^+ (*m*/*z* 78), Cr^+ (*m*/*z* 52)
Analyte species	As^{+3}, As^{+5}, Cr^{+3}, Cr^{+6}, Se^{+4}, Se^{+6}, $SeCN^-$
Dwell time	330 ms (per analyte)
Analysis time	5.5 min

Source: Neubauer, K. R., et al., *Curr. Trends Mass Spectrosc.*, May 2006.

It was used to react with the arsenic ion to form the arsenic-oxygen molecular ion ($^{75}As^{16}O$) at mass 91, and move it away from the argon-chloride interference at mass 75. This novel approach has been reported many times in the literature[22,24] and has in fact been approved by the EPA in the recent ILM05.4 analytical procedure update for its Superfund Contract Laboratory.

The instrumental conditions for the speciation analysis are shown in Table 21.3, and a plot of signal intensity versus time of the simultaneous separation of a 1 µg/L standard of As, Cr, and Se is shown in Figure 21.7.

It should be emphasized that even though oxygen was used as the reaction gas in this study, many other collision/reaction strategies using inert gases, such as helium, and low-reactivity gases, such as hydrogen, have been successfully used for the determination of As, Cr, and Se. The question is whether a single gas can be used to remove all the interferences to an acceptable level and allow quantitation at the trace level. If more than one gas is required, that is not such a major hardship, especially as all gas flows are under computer control and can be changed in a multielement run. However, if the method needs to be automated for a high-sample-throughput environment, it will be significantly faster if only one gas is used for interference removal, so that the collision/reaction cell conditions need not be changed for each element.

Optimization of Peak Measurement Protocol

It can be seen from the chromatogram that the total separation time for the seven inorganic species is on the order of 2 min for the common oxidation states of the elements, and about 5 min if there is any selenocyanide in the sample. This means that all the peaks have to be integrated and quantified in a transient signal lasting 2–5 min. So, even though this is not considered a short transient, such as an ETV or small-spot laser ablation work that typically lasts 2–5 s, it is not a continuous signal generated by a pneumatic nebulizer. For that reason, it is critical to optimize the

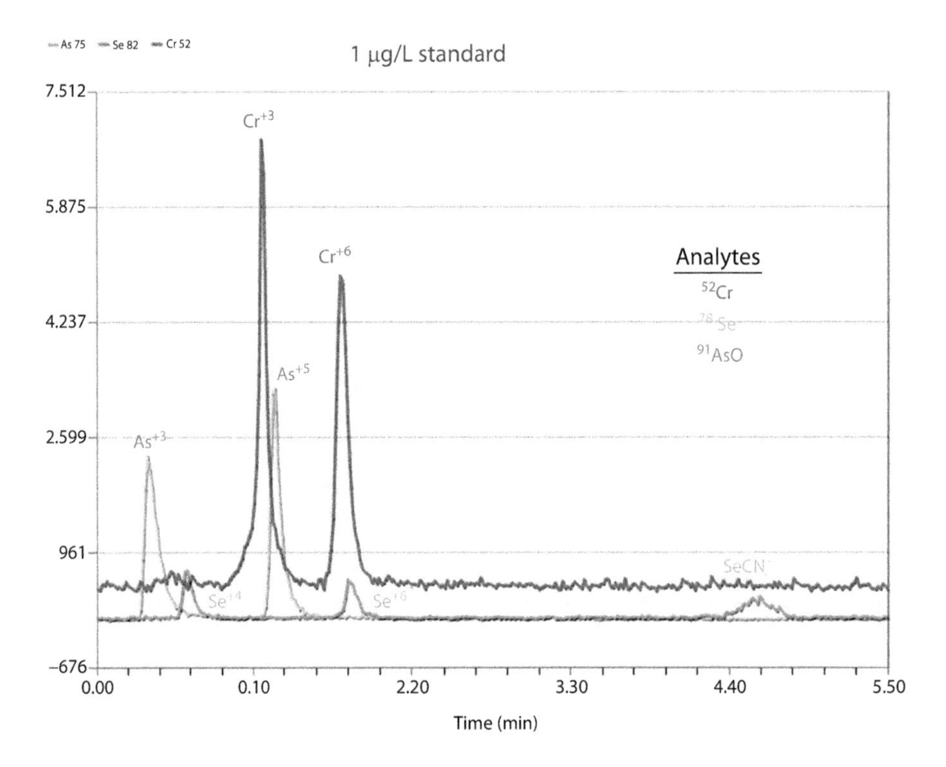

FIGURE 21.7 A plot of signal intensity versus time of the simultaneous separation of a 1 µg/L standard of As, Cr, and Se species. (From K. R. Neubauer, P. A. Perrone, W. Reuter, R. Thomas, *Current Trends in Mass Spectroscopy*, May 2006.)

measurement time in order to achieve the best multielement signal-to-noise ratio in the sampling time available. This is demonstrated in Figure 21.8, which shows the temporal separation of a group of elemental species in a chromatographic transient peak. The plot represents signal intensity against mass over the time period of the chromatogram. To get the best detection limits for this group of elements or species, it is very important to spend all the available time quantifying the peaks of interest.

For this reason, the quadrupole scanning or settling time and the time spent measuring the analyte peaks must be optimized to achieve the highest signal quality. This is described in greater detail in Chapter 14 and basically involves optimizing the number of sweeps, selecting the best dwell time, and using short settling times to achieve the highest measurement duty cycle and maximize the peak signal-to-noise ratio over the duration of the transient event. If this is not done, there is a strong likelihood that the quality of the speciation data could be compromised. In addition, if the extended dynamic range is used to determine higher concentrations of elemental species, the scanning and settling time of the detector will also have an impact on the quality of the signal. For that reason, detectors that require two scans to characterize an unknown sample will use up valuable time in the quantitation process. This is somewhat of a disadvantage when doing multi-element speciation on a chromatographic transient signal, especially if you have limited knowledge of the analyte concentration levels in your samples.

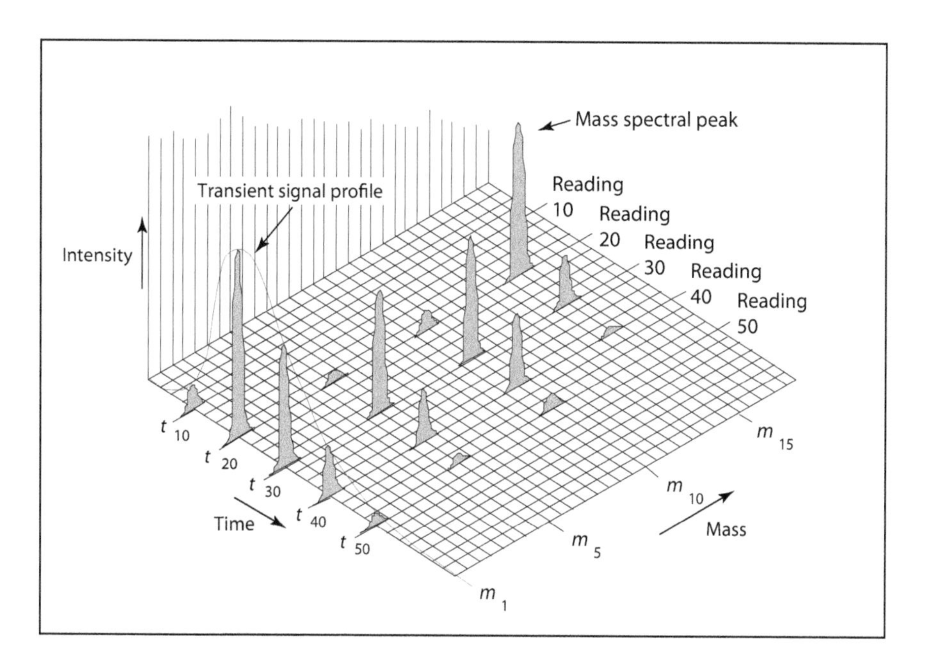

FIGURE 21.8 The temporal separation of a group of elemental species in a chromatographic transient peak.

Full Software Control and Integration

In the early days of coupling HPLC components with ICP-MS, there were very few sophisticated communication protocols between the two devices. The sample was injected into the chromatographic separation system, and when the analyte species was close to being eluted off the column, the read cycle of the ICP mass spectrometer was initiated manually to capture the data using the instrument's time-resolved software. Processing and manipulation of the data was then carried out after the chromatogram had been captured, sometimes by a completely different software program. The HPLC and the ICP-MS were considered almost two distinctly different hardware and software devices, with very little communication between them. This was even more surprising considering that many of the ICP-MS vendors also offered LC equipment. As you can imagine, this was not the ideal scenario for a laboratory that wanted to carry out automated speciated analysis. As a result of this demand, manufacturers realized that unless they offered fully integrated hardware and software solutions for trace element speciation work, it was never going to be accepted as a routine analytical tool. Today, just about all the ICP mass spectrograph manufacturers offer fully integrated HPLC-ICP-MS systems. Some of the features available on today's instrumentation include the following:

- Full software and hardware control of both the chromatograph and the ICP-MS system from one computer
- Ability to check the status of the ICP-MS system when setting up the chromatography method or vice versa

- Computer control of the switching valve to allow the HPLC to be used in tandem with the ICP-MS normal sample introduction system
- Real-time display of spectral data, including peak identification and quantitation, while the chromatogram is being generated
- Full data handling capability, including manipulation of spectral peaks and calibration curves
- Comprehensive reporting options or exportable data to third-party programs

It is also important to emphasize that most vendors also offer integrated systems with turnkey application methods that are almost ready to run samples as soon the instrument is installed. Although this is not available for all applications, it is becoming a standard offering for some of the more routine environmental applications. Manufacturers are also realizing that most analytical chemists who are experienced in trace element analysis may have very little expertise in chromatography. For that reason, they are providing full backup and customer support with HPLC application specialists, as well as with the traditional ICP-MS product specialists.

FINAL THOUGHTS

Combining chromatography with ICP-MS has revolutionized trace metal speciation analysis. In particular, when HPLC or even low-pressure LC systems are coupled with the selectivity and sensitivity of ICP-MS, many elemental species at sub-parts-per-billion levels can now be determined in a single sample injection. When interference reduction methods, such as collision/reaction cell or interface technology, are available, it is possible to separate and detect different inorganic species of environmental significance in one automated run. By using a fully integrated, computer-controlled HPLC-ICP-MS system and optimizing the chromatographic separation, sample introduction, and ICP-MS detection parameters, simultaneous quanitation can be carried out in a few minutes. There is no question that this kind of sample throughput will arouse the interest of the environmental, biomedical, nutritional, and other application communities interested in trace element speciation studies and help them realize that it is feasible to carry out this kind of analysis in a truly routine manner.

However, it should be emphasized that historically there has been limited demand for the determination of elemental species in pharmaceutical materials or dietary supplements. As a result, there is very little application material in the public domain describing the separation and quanitation of inorganic and organic forms of arsenic and mercury in these kinds of sample matrices. For that reason, if there is a need to determine these species, it is strongly advised that a person or a laboratory with expertise in speciated techniques initially develop the method, as many interlaboratory round-robin studies have indicated that an optimized sample preparation procedure and chromatographic separation technology are critical to get high-quality data.[25]

REFERENCES

1. R. Lobinski, I. R. Pereiro, H. Chassaigne, A. Wasik, and J. Szpunar. *Journal of Analytical Atomic Spectrometry*, 13, 860–867, 1998.

2. G. Cox and C. W. McLeod. *Mikrochimica Acta*, 109, 161–164, 1992.
3. S. Branch, L. Ebdon, and P. O'Neill. *Journal of Analytical Atomic Spectrometry*, 9, 33–37, 1994.
4. J. R. Dean, L. Ebdon, M. E. Foulkes, H. M. Crews, and R. C. Massey. *Journal of Analytical Atomic Spectrometry*, 9, 615–618, 1994.
5. Z. A. Grosser and K. Neubauer. *Today's Chemist at Work*, 43–46, May 2004.
6. Y. L. Chang and S. J. Jiang. *Journal of Analytical Atomic Spectrometry*, 9, 858, 2001.
7. K. E. Lokits, D. D. Richardson, and J. A. Caruso. *Handbook of Hyphenated ICP-MS Applications*. Santa Clara, CA: Agilent Technologies, 2007.
8. K. DeNicola, D. D. Richardson, and J. A. Caruso. *Spectroscopy*, 21(2), 18–24, 2006.
9. D. Wallschlager and N. Bloom. *Journal of Analytical Atomic Spectrometry*, 16, 1322, 2001.
10. J. J. Sloth, E. H. Larsen, and K. Julshamn. *Journal of Analytical Atomic Spectrometry*, 18, 452–459, 2003.
11. C. Rosal, G. Momplaisir, and E. Heithmar, *Electrophoresis*, 26, 1606–1614, 2005.
12. L. Yang, Z. Mester, and R. E. Sturgeon. *Analytical Chemistry*, 74, 2968, 2002.
13. V. Vacchina, L. Torti, C. Allievi, and R. Lobinski. *Journal of Analytical Atomic Spectrometry*, 18, 884–890, 2003.
14. S. Mounicou, J. Szpunar, R. Lobinski, D. Andrey, and C. J. Blake. *Journal of Analytical Atomic Spectrometry*, 17, 880–886, 2002.
15. B. Bouyssiere, P. Leonhard, D. Pröfrock, F. Baco, C. Lopez Garcia, S. Wilbur, and A. Prange. *Journal of Analytical Atomic Spectrometry*, 5, 700–702, 2004.
16. D. Chen, M. Jing, and S. Wang. *Handbook of Hyphenated ICP-MS Applications*. Santa Clara, CA: Agilent Technologies, 2007.
17. B. Avula, Y. H. Wang, M. Wang, and A. Khan. Analysis of arsenic and mercury species from botanicals and dietary supplements using LC-ICP-MS. *Planta Medica*, 78, 136–144, 2012.
18. K. R. Neubauer, P. A. Perrone, W. Reuter, and R. Thomas. Current Trends in Mass Spectroscopy, 43–50 May 2006.
19. McElroy, A. Mennito, E. Debrah, and R. Thomas. *Spectroscopy*, 13(2), 42–53, 1998.
20. M. Bueno, F. Pannier, M. Potin-Gautier, and J. Darrouzes. Determination of organic and inorganic selenium using HPLC-ICP-MS. Application Note 5989-7073EN. Agilent Technologies, 2007.
21. S. McSheehy and M. Nash. Determination of selenomethionine in nutritional supplements using HPLC coupled to the XSeriesII ICP-MS with CCT. Application Note 40745. Thermo Scientific, 2005. http://www.thermo.com/eThermo/CMA/PDFs/Articles/articlesFile_26474.pdf.
22. J. Di Bussolo, W.Reuter, L. Davidowski, and K. Neubauer. Speciation of five arsenic compounds in urine by HPLC-ICP-MS. LAS Application Note D-6736. PerkinElmer, 2004.
23. M. Leist and A. Toms. Low level speciation of chromium in drinking waters using LC-ICP-MS. Application Note 29. http://alfresco.ubm-us.net/alfresco_images/pharma/2014/08/22/e24804f7-20db-4068-97f8-13061098e05a/article-352912.pdf
24. D. S. Bollinger and A. J. Schleisman. *Atomic Spectroscopy*, 20(2), 60–63, 1999.
25. M. L. Briscoe, T. M. Ugrai, J. Creswell, and A. T. Carter. An interlaboratory comparison study for the determination of arsenic and arsenic species in rice, kelp, and apple juice. *Spectroscopy Magazine*, 30(5), 48–61, 2015.

22 Fundamental Principles, Method Development, and Operational Requirements of ICP-OES

Since its introduction more than 40 years ago, inductively coupled plasma optical emission spectroscopy (ICP-OES) has significantly changed the capabilities of elemental analysis. This technique combined the energy of an argon-based plasma with an optical spectrometer and detection system capable of measuring low-level emission signals, which allowed laboratories to perform rapid, automated, multielement analyses at trace concentrations.[1] This was approximately 10 years before the introduction of the first commercial inductively coupled plasma mass spectrometer (ICP-MS), so ICP-OES became the workhorse instrument in many laboratories required to perform elemental analysis at trace-level concentrations. This chapter, written by Maura Rury from Thermo Fisher Scientific, gives a detailed description of the fundamental principles, together with method development optimization procedures and the operational requirements of this technique.

The advantage in using an atmospheric pressure ICP source for making optical emission measurements was first published in 1964,[2] and the sensitivity, speed of analysis, ease of use, and tolerance to high levels of dissolved solids are advantages that laboratories continue to rely on more than half a century later.[3,4] The success of the technique itself can be measured by the fact that more than 50,000 ICP-OES instruments have been installed between 1983 and 2016, which has resulted in approximately 59,000 publications, with more than 28,000 published since 2012 (results courtesy of Google Scholar search). That published literature features elemental determinations in a variety of sample matrices in industries, including environmental, nuclear, mining and geochemistry, materials testing, semiconductor, industrial, petrochemical, clinical and toxicological, food safety, and pharmaceutical.

BASIC DEFINITIONS

A full glossary of ICP-OES and ICP-MS terms is given at the end of this book. However, several ICP-OES terms are defined here to ensure clarity while reading this chapter. Several optical emission techniques exist, based on atmospheric discharges, which include inductively coupled plasmas (ICPs), direct coupled plasmas (DCPs), microwave-induced plasmas (MIPs), direct current (DC) arcs, and alternating current (AC) sparks. Each discharge is generated via a different mechanism and has its own inherent advantages and disadvantages; however, a comparative discussion of these techniques is outside the scope of this text. The remainder of this chapter focuses solely on ICP-OES.

It should be noted that *ICP-AES* and *ICP-OES* are terms that are sometimes used interchangeably; however, the former term can be a source of error and confusion. The term *ICP-AES* refers to atomic emission spectroscopy, which nominally excludes emission contributions from other species, such as ions and molecules. The latter term refers to optical emission spectroscopy and is more commonly used, as it includes emission from multiple contributors. Only the term *ICP-OES* is used in this chapter.

PRINCIPLES OF EMISSION

For most ICP-OES applications, a sample is delivered to the instrument's plasma in the form of an aerosol. As the aerosol travels from the base of the plasma to its tail, it travels through a variety of heated zones where it gets desolvated (unless delivered as a dry aerosol), vaporized, atomized, and ionized. Further time spent in the plasma allows the atoms and ions to absorb additional energy, which excites an outer electron and produces excited state species. Relaxation back to a ground-state atom produces energy in the form of a photon. This production of photons from excited atoms and ions forms the basis for atomic emission measurements. There are many species in a sample that may absorb energy from the plasma and produce emission spectra. These species include atoms, ions, and molecules. For the purpose of this section, the contribution from molecular emission is excluded. All references to emission include the contribution from atoms and ions only.

ATOMIC AND IONIC EMISSION

Elemental analysis by ICP-OES relies on the emission from excited atoms and ions within a sample. Argon plasmas contain ~15.8 eV of energy, which is sufficient to remove one or two electrons from the outer orbital of most atoms. This results in the presence of both atoms and ions in the plasma, all of which are in their ground (lowest-level) energy state. Excitation, and subsequent emission, occurs when a species' absorbed energy from the plasma is released in the form of wavelength-specific photons.

A simplified schematic of atomic absorption and emission is illustrated in Figure 22.1. The horizontal lines represent energy levels in an atom. The lowest horizontal line and the four remaining horizontal lines represent the ground and excited states, respectively. If included in the schematic, additional horizontal lines to represent ionic ground and excited states are illustrated above the atomic excited states. The vertical arrows represent an energy transition for an electron, following the absorption or emission of a photon. The length of each vertical arrow correlates to the amount of energy involved in the transition.

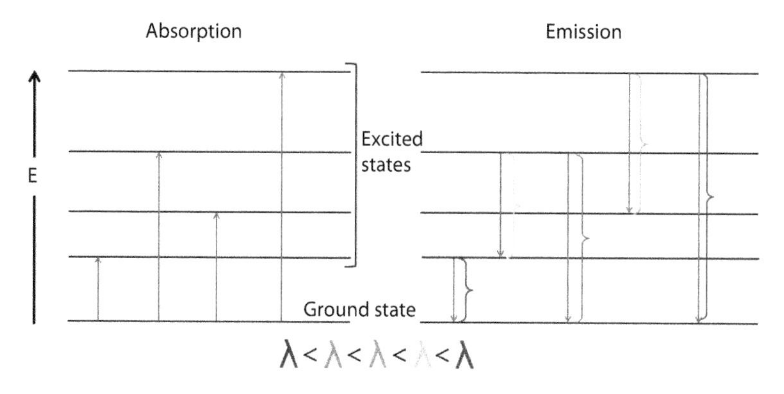

FIGURE 22.1 Diagram depicting energy transitions involved in an atom's absorption and emission of energy.

As the schematic indicates, absorbed energy can shift electrons to different excited states, both atomic and ionic. Relaxation of these excited electrons produces energy in the form of photons. Photons vary in energy and can be correlated to their associated emission wavelength using Einstein's equation,[5] which relates the energy of light and its frequency according to

$$E = h\nu$$

where E represents the energy of light, h represents Planck's constant, and ν represents the frequency of light. In optical emission spectroscopy, it is more practical to speak in terms of wavelength, so the term c/λ can be substituted to yield

$$E = hc/\lambda$$

where E represents energy in joules, h represents Planck's constant in units of joule seconds, c represents the speed of light in meters per second, and λ represents the wavelength in units of meters. From this equation, it becomes clear that each emitted photon is wavelength specific and represents the inverse relationship between energy and wavelength. These emission wavelengths represent the energy levels that are characteristic to each element, thus making optical emission spectroscopy a useful technique for identifying and quantifying elements in unknown samples.

INSTRUMENTATION

Commercially available spectrometers collect emission at wavelengths from 165 to 1100 nm. Early elemental analysis did not include the collection of optical emission in the vacuum ultraviolet (VUV) region. Wavelengths below 190 nm are absorbed by oxygen, water vapor, and other components in the ambient atmosphere, and early instrument designs did not sufficiently purge the optical path between the torch and the entrance slit of the spectrometer. The use of a completely purged optical path was first proposed in the early 1970s, where the use of high-purity nitrogen made it feasible to collect optical emission in the VUV wavelength region.[6,7]

SAMPLE INTRODUCTION

The "Achilles' heel" of most plasma spectrochemical techniques lies with sample introduction.[8] This fundamental drawback was noted in the literature in the 1980s; however, it is an issue that holds true even decades later. Improving sample introduction systems is a research topic that has been investigated for many years, the outcome of which has produced many nebulizer and spray chamber designs, as well as introduction systems that incorporate flow injection and internal standardization techniques. Research in this area is ongoing, as there are still a number of areas in which sample introduction falls short. Introduction system designs have made little progress in the way of improving sample throughput and reducing carryover from memory-prone elements. A number of systems allow for online addition of internal standards; however, many of those systems suffer from inconsistent mixing between the internal standard and carrier solutions. Flow injection techniques have been incorporated into a number of introduction systems. However, the use of a relatively small, defined sample volume at the sample flow rates typically used with ICP-MS and ICP-OES results in short-lived transient signals that do not allow for the measurement of more than a few analytes at a time.

As with all analytical atomic spectroscopic techniques, sample introduction is a critical step that strongly influences, and often dictates, analytical figures of merit, such as sensitivity, precision, and stability.[9] Improving the efficiency of sample introduction is an ongoing research topic; however, it has proven to be a nontrivial task, as the processes involved within both the nebulizer and spray chamber are numerous and complicated.

Alternative methods of sample introduction, including flow injection, flow injection with hydride generation, and laser ablation, have been employed with the intent to improve plasma spectrochemical measurements. Flow injection is a technique that provides a continuous flow of solution to the nebulizer, which allows sample uptake and stabilization in the plasma to take place more rapidly, thereby increasing sample throughput. Furthermore, injection of small sample volumes into a relatively clean carrier stream reduces the amount of salt introduced into the instrument, which can reduce matrix effects and improve limits of detection (LODs).[10] A drawback to this setup is that the introduced sample is of finite volume and the measured signal becomes transient. This is an undesirable situation when measuring a large suite of elements with a sequential instrument, as many of the elements will be measured off the peak center, which degrades the signal-to-noise ratio, and therefore the sensitivity of the measurement.[11,12] In an effort to retain both steady-state signal analysis and rapid wash-in and washout, a relatively large-volume sample loop should be employed for analysis.

Hydride generation provides a way to remove interfering matrix components to allow only the analytes of interest to be presented to the analytical instrument.

Hydride generation methods are well established[13–16] and rely on the generation of volatile gaseous hydrides, following a reaction with a reducing agent at an appropriate pH. Hydride generation can be performed online or offline prior to analysis, and it is a technique that can be used for preconcentration as well as matrix removal for the determination of hydride-active elements.[17,18]

Laser ablation, when combined with ICP-OES or ICP-MS, is a sampling technique that allows elemental impurities to be quantified in their native solid sample. In this technique, a laser is focused on a prepared sample to remove (ablate) surface material in the form of fine particles. The particles are swept into the instrument using an argon or helium carrier gas where they are atomized or ionized and excited, similar to the processes that would occur with an aerosol from a liquid sample.[19]

The popularity of this technique stems from its ability to provide a sample aerosol from materials that are difficult to dissolve, such as metals, refractory materials, glasses, and insoluble alloys. Additional advantages include simplified sample preparation procedures with a reduced risk for contamination or loss, a reduced sample size requirement, and the ability to determine the spatial distribution of the quantified elements.[20,21] A significant challenge with laser ablation is in obtaining calibration standards that are matrix matched to the unknown samples and which contain elements at concentrations that bracket those of the analytes of interest.[22] The development of synthetic liquid standards is relatively straightforward; however, this task is nontrivial for solid-based techniques. Note: A more detailed description of laser sampling for ICP-MS is given in Chapter 20.

The driving forces behind the design of sample introduction systems include improved sensitivity detection limits, and precision across the working range of the instrument, together with fewer interferences from matrix effects. Much of the design efforts have centered around improvements in nebulizers and spray chambers; however, significant progress has diminished in the recent years.[23]

AEROSOL GENERATION

Most ICP-OES applications require the delivery of a sample to the instrument's plasma in the form of an aerosol. The components in the sample introduction system, which typically include a pump, nebulizer, spray chamber, and torch, must work in a complementary fashion to convert a bulk sample solution into a body of micrometer-sized droplets. The pump (peristaltic or syringe) consistently delivers solution to the nebulizer, which uses a high-velocity gas stream to break the solution into small droplets. The spray chamber then filters the droplets by diameter, allowing only the smallest droplets to be transferred to the torch, where they are desolvated, vaporized, and atomized in the plasma. A proper aerosol should consist of as many small (<10 μm diameter) droplets as possible, to maximize the efficiency of desolvation once the sample reaches the plasma.

The generation of an optimized aerosol is a significant factor in determining the quality of the resulting emission signals and data.[24] Sample aerosols may vary in their density of droplets; however, a proper aerosol should maintain a consistent shape and density during data collection to promote analytical measurements with optimal precision. Two example aerosols are pictured in Figure 22.2 for illustrative purposes. They are generated from nebulizers with two different structural designs that produce differences in their shape and droplet density.

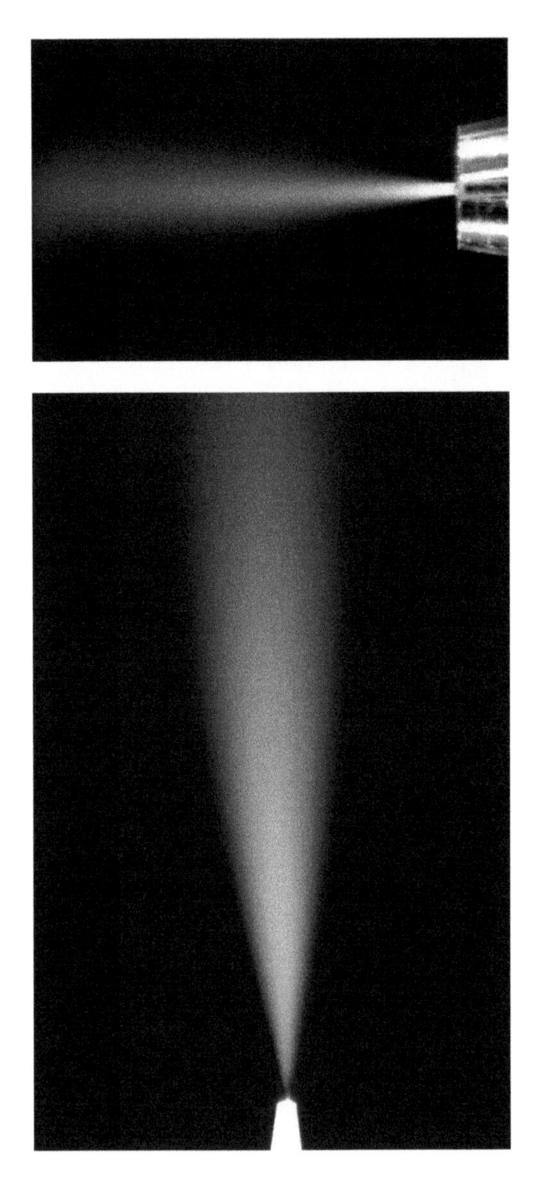

FIGURE 22.2 Examples of proper sample aerosols, generated by a concentric nebulizer (a) (with permission from Meinhard Glass Products, part of Elemental Scientific Inc., Omaha, NE) and a perfluoroalkoxy microflow concentric nebulizer (b). (With permission from Elemental Scientific Inc., Omaha, NE.)

NEBULIZERS

The main role of the nebulizer is to generate an aerosol of small droplets from the interaction between a sample stream and a high-velocity gas stream.[25] Ideally, the nebulizer will transport an aerosol with a narrow drop size distribution into the spray chamber to allow the spray chamber to more efficiently filter out the large aerosol droplets.[26] Since the droplets must be desolvated, vaporized, and atomized during their short residence time in the plasma, the spray chamber allows passage of only small droplets (typical diameter <10 μm).[27,28]

In an effort to overcoming the limits associated with sample introduction, much effort has been put into the design of nebulizers and spray chambers. Typical introduction systems consist of a nebulizer–spray chamber arrangement in which the nebulizer is pneumatic.[29] The relative simplicity and low cost of this setup make it the preferred choice for ICP sample introduction; however, its associated drawbacks include low analyte transport efficiency (~1%–2%), high sample consumption (1–2 mL/min), and relatively high retention of some elements.[30] An image of a typical pneumatic nebulizer is illustrated in Figure 22.3.

Microconcentric nebulizers (MCNs)[31–36] were designed to improve the gas–liquid interaction and reduce the size distribution of droplets formed in the aerosol.[37,38] These nebulizers, which operate at lower sample flow rates than those typically used with pneumatic nebulizers, include high-efficiency nebulizers (HENs),[39–42] oscillating capillary nebulizers (OCNs),[43–45] and sonic spray nebulizers (SSNs).[46] These nebulizers have a relatively low dead volume and operate at normal nebulizer gas pressures, which improves analyte transport efficiency regardless of whether organic or aqueous solvents are used.[47,48]

Ultrasonic nebulizers (USNs)[49–53] were developed in an effort to improve the efficiency of aerosol generation. These nebulizers are highly efficient at producing a large volume of small droplets with a narrow size distribution.[54] The aerosol volume is sufficiently large that most USNs are used with desolvation systems to reduce the introduction of water vapor in the case of aqueous sample analyses,[55,56] and to reduce solvent loading and carbon deposition in the analysis of samples containing organic solvents or high concentrations of dissolved salts.[57,58] A schematic of a USN is shown in Figure 22.4.

Nebulizers have also been designed in which no spray chamber is present and samples are injected directly into the plasma. These direct injection nebulizers (DINs)[59–63] and direct injection high-efficiency nebulizers (DIHENs)[64–67] improve

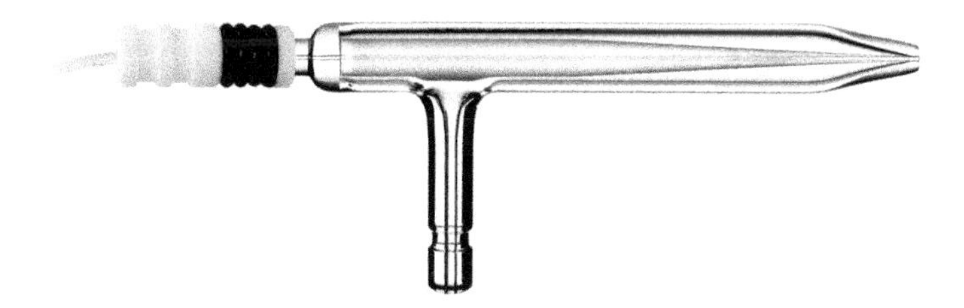

FIGURE 22.3 Image of a typical pneumatic nebulizer. (With permission from Glass Expansion, Pocasset, MA.)

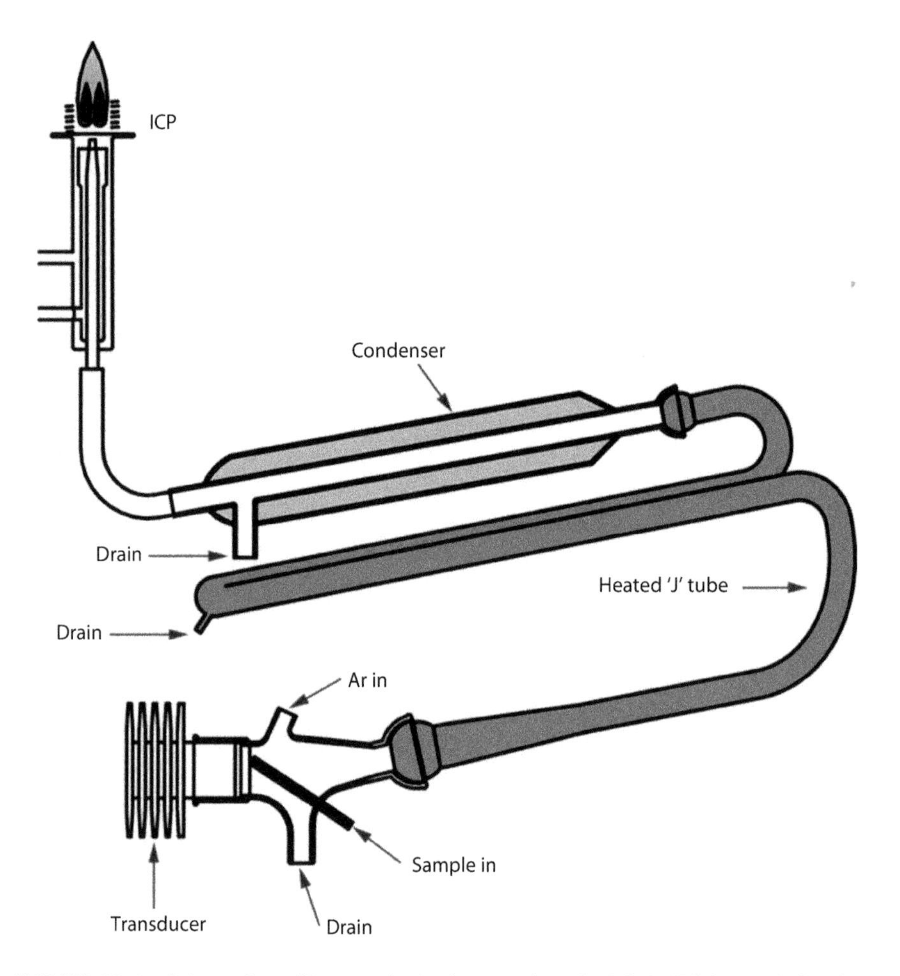

FIGURE 22.4 Schematic to illustrate the basic operation of a USN. (With permission from Teledyne Cetac Technologies, Omaha, NE.)

the sample transport efficiency to 100%, even at relatively low flow rates. Direct injection also reduces the dead volume associated with spray chambers, which increases the response time of the measurement and reduces memory effects.[68–71] A significant drawback of DINs is their vulnerability toward samples containing high concentrations of dissolved salts or volatile solvents that cause plasma instability and tip clogging.[72] The large-bore direct injection high-efficiency nebulizer (LB-DIHEN)[73] was designed to reduce its susceptibility to blockage; however, the larger inside diameter of the nebulizer capillary produces a relatively large droplet size distribution, which degrades both the precision and detection limits.[74] An example of a direct injection nebulizer, specifically a quartz DIHEN, is pictured in Figure 22.5.

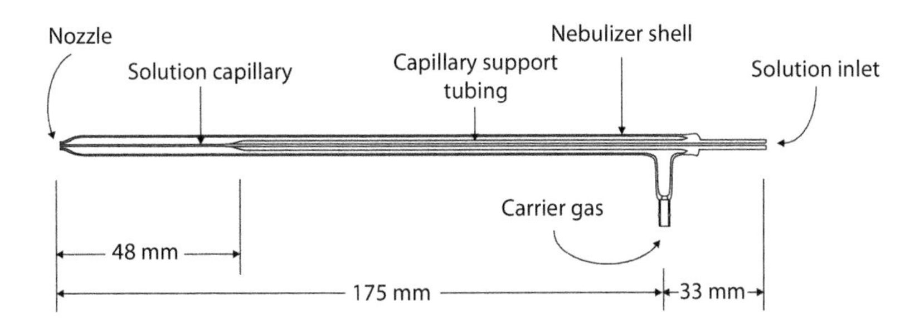

FIGURE 22.5 Schematic DIHEN nebulizer. (With permission from Meinhard, partners with Elemental Scientific Inc., Omaha, NE.)

SPRAY CHAMBERS

Much effort has been put into spray chamber design, as spray chambers are responsible for the loss of >90% of the aerosol produced by the nebulizer.[75] Improvements have focused on directing the flow of aerosol from the nebulizer to maximize the efficiency with which larger droplets are filtered out.[76] Reverse-flow, commonly called Scott-type double-pass,[77,78] spray chambers are the popular choice for relatively simple, low-cost introduction. The double-pass design is particularly useful for samples containing high concentrations of dissolved salts, as the transport efficiency is relatively low compared with other spray chamber designs.[79] An example double-pass spray chamber is illustrated in Figure 22.6.

Cyclonic designs have been employed to improve transport efficiency, precision, and detection limits.[80,81] Popular cyclonic spray chamber designs include a flow spoiler or dimple[82] to disrupt the flow of aerosol within the spray chamber. Computer modeling of the fluid dynamics within cyclonic spray chambers suggests that the

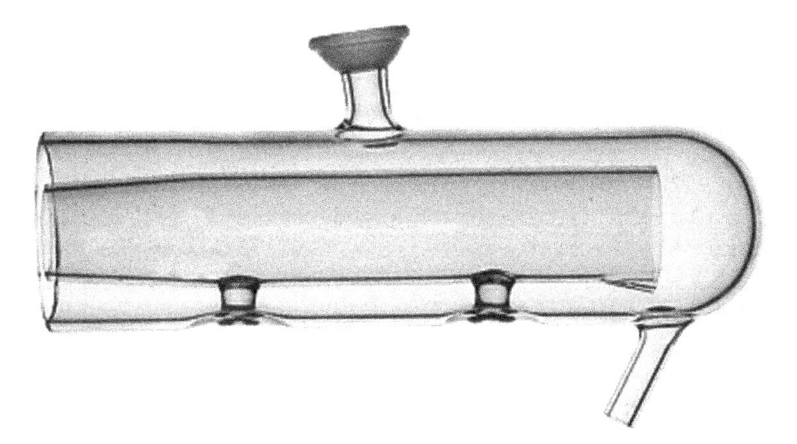

FIGURE 22.6 Image and schematic of a typical Scott-type double-pass spray chamber. (With permission from Precision Glassblowing, Centennial, CO.)

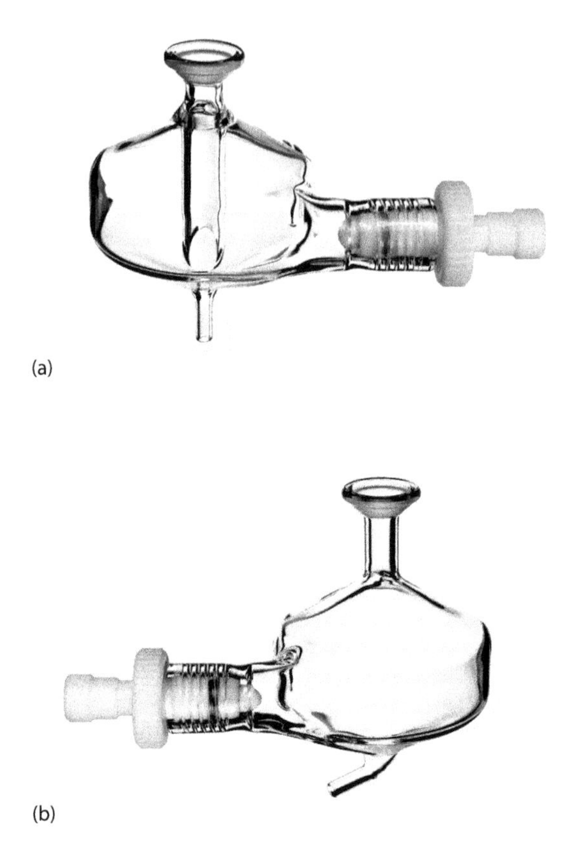

FIGURE 22.7 Cyclonic spray chamber with (a) and without (b) a knockout tube. (With permission from Glass Expansion, Pocasset, MA.)

presence of three spoilers creates a "virtual cyclone,"[83] which reduces the interaction between the aerosol and the walls of the spray chamber, thereby improving transport efficiency and reducing memory effects. A commonly employed cyclonic spray chamber is illustrated in Figure 22.7. Spray chambers, with (Figure 22.7a) and without (Figure 22.7b) a flow spoiler, are illustrated for comparison.

Single-pass or cylindrical-type spray chambers have been designed for use in low-flow introduction.[84–86] These spray chambers provide high efficiency and reduced memory effects;[87] however, this design is unsuitable for conventional ICP analysis and is typically used when electrophoretic or chromatographic separations are employed in conjunction with ICP detection.[88,89] Spray chambers have also been blamed for the retention of elements such as B and Hg.[90–93] As discussed above, one approach to solving this problem has been in the removal of the spray chamber. Other approaches have involved the use of a thermostated spray chamber. Water-cooled spray chambers have been used to reduce aerosol desolvation and deposition along the walls of the spray chamber, thereby reducing memory effects and reducing oxide formation in the plasma.[94–96] This logic has been used in the application of heated spray chambers for desolvating and therefore improving the efficiency of aerosol generation in aqueous samples.[97–99]

TORCHES

The torch is the final component in the sample's journey from sample container to plasma and often has a major influence on the analytical performance of the instrument in terms of sensitivity, detection limits, and plasma robustness. Torch design for ICPs has been investigated for more than 50 years. The original ICP torch was based on Reed's design[100–102] and introduced the coolant and auxiliary gas flows in a tangential direction. This method of gas introduction was thought to produce fluid dynamics with maximum stability and ion density.[103]

Reed's torch design was modified by Greenfield and colleagues,[104,105] who constructed a torch containing three concentric tubes to generate and sustain an annular-shaped plasma. The outer and intermediate tubes were made out of silica, and the center tube was constructed out of borosilicate glass. High-purity gas was introduced tangentially, similar to the Reed torch; however, gas was not introduced into the outer tube until after the plasma had formed. Only the intermediate tube was purged during ignition.

Fassel made minor modifications, and the resulting torch design[106] is used by most commercial instrument manufacturers even today. The Fassel torch design evolved to a design similar to that shown in Figure 22.8. The torch consists of three concentric quartz tubes: the outer tube, the intermediate tube, and the center tube (or injector). The geometry of these tubes and their positioning with respect to each other and to the load coil, along with the gas flow settings, directly affect the formation, stability, and sustainability of the plasma.[107] This makes torch design critical to data quality in elemental analysis.

The inner tube, or injector, provides an inlet for introducing the sample aerosol into the plasma. The center tube provides a structured path for the auxiliary gas, which is used to shift the base of the plasma farther away from the injector. The outer tube, or coolant, houses the largest volume of gas, which is used to generate and sustain the plasma. This gas also provides sufficient cooling to prevent the torch from melting or contributing background emission signals.

Since the introduction of the Fassel torch, minor changes have been made to the torch design. Injector tubes have been constructed with both a tapered and a parallel end, and they are available with a variety of inside diameters. Small-diameter injectors are typically used for applications involving solvents that have a relatively high vapor pressure and are likely to overload and extinguish the plasma. Larger-diameter injectors are typically used for the analysis of relatively clean samples to increase the aerosol volume delivered to the plasma, which increases the resulting emission signals.

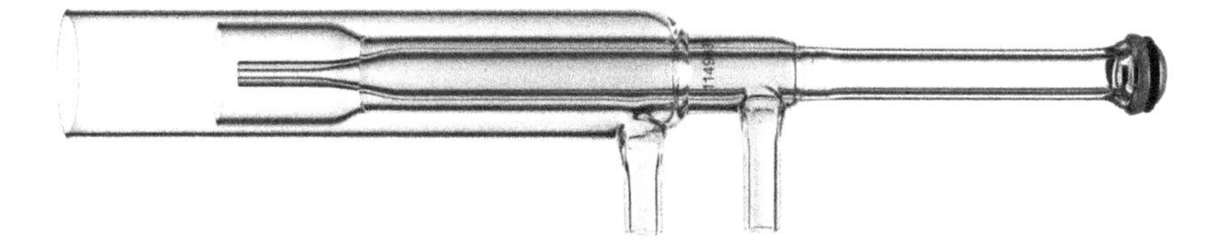

FIGURE 22.8 Schematic for a standard, commercially available ICP-OES torch. (With permission from Glass Expansion, Pocasset, MA.)

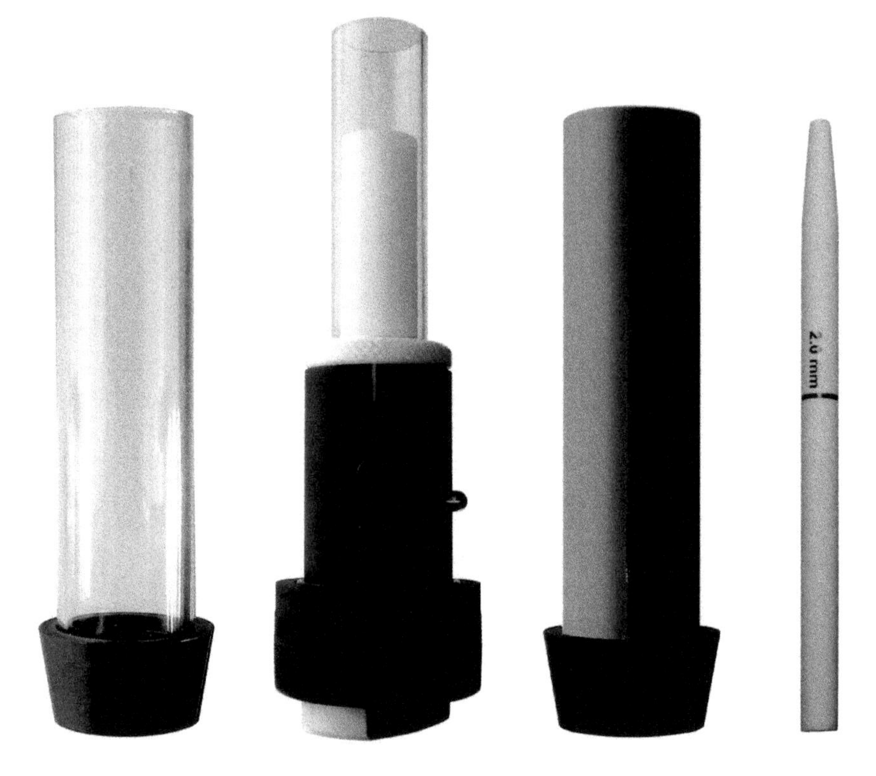

FIGURE 22.9 Demountable ceramic torch parts, showing, from left to right, a quartz outer tube, base and inner tube, ceramic outer tube, and ceramic injector. (With permission from Glass Expansion, Pocasset, MA.)

The intermediate tube has been produced with both a parallel and a tulip-shaped design. Research has been done to investigate the effect of the intermediate tube shape on the shape and stability of the plasma; however, both designs are still employed.[108] A parallel intermediate tube allows the auxiliary gas flow to remain constant along the length of the torch to reduce turbulence at the base of the plasma and to allow the auxiliary gas to reach a true laminar flow. This reduction in turbulence produces a plasma with a relatively flat base and aids in the penetration of the nebulizer gas, and therefore the sample, into the plasma.

Torches are available as a single piece or in a demountable design. The single-piece torch (e.g., see Figure 22.8) is generally easier for operator use and ensures that the injector will be fully centered in the intermediate tube; however, the individual parts of the torch cannot be replaced and the injector diameter cannot be changed. The demountable design is advantageous and more cost-effective for applications that benefit from the use of multiple injector diameters, or for those that are likely to clog the torch injector. The disadvantage to this style of torch is the requirement that it be manually assembled and aligned to ensure that the injector is centered in the torch body.

Ceramic torches have been designed for use in measuring organic solvents, extremely high levels of dissolved solids, and other samples that devitrify a standard quartz torch.[109] Commercial ceramic torches are available in fully demountable designs, which allows quartz and ceramic pieces to be combined to meet the needs of each specific application. An example of commercially available ceramic torch components is illustrated in Figure 22.9.

SPECTROMETERS

The optical spectrometer is the heart of the modern ICP instrument and is comprised of the fore-optics and the polychromator, with the detector attached. The purpose of the optical system is to separate the ICP source emission into element-specific wavelengths and to focus the resolved light onto the detector with high efficiency and with minimal stray light contributions.[110] This should be accomplished with a minimum amount of absorption, scattering, and optical aberration to maximize the light throughput to the detector.

FORE-OPTICS

The transfer of light from the plasma into the spectrometer is a critical first step in obtaining results with maximum sensitivity and signal-to-background ratios. Of equal importance is the transfer of emission with minimal contributions from molecular emission and stray light sources. The gap between the plasma and the spectrometer's entrance optics can be a challenging environment to control. In addition to protecting the entrance optics from the plasma's intense heat, the plasma–spectrometer interface must effectively replace the ambient air to prevent it from absorbing emission wavelengths below ~200 nm. For a horizontally mounted torch, the interface must also remove the relatively cool portion of the plasma ("tail") to minimize interferences from self-absorbed analytes and atoms or ions that recombine and produce molecular emission.[111]

The purpose of the fore-optics is to focus emitted light from the plasma onto the entrance slit as efficiently as possible and to provide a light path suitable for transmitting in either radial or axial plasma views, or both, for systems that have that capability. In dual-view systems, changing plasma views should be a fast, automated process to avoid undesirable increases in sample analysis time and subsequent degradation of productivity and cost efficiency.

Properly designed fore-optics should exhibit a number of characteristics to ensure optimal instrument performance. The light from the plasma should be transferred into the instrument with a high level of efficiency to maximize the sensitivity of the instrument. This requires a minimum number of optical components, as each light-reflecting surface will absorb or scatter a small amount of light. Each additional optic reduces the transfer efficiency, which reduces the instrument's achievable detection limits.[112]

The fore-optics should have optimum focus for better sensitivity and detection limits. Photons must be collected from the analytical zone of the plasma—the area of the plasma where the maximum amount of analyte emission and the minimum amount of emission from the plasma exist. Once collected, the light must be properly focused onto the entrance slit of the spectrometer to maximize the amount of light that travels through the spectrometer.

There should be a minimal number of moving parts for greater stability. A small amount of movement is required to optimize the viewing height of a radially configured plasma, and to switch between axial and radial light collection in a dual-view instrument. The movement needs to be rapid and highly reproducible to minimize the degradation of sample throughput and precision.

The fore-optics should be robust and provide thermal protection from external heat sources, such as the plasma and the atmosphere in the torch box. Excessive heat transfer to the optics will produce instrumental thermal drift, resulting in quality control (QC) failures and frequent recalibration.

OPTICAL DESIGNS

The earliest ICP spectrometers are based on the Paschen–Rünge design.[113] In this configuration, all the optics (diffraction grating, entrance slit, and exit slit) are permanently attached and the exit slits are mounted along a portion of a Rowland circle.[114] In its original design, this type of spectrometer was referred to as a direct reader and consisted of a polychromator with photomultiplier tube (PMT) detectors positioned on the focal plane behind a series of exit slits. One exit slit and PMT was used for each analytical wavelength measured.[115] This optical layout, which is illustrated in Figure 22.10, allowed several emission wavelengths to be measured simultaneously in a matter of a few seconds, depending on the integration time used.

Despite the ability to acquire emission signals with remarkable speed, early spectrometers with this optical configuration had relatively poor spectral resolution.

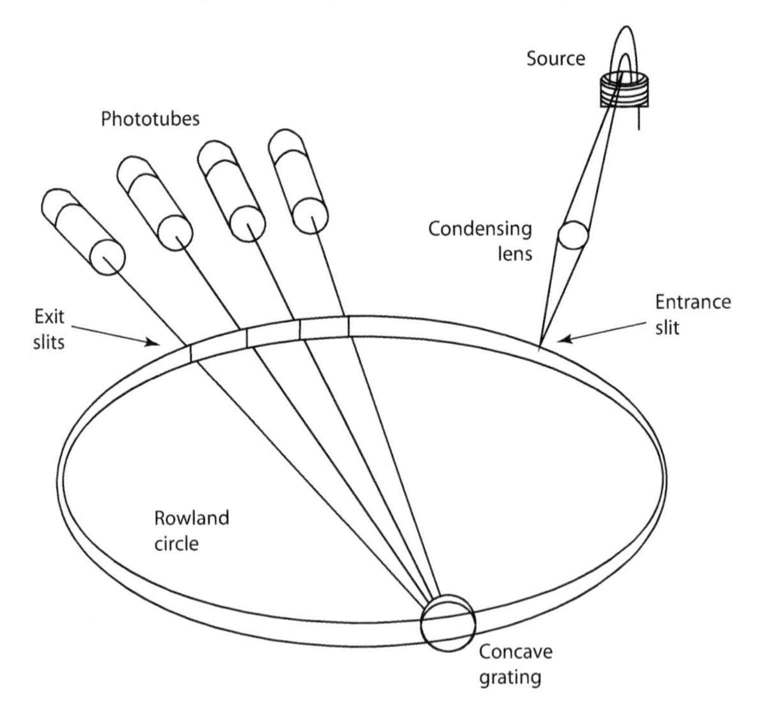

FIGURE 22.10 Schematic of a Paschen–Rünge optical design. (From C. B. Boss and K. J. Fredeen. *Concepts, Instrumentation, and Techniques in Inductively Coupled Plasma Optical Emission Spectrometry*. 2nd ed. Waltham, MA: PerkinElmer, 1997.)

Furthermore, most spectrometers utilized PMT detectors at that time, which were responsive over narrow wavelength ranges. This restriction required the end user's desired suite of elements to be selected prior to the instrument being built and limited the instrument to a maximum of 20–30 emission lines due to the space required for having an exit slit and PMT for each wavelength.[116] This optical design poses additional challenges with internal standardization and interference corrections. The wavelengths being used for internal standard measurements and for interference calculations must have been included in the instrument at the time of manufacture.

The interference challenges and wavelength restrictions associated with the Paschen–Rünge spectrometer led to the development of the Czerny–Turner and Ebert designs. The fixed, concave grating layout of the Paschen–Rünge spectrometer is replaced with a scanning, dispersive optic in the Czerny–Turner and Ebert designs.[117–120] In this arrangement, a dispersive optic such as a plane grating diffracts light onto either one or two collimating mirrors (Ebert and Czerny–Turner, respectively). The focused light is then directed onto a single exit slit.

The wavelength flexibility of these monochromator-based systems is an advantage over polychromators; however, their measurement versatility was offset by the relatively long sample analysis times. The use of a scanning optic dictates that wavelengths be accessed and measured sequentially, which means a single sample could require several minutes to analyze.

An additional challenge with this spectrometer design was in the spectral aberrations resulting from the use of spherical collimating mirrors.[121] Image distortions such as astigmatism, spherical aberration, coma aberration, or elongated slit images are inherent to the traditional design.[122] Light from the grating strikes the mirror at an angle, which prevents the rays from meeting at a common focal point. To prevent these image distortions from creating errors in the spectral results, compensating optics are often used. Example schematics of typical Ebert and Czerny–Turner spectrometers are illustrated in Figure 22.11.

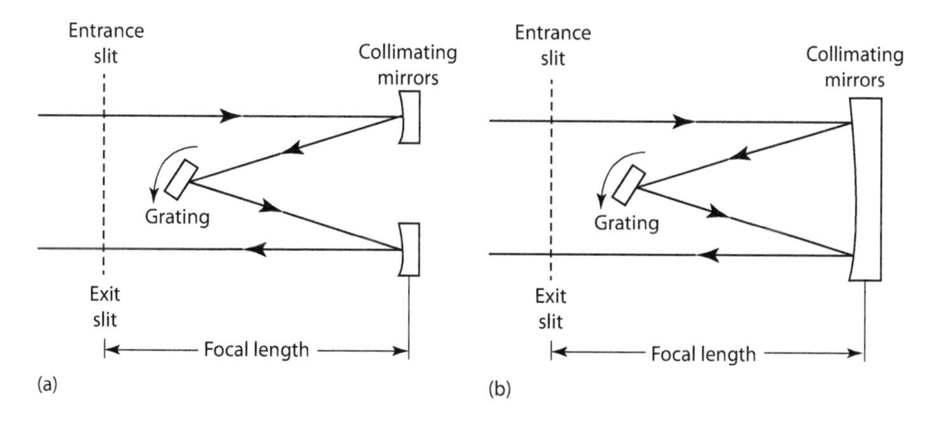

FIGURE 22.11 Schematic of Czerny–Turner (a) and Ebert (b) optical designs. (From C. B. Boss and K. J. Fredeen. *Concepts, Instrumentation, and Techniques in Inductively Coupled Plasma Optical Emission Spectrometry*. 2nd ed. Waltham, MA: PerkinElmer, 1997.)

The early 1980s saw the development of an echelle-based optical spectrometer.[123] This design utilizes a high-dispersion echelle grating in addition to a second dispersive optic (such as a grating or a prism) to separate incident polychromatic light into a two-dimensional spectrum, known as an echellogram. Unlike conventional diffraction gratings, echelle gratings are blazed with a relatively low groove density and relatively high angles of incidence to disperse light into its constituent wavelengths at higher spectral orders. While this dispersive behavior produces exceptional resolution, the resulting spectral orders overlap and are of no use without a second, cross-dispersive optic.

In 1983, the first commercially available benchtop echelle optical system was released, significantly improving the optical dispersion and resolution capabilities of optical emission spectrometers.[124] Due to the high resolution of these systems over relatively short focal lengths, echellograms could be enlarged to fill the relatively large surface area of a PMT detector, or reduced to fill the small active surface area of a solid-state array detector. Further advances in instrument design resulted in spectrometers that could access more than one wavelength at the same time, allowing simultaneous and sequential measurements, or a combination of both, using a single optical platform.[125] Examples of simultaneous and sequential echelle-based optical spectrometers are illustrated in Figure 22.12.

DETECTORS

Detectors are a crucial component to the success of optical emission measurements. Once the plasma transfers sufficient energy to the sample to generate photons, the emitted light is collected, collimated, and diffracted by the spectrometer, where it is then transferred to the detector. The efficiency and performance of the detector then determine the quality and sensitivity of the resulting measured signals.

HISTORICAL PERSPECTIVE

The earliest published research involving optical emission measurements utilized photographic plate detectors. Photographic plate detectors operate via the photoelectric effect, so-called because the light emitted from the sample is of sufficient energy to eject electrons from a photoreactive solid surface. The resulting image contains a series of dark bands and represents the emission spectrum for the sample being analyzed. The position and width of the dark bands can be used to determine which elements, and at what concentrations, were present in the original sample.[126]

Photographic plate spectrometers have been utilized since the early 1900s, and were successful in analyzing complex spectra, such as steel. Spectrographs were designed with a variety of optical systems to achieve various levels of resolution and to accommodate photographic plates of various sizes.[127]

Photographic plates offered the first permanent record of emission spectra from a variety of excitation sources, including arc, spark, and ICP. The photographic emulsion coated on its surface reacts over a wide energy range, allowing a single plate to record emission from most (or all) analytes of interest in a single sample run. A drawback to the detector technology, however, was its nonlinear response and its relatively low quantum efficiency (1%–3%).[128]

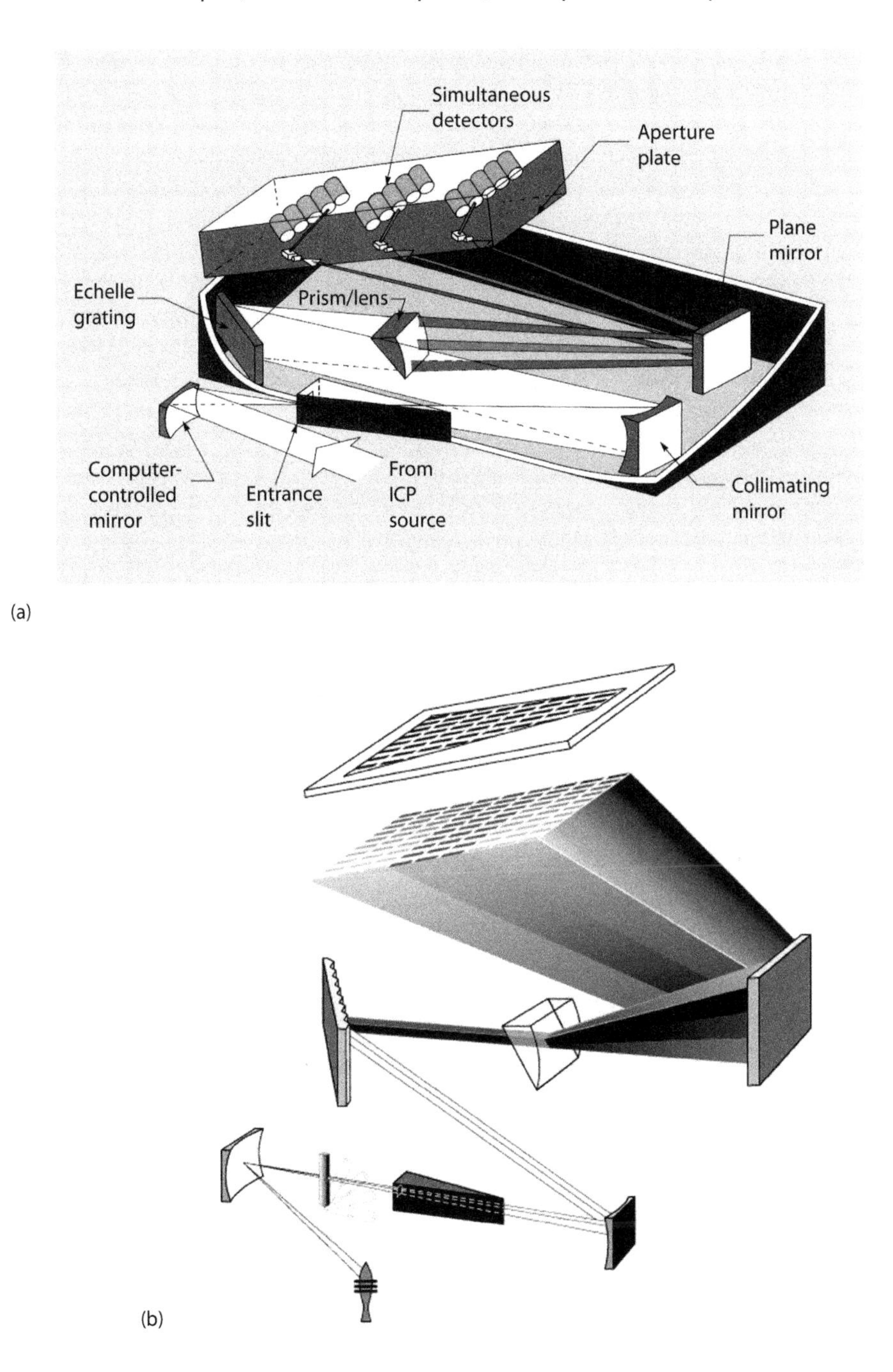

FIGURE 22.12 Schematic of a simultaneous (a) and sequential (b) echelle optical system. (With permission from Teledyne Leeman Labs, Hudson, NH.)

The early 1930s saw the development of PMT detectors, and by the late 1970s, PMTs were being used fairly routinely in atomic emission spectrometers.[129–131] By the early 1980s, technological advances were made to produce photodiode array (PDA) detectors, which possessed multichannel measurement capabilities. These photodiodes were responsive in both the ultraviolet and visible wavelength ranges and had a wider dynamic range than their photomultiplier predecessors. This technology made a fundamental deviation in the way in which it measured spectral emission. Unlike its PMT predecessor, which produces a measureable current from incoming radiation, the PDA is able to generate and store that photo-generated charge.[132,133]

During the late 1980s and early 1990s, charge transfer devices (CTDs), such as charge-coupled devices (CCDs) and charge injection devices (CIDs), began to replace PMTs and PDAs in analytical spectroscopy. These devices respond more uniformly across relatively wide wavelength ranges, and they have the ability to perform simultaneous background measurements. Today, CTD detectors are used almost exclusively in optical spectrometers to quantify spectral emission from a variety of analytes over a wide range of wavelengths.[134–137]

PHOTOMULTIPLIER TUBES

Over the years, PMTs have been the most commonly used detectors for ICP-OES instruments. These detectors typically consist of a sealed vessel, with a series of emission dynodes placed between an anode and a large-surface-area cathode. The dynodes are mounted in series and kept at voltage potentials that get progressively more positive than that of the cathode. A photon that strikes the cathode with a sufficient amount of energy dislodges an electron. The ejected electrons accelerate toward the first dynode in the series, which causes a cascade of electrons to be

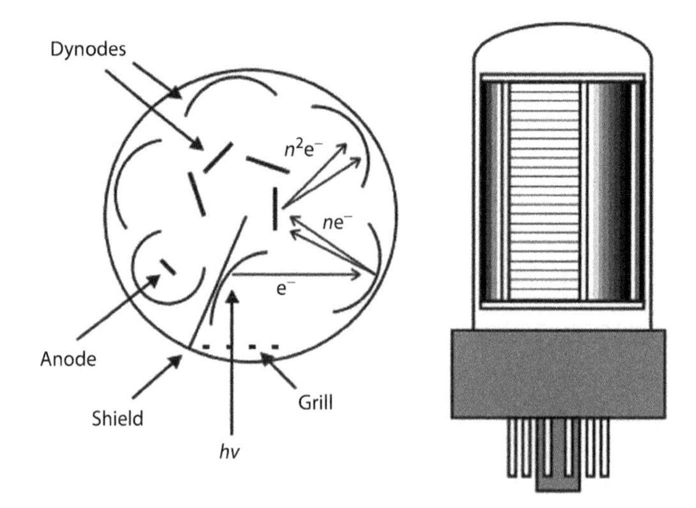

FIGURE 22.13 Schematic and basic function of a PMT. (From X. Hou and B. T. Jones. *Encyclopedia of Analytical Chemistry*, ed. R. A. Meyers. Chichester, UK: John Wiley & Sons, 2000, 9468–9485.)

ejected and accelerated toward the next dynode. The process repeats until the multitude of electrons reaches the anode, which generates a measurable electrical pulse.[138] An example of a PMT detector is illustrated in Figure 22.13.

This type of detector is well-known for its long linear working range and its extremely low background noise. The detector's design gives it the ability to generate more than a million electrons from a single photon, which results in a signal gain of 10^6 to 10^8. The gain is achieved with almost no measureable dark current or background noise, making it well suited for small signals, such as those from trace-level concentrations of analytes.

These detectors lack simultaneous multielement collection capability; however, their single-channel design allows the signal-to-noise ratio to be optimized for each analyte that is being measured. If simultaneous emission detection is desired, multiple PMT detectors can be positioned behind multiple exit slits and used in parallel. The metal oxide construction of the cathode makes the PMT responsive over a relatively short wavelength range; however, a combination of materials can be used to widen the response range.

PHOTODIODE ARRAYS

Photodiode arrays were available for use in making spectrochemical measurements by 1980.[139] PDA devices do not have the same gain and signal response as PMTs; however, they offer the advantage of being able to store photo-generated charge, which allows them to integrate emission signals.[140]

PDAs do not have the sensitivity and signal gain achievable with PMTs; however, their responses are linear over a relatively wide range, and they can be operated in a high-speed mode that allows for parallel readout (similar to the random access readout feature of CID detectors).[141] These devices also require cooling to reduce contributions from dark current and fixed pattern noises; however, signal-to-noise ratios could be improved if measurements are made with longer integration times and fewer signal averages.[142]

CHARGE TRANSFER DEVICES

CTD detectors are solid-state devices, consisting of an array of photoactive elements, which convert incident radiation to an electrical charge that is stored and measured. These photoactive elements, also known as pixels, can be configured as a linear or two-dimensional array, and can be manufactured in a variety of sizes, to meet the requirements of the instrument.[143] Among other benefits, the structure of these array-based devices provides them with the ability to measure more than one wavelength simultaneously, which allows for true simultaneous background correction and real-time internal standardization.[144] In some applications, the efficiency of the background subtraction is sufficient to remove the contribution from background emission that is significantly larger than that for the analytes being measured.

As the name implies, these devices employ a charge transfer process to read out the accumulated charge. Charge transfer is executed via one of two methods: (1)

intercell charge transfer, which shifts charge from where the charge accumulated to a single-output amplifier, or (2) intracell charge transfer, which measures the voltage change induced by shifting charge within the detector element where it was first accumulated.[145] CTDs generate dark current when in use; however, operating them under cooled conditions reduces the dark current contribution to a level that is almost immeasurable, even when emission signals are integrated for long periods of time.[146]

F. L. J. Sangster and K. Teer invented the first CTD at Philips Research Laboratories almost 50 years ago. Unbeknownst to them, it would become a device that had a significant impact on future developments and advancements in atomic spectroscopy.[147] These devices were originally referred to as bucket brigade devices (BBDs) to analogize the method of storing and transferring signals to a line of people passing buckets of water to fill a storage container or to put out a small fire. The potential for using these devices in imaging sensors was quickly realized, and they became a memory storage device with significant use.[148]

In less than a year, Willard Boyle and George Smith, working at Bell Laboratories at the time, improved the BBD's method for processing signals and introduced the CCD.[149] The BBD, which used transistors to transfer charge between a series of capacitors, was improved with the CCD, which transferred charge between capacitive bins (also known as "wells") across the surface of a metal oxide semiconductor (MOS). This invention would earn them a Nobel Prize for Physics and would find mainstream use in imaging devices, such as telescopes, digital cameras, and video cameras.[150,151]

By 1973, Hubert Burke and Gerald Michon, working at General Electric, had introduced the CID detector. For the next two decades, CIDs were used as imagers in machine vision applications where digital image processing capabilities were required. By 1990, CID devices were adapted for use in astronomy, astrophysics, and microscopy, where detector requirements included high sensitivity and spectral accuracy, large dynamic range, and low dark current levels, even during integration periods lasting several hours.[152] The most significant difference between CCDs and CIDs is their readout mechanism, which leads to unique benefits and challenges in atomic spectroscopy applications. Each type of device will be discussed separately in the following sections.

CHARGE-COUPLED DEVICES

CCDs are compact CTDs that have found use as solid-state imagers in a variety of applications. In its original design, the CCD was constructed for the storage of digital information for use in devices such as digital cameras and video recorders. Once in use for routine consumer applications, it became clear that the CCD possessed the ability to convert digital charges to analog signals, broadening its utility to include industrial and scientific applications.[153,154] Figure 22.14 illustrates the structure of a typical CCD device.

The mechanism for transferring charge through a typical CCD detector is illustrated in Figure 22.15. When the device is read out, charge that has accumulated across the entire device must be transferred and measured, even if only a small portion of the array needs to be read out. To accomplish this, a small bias voltage is

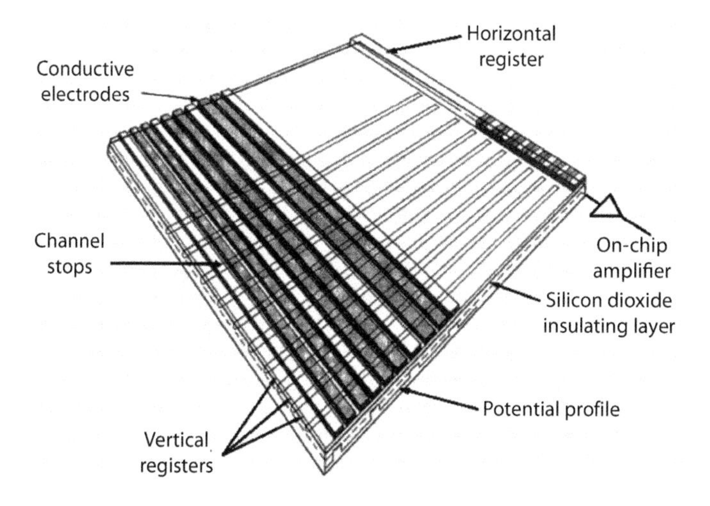

FIGURE 22.14 Basic schematic of a CCD. (From J. R. *Janesick. Scientific Charge-Coupled Devices*. Bellingham, WA: International Society for Optical Engineering, 2001.)

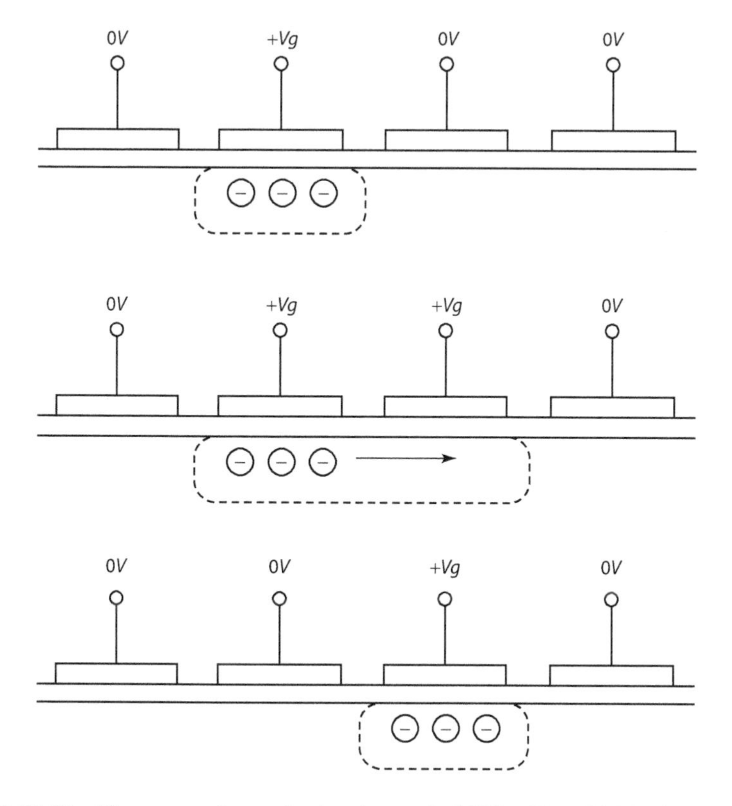

FIGURE 22.15 Charge transfer mechanism in a typical CCD. (From N. Waltham. CCD and CMOS sensors. In *Observing Photons in Space: A Guide to Experimental Space Astronomy*, ed. M. C. E. Huber, A. Pauluhn, J. L. Culhane, J. G. Timothy, K. Wilhelm, and A. Zehnder. New York: Springer Science & Business Media, 2013, 423– 442.)

applied to one of the pixels, which creates an area of depleted charge in the substrate directly below. A greater bias voltage is applied to the adjacent pixel, which creates a similar depletion area and promotes the transfer of charge. The process repeats itself until the accumulated charge is transferred through the entire device until the charge is shifted to the output amplifier for analog-to-digital conversion and measurement with a digital processor. This readout process destroys the stored charge and prevents it from being measured again in the future.[156]

The structure of a CCD, combined with its readout mechanism, allows it to read its stored charge with a high level of uniformity, which translates to high-quality spectral images. However, it is this structure and readout that make the CCD susceptible to a phenomenon known as blooming. Blooming occurs when one of the device's detector elements (pixels) accumulates more charge than it can store and spills the excess charge into nearby pixels. This occurrence can significantly impact measurements for emission signals that are being measured simultaneously with intense emission from nearby wavelengths.[157]

To reduce the risk of blooming, a variety of modifications have been implemented. Some CCD devices have been assembled to include structures that are designed to remove and drain excess charge from pixels before they saturate and spill that charge into nearby pixels. These structures, often referred to as charge drains or antiblooming gates, surround and isolate active pixels to allow low-level and high-level emission signals to be accurately measured, even if this requires storage in nearby pixels on the same detector.[158] These charge drains have been shown to be effective enough to prevent blooming, even when reading charge that's adjacent to pixels that accumulated charge 1000 times their saturation level.[159]

Another method employed to reduce the likelihood of charge spillage was to use a series of separate linear CCD arrays, working as a cohesive unit. This device, known as a segmented charge-coupled device (SCD) detector, is constructed such that a separate linear array is used for the measurement of one wavelength (up to three wavelengths if they occur in close proximity) for each of the elements on the periodic table. Using arrays that are physically separate from one another mimics the structure of a single CCD detector with built-in charge drains.[160,161] Size constraints limit the number of arrays that can be used; however, more than 200 arrays can be combined for use in covering important analytical wavelengths across both the ultraviolet and visible wavelength regions.[162]

In addition to overcoming potential effects from blooming, the structure of an SCD allows it to measure a small number of wavelengths (or even a single wavelength) without reading the charge across the entire detector, in a method known as random access integration (RAI). This feature increases the readout speed and dynamic range, compared with a single (nonsegmented) CCD detector.[163] Figure 22.16 is an image of an SCD detector showing a photomicrograph close-up of the separate photosensitive linear arrays.

CHARGE INJECTION DEVICES

CID detectors fall into the category of complementary metal oxide semiconductor (CMOS) technology. CIDs contain an array of detector elements that can store

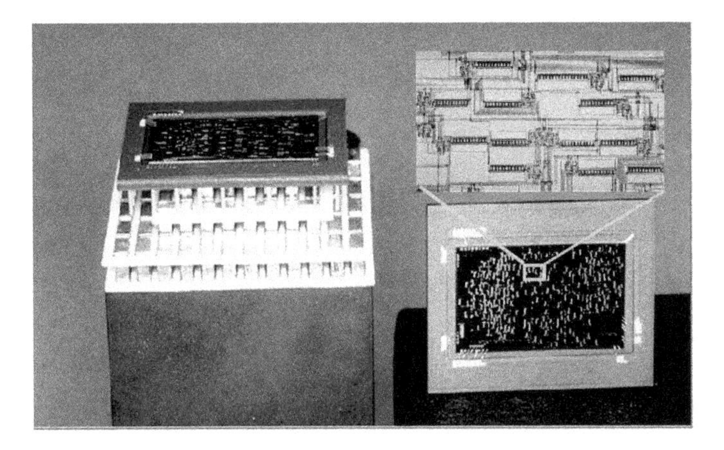

FIGURE 22.16 SCD detector showing a photomicrograph close-up of the separate photo-sensitive linear arrays. (From T. W. Barnhard, M. I. Crockett, J. C. Ivaldi, P. L. Lundberg, D. A. Yates, P. A. Levine, and D. J. Sauer. *Analytical Chemistry*, 65, 1231–1239, 1993.)

photo-generated charge. Each detector contains a light-sensitive area, as well as row and column electrodes that are addressable for the purposes of storing and reading out charge. The electrodes are constructed from a conductive silicon material laid over an insulating layer, which provides a region for charge to be stored.[164] An example of a CID detector element is illustrated in Figure 22.17.

The mechanism for transferring charge through a typical CID detector is illustrated in Figure 22.18a–d. The four images represent a single pixel during the four-step charge readout process. The pixel contains two photogates (labeled "Row" and "Column") that are used for storing and measuring photo-generated charge. When an analytical measurement commences, the detector starts collecting charge, based on the integration time set by the user (Figure 22.18a). During this step, the incident radiation is converted to charge and collected in the column photogate. Once integration has begun, the detector measures the accumulated charge by shifting it between the row and column photogates (Figure 22.18b and c).

If the pixel has accumulated enough charge to approach saturation, the charge can be cleared by injecting it into the substrate (Figure 22.18d). This is a destructive readout (DRO) and resets the pixel to allow it to resume charge collection. If the pixel's charge is measured and is far from reaching its saturation point, the charge can be transferred back to the column photogate, where the pixel can continue collecting charge. This process is known as a nondestructive readout (NDRO), as the detector shifts the charge without destroying it.[165]

The structure of each pixel allows CID detectors to perform NDROs, which is the main functional difference between CID and CCD detectors. Measuring charge without destroying it allows each pixel to perform a NDRO repeatedly during an analytical measurement, as needed. Since each pixel can sense and measure its own charge, the pixels within a CID can be randomly accessed for charge integration. These NDRO and RAI features make these detectors well suited for a number of applications in spectroscopy.[166,167]

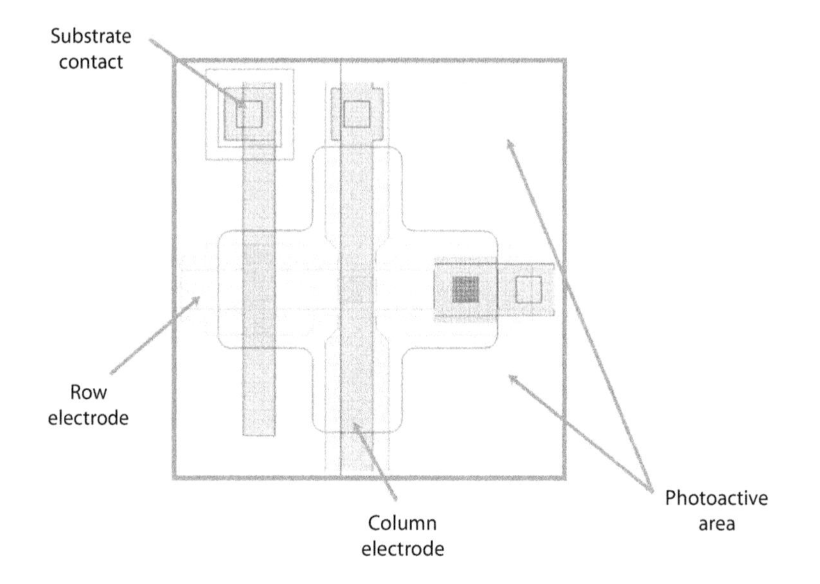

FIGURE 22.17 Schematic of an example detector element (pixel) within a CID detector. (With permission from Teledyne Leeman Labs, Hudson, NH.)

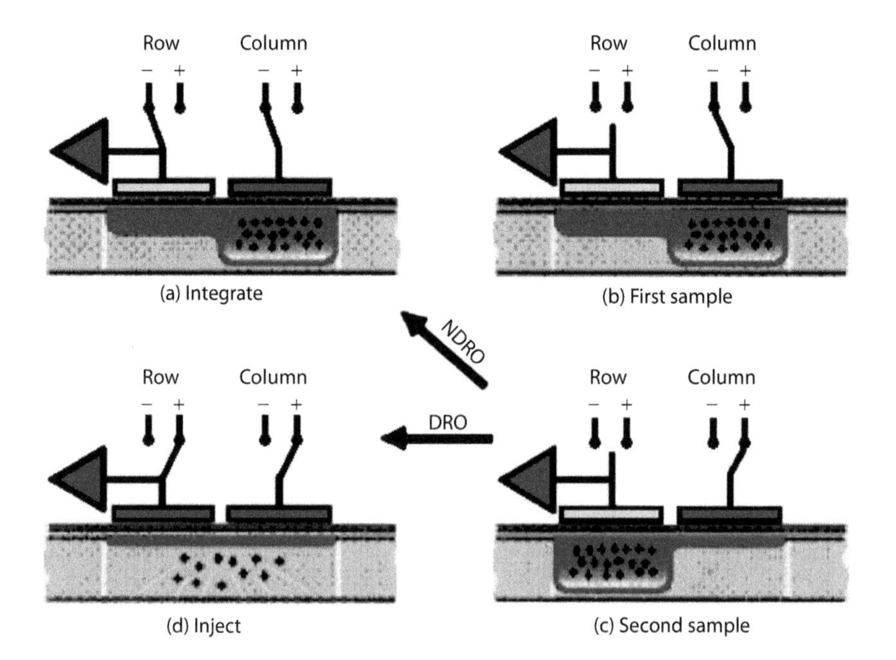

FIGURE 22.18 Schematic and operation of a CID pixel. (From J. M. Harnly and R. E. Fields. *Applied Spectroscopy*, 51, 334A–351A, 1997.)

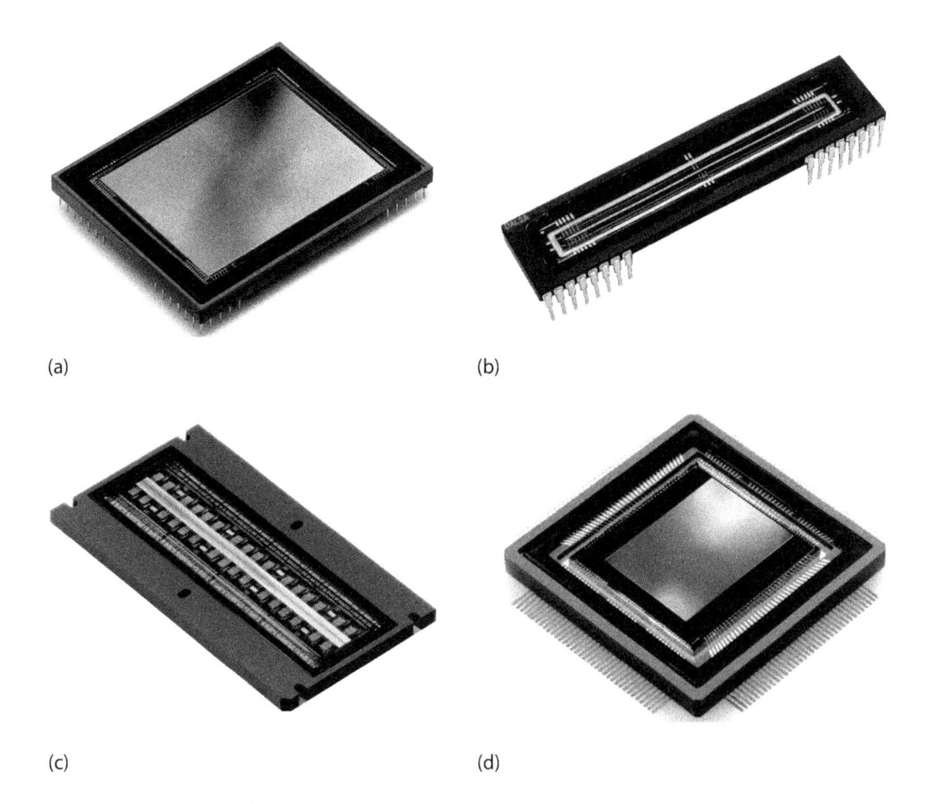

(a) (b) (c) (d)

FIGURE 22.19 CCD (a and b) and CMOS (c and d) imagers. (With permission from Teledyne DALSA, Billerica, MA.)

The presence of photogates on each detector element reduces the size of the photoactive area of each pixel, which decreases the total amount of charge that each pixel can accumulate before reaching saturation. This construction also decreases the uniformity of charge storage and readout across the array. Benefits to the structure of this detector are that it is inherently resistant to charge blooming, and it allows for simultaneous wavelength measurements and background correction, plus simultaneous internal standardization for real-time drift correction.[168]

These devices can rapidly access all the addressable pixels within the array, which allows the detector to optimize the signal-to-noise ratio for each wavelength being measured. This optimization also increases the device's dynamic range to make it functional over several orders of magnitude. This feature is beneficial for applications in which both low- and high-intensity wavelengths must be measured simultaneously.[169] Figure 22.19 illustrates examples of typical CCD (a and b) and CID (c and d) devices.

ANALYTICAL PERFORMANCE

The quality of any developed and optimized analytical method can be assessed by a number of different figures of merit, including accuracy, precision, sensitivity, speed, detection limits, working range, ease of use, and reproducibility. Both the instrument operating conditions and method parameters can affect these performance characteristics, so it is important to be familiar with these factors, to understand their effects, on the analytical data being collected.

DEPENDENCE ON ENVIRONMENTAL OPERATING CONDITIONS

All commercially available ICP-OES instruments have cooling, venting, electrical, and gas requirements for proper and safe operation. These instruments also prefer a certain set of room conditions with specified temperature, pressure, humidity, and room vibration requirements. The environmental conditions will vary slightly between instruments; however, they should generally be operated under the following range of conditions.

The laboratory should be maintained at a temperature between 15°C and 35°C. Ideally, the temperature within the room will not change, particularly when the instrument is being used for analysis. If the room temperature cannot be maintained over the course of a day, the rate of temperature change should be minimal. Most instruments can tolerate temperature changes of 1°–3°C per hour; however, significant temperature changes will produce wavelength drift.

The relative room humidity should be kept between 20% and 80% in a noncondensing and noncorrosive atmosphere. Most spectrometers remain constantly purged with a low flow of dry, high-purity gas; however, if the purge is disrupted or stopped and the gas inside the spectrometer exchanges with the air in the room, moisture in the air could deposit on the components inside the spectrometer and fog the optics.

Although ICP-OES instruments are designed with a certain level of robustness built in, the manufacturer's recommendations for environmental conditions should be followed for optimal instrument performance.

EXHAUST REQUIREMENTS

An exhaust system is used for removing heat and gases from the instrument for both safety and stability reasons. Heat generated by the system electronics must be removed to prevent them from overheating and malfunctioning, while excess heat from the plasma must be withdrawn to prevent the temperature inside the torch box from rising and producing background emission from the torch.

The torch box area must be properly vented to remove gases generated in the plasma, which may contain ozone and other noxious substances. The positioning of the exhaust vents, the minimum and maximum draw required, and the diameter of the vent tubing vary between instruments, so it is important to follow manufacturer recommendations.

ELECTRICAL REQUIREMENTS

Optical emission instruments require high-voltage power to operate properly. If the instrument operates with a water-cooled load coil, a chiller or water recirculator may be used that has a voltage requirement that is different from that for the instrument. A computer, which likely controls the instrument, may have its own set of electrical requirements. The power requirements, including voltage, frequency, and phase, will vary slightly between instruments, and the manufacturer requirements should be closely followed.

The performance and longevity of an ICP-OES can be affected by the quality of the power being delivered to the instrument. In severe cases, power fluctuations can cause significant damage to electronic components in the instrument. A power conditioner may be employed to protect the instrument from voltage sags, transient voltage issues, and general regulation problems. An uninterruptible power supply (UPS) will allow the instrument to be properly shut down in the event of a power outage.

TEMPERATURE AND PRESSURE REQUIREMENTS

The spectrometer must be maintained at a constant temperature and pressure to prevent excessive wavelength drift during analysis. Spectrometers are typically heated to a temperature between 35°C and 38°C, and the set point is maintained to a precision of approximately ±0.1°C. Maintaining the spectrometer at a temperature well above that in the laboratory reduces the effect that small temperature changes in the room have on wavelength drift.

The spectrometer is typically purged with a relatively dry, high-purity gas to maintain a constant environment inside the spectrometer. A wide range of flow rates can be used for the purge gas, from 0.1 to 15 L/min; however, the pressure inside the spectrometer should be at equilibrium prior to starting a sample run to avoid wavelength drift due to pressure changes.

MAINTENANCE

Routine preventive maintenance is one of the best ways to ensure proper instrument performance, as poor-quality data can often be traced back to the sample introduction system. General good laboratory practices should be followed in terms of maintaining a clean laboratory, and any spills should be cleaned up immediately. Solutions should be properly stored and labeled, and standards and samples should be kept in covered vials or containers at the completion of a sample run to minimize spills, contamination, or loss from volatilization. A clean rinse solution should be run through the sample introduction system for a suitable length of time at the end of each sample run to help reduce the frequency with which the sample introduction system must be disassembled and thoroughly cleaned.

If a peristaltic pump is being used, all pump tubing should be regularly checked. Solution must move smoothly through the tubing, both into and out of the instrument, to minimize pulsations in the plasma. Tubing should be checked for distortions, flat spots, and tears and should be replaced if any are discovered. The platens on the pump should be properly tightened down on the pump tubing to allow solution to move through the tubing without any jerking movements.

The nebulizer should periodically be checked for cracks, blockages, or leakages. The performance should be checked by pumping a clean solution through the nebulizer and visually inspecting the resulting aerosol to ensure that it has the proper shape and lacks irregular pulsations. The nebulizer should be cleaned according to the manufacturer instructions and should be carefully stored when not in use to prevent breakages.

The spray chamber requires little to no maintenance; however, it should be visually checked for cracks on occasion. The spray chamber should be viewed during aerosol generation to ensure that the inside surface is being properly wetted and large droplets are not beading up and disrupting the aerosol transfer. The drain tubing should fit tightly onto the spray chamber to allow waste solution to be properly removed.

The torch should be checked for leaks, cracks, or other physical damage. The injector should be checked for deposits or blockages and thoroughly cleaned if any are present. If the torch has gas connections or O-rings, those should be inspected for damage or leakages. Torches will accumulate deposits during normal sample analysis and should be cleaned as necessary to remove them. Over time, small deposits will accumulate that do not get removed during cleaning. These deposits should not be of concern unless the instrument's performance is affected.

Most of the remaining components of the instrument, including the radio-frequency (RF) generator, the spectrometer, and the detector, require little or no maintenance other than an optical entrance window that requires occasional cleaning or air filters that need to be cleaned or replaced.

DEPENDENCE ON PLASMA OPERATING CONDITIONS

Optimum plasma operating conditions are critical in obtaining the highest-quality data. Plasma parameters will impact the stability and energy of the plasma, and therefore, the behavior of the elements. To minimize these effects, plasma parameters are typically optimized to maximize the emission intensity for the analyte wavelengths in the method. Plasma conditions that are optimal for one type of sample are often different from those for another sample type, so the operating conditions should be carefully evaluated for each application. A few important points to consider when optimizing plasma conditions are outlined in the subsequent sections.

Plasma parameters to consider for optimization are

- RF power
- Nebulizer gas flow
- Auxiliary gas flow
- Coolant gas flow
- Pump settings
- Radial viewing height

It should be noted that, while important, plasma optimization is not as critical as the other steps involved in developing an optimized analytical method. ICP-OES methods typically are developed for the analysis of multiple elements. Each element prefers a slightly different set of plasma parameters to achieve its maximum intensity for the wavelength selected. Since ICP-OES instruments collect emission intensities from groups of elements simultaneously, it is not practical to select different plasma settings for each individual element. Therefore, the plasma settings for multielement analysis are often of a compromise for most of the elements in the method. Most modern ICP-OES instrumentation provides an automated optimization feature to rapidly and automatically select the best plasma parameters to maximize the emission intensity for as many elements as is possible in a multielement method.[170]

RF POWER

The quality of the RF generator and the RF power selected for use during analysis will determine the quality and precision of the resulting analytical measurements. While there may be more than one RF power setting that will produce high-quality data, this is an important plasma parameter because the RF field, in combination with the argon gas, is responsible for generating and maintaining a stable plasma source.

RF generators typically operate at a frequency of either 27.12 or 40.68 MHz (although a new 33.90 MHz generator has recently been commercialized for use in ICP-MS). These are the only frequencies approved by the Federal Communications Commission (FCC) for RF generator operation, as other frequencies will interfere with established communication activities. There has been much debate over the relationship between an RF generator's operating frequency and its resulting performance;[171,172] however, there is very little evidence to indicate that one frequency is beneficial over the other.

The stability of the plasma largely relies on the generator's ability to adjust to changing plasma conditions. Each time a new sample is introduced, the composition of the plasma rapidly changes. As the composition changes, the power required to adjust to the new plasma conditions changes and the RF generator must react accordingly. Matching the power output of the generator to that required by the new plasma conditions is known as impedance matching and can be accomplished with two basic oscillator designs: (1) crystal controlled (fixed frequency) and (2) free running (variable frequency).[173] There are benefits and detriments to both designs; however, the variable-frequency design of free-running oscillators makes them better able to generate a plasma from a cold start and adapt to changes incurred by challenging sample matrices. A stable, successfully formed plasma is illustrated in Figure 22.20.

FIGURE 22.20 A close-up of a Thermo Jarrell Ash Atomscan 16 inductively coupled argon plasma torch. (With permission from Wblanchard.)

In addition to having a well-designed generator, an appropriate RF power must be selected for the application. Generally speaking, RF power is proportional to plasma temperature. Given the same set of gas flows and sample uptake parameters, a higher RF power produces a hotter, more energetic plasma. This is particularly advantageous for sample matrices that contain organic solvents or high levels of dissolved solids that are notorious for causing plasma instabilities or blowouts. While a higher RF power produces a more robust plasma, this setting should still be carefully selected to avoid using a power that is higher than necessary. Increasing the RF power increases the amount of emission that gets produced from all components in the sample solution, which can increase the level of background emission, degrade the signal-to-noise ratio for the analytes of interest, and degrade the background equivalent concentration (BEC).[174,175]

PLASMA GASES

While the plasma gases play an important role in generating and sustaining a stable plasma, the individual gas flow settings can have an effect on the emission of some elements and on the overall data quality. The coolant gas serves two major purposes when the plasma is being operated: plasma generation and torch cooling. The coolant gas flows through the outer tube in a swirled pattern to help the plasma maintain its annular shape. A relatively high flow rate is used to sustain a stable plasma and to prevent the torch from overheating. Since the main function of the coolant gas is related to the operation of the plasma, the flow rate does not have a significant effect on the emission of most elements. However, it should be pointed out that the flow rate should be chosen with logic and care. A setting that is too low will provide insufficient cooling of the torch. A slight overheating can cause an increase in background emission as blackbody radiation from the torch itself is emitted. A severely overheated torch will melt and will require replacement. A flow setting that is too high is wasteful of the coolant gas and may provide more turbulence in the plasma than is necessary. Flow rates for the coolant gas can range from 8 to 20 L/min; however, flow rates between 10 and 14 L/min are appropriate for most ICP-OES applications.

The auxiliary gas flows through the intermediate tube of the torch and supplements the coolant gas to shift the base of the plasma farther away from the end of the inner tube (injector). Not all applications or torch configurations require the use of an auxiliary gas; however, it is useful during the analysis of sample matrices with high levels of dissolved solids, organic solvents, or other matrix components that are known to cause a buildup of material inside the injector. The auxiliary gas adds turbulence to the plasma, which degrades the penetration of the nebulizer gas into the plasma. Therefore, ideal circumstances would dictate that minimal or no auxiliary gas is used. However, if the application requires an auxiliary gas, the flow rate should be set such that the plasma is far enough from the injector to avoid a blockage from forming, yet not so far that the plasma becomes unstable and at risk for being extinguished. Flow rates for the auxiliary gas, if being used, are typically between 0.1 and 1.5 L/min.

Of the three plasma gases, the nebulizer gas typically has the most significant effect on analyte emission and should be optimized carefully. The nebulizer gas flows through the center tube (injector) of the torch and carries the sample into the plasma. The gas must be able to penetrate the plasma and travel along its central channel to maximize the efficiency with which the sample is desolvated, vaporized, atomized, ionized, and excited. Since the nebulizer gas carries the sample into the plasma, it stands to reason that increasing the nebulizer gas will transport more sample into the plasma, which will produce larger emission signals and more sensitive measurements. This is not the case, however. The nebulizer gas must be carefully selected based on the challenge of the application and analytes being measured. In addition to generating some turbulence that affects plasma stability, the nebulizer gas transports the nebulized sample aerosol, which affects the load on the plasma. If the change in plasma load is too severe, the RF generator will not be able to adjust its power output to compensate and severe plasma instability will result. In severe cases, the plasma load will be great enough to extinguish the plasma completely. Conversely, if the nebulizer gas flow is too low, the sample will not adequately penetrate the plasma and emission signals will significantly degrade. Most instruments will allow the nebulizer gas flow to be set between 0.1 and 2 L/min; however, a setting of approximately 1 L/min (or the equivalent pressure if the nebulizer is not mass flow controlled) is typically used for most applications. Elements from different groups on the periodic table will produce optimal emission with slightly different nebulizer gas settings. Since most applications involve the analysis of multiple elements, a compromise must be made and a nebulizer setting should be selected that will produce optimal sensitivity and detection limits for the overall method.

PUMP SETTINGS

The sample solution is delivered to the instrument via either a peristaltic or a syringe pump. The speed and precision with which the solution is introduced to the nebulizer significantly affect the quality of the resulting aerosol. The pump speed, combined with the inside diameter of the pump tubing, dictates the rate of solution that is delivered to the nebulizer. While each nebulizer will have a recommended flow rate for sample introduction, an optimal pump speed should be selected such that a consistent, stable aerosol is produced. If the pump speed is set such that the nebulizer aspirates solution faster than it is supplied via the pump, the nebulizer is being underfed ("starved"). Conversely, if the pump delivers solution faster than the nebulizer can aspirate it and produce an aerosol, the nebulizer is being flooded. Most nebulizers operate optimally when they are being starved for solution. Flooded nebulizers will produce an aerosol with a pulsing stream of large droplets that spill out the end. This "spitting" is from the excess solution that was pumped into the nebulizer but not converted to an aerosol. Some nebulizers, known as self-aspirating nebulizers, are designed to operate by drawing the solution through the pump tubing without assistance from a pump. Since these nebulizers do not require a pump for sample uptake, self-aspirating nebulizers are excluded from the text in this section.

The precision of the delivered solution will affect the precision of the aerosol. This is dictated by the style of the pump. A peristaltic pump contains a number of rollers that push solution through a flexible piece of tubing by alternatively compressing and relaxing the walls of the tubing. Each time the tubing is decompressed,

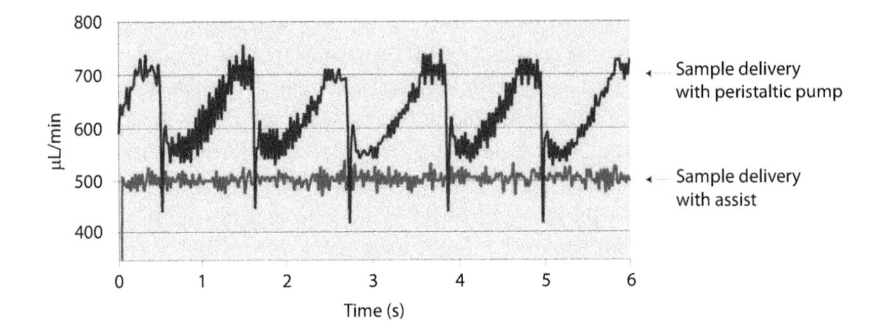

FIGURE 22.21 Signal fluctuation produced with a peristaltic pump vs. a syringe pump. (With permission from Glass Expansion, Pocasset, MA.)

a vacuum is created and a miniscule amount of solution is drawn backward into the pump tubing. This causes a fluctuation in the volume of solution being delivered to the nebulizer, which produces an aerosol and resulting emission signals with a measurable pulsation.

A syringe pump can also be used for solution delivery to the nebulizer. A syringe pump is a type of infusion pump that delivers solution via small, rapid pulses. The pulsations are smaller and more frequent than those incurred with a peristaltic pump, which results in an aerosol and emission signals with smaller pump-related fluctuations. This concept is illustrated in Figure 22.21 A syringe pump will produce an aerosol with fewer fluctuations; however, pulsations from a peristaltic pump can be minimized with the use of a small-diameter pump and small pump rollers.

PLASMA VIEWING HEIGHT

Emission signals should be collected from a specific part of the plasma to maximize the signal-to-background ratio and sensitivity of each measurement. Often referred to as the "normal analytical zone," this section of the plasma consists of a temperature zone that is optimal for atomic and ionic emission. Axial (end-on) measurements are made along the central channel of the plasma, which prevents emission from being collected from only one of the temperature zones. Radial (side-on) measurements are made perpendicularly to the direction of the plasma so that the observation height, also known as the viewing height, should be carefully optimized.[176]

Viewing height affects the emission intensity of different elements in different ways. Some would prefer a hotter plasma, which would render an optimal viewing height that is closer to the load coil. Other elements would prefer slightly cooler conditions, which would put the optimal viewing height farther from the load coil and closer to the tail of the plasma. Unless the instrument is being used for single-element analysis, the plasma viewing height will be a compromise. This process is computer controlled through the instrument's software; however, manual adjustments can sometimes be made to fully optimize the observation height of the measurement.

PRECISION AND ACCURACY

When discussing analytical performance, precision and accuracy are figures of merit that indicate the robustness and reliability of the instrument and of the method developed for analysis. The terms are often used together; however, it is important not to use them incorrectly or interchangeably. *Precision* is a term that describes the agreement between replicate results, which provides an indication of the reproducibility of the method. It is a figure of merit that can be obtained by repeating the measurement of a particular standard or sample and calculating the variation between the replicate results. To properly calculate precision, data for the same elements must be collected the exact same way each time.

Both short- and long-term precision can be calculated for a method. Short-term precision typically represents the variation between back-to-back measurements of a small number of standards and/or samples. This represents the reliability of the method for a particular application, and depending on which solutions are included in the measurement, this can include the reliability of the sample preparation procedure. Long-term precision, which typically represents the robustness and/or transferability of a method or procedure, can be calculated using a number of different approaches. Results can be collected for a chosen standard or sample periodically over the course of a sample run to determine the reliability of the method during a typical 8-hour day. Results can also be collected on nonconsecutive days, with two different laboratory technicians or in two different laboratories (as long as both laboratories are using the same instrumentation). Results collected under these conditions will determine the robustness and transferability of the method.

Accuracy is a term that describes how close the experimental results are to the known ("true") results. In other words, the accuracy of the method determines the correctness of the results. This is expressed as either an absolute or relative error, and the calculation should be based on the measurement of a certified or other well-characterized standard. The acceptable level of inaccuracy should be determined during the method development and optimization stage. All calculated results are going to possess a small amount of inaccuracy due to basic errors and uncertainties in preparing and analyzing samples. If the calculated accuracy does not meet the requirements of the method, the source of the error should be investigated. Errors can consist of two different types: random or systematic. Each type of error could be due to a number of sources, and a combination of errors could be contributing to the overall issue, so troubleshooting inaccuracies should be approached with care.

Precision and accuracy should not be used interchangeably, as they describe significantly different figures of merit. While it's desirable for a method to have excellent precision and accuracy, it is possible to have one without the other. Results that are precise but inaccurate could indicate that the instrument settings are properly chosen, but an issue exists with the calibration curve or there are interferences that have not been properly accounted for.

If a method has good accuracy but poor precision, it is sometimes referred to as being "accurate in the mean." In this situation, the accuracy is a result of systematic imprecision and typically indicates an issue with the sample introduction system (steering mirror is moving too slowly [dual-view methods only], sample uptake time is too short, sample flow rate is too high, or platens on the peristaltic pump are not tightened properly).

DETECTION LIMITS

One of the most important, yet one of the more highly debated figures of merit for an analytical method is the LOD. For a given analyte, the International Union of Pure and Applied Chemistry (IUPAC) defines the detection limit as the smallest change in the emission signal (x_L) that can be detected with statistical certainty.[177] In other words, this is the lowest signal that can be detected above the blank and is defined according to the following equation:

$$x_L = k\sigma_B$$

where x_L is the net intensity, k represents a numerical multiplier that is selected according to the desired statistical confidence level, and σ_B represents the standard deviation of replicate measurements of the blank solution. The net intensity can be converted to a detection limit concentration using the following equation:[178]

$$LOD = \left(k\sigma_B\right) / S$$

where k represents the same multiplier as that in the previous equation, σ_B represents the standard deviation of the blank, and S is the slope of the calibration curve for that analyte (i.e., the sensitivity). Kaiser[179] asserts that an appropriate value for k is 3, as that will represent a 95% confidence interval for the calculated detection limit for most applications, so 3 is the most commonly used value in this calculation.

The debate over the formula for calculating detection limits stems from its definition, which indicates that the concentrations must be calculated with statistical certainty.[180] Most accepted formulas for calculating detection limits are based on the IUPAC equation listed above. The resulting concentrations are typically referred to as "instrumental detection limits" and allow detection limits to be compared between different instruments or between different conditions on the same instrument. While useful for comparison purposes, detection limits based on this formula are unrealistic and do not reflect those obtained when measuring analyte signals in the presence of the sample matrix under investigation.

Suggested modifications have been made to reflect more conservative estimates of detection limits. The Environmental Protection Agency (EPA) suggests the calculation of a "method detection limit" based on seven replicate analyses of a blank fortified with the analytes of interest at two or three times the concentration of the calculated instrument detection limit.[181] This is a more conservative estimate of the detection limits and accounts for any signal degradation caused by the sample matrix. Currie[182] suggests calculating a "determination limit" by using the IUPAC definition of LOD with a multiplier of 10 instead of 3. Currie defines this as "a determination limit at which a given procedure will be sufficiently precise to yield a satisfactory quantitative estimate." Long and Winefordner[183] recommend multiplying the standard deviation of the blank by 6 to calculate a "limit of guarantee."

LIMIT OF QUANTITATION

Limits of quantitation (LOQs) are often calculated along with the LODs. The LOQ is a conservative estimate of the LOD and is sometimes used as the reporting limit for a method.[184] For each analyte, the LOQ represents the lowest concentration that can be measured and reported with a sufficient amount of precision and accuracy. The desirable level of precision and accuracy is somewhat arbitrarily defined so that there is no concrete formula for calculating the LOQ. However, the desired precision level is often around 10% (reported as a relative standard deviation), which means the LOQ is calculated using 10 times the standard deviation of the blank.[185] By definition, the LOQ cannot be lower than the LOD, and it should not be equal to the LOD.

BACKGROUND EQUIVALENT CONCENTRATION

A figure of merit that relates to the instrument detection limit is the BEC. For a given analyte, the BEC is the concentration of that analyte that produces a net signal (emission signal minus the contribution from the background) equal to the background signal at that wavelength. In other words, this is the analyte concentration that yields a signal-to-background ratio of 1.[186] There are a number of different formulas that exist for calculating BEC values; however, they can easily be determined using the calculated calibration curves. If the curves are plotted as measured emission intensity (x-axis) versus concentration (y-axis), the BEC for each analyte is represented by the y-intercept value. Since the BEC is based on the noise associated with the background signal (which is typically ~1% of the signal intensity of the background), an approximate BEC calculation can be performed by multiplying the instrument detection limit by a factor of 30.

SENSITIVITY

Sensitivity is generally referred to as the ability to confidently measure small differences in analyte concentrations. The acceptable definition of sensitivity is outlined by IUPAC and refers to calibration sensitivity. Sensitivity can vary by element and is represented by the calculated calibration curve. Most calibrations that are used with ICP-OES applications are linear and can be represented using the following equation:

$$y = mx + b$$

where y represents the instrument response for a given analyte at a given concentration, m represents the calibration sensitivity (i.e., the slope of the calibration curve), x represents the analyte concentration, and b is the measured signal for a blank solution. If the concentration of the signal is zero, the equation becomes $y = mx$ and the calibration sensitivity becomes independent of concentration.

Mandel and Stiehler[187] published a slightly different definition of sensitivity. Referred to as analytical sensitivity, this definition includes the precision of the measurement according to the following equation:

$$\gamma = m/s_S$$

where γ represents the analytical sensitivity, m represents the slope of the calibration curve, and s_S represents the standard deviation of the measured signal. Since this term is based on the slope and standard deviation of the measurement, the sensitivity value is independent of the units that were used to measure the emission signals. A drawback to this definition is the inclusion of the signal standard deviation. Since s_S can vary with analyte concentration, analytical sensitivity is typically concentration dependent.

METHOD DEVELOPMENT CONSIDERATIONS

As with many techniques, its success in the laboratory depends on the quality of the methods that are developed for the application work being conducted. While ICP-OES instruments provide automated, intuitive operation with parts per billion–level detection limits and working ranges that extend over several orders of magnitude, developing a useful, well-optimized analytical method can be a manual, labor-intensive, and time-consuming process.

Whether a method is being developed from scratch or an existing method is being further optimized, there are a number of parameters that should be taken into consideration to ensure that the method is best suited for the intended application. These include

- Analytical wavelengths
- Interferences
- Plasma parameters
- Data acquisition parameters
- Validation of method

Let's take a look at each of these considerations, in turn, to understand how each factor plays a role in the quality of the final optimized method. It should be noted that the theory behind this approach is applicable to the development of an analytical method for any elemental analysis technique; however, some of the specific considerations will change. For example, if developing an analytical method for an ICP-MS application, one would be choosing isotopes instead of emission wavelengths for measuring the analytes and internal standards.

ANALYTICAL WAVELENGTH CONSIDERATIONS

When choosing appropriate wavelengths for the elements of interest, it's important to keep in mind the method's desired performance attributes. For example, if method development is taking place for quantifying trace elements in drinking water samples, the main performance attributes will likely be detection capability and accuracy. Therefore, the most sensitive wavelength should be chosen for each element, and all potential interferences should be identified and carefully corrected. Alternatively, if a method is being developed to quantify elements in oil additives, a wider working range may be desired. In this case, wavelengths with lower sensitivity may need to be chosen to maximize the concentration range over which the elements can be calibrated.

A critical factor in determining the suitability of a wavelength is whether it suffers from interferences. Several types of interferences exist in ICP-OES, and their severity is dependent on the analyte wavelength, other elements present in the sample, and the sample matrix itself.[188] Evaluating interferences is not a trivial task and must be carried out with a great deal of care and thought. Interferences that are erroneously identified or not compensated for correctly will result in poor-quality data. With the number of interference types and the variety of correction approaches possible, interference correction can quickly become an overwhelming task, particularly to an inexperienced operator. For this reason, when addressing interferences the following wavelength selection procedure is recommended:

1. Choose several wavelengths for each element in the method.
2. Collect analyte wavelength scans.
3. Visually inspect the peaks for the data collected.
4. Review data.
5. Eliminate unsuitable wavelengths.

Let's review this process in more detail. In the first step, it is recommended to choose two or three wavelengths for each element of interest. This includes both the analyte elements and any applicable internal standard elements. Typically, the most sensitive wavelength for each element would be chosen, along with two wavelengths of slightly lower sensitivity. If a wide range of concentrations is expected to be encountered in the samples, a high-sensitivity wavelength might be chosen, along with a low-sensitivity wavelength. In some cases, the sample matrix will produce an emission spectrum that is so complex that there may be only one or two suitable wavelengths (interference-free and providing the desired working concentration range) to choose for the analysis.

Once a set of wavelengths has been chosen for evaluation, data should be collected to determine which wavelengths are best suited, based on the requirements of the application. It is recommended that three wavelength measurements be collected for each of the following solutions:

- A blank (calibration and method blanks, where applicable)
- A low-concentration calibration standard
- A high-concentration calibration standard
- A sample that represents each type of sample matrix to be analyzed

Data from the blank provides an emission profile of the matrix in the absence of analytes. Data from the calibration standards provides profiles for the elements of interest at low and high concentrations. Both sets of measurements are required for the data inspection outlined in steps 3 and 4 described above. The highest-concentration calibration standard should be used for this data collection to ensure that none of the selected wavelengths suffer from peak broadening or self-absorption.[189] Sample measurements produce emission profiles for the elements of interest in the presence of the sample matrix. One sample for each sample matrix type should be collected to ensure emission profiles are examined in every applicable sample matrix. For laboratories

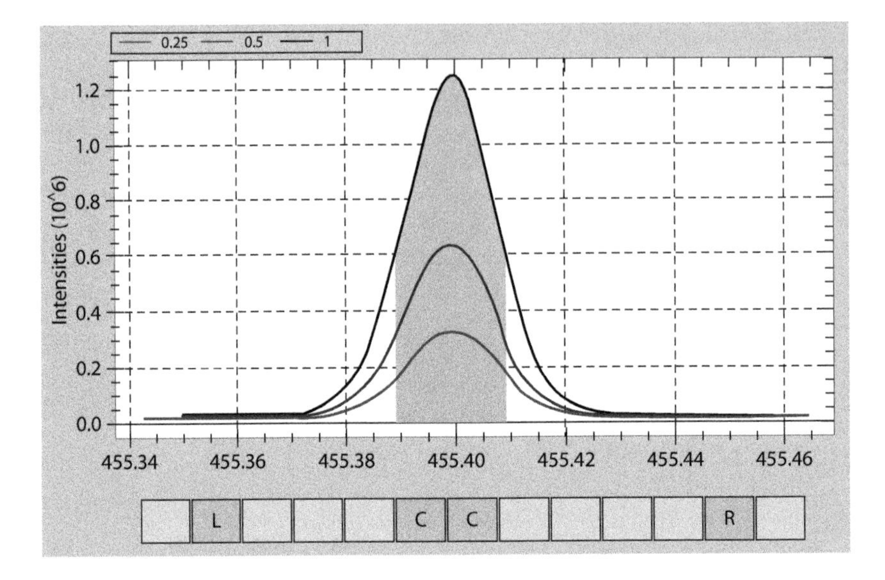

FIGURE 22.22 Examples of emission profiles for three calibration standards and a blank. (With permission from Thermo Fisher Scientific, Waltham, MA.)

analyzing a wide range of sample matrices, collecting wavelength measurement scans for each matrix type could require data collection for a large number of samples.

After wavelength data has been collected, it should be visually inspected to determine that the emission peaks have the correct shape and size. An ideal wavelength would produce the series of emission signals illustrated in Figure 22.22. In this example, the emission from the blank produces a flat emission signal with a relatively low intensity. The emission signals for the calibration standards and representative samples should produce a symmetric, Gaussian-like shape with a single peak. Each peak should exhibit a proportional change in magnitude to match the difference in concentrations. For example, data from a standard containing elements at 1.0 ppm would be expected to have roughly twice the signal of that for a 0.5 ppm standard, while maintaining approximately the same shape.

Numerical data should be inspected to determine the intensity and precision at each wavelength. The intensity for each standard should increase at a rate that is proportional to the increase in its concentration. If a set of three wavelength measurements was collected for each solution, the measured intensities can be used to calculate the approximate precision for each measured standard. Calculated precision values should be within the acceptable limits for the application.

INTERFERENCES

Interferences are common with any plasma-based technique, particularly when trying to measure trace-level concentrations in a sample matrix that contains high concentrations of elements that produce line-rich spectra, such as Fe, Ca, and Si.[190] Sample matrices that are known to generate these types of interferences include geological, metallurgical, and high matrix environmental samples, such as soils or wastewaters. The three common types of interferences in ICP-OES are physical, chemical, and spectral.

PHYSICAL INTERFERENCES

Physical interferences occur when the nebulization and/or transport efficiency of the standards differs from that of the samples. These differences in the physical characteristics of the matrices (density, viscosity, and level of dissolved solids) can produce errors in the measured sample concentrations. Physical interferences are not wavelength specific and can be overcome by utilizing internal standards and preparing the calibration standards in a matrix that matches that of the samples.[191]

CHEMICAL INTERFERENCES

Chemical interferences occur when the standards behave differently from the samples as they enter the plasma. These types of interferences typically result from changes in temperature within the plasma and include easily ionized element (EIE) effects, molecular emission, and plasma loading. Referring to Figure 22.23,[192] the plasma consists of several temperature zones, which translate to varying amounts of energy available for excitation of the ground-state atoms. The plasma is hottest and contains the greatest amount of energy at its base, which is where the sample is introduced. In this region, the plasma contains sufficient energy to atomize and ionize elements before they are excited. Ionization is particularly prevalent for elements in the first two groups of the periodic table, which all have relatively low first ionization potentials. These are often referred to as EIEs and include elements like Na, K, and Li.

When EIEs are present at low concentrations, very few of the atoms become ionized, which means that emission will take place from excited atoms. At high concentrations, a significant number of the atoms will become ionized, which shifts the emission wavelength such that it will predominantly occur from the excited ions. If left uncorrected, this interference will reduce the linear dynamic range of the calibration curve and can potentially produce highly inaccurate results.[193]

This interference can be corrected by adding an ionization suppressant (also known as an ionization buffer) to all the solutions prior to analysis. An ionization suppressant is a solution containing a high concentration of an EIE (e.g., 1000 ppm Cs). The solution will produce an extremely high concentration of ions in the plasma, which, according to Le Châtelier's principle,[194] will reduce the concentration of ions and increase the concentration of atoms to provide a chemical equilibrium balance in the plasma. Therefore, the element of interest will remain in its atomic state in the plasma.

If a dual-view instrument is being used, an additional step to take in correcting for this type of interference is to measure emission from a radially configured plasma. EIE effects are much more significant for axial plasmas, as emission is being measured from species that are present in all temperature zones within the central channel of the plasma. If emission is measured perpendicularly to the direction of the plasma (radial view), emission will be measured from a single temperature zone.[195]

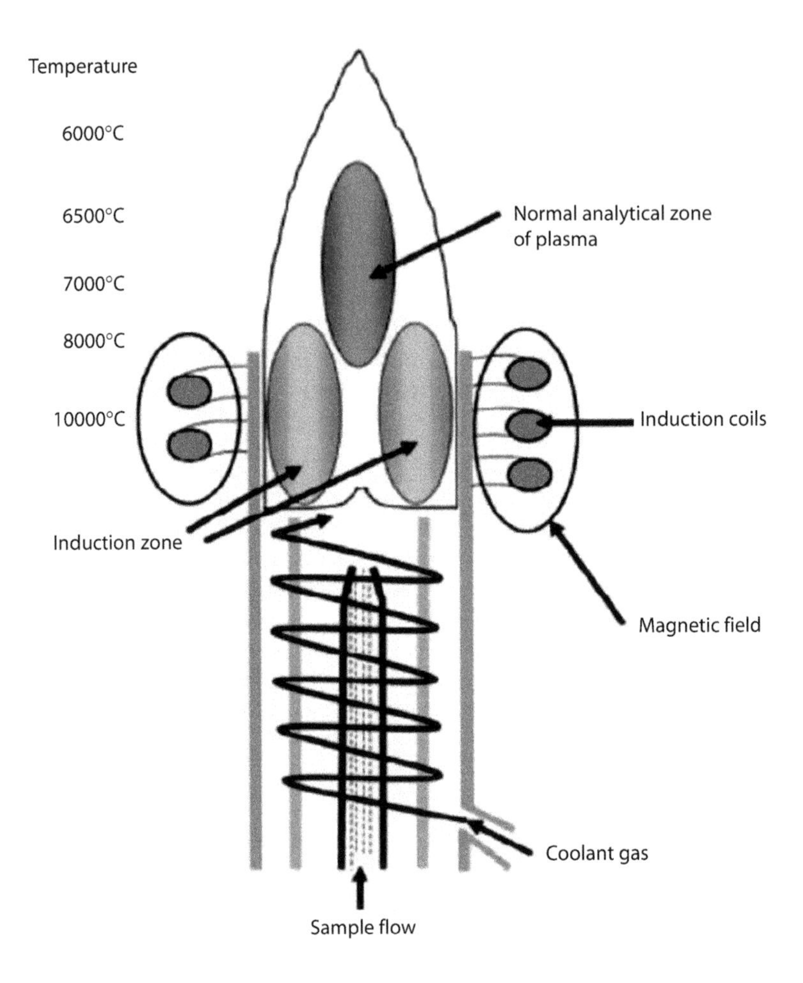

FIGURE 22.23 Different temperature zones within a plasma discharge. (From C. B. Boss and K. J. Fredeen. *Concepts, Instrumentation, and Techniques in Inductively Coupled Plasma Optical Emission Spectrometry*. 2nd ed. Waltham, MA: PerkinElmer, 1997.)

Converse to the formation of EIE interferences is the formation of molecular interferences. If we refer again to Figure 22.23, the tip (also known as the tail) of the plasma is the coolest and least energetic part of the plasma.[196] In this region, atoms and ions that were formed in the hotter part of the plasma can combine to form molecules. These molecules can become excited and emit light, which will produce broadband, molecular emission spectra.

This type of interference is relatively straightforward to correct and requires no intervention from the analyst. Since the coolest part of the plasma is in its tail, removing that portion of the plasma prevents atoms and ions from traveling through a zone that's cool enough to allow them to combine and form molecular species. All modern ICP-OES instruments are set up to automatically and effectively remove the tail of the plasma. There are a variety of methods in which this is accomplished—a flow of

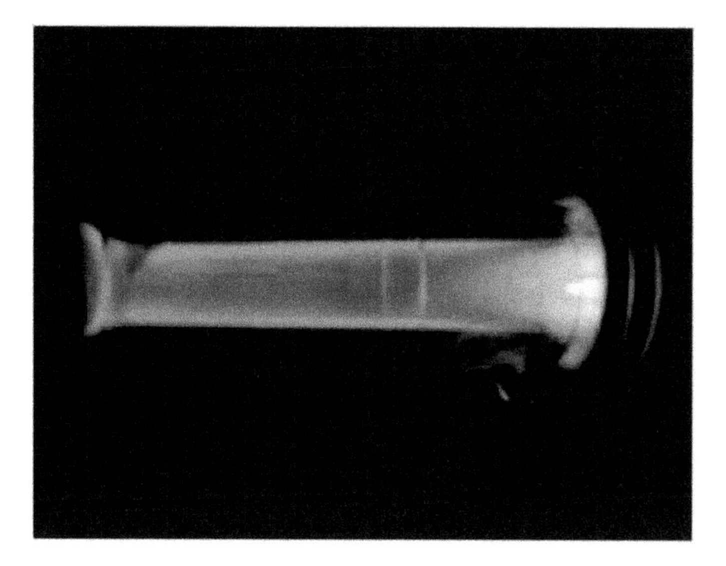

FIGURE 22.24 Image of a dual-view plasma to demonstrate removal of the plasma tail. (With permission from Thermo Fisher Scientific, Waltham, MA.)

inert gas that is counter to the direction of the plasma, a high-velocity shear gas, or a cooled cone interface; however, all are effective in cutting off the tail of the plasma. An example of the way in which the tail is removed is illustrated in Figure 22.24.

As with chemical interferences, another option for eliminating molecular interferences is to measure emission from a radially configured plasma. If emission is collected at 90° to the plasma and the observation height is optimized, emission should be collected from the normal analytical zone, which is well-separated from the cooler tail of the plasma.

Plasma loading is an interference that is sometimes experienced when samples contain organic solvents or high levels of dissolved solids.[197] When a significant amount of material is introduced into the plasma, it cools the plasma and, in severe cases, can extinguish it completely. Cooling the plasma reduces the amount of available energy, which affects the emission intensity of the elements present. As a result, if samples contain concentrations of dissolved solids that are different from those in the standards, the emission intensity for the same analyte concentration will be different in the samples and standards, which could produce data inaccuracies.

This interference can usually be addressed by preparing standards in a matrix that matches that of the samples, utilizing proper internal standards, choosing sample introduction components that were designed for the application, and selecting sample introduction conditions that minimize plasma loading.

SPECTRAL INTERFERENCES

The third, and sometimes most challenging, interferences are spectral. These interferences occur when emission from one or more species in the sample matrix overlaps with the emission for the analyte of interest. These interferences can result in a background shift, a partial peak overlap, or a direct peak overlap. If the analyte is suffering from a direct spectral overlap, either the interfering element must be removed from all solutions prior to analysis, or an alternative wavelength must be selected. If the interfering element produces a simple background shift (either a flat, raised baseline or a sloping background), the analyte can be accurately quantified by the use of optimized background correction points.

If the analyte suffers from partial spectral overlap, a combination of background correction and mathematical correction may need to be employed. An example of partial spectral overlap is illustrated in Figure 22.25. The figure depicts data for a solution containing 1 ppm B in a matrix that contains 100 ppm Fe.

A shoulder is clearly visible on the right side of the main peak, indicating the presence of a spectral overlap on B at 249.773 nm. If the emission profile for the sample solution is overlaid with that for single-element solutions containing 1 ppm B and 100 ppm Fe, the image in Figure 22.26 is produced.

In this figure, the red peak represents the emission profile for 1 ppm B at 249.773 nm, whereas the blue trace represents the emission peak for 100 ppm Fe at 249.782 nm. The black outline represents the combined emission from a sample containing 1 ppm B and 100 ppm Fe. The center green-shaded region represents the area under the peak, which will be included for data collection. This area includes a portion of the emission peak from Fe, and if left uncorrected, this analytical measurement might produce erroneous data.

The best way to correct for this interference is to select an alternative emission wavelength for B that is free from interferences. If an alternative wavelength cannot be used, an interelement correction (IEC) factor should be calculated to correct for the overlap of Fe on B. If IEC factors are to be used, they must be carefully calculated and applied to ensure that spectral overlaps are being accurately corrected. Improperly calculated IEC factors can produce data that is more inaccurate than if no IEC factors were used.[199]

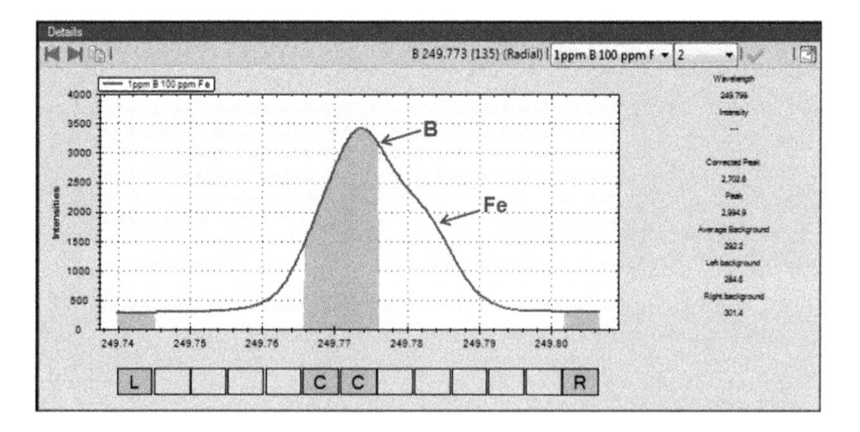

FIGURE 22.25 Emission profile of 1 ppm B in a matrix containing 100 ppm Fe. (From M. Rury, Spectroscopy, 31, 5, 16–32, 2016. With permission from Thermo Fisher Scientific.

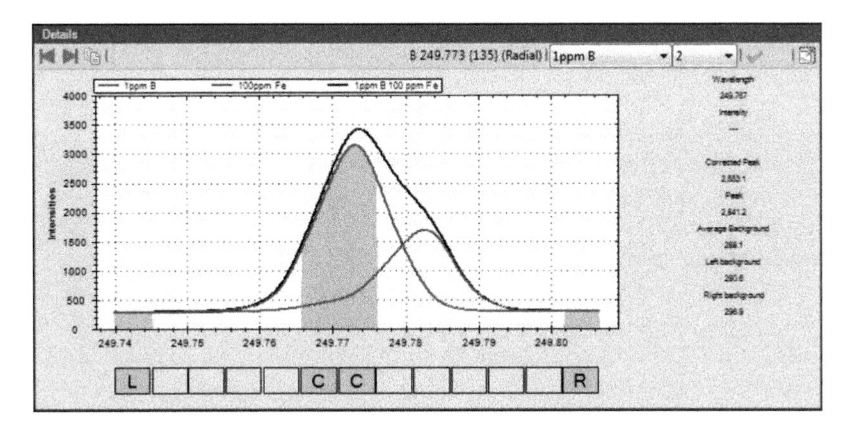

FIGURE 22.26 Emission profiles to demonstrate spectral overlap. (From M. Rury, Spectroscopy, 31, 5, 16-32, 2016. With permission from Thermo Fisher Scientific.)

DATA ACQUISITION

Data acquisition parameters play a crucial role in determining the quality of the resulting data. Even if the instrument is performing optimally, sensitive wavelengths have been selected for all analytes of interest, and all interferences have been identified and removed, the analytical data will suffer significantly if the proper data acquisition parameters are not chosen. For the purposes of this text, data acquisition parameters will be restricted to the choice of axial or radial plasma viewing (if using a dual-view instrument), the integration time, and the number of integrations. In general, an axial plasma is used for determining low parts per billion–level concentrations, while a radial plasma view is used for higher (parts per million–level) concentrations or to help compensate for chemical interferences.[200] The integration time for each plasma viewing configuration must be long enough to collect sufficient emission from all elements being measured. This would usually be dictated by the element with the weakest emission line or the element that is present at the lowest concentration. An axial integration time of 15 s and a radial integration time of 5 s are fairly typical to use. Finally, the number of integrations must be set. Keep in mind that if every sample is collected using a single integration, statistical analysis, precision data, and LOQs cannot be calculated and reported. Therefore, two or more integrations must be collected for each sample in order to report this kind of information.

METHOD VALIDATION

The final step in the method development process is to validate that the method will produce results that meet the figures of merit required for the application. Methodology that achieves good detection limits, yet cannot obtain the required accuracy, is unlikely to meet the data quality objectives of the analysis.

A detection limit study is always encouraged for any analytical method. Calculated detection limits and LOQs allow reporting limits to be calculated, in order to determine the lowest concentrations that can be measured and reported for a suite of elements in a given sample matrix.[201]

Regardless of whether a detection limit study is performed, the developed method needs to be validated prior to use with real samples. The best approach to validating an analytical method is to obtain a certified reference material (CRM) that is appropriate for the application. If an analytical method is valid for the application, data acquired for a suitable CRM should match the certified concentrations that accompany the CRM (within defined error limits).

If a suitable CRM is not available for a given application, a number of other check standards can be utilized. A duplicate calibration standard can be prepared and analyzed as a sample to verify the accuracy of the calibration curve. A calibration or QC standard can be purchased from an external source and analyzed as a sample. Samples can be prepared and measured in duplicate to determine the precision of the method. Samples can be spiked with a known concentration of the elements of interest to determine whether the spiked elements can be measured and recovered at their expected concentrations. If interference corrections are being used, interference check standards can be prepared or purchased and analyzed to ensure that the correct results are obtained.

FINAL THOUGHTS

It should be emphasized that even though ICP-OES is a very useful analytical technique for measuring sub-parts per million and parts per billion levels in different sample matrices, it might have limited use in determining low-level elemental impurities in many pharmaceutical materials. This is mainly due to the fact that the sample digestion procedure and dilution factors involved can reduce the analyte levels in solution below the detection capability of the technique. If a drug product is a liquid or already in solution, it could be a viable approach to use. However, with parenteral or inhalation drug products, which can have permissible daily exposure (PDE) limits one or two orders of magnitude lower than those of oral medications, ICP-MS is probably the best choice to use. Chapter 23 gives this a little more clarity, by comparing the detection capabilities of ICP-MS, ICP-OES, and AA with regard to the J-values for all 24 elemental impurities defined in Chapter <232>.

REFERENCES

1. K. Ohls and B. Bogdain. *Journal of Analytical Atomic Spectrometry*, 31, 22–31, 2016.
2. K. Ohls and B. Bogdain. *Journal of Analytical Atomic Spectrometry*, 31, 22–31, 2016.
3. S. Greenfield, I. L. I. Jones, and C. T. Berry. *Analyst*, 89, 713–720, 1964.
4. R. H. Wendt and V. A. Fassel. *Analytical Chemistry*, 37(7), 920–922, 1965.
5. D. A. Skoog, F. J. Holler, and T. A. Nieman. *Principles of Instrumental Analysis*. FL: Harcourt Brace & Company, 1998.
6. G. F. Kirkbright, A. F. Ward, and T. S. West. *Analytica Chimica Acta*, 62(2), 241–251, Orlando, FL, 1972.

7. G. F. Kirkbright, A. F. Ward, and T. S. West. *Analytica Chimica Acta*, 64(3), 353–362, 1973.
8. R. F. Browner and A. W. Boorn. *Analytical Chemistry*, 56, A786–A798, 1984.
9. A. Montaser, M. G. Minnich, J. A. McLean, H. Liu, J. A. Caruso, and C. W. McLeod. *Sample Introduction in ICP-MS*. New York: Wiley-VCH, 1998.
10. J. F. Tyson. *Analytica Chimica Acta*, 234, 3–12, 1990.
11. E. R. Denoyer. *Atomic Spectroscopy*, 13, 93–98, 1992.
12. E. R. Denoyer. *Atomic Spectroscopy*, 15, 7–16, 1994.
13. G. G. Bortoleto and S. Cadore. *Talanta*, 67, 169–174, 2005.
14. J. Muller. *Fresenius' Journal of Analytical Chemistry*, 363, 572–576, 1999.
15. A. U. Shaikh and D. E. Tallman. *Analytical Chemistry*, 49, 1093–1096, 1977.
16. M. A. Wahed, D. Chowdhury, B. Nermell, S. I. Khan, M. Ilias, M. Rahman, L. A. Persson, and M. Vahter. Journal of Health, *Population and Nutrition*, 24, 36–41, 2006.
17. E. H. Evans, J. A. Day, C. Palmer, W. J. Price, C. M. M. Smith, and J. F. Tyson. *Journal of Analytical Atomic Spectrometry*, 22, 663–696, 2007.
18. Z. Long, Y. Luo, C. Zheng, P. Deng, and X. Hou. *Applied Spectroscopy Reviews*, 47, 382–413, 2012.
19. R. E. Russo, X. Mao, H. Liu, J. Gonzalez, and S. S. Mao. *Talanta*, 57, 425–451, 2002.
20. J. R. Bacon, K. L. Linge, R. R. Parrish, and L. Van Vaeck. *Journal of Analytical Atomic Spectrometry*, 21(8), 785–818, 2006,
21. O. T. Butler, J. M. Cook, C. F. Harrington, S. J. Hill, J. Rieuwerts, and D. L. Miles. *Journal of Analytical Atomic Spectrometry*, 21(2), 217–243, 2006.
22. S. A. Wilson, W. I. Ridley, and A. E. Koenig. *Journal of Analytical Atomic Spectrometry*, 17, 406–409, 2002.
23. E. H. Evans, J. A. Day, C, Palmer, W. J. Price, C. M. M. Smith, and J. F. Tyson. *Journal of Analytical Atomic Spectrometry*, 22, 663–696, 2007.
24. J. Mora, S. Maestre, V. Hernandis, and J. L. Todoli. *Trends in Analytical Chemistry*, 22(3), 123–132, 2003.
25. B. L. Sharp. *Journal of Analytical Atomic Spectrometry*, 3, 613–652, 1988.
26. R. F. Browner and A. W. Boorn. *Analytical Chemistry*, 56, A875–A888, 1984.
27. A. Montaser, M. G. Minnich, H. Liu, A. G. T. Gustavsson, and R. F. Browner. *Fundamental Aspects of Sample Introduction in ICP Spectrometry*. New York: Wiley-VCH, 1998.
28. G. Schaldach, L. Berger, I. Razilov, and H. Berndt. *Journal of Analytical Atomic Spectrometry*, 17, 334–344, 2002.
29. J. W. Olesik and L. C. Bates. Spectrochimica Acta, *Part B*, 50, 285–303, 1995.
30. L. Ebdon and M. R. Cave. *Analyst*, 107, 172–178, 1982.
31. S. Augagneur, B. Medina, J. Szpunar, and R. Lobinski. *Journal of Analytical Atomic Spectrometry*, 11, 713–721, 1996.
32. E. Debrah, S. A. Beres, T. J. Gluodenis, R. J. Thomas, and E. R. Denoyer. *Atomic Spectroscopy*, 16, 197–202, 1995.
33. F. Vanhaecke, M. VanHolderbeke, L. Moens, and R. Dams. *Journal of Analytical Atomic Spectrometry*, 11, 543–548, 1996.
34. J. W. Olesik and S. E. Hobbs. *Analytical Chemistry*, 66, 3371–3378, 1994.
35. K. E. Lawrence, G. W. Rice, and V. A. Fassel. *Analytical Chemistry*, 56, 289–292, 1984.
36. J. L. Todoli and J. M. Mermet. *Journal of Analytical Atomic Spectrometry*, 13, 727–734, 1998.
37. A. Gustavsson. *Spectrochimica Acta, Part B*, 39, 743–746, 1984.
38. A. Gustavsson. *Spectrochimica Acta, Part B*, 39, 85–94, 1984.
39. H. Y. Liu and A. Montaser. *Analytical Chemistry*, 66, 3233–3242, 1994.

40. H. Y. Liu, R. H. Clifford, S. P. Dolan, and A. Montaser. *Spectrochimica Acta, Part B*, 51, 27–40, 1996.
41. S. H. Nam, J. S. Lim, and A. Montaser. *Journal of Analytical Atomic Spectrometry*, 9, 1357–1362, 1994.
42. H. Y. Liu, A. Montaser, S. P. Dolan, and R. S. Schwartz. *Journal of Analytical Atomic Spectrometry*, 11, 307–311, 1996.
43. T. T. Hoang, S. W. May, and R. F. Browner. *Journal of Analytical Atomic Spectrometry*, 17, 1575–1581, 2002.
44. P. W. Kirlew and J. A. Caruso. *Applied Spectroscopy*, 52, 770–772, 1998.
45. L. Q. Wang, S. W. May, R. F. Browner, and S. H. Pollock. *Journal of Analytical Atomic Spectrometry*, 11, 1137–1146, 1996.
46. M. Huang, H. Kojima, A. Hirabayashi, and H. Koizumi. *Analytical Sciences*, 15, 265–268, 1999.
47. E. Debrah, S. A. Beres, T. J. Gluodenis, R. J. Thomas, and E. R. Denoyer. *Atomic Spectroscopy*, 16, 197–202, 1995.
48. J. L. Todoli and V. Hernandis. *Journal of Analytical Atomic Spectrometry*, 14, 1289–1295, 1999.
49. R. I. Botto and J. J. Zhu. *Journal of Analytical Atomic Spectrometry*, 9, 905–912, 1994.
50. B. Budic. *Journal of Analytical Atomic Spectrometry*, 16, 129–134, 2001.
51. P. Masson, A. Vives, D. Orignac, and T. Prunet. *Journal of Analytical Atomic Spectrometry*, 15, 543–547, 2000.
52. M. A. Tarr, G. X. Zhu, and R. F. Browner. *Applied Spectroscopy*, 45, 1424–1432, 1991.
53. R. J. Thomas and C. Anderau. *Atomic Spectroscopy*, 10, 71–73, 1989.
54. Q. H. Jin, F. Liang, Y. F. Huan, Y. B. Cao, J. G. Zhou, H. Q. Zhang, and W. Yang. *Laboratory Robotics and Automation*, 12, 76–80, 2000.
55. S. Yamasaki and A. Tsumura, *Water Science & Technology*, 25, 205–212, 1992.
56. T. T. Nham. *American Laboratory*, 27, 48L–48V, 1995.
57. I. B. Brenner, J. Zhu, and A. Zander. *Fresenius' Journal of Analytical Chemistry*, 355, 774–777, 1996.
58. J. Kunze, S. Koelling, M. Reich, and M. A. Wimmer. *Atomic Spectroscopy*, 19, 164–167, 1998.
59. S. C. K. Shum and R. S. Houk. *Analytical Chemistry*, 65, 2972–2976, 1993.
60. S. C. K. Shum, R. Neddersen, and R. S. Houk. *Analyst*, 117, 577–582, 1992.
61. S. C. K. Shum, H. M. Pang, and R. S. Houk. *Analytical Chemistry*, 64, 2444–2450, 1992.
62. T. W. Avery, C. Chakrabarty, and J. J. Thompson. *Applied Spectroscopy*, 44, 1690–1698, 1990.
63. D. R. Wiederin, F. G. Smith, and R. S. Houk. *Analytical Chemistry*, 63, 219–225, 1991.
64. J. A. McLean, H. Zhang, and A. Montaser. *Analytical Chemistry*, 70, 1012–1020, 1998.
65. M. G. Minnich and A. Montaser. *Applied Spectroscopy*, 54, 1261–1269, 2000.
66. J. L. Todoli and J. M. Mermet. *Journal of Analytical Atomic Spectrometry*, 16, 514–520, 2001.
67. E. Bjorn and W. Frech. *Journal of Analytical Atomic Spectrometry*, 16, 4–11, 2001.
68. M. G. Minnich and A. Montaser. *Applied Spectroscopy*, 54, 1261–1269, 2000.
69. A. C. S. Bellato, M. F. Gine, and A. A. Menegario. *Microchemical Journal*, 77, 119–122, 2004.
70. M. J. Powell, E. S. K. Quan, D. W. Boomer, and D. R. Wiederin. *Analytical Chemistry*, 64, 2253–2257, 1992.
71. S. E. O'Brien, J. A. McLean, B. W. Acon, B. J. Eshelman, W. F. Bauer, and A. Montaser. *Applied Spectroscopy*, 56, 1006–1012, 2002.
72. J. A. McLean, M. G. Minnich, L. A. Iacone, H. Y. Liu, and A. Montaser. *Journal of Analytical Atomic Spectrometry*, 13, 829–842, 1998.

73. B. W. Acon, J. A. McLean, and A. Montaser. *Analytical Chemistry*, 72, 1885–1893, 2000.
74. C. S. Westphal, K. Kahen, W. E. Rutkowski, B. W. Acon, and A. Montaser. *Spectrochimica Acta, Part B*, 59, 353–368, 2004.
75. G. Schaldach, L. Berger, I. Razilov, and H. Berndt. *Journal of Analytical Atomic Spectrometry*, 17, 334–344, 2002.
76. B. L. Sharp. *Journal of Analytical Atomic Spectrometry*, 3, 939–963, 1988.
77. C. Rivas, L. Ebdon, and S. J. Hill. *Journal of Analytical Atomic Spectrometry*, 11, 1147–1150, 1996.
78. R. H. Scott, V. A. Fassel, R. N. Kniseley, and D. E. Nixon. *Analytical Chemistry*, 46, 75–81, 1974.
79. D. R. Luffer and E. D. Salin. *Analytical Chemistry*, 58, 654–656, 1986.
80. J. L. Todoli, S. Maestre, J. Mora, A. Canals, and V. Hernandis. *Fresenius' Journal of Analytical Chemistry*, 368, 773–779, 2000.
81. X. H. Zhang, H. F. Li, and Y. F. Yang. *Talanta*, 42, 1959–1963, 1995.
82. M. Wu and G. M. Hieftje. *Applied Spectroscopy*, 46, 1912–1918, 1992.
83. G. Schaldach, H. Berndt, and B. L. Sharp. *Journal of Analytical Atomic Spectrometry*, 18, 742–750, 2003.
84. H. Isoyama, T. Uchida, C. Iida, and G. Nakagawa. *Journal of Analytical Atomic Spectrometry*, 5, 307–310, 1990.
85. H. Isoyama, T. Uchida, T. Niwa, C. Iida, and G. Nakagawa. *Journal of Analytical Atomic Spectrometry*, 4, 351–355, 1989.
86. B. Bouyssiere, Y. N. Ordonez, C. P. Lienemann, D. Schaumloffel, and R. Lobinski. *Spectrochimica Acta, Part B*, 61, 1063–1068, 2006.
87. B. Bouyssiere, Y. N. Ordonez, C. P. Lienemann, D. Schaumloffel, and R. Lobinski. *Spectrochimica Acta, Part B*, 61, 1063–1068, 2006.
88. A. Prange and D. Schaumloffel. *Journal of Analytical Atomic Spectrometry*, 14, 1329–1332, 1999.
89. D. Schaumloffel, J. R. Encinar, and R. Lobinski. *Analytical Chemistry*, 75, 6837–6842, 2003.
90. A. Woller, H. Garraud, F. Martin, O. F. X. Donard, and P. Fodor. *Journal of Analytical Atomic Spectrometry*, 12, 53–56, 1997.
91. Y. F. Li, C. Y. Chen, B. Li, J. Sun, J. X. Wang, Y. X. Gao, Y. L. Zhao, and Z. F. Chai. *Journal of Analytical Atomic Spectrometry*, 21, 94–96, 2006.
92. A. Al-Ammar, R. K. Gupta, and R. M. Barnes. *Spectrochimica Acta, Part B*, 54, 1077–1084, 1999.
93. A. Al-Ammar, R. K. Gupta, and R. M. Barnes. Spectrochimica Acta, *Part B*, 55, 629–635, 2000.
94. R. L. Sutton. *Journal of Analytical Atomic Spectrometry*, 9, 1079–1083, 1994.
95. H. Naka and H. Kurayasu. *Bunseki Kagaku*, 45, 1139–1144, 1996.
96. P. Schramel. *Fresenius' Journal of Analytical Chemistry*, 320, 233–236, 1985.
97. J. H. D. Hartley, S. J. Hill, and L. Ebdon. *Spectrochimica Acta, Part B*, 48, 1421–1433, 1993.
98. W. Schron and U. Muller. *Fresenius' Journal of Analytical Chemistry*, 357, 22–26, 1997.
99. A. R. Eastgate, R. C. Fry, and G. H. Gower. *Journal of Analytical Atomic Spectrometry*, 8, 305–308, 1993.
100. T. B. Reed. *Journal of Applied Physics*, 32, 821–824, 1961.
101. T. B. Reed. *Journal of Applied Physics*, 32, 2534–2535, 1961.
102. T. B. Reed. Presented at 57th Annual Meeting of the American Institute of Chemical Engineers, December 1964.
103. C. D. Allemand and R. M. Barnes. *Applied Spectroscopy*, 31, 434–443, 1977.

104. S. Greenfield, I. L. I. Jones, and C. T. Berry. *Analyst*, 89, 713–720, 1964.
105. S. Greenfield. U.S. Patent 3,467,471, September 16, 1969.
106. R. H. Scott, V. A. Fassel, R. N. Kniseley, and D. E. Nixon. Plasma light source for spectroscopic investigation. *Analytical Chemistry*, 46, 75–80, 1975.
107. R. Rezaaiyaan, G. M. Hieftje, H. Anderson, H. Kaiser, and B. Meddings. *Applied Spectroscopy*, 36, 627–631, 1982.
108. P. W. J. M. Boumans. *Fresenius' Journal of Analytical Chemistry*, 299, 337–361, 1979.
109. Thermo Scientific. Radial demountable ceramic torch for the Thermo Scientific iCAP 6000 Series ICP spectrometer. Thermo Fisher Product Technical Note 43053. 2010. http://tools.thermofisher.com/content/sfs/brochures/D01563~.pdf.
110. G. F. Larson, V. A. Fassel, R. K. Winge, and R. N. Kniseley. *Applied Spectroscopy*, 30, 384–391, 1976.
111. F. V. Silva, L. C. Trevizan, C. S. Silva, A. R. A. Nogueira, and J. A. Nóbrega. *Spectrochimica Acta, Part B*, 57, 1905–1913, 2002.
112. Thermo Scientific. Thermo Scientific iCAP 7000 Plus Series ICP-OES: Innovative ICP-OES optical design. Thermo Fisher Scientific Product Technical Note 43333. 2016. https://tools.thermofisher.com/content/sfs/brochures/TN-43333-ICP-OES-Optical-Design-iCAP-7000-Plus-Series-TN43333-EN.pdf.
113. C. R. Runge and F. Paschen. *Abh. K. Akad. Wiss. Berlin*, 1902.
114. G. L. Clark, ed. *The Encyclopedia of Spectroscopy*. New York: Reinhold Publishing Corporation, 1960.
115. J. F. James and R. S. Sternberg. *The Design of Optical Spectrometers*. London: Chapman & Hall, 1969.
116. C. B. Boss and K. J. Fredeen. *Concepts, Instrumentation, and Techniques in Inductively Coupled Plasma Optical Emission Spectrometry*. 2nd ed. Waltham, MA: PerkinElmer, 1997.
117. K. Jankowski, A. Jackowska, A. P. Ramsza, and E. Reszke. *Journal of Analytical Atomic Spectrometry*, 23, 1234–1238, 2008.
118. Y. Okamoto. *Analytical Sciences*, 7, 283–288, 1991.
119. T. Maeda and K. Wagatsuma. *Microchemical Journal*, 76, 53–60, 2004.
120. U. Engel, C. Prokisch, E. Voges, G. M. Hieftje, and J. A. C. Broekaert. *Journal of Analytical Atomic Spectrometry*, 13, 955–961, 1998.
121. Q. Xue. *Applied Optics*, 50, 1338–1344, 2011.
122. Q. Xue, S. Wang, and F. Lu. *Applied Optics*, 48, 11–16, 2009.
123. J. A. C. Broekaert. Instrument column. *Spectrochimica Acta*, 37B, 359–369, 1982.
124. Plasma-Spec. The next generation in plasma spectrometry. Technical Bulletin of Leeman Labs, Inc., Lowell, MA.
125. X. Hou and B. T. Jones. In *Encyclopedia of Analytical Chemistry: Applications, Theory and Instrumentation, Supplementary Volumes S1-S3*, ed. R. A. Meyers. Chichester, UK: John Wiley & Sons, 2000, pp. 9468–9485.
126. K. Paech and M. V. Tracey. *Modern Methods of Plant Analysis*. Vol. 1. Berlin: Springer-Verlag, 1956.
127. R. F. Jarrell, F. Brech, and M. J. Gustafson. *Journal of Chemical Education*, 77, 592–598, 2000.
128. A. Scheeline, C. A. Bye, D. L. Miller, S. W. Rynders, and R. C. Owen Jr. *Applied Spectroscopy*, 45, 334–346, 1991.
129. A. T. Zander and P. N. Keliher. *Applied Spectroscopy*, 33, 499–502, 1979.
130. C. Allemand. *ICP Information Newsletter*, 2(1), 1976.
131. C. Allemand. *ICP Information Newsletter*, 4(44), 1978.
132. R. B. Bilhorn, P. M. Epperson, J. V. Sweedler, and M. B. Denton. *Applied Spectroscopy*, 41, 1125–1136, 1987.
133. Y. Talmi and R. W. Simpson. *Applied Optics*, 19, 1401–1414, 1980.

134. F. M. Pennebaker, D. A. Jones, C. A. Gresham, R. H. Williams, R. E. Simon, M. F. Schappert, and M. B. Denton. *Journal of Analytical Atomic Spectrometry*, 13, 821–827, 1998.
135. J. Marshall, A. Fisher, S. Chenery, and S. T. Sparkes. *Journal of Analytical Atomic Spectrometry*, 11, 213R–238R, 1996.
136. Q. S. Hanley, C. W. Earle, F. M. Pennebaker, S. P. Madden, and M. B. Denton. *Analytical Chemistry*, 68, 661A–667A, 1996.
137. J. V. Sweedler, K. L. Ratzlaff, and B. M. Denton, eds. *Charge Transfer Devices in Spectroscopy*. New York: VCH Publishers, 1994.
138. X. Hou and B. T. Jones. *Encyclopedia of Analytical Chemistry*, ed. R. A. Meyers. Chichester, UK: John Wiley & Sons, 2000, pp. 9468–9485.
139. Y. Talmi and R. W. Simpson. *Applied Optics*, 19, 1401–1414, 1980.
140. R. B. Bilhorn, P. M. Epperson, J. V. Sweedler, and M. B. Denton. *Applied Spectroscopy*, 41, 1125–1136, 1987.
141. Y. Talmi. *Applied Spectroscopy*, 36, 1–18, 1982.
142. E. D. Salin and G. Horlick. *Analytical Chemistry*, 52, 1578–1582, 1980.
143. J. M. Harnly and R. E. Fields. *Applied Spectroscopy*, 51, 334A–351A, 1997.
144. J. V. Sweedler, R. D. Jalkian, R. S. Pomeroy, and M. B. Denton. *Spectrochimica Acta, Part B*, 44B, 683–692, 1989.
145. R. B. Bilhorn, J. V. Sweedler, P. M. Epperson, and M. B. Denton. *Applied Spectroscopy*, 41, 1114–1125, 1987.
146. Q. Xue, S. Wang, and F. Lu, *Applied Optics*, 48, 11–16, 2009.
147. F. L. J. Sangster and K. Teer. *IEEE Journal of Solid-State Circuits*, SC-4, 131–136, 1969.
148. A. J. P. Theuwissen. *Solid-State Imaging with Charge-Coupled Devices*. New York: Kluwer Academic Publishers, 2002.
149. W. Boyle and G. Smith. *Bell System Technical Journal*, 49, 587–593, 1970.
150. W. Boyle and G. Smith. Buried channel charge coupled devices. U.S. Patent 3792322, February 12, 1974.
151. W. Boyle and G. Smith. Three dimensional charge coupled devices. U.S. Patent 3796927, March 12, 1974.
152. P. M. Epperson, J. V. Sweedler, R. B. Bilhorn, G. R. Sims, and M. B. Denton. *Analytical Chemistry*, 60, 327A–335A, 1988.
153. A. G. Milnes. Charge-transfer devices. In *Semiconductor Devices and Integrated Electronics*. New York: Van Nostrand Reinhold, 1980, pp. 590–642.
154. G. C. Holst and T. S. Lomheim. *CMOS/CCD Sensors and Camera Systems*. Oviedo, FL: JCD Publishing, 2007.
155. J. R. Janesick. *Scientific Charge-Coupled Devices*. Bellingham, WA: International Society for Optical Engineering, 2001.
156. N. Waltham. CCD and CMOS sensors. In *Observing Photons in Space: A Guide to Experimental Space Astronomy*, ed. M. C. E. Huber, A. Pauluhn, J. L. Culhane, J. G. Timothy, K. Wilhelm, and A. Zehnder. New York: Springer Science & Business Media, 2013, pp. 423–442.
157. J. V. Sweedler, R. D. Jalkian, R. S. Pomeroy, and M. B. Denton. *Spectrochimica Acta, Part B*, 44, 683–692, 1989.
158. J. M. Mermet, A. Cosnier, Y. Danthez, C. Dubuisson, E. Fretel, O. Rogerieux, and S. Vélasquez. Tech note: Design criteria for ICP spectrometry using advanced optical and CCD technology. *Spectroscopy*, 20, 60–66, 2005.
159. J. Marshall, A. Fisher, S. Chenery, and S. T. Sparkes. *Journal of Analytical Atomic Spectrometry*, 11, 213R–238R, 1996.
160. T. W. Barnhard, M. I. Crockett, J. C. Ivaldi, and P. L. Lundberg. *Analytical Chemistry*, 65, 1225–1230, 1993.

161. T. W. Barnhard, M. I. Crockett, J. C. Ivaldi, P. L. Lundberg, D. A. Yates, P. A. Levine, and D. J. Sauer. *Analytical Chemistry*, 65, 1231–1239, 1993.
162. I. B. Brenner and A. T. Zander. *Spectrochimica Acta, Part B*, 55, 1195–1240, 2000.
163. J. M. Harnly and R. E. Fields. *Applied Spectroscopy*, 51, 334A–351A, 1997.
164. R. B. Bilhorn, J. V. Sweedler, P. M. Epperson, and M. B. Denton. *Applied Spectroscopy*, 41, 1114–1125, 1987.
165. J. M. Harnly and R. E. Fields. *Applied Spectroscopy*, 51, 334A–351A, 1997.
166. P. M. Epperson, J. V. Sweedler, R. B. Bilhorn, G. R. Sims, and M. B. Denton. *Analytical Chemistry*, 60, 327A–335A, 1988.
167. G. R. Sims and M. B. Denton. In *Multichannel Image Detectors*, ed. Y. Talmi. ACS Symposium Series No. 236, vol. 2. Washington, DC: American Chemical Society, 1983, chap. 5.
168. S. Bhaskaran, T. Chapman, M. Pilon, and S. VanGorden. In *SPIE Proceedings, Infrared Systems and Photoelectronic Technology III*, 2008, p. 70561R.
169. R. B. Bilhorn and M. B. Denton. *Applied Spectroscopy*, 44, 1538–1546, 1990.
170. Thermo Scientific. Overcoming interferences with the Thermo Scientific iCAP 7000 Plus Series ICP-OES. Thermo Fisher Scientific Product Technical Note 43332. 2016. https://tools.thermofisher.com/content/sfs/brochures/TN-43332-ICP-OES-Overcoming-Interferences-iCAP-7000-Plus-Series-TN43332-EN.pdf.
171. K. E. Jarvis, P. Mason, T. Platzner, and J. G. Williams. *Journal of Analytical Atomic Spectrometry*, 13, 689–696, 1998.
172. G. H. Vickers, D. A. Wilson, and G. M. Hieftje. *Journal of Analytical Atomic Spectrometry*, 4, 749–754, 1989.
173. H. E. Taylor. *Inductively Coupled Plasma-Mass Spectrometry: Practices and Techniques*. San Diego: Academic Press, 2001.
174. I. B. Brenner, A. Zander, M. Cole, and A. Wiseman. *Journal of Analytical Atomic Spectrometry*, 12, 897–906, 1997.
175. I. B. Brenner, M. Zischka, B. Maichin, and G. Knapp. *Journal of Analytical Atomic Spectrometry*, 13, 1257–1264, 1998.
176. L. C. Trevizan and J. A. Nobrega. *Journal of the Brazilian Chemical Society*, 18, 1678–4790, 2007.
177. G. L. Long and J. D. Winefordner. *Analytical Chemistry*, 55, 712A–724A, 1983.
178. J. Mermet and E. Poussel. *Applied Spectroscopy*, 49, 12A–18A, 1995.
179. H. Kaiser. *Analytical Chemistry*, 42, 53A, 1987.
180. V. Thomsen, D. Schatzlein, and David Mercuro. *Spectroscopy*, 18, 112–114, 2003.
181. Environmental Protection Agency. EPA Method 200.7, Revision 4.4. 1994. https://www.epa.gov/sites/production/files/2015-08/documents/method_200-7_rev_4-4_1994.pdf.
182. L. A. Currie. *Analytical Chemistry*, 40, 586–593, 1968.
183. G. L. Long and J. D. Winefordner. *Analytical Chemistry*, 55, 713A–724A, 1983.
184. J. M. Mermet, G. Granier, and P. Fichet. *Spectrochimica Acta, Part B*, 76, 221–225, 2012.
185. F. C. Garner and G. L. Robertson. *Chemometrics and Intelligent Laboratory Systems*, 3, 53–59, 1988.
186. V. Thompsen. *Spectroscopy*, 27(3), 2012.
187. J. Mandel and R. D. Stiehler. *Journal of Research of the National Bureau of Standards*, 155, A53, 1964.
188. G. F. Larson, V. A. Fassel, R. H. Scott, and R. N. Kniseley. *Analytical Chemistry*, 47, 238–243, 1975.
189. Thermo Scientific. High performance radio frequency generator technology for the Thermo Scientific iCAP 7000 Plus Series ICP-OES. Thermo Fisher Scientific Product Technical Note 43334. 2016. https://tools.thermofisher.com/content/sfs/brochures/TN-43334-ICP-OES-RF-Generator-iCAP-7000-Plus-Series-TN43334-EN.pdf.

190. H. G. C. Human and R. H. Scott. *Spectrochimica Acta, Part B*, 31, 459–473, 1976.
191. Thermo Scientific. Overcoming interferences with the Thermo Scientific iCAP 7000 Plus Series ICP-OES. Thermo Fisher Scientific Product Technical Note 43332. 2016. https://tools.thermofisher.com/content/sfs/brochures/TN-43332-ICP-OES-Overcoming-Interferences-iCAP-7000-Plus-Series-TN43332-EN.pdf.
192. R. F. Browner, Fundamental aspects of aerosol generation and transport. In *Inductively Coupled Plasma Emission Spectrometry, Part II: Applications and Fundamentals*, ed. P. W. J. M. Boumans. New York: Wiley-Interscience, 244–288 1987.
193. M. W. Blades and G. Horlick. *Spectrochimica Acta, Part B*, 36, 881–900, 1981.
194. A. J. Miller. *Journal of Chemical Education*, 31, 455, 1954.
195. M. H. Abdallah, R. Diemiaszonek, J. Jarosz, J. M. Mermet, J. Robin, and C. Trassy. *Analytica Chimica Acta*, 84, 271–282, 1976.
196. I. B. Brenner and A. T. Zander. *Spectrochimica Acta, Part B*, 55, 1195–1240, 2000.
197. A. W. Boorn and R. F. Browner. *Analytical Chemistry*, 54, 1402–1410, 1982.
198. Thermo Scientific. Overcoming interferences with the Thermo Scientific iCAP 7000 Plus Series ICP-OES. Thermo Fisher Scientific Product Technical Note 43332. 2016. https://tools.thermofisher.com/content/sfs/brochures/TN-43332-ICP-OES-Overcoming-Interferences-iCAP-7000-Plus-Series-TN43332-EN.pdf.
199. V. Thomsen, D. Mercuro, and D. Schatzlein. *Spectroscopy*, 21(7), 2006. http://www.spectroscopyonline.com/interelement-corrections-spectrochemistry.
200. F. V. Silva, L. C. Trevizan, C. S. Silva, A. R. A. Nogueira, and J. A. Nóbrega. *Spectrochimica Acta, Part B*, 57, 1905–1913, 2002.
201. T. R. Dulski. Statistics and specifications. In *A Manual for the Chemical Analysis of Metals*. ASTM, 1996, pp. 200–201.

23 What Atomic Spectroscopic Technique is Right for Your Lab?

Since the introduction of the first commercially available atomic absorption spectrophotometer (AAS) in the early 1960s, there has been an increasing demand for better, faster, easier-to-use, and more flexible trace element instrumentation. A conservative estimate shows that in 2017, the market for atomic spectroscopy (AS)–based instruments will be well in excess of $1 billion in annual revenue—and that doesn't include aftermarket sales and service costs, which could increase this number by an additional $500 million. As a result of this growth, we have seen a rapid emergence of more sophisticated equipment and easier-to-use software. Moreover, with an increase in the number of manufacturers of AS instrumentation and its sampling accessories, together with the availability of traditional solid sampling techniques, such as x-ray fluorescence (XRF), and newer techniques, like microwave-induced optical emission spectroscopy (MIP-OES) and laser-induced breakdown spectroscopy (LIBS), the choice of which technique to use is often unclear. It also becomes even more complicated when budgetary restrictions allow only one analytical technique to be purchased to solve a particular application problem. Chapter 23 takes a look at the major AS techniques and compares their performance characteristics, and in particular examines their respective limits of quantitation (LOQs) for pharmaceutical- and dietary supplement–type samples.

In order to select the best technique for a particular analytical problem, it is important to understand exactly what the problem is and how it is going to be solved. For example, if the requirement is to monitor copper at percentage levels in a copper plating bath and it is only going to be done once per shift, flame atomic absorption (FAA) would adequately fill this role. Alternatively, when selecting an instrument to determine 24 elemental impurities in pharmaceutical products according to United States Pharmacopeia (USP) chapters <232> and <233>, directives, or International Conference on Harmonisation of Technical Requirements for Registration of Pharmaceuticals for Human Use (ICH) Q3D Step 4 guidelines, this application is clearly better suited for a multielement technique, such as inductively coupled plasma optical emission spectrometry (ICP-OES), or perhaps inductively coupled plasma mass spectrometry (ICP-MS), depending on the materials being analyzed.

So when choosing a technique, it is important to understand not only the application problem, but also the strengths and weaknesses of the technology being applied to solve the problem. However, there are many overlapping areas between the major AS techniques, so it is highly likely that for some applications, more than one technique would be suitable. For that reason, it is important to go through a carefully thought-out evaluation process when selecting a piece of equipment.[1]

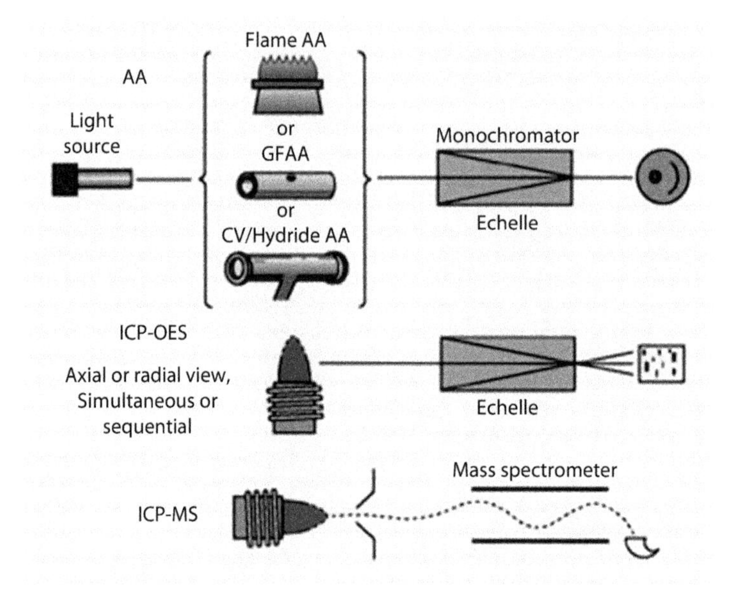

FIGURE 23.1 Simplified schematic of AA, ICP-OES, and ICP-MS.

The main intent of this chapter is therefore to look at the application requirements for the determination of 24 elemental impurities in pharmaceutical materials and the big four heavy metals in dietary supplements to offer some insight as to which might be the most suitable atomic spectroscopic approach. We will not be discussing XRF, MIP-OES, or LIBS techniques in this evaluation. They are all very useful techniques for many sample types and have been described at length in the open literature.[2] Chapter <233> does allow the use of an alternative technique, as long as its meet the validation and verification protocols described in the analytical procedure section. In fact, it has been shown that XRF is a valid technique to use as a screening tool to reduce the number of samples that need to be analyzed by one of the spectrochemical techniques.[3] However, for the purpose of this comparison, we focus on the four most commonly used AS techniques—FAA, electrothermal atomization (ETA), ICP-OES, and ICP-MS.

In order to set the stage, let's first take a brief look at their fundamental principles. We have covered ICP-OES and ICP-MS in previous chapters, but it's useful to briefy compare the performance of these two plasma spectrochemical techniques with FAA and ETA. Simplified schematics of each of these techniques are shown in Figure 23.1.

FLAME ATOMIC ABSORPTION

This is predominantly a single-element technique for the analysis of liquid samples that uses a flame to generate ground-state atoms. The sample is aspirated into the flame via a nebulizer and a spray chamber. The ground-state atoms of the sample absorb light of a particular wavelength, from either an element-specific, hollow cathode lamp or a continuum source lamp. The amount of light absorbed is measured by a monochromator (optical system) and detected by a photomultiplier tube or solid-state detector, which converts the photons into an electrical signal. As in all AS techniques, this signal is used to determine the concentration of that element in the sample, by comparing it with calibration or reference standards.

FAA typically uses about 2–5 mL/min of liquid sample and is capable of handling in excess of 10% total dissolved solids, although for optimum performance it is best to keep the solids down below 2%. For the majority of elements, it's detection capability is in the order of 1–100 parts per billion (ppb) levels, with an analytical range up to 10–1000 parts per million (ppm), depending on the absorption wavelength used. However, it is not really suitable for the determination of the halogens, nonmetals like carbon, sulfur, and phosphorus, and has very poor detection limits for the refractory, rare earth, and transuranic elements. Sample throughput for 15 elements per sample is in the order of 10 samples an hour.

ELECTROTHERMAL ATOMIZATION

This is also mainly a single-element technique, although multielement instrumentation is now available. It works on the same principle as FAA, except that the flame is replaced by a small heated tungsten filament or graphite tube. The other major difference is that in ETA, a very small sample (typically 50 μL) is injected onto the filament or into the tube, and not aspirated via a nebulizer and a spray chamber. Because the ground-state atoms are concentrated in a smaller area, more absorption takes place. The result is that ETA offers detection capability at the 0.01–1 ppb level, with an analytical range up to 10–100 ppb.

The elemental coverage limitations of the technique are similar to those of the FAA technique. However, because a heated graphite tube is used for atomization in most commercial instruments, it cannot determine the refractory, rare earth, and transuranic elements, because they tend to form stable carbides that cannot be readily atomized. One of the added benefits is that ETA can also analyze slurries and some solids due to the fact that no nebulization process is involved in introducing the sample. This technique is not ideally suited for multielement analysis, because it takes 3–4 min to determine one element per sample. As a result, sample throughput for 15 elements is in the order of one sample per hour.

HGAA

Hydride generation (HG) is a very useful analytical technique to determine the hydride-forming metals, such as As, Bi, Sb, Se, and Te, usually by AA (although detection by either ICP-OES or ICP-MS can also be used). In this technique, the analytes in the sample matrix are first reacted with a very strong reducing agent, such as sodium borohydride, to release their volatile hydrides, which are then swept into a heated quartz tube for atomization. The tube is heated by either a flame or a small oven, which creates the ground-state atoms of the element of interest, and then measured by AA. When used with ICP-OES or ICP-MS, the volatile hydrides are passed directly in the plasma for excitation or ionization.

By choosing the optimum chemistry, mercury can also be reduced in solution in this way to generate elemental mercury. This is known as the cold vapor (CV) technique. HGAA and CVAA can improve the detection for these elements over FAA by up to three orders of magnitude, achieving detection capability in the order of 0.005–0.1 ppb levels, with an analytical range of 5–100 ppb, depending on the element of interest. It should also be pointed out that dedicated mercury analyzers (some using gold amalgamation techniques), coupled with AA or atomic fluorescence, are capable of better detection limits. Because of the online chemistry involved, these techniques are very time-consuming and are normally used in conjunction with FAA, so they will most likely impact the overall sample throughput.

RADIAL ICP-OES

Radially viewed ICP-OES is a multielement technique that uses a traditional radial (side-view) ICP to excite ground-state atoms to the point where they emit wavelength-specific photons of light that are characteristic of a particular element. The number of photons produced at an element-specific wavelength is measured by high-resolution optics and a photon-sensitive device, such as a photomultiplier tube or a solid-state detector. This emission signal is directly related to the concentration of that element in the sample. The analytical temperature of an ICP is about 6000–7000 K, compared with that of a flame or a graphite furnace (GF), which is typically 2500–3500 K.

For the majority of elements, a radial ICP instrument can achieve a detection capability in the order of 0.1–100 ppb levels, with an analytical range up to 10–1000 ppm, depending on the emission wavelength used. The technique can determine a similar number of elements as FAA, but has the advantage of offering the capability of nonmetals, like sulfur and phosphorus, together much better performance for the refractory, rare earth, and transuranic elements. The sample requirement for ICP-OES is approximately 1 mL/min, and is capable of aspirating samples containing up to 10% total dissolved solids, but for optimum performance, it is usually kept below 2%. ICP-OES is a rapid multielement technique, so sample throughput for 15 elements per sample is in the order of 20 samples an hour.

AXIAL ICP-OES

The principle is exactly the same as for radial ICP-OES, except that in the axial view, the plasma is viewed horizontally (end on). The benefit is that more photons are seen by the detector, and for this reason, detection limits can be as much as an order of magnitude lower, depending on the design of the instrument. The disadvantage is that the working range is also reduced by an order of magnitude. As a result, for the majority of elements, an axial ICP instrument can achieve detection capability in the order of 0.01–10 ppb levels, with an analytical range up to 1–100 ppm, depending on the emission wavelength used. The other disadvantage of viewing axially is that more severe matrix interferences are observed, which means that the total dissolved solids content of the sample needs to be kept much lower. Sample flow requirements are the same as for radial ICP-OES.

It's also important to point out that most commercial ICP-OES instruments have both radial and axial capability built in. However, because of the different hardware configurations available, some instruments work by carrying out the analysis using either the radial or the axial view (sequentially). Others can use both the radial and the axial view at exactly the same time (simultaneously), which for some applications may be advantageous. Strategic use of both radial and axial views, together with the optimum wavelength selection, can extend the analytical range by two to three orders of magnitude. Sample throughput will be approximately 20 samples per hour, the same as for radial ICP-OES.

ICP-MS

The fundamental difference between ICP-OES and ICP-MS is that in ICP-MS, the plasma is not used to generate photons, but to generate positively charged ions. The ions produced are transported and separated by their atomic mass-to-charge ratio using a mass filtering device such as a quadrupole or a magnetic sector. The generation of such large numbers of positively charged ions allows ICP-MS to achieve detection limits approximately three orders of magnitude lower than those for ICP-OES, even for the refractory, rare earth, and transuranic elements. As a result, for the majority of elements, an ICP-MS instrument can achieve detection capability in the order of 0.0001–1 ppb, with an analytical range up to 0.1–100 ppm, using pulse-counting measurement, but can be extended even further up to 100–100,000 ppm by using analog counting techniques. However, it should be emphasized that if such large analyte concentrations are being measured, expectations should be realistic about also carrying out ultratrace determinations of the same element in the same sample run.

The sample requirement for ICP-MS is approximately 1 mL/min, and is capable of aspirating samples containing up to 5% total dissolved solids for short periods with the use of specialized sampling accessories. However, because the sample is being aspirated into the mass spectrometer, for optimum performance, matrix components should ideally be kept below 0.2%. This is particularly relevant for laboratories that experience a high sample workload. Sample throughput will be approximately 15 samples per hour, for the determination of 24 elements in a sample.

COMPARISON HIGHLIGHTS

There is no question that the multielement techniques like ICP-OES and ICP-MS are better suited for the determination of the elements defined in USP chapters <232> and <2232>, especially in labs that are expected to be carrying out the analysis in a high-throughput, routine environment. If the best detection limits are required, ICP-MS offers the best choice, followed by GFAA (ETA). Axial ICP-OES offers very good detection limits for most elements, but generally not as good as ETA. Radial ICP-OES and FAA show approximately the same detection limits performance, except for the refractory, rare earth, and transuranic elements, for which performance is much better by ICP-OES. For mercury and those elements that form volatile hydrides, such as As, Bi, Sb, Se, and Te, the CV or HG techniques offer

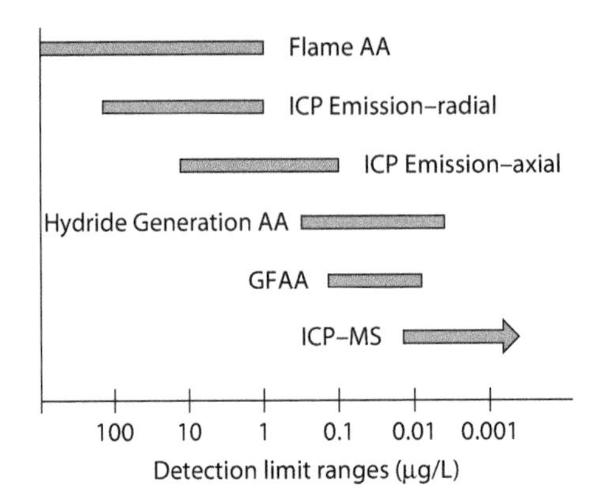

FIGURE 23.2 Typical AS detection limit ranges.

exceptional detection limits. Figures 23.2 and 23.3 and Table 23.1 show an overview of the detection capability, working analytical range, and sample throughput of the major AS approaches, which are three of the metrics most commonly used to select a suitable technique. They should be considered an approximation and only used for guidance purposes.

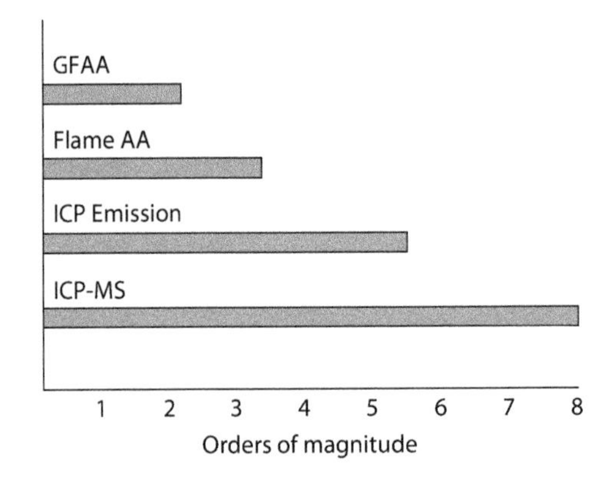

FIGURE 23.3 Typical AS analytical working ranges.

TABLE 23.1
Approximate Sample Throughput Capability of AS Techniques

Technique	Elements at a Time	Duplicate Analysis (min)	Samples per Hour (1 element)	Samples per Hour (5 elements)	Samples per Hour (15 elements)
FAA	1	0.3	150	30	10
ETA	1	5	12	3	~1
ICP-OES	Up to 70	3	20	20	20
ICP-MS	Up to 70	3	20	20	20

It is also worth mentioning that the detection limits of quadrupole-based ICP-MS, when used in conjunction with collision/reaction cell and interface and/or multiple mass separation and selection devices, is now capable of sub-parts-per-trillion (ppt) detection limits, even for elements such as Fe, K, Ca, Se, As, Cr, Mg, V, and Mn, which traditionally suffer from plasma- and solvent-based polyatomic interference.

This is not meant to be a detailed description of each technique, but a basic understanding as to how they differ from each other. For a more detailed comparison of the fundamental principles and application strengths and weaknesses of ICP-MS and ICP-OES, refer to the previous chapters. Let's now turn our attention to selecting the best technique based on the demands of pharmaceutical application.

PHARMACEUTICAL APPLICATION DEMANDS

Comparing an instrumental technique based on its performance specifications with simple standards is important, but it bears little relevance to how that instrument is going to be used in a real-world situation. For example, instrument detection limits (IDLs) are important to know, but how are they impacted by the matrix and sample preparation procedure? So what are the real-world method detection limits (MDLs), and is the technique capable of quantifying the maximum concentration values expected for this analysis? And what kind of precision and accuracy can be expected if working close to the LOQ for the overall methodology being used. Additionally, on the sample throughput side, how many samples are expected, and at what frequency will they be coming into the lab? How much time can be spent on sample preparation, and how quickly must they be analyzed? Sometimes the level of interferences from the matrix will have a major impact on the selection process or even the amount or volume of sample available for analysis.

Understanding the demands of an application is therefore of critical importance when a technique is being purchased, and particularly if there is a minimum amount of expertise or experience in-house on how best to use it to solve a particular application problem. This could be the likely scenario in a pharmaceutical or dietary supplement production laboratory that is being asked to check the elemental impurities of incoming raw materials used in the manufacturing process. They will now have to follow the new USP chapters <232> (or Chapter <2232>) and <233>, which recommend the use of a plasma-based spectroscopic technique to carry out the analysis—or any other AS technique, as long as the validation protocols are met.

The expertise of the operator should never be underestimated, because if ICP-MS is being seriously considered, it generally requires an analyst with a higher skill level to develop rugged methodology free of interferences that can eventually be put in the hands of an inexperienced user to operate on a routine basis. This again is a real concern if the technique is being used by novice users who have limited expertise in running analytical instrumentation, which may be the case in the pharmaceutical or nutraceutical industries. So let's take a closer look at the new USP chapters to understand what AS techniques might best meet the demands of this application.

SUITABILITY OF TECHNIQUE

To get a better understanding of the suitability of the technique being used and whether its detection capability is appropriate for pharmaceutical materials, it's important to know the permissible daily exposure (PDE) limit for each target element and, in particular, what the USP calls the J-value, as described in Chapter <233>. As mentioned, the J-value is defined as the PDE concentration of the element of interest, appropriately diluted to the working range of the instrument, after the sample preparation procedure to get the sample into solution is completed. So let's take Pb as an example. The PDE limit for Pb in an oral medication defined in Chapter <232> is 5 µg/day.

Based on a suggested dosage of 10 g of the drug product per day, that's equivalent to 0.5 µg/g Pb. If 1.0 g of sample is digested or dissolved and made up to 500 mL, that's a 500-fold dilution, which is equivalent to 1.0 µg/L. So the J-value for Pb in this example is equal to1.0 µg/L.

The method then suggests using a calibration made up of 2 standards: Standard 1 = 1.5J and Standard 2 = 0.5J. So for Pb, that's equivalent to 1.5 µg/L for Standard 1 and 0.5 µg/L for Standard 2.

The suitability of a technique is then determined by measuring the calibration drift and comparing the results for Standard 1 before and after the analysis of all the sample solutions under test. This calibration drift should be <20% for each target element. However, once the suitability of the technique has been determined, further validation protocols, described previously, must be carried out to show compliance to the regulatory agency if required.

It should also be pointed out that no specific instrumental parameters are suggested in Chapter <233>, but only to analyze according to the manufacturer's suggested conditions and to calculate and report results based on the original sample size. However, it does say that appropriate measures must be taken to correct for interferences, such as matrix-induced wavelength overlaps in ICP-OES and argon-based polyatomic interference with ICP-MS.

Let's exemplify this by taking two analytical scenarios: (1) measuring 24 elemental impurities in an oral drug according to Chapter <232> and (2) measuring the big four heavy metal contaminants in a dietary supplement according to Chapter (2232>. Next, calculate the J-values for each elemental impurity and compare them with the LOQs for each technique to give us an assessment of their suitability. For this analytical scenario, we'll take the LOQ for the technique as 10 times the IDL. These LOQs were calculated by taking the average of published IDLs from three instrument vendors' application materials and multiplying them by 10 to get an approximation of LOQ. In practice, a method LOQ is typically determined by processing the matrix blank through the entire sample preparation procedure and taking 10 replicate measurements. The method LOQ, sometimes referred to as the MDLs, is then calculated as three to seven times the standard deviations of these 10 measurements, depending on the percent confidence level required.

To make this comparison valid, the sample weight was adjusted for each technique, based on the detection limit and analytical working range. So for AA and ICP-OES, we used a sample dilution of 2 g/100 mL, whereas for ICP-MS we used

0.2 g/100 mL. AA/ICP-OES could definitely use larger sample weights, but for high-throughput routine analysis, we are probably at the optimum dilution for ICP-MS. Tables 23.2 through 23.4 show the comparison of AA (FAA and ETA), ICP-OES, and ICP-MS, respectively, for 24 elemental impurities in an oral drug according to Chapter <232>. Table 23.5 shows equivalent data for Cd, Pb, As, and Hg in a supplement according to Chapter <2232>. The important data to consider is in the final column, labeled "Factor Improvement," which is the J-value divided by the LOQ. Generally speaking, the higher this number, the more suitable is the technique.

TABLE 23.2
USP Chapter <232> J-Values Compared with Limits of Quantitation for FAA and ETA

Element	Concentration Limits for an Oral Drug with a Maximum Daily Dose of ≤10 g/day(μg/g)	J-Value with a Sample Dilution of 2 g/100 mL(μg/L)	~AA LOQ (IDL × 10)(μg/L)		Factor Improvement(J-Value/LOQ)	
			FAA	ETA	FAA	ETA
Cadmium	0.5	10	7	0.02	1.4	500
Lead	0.5	10	140	0.5	0.07	20
Arsenic	1.5	30	0.2[a]	0.5	150[a]	60
Mercury	3	60	0.1[a]	5	600[a]	12
Cobalt	5	100	100	2	1	50
Vanadium	10	200	600	1	0.3	200
Nickel	20	400	60	0.8	7	500
Thallium	0.8	16	150	1	0.1	16
Gold	10	200	100	1.5	2	133
Palladium	10	200	300	1	0.7	200
Iridium	10	200	300	NA	0.7	NA
Osmium	10	200	NA	NA	NA	NA
Rhodium	10	200	60	NA	3.3	NA
Ruthenium	10	200	1,000	10	0.2	20
Selenium	15	300	0.3[a]	0.5	1,000[a]	600
Silver	15	300	20	0.05	15	6,000
Platinum	10	200	600	20	0.3	10
Lithium	55	1,100	10	0.6	110	1833
Antimony	120	2,400	500	0.5	4.8	4,800
Barium	140	2,800	200	4	14	700
Molybdenum	300	6,000	500	0.5	12	12,000
Copper	300	6,000	420	15	14.3	400
Tin	600	12,000	1,500	1	8	12,000
Chromium	1,100	22,000	30	0.05	733	440,000

Note: NA = not appropriate by the technique.

[a] These values are based on HGAA for arsenic and CVAA for mercury.

TABLE 23.3
USP Chapter <232> J-Values Compared with Limits of Quantitation for Axially Viewed ICP-OES

Element	Concentration Limits for an Oral Drug with a Maximum Daily Dose of ≤10 g/day(μg/g)	J-Value with a Sample Dilution of 2 g/100 mL(μg/L)	~Axial ICP-OES LOQ (IDL × 10) (μg/L)	Factor Improvement (J-Value/LOQ)
Cadmium	0.5	10	0.02	500
Lead	0.5	10	10	1
Arsenic	1.5	30	10	3
Mercury	3	60	10	6
Cobalt	5	100	2	50
Vanadium	10	200	5	40
Nickel	20	1,200	5	240
Thallium	0.8	16	20	0.8
Gold	10	200	10	20
Palladium	10	200	20	10
Iridium	10	200	10	20
Osmium	10	200	50	4
Rhodium	10	200	50	4
Ruthenium	10	200	10	20
Selenium	15	300	20	15
Silver	15	300	10	30
Platinum	10	200	10	20
Lithium	55	1,100	4	275
Antimony	120	2,400	20	120
Barium	140	2,800	0.4	7,000
Molybdenum	300	6,000	5	1,200
Copper	300	6,000	5	1,200
Tin	600	12,000	20	600
Chromium	1,100	22,000	2	11,000

RELATIONSHIP BETWEEN LOQ AND J-VALUE

It should be emphasized again that LOQ in these examples is just a guideline as to the real-world detection capability of the technique for this method. However, it does offer a very good approximation as to whether the technique is suitable based on the factor improvement number compared with the J-values for each elemental impurity. Clearly, if this improvement number is close to or less than 1, as it is with the majority of elements by FAA, the technique is just not going to be suitable, particularly for the big four heavy metals, which are the most critical. On the other hand, ETA would be suitable for the majority of the impurities (including the heavy metals), except for a few of the catalyst-based elements, which are not ideally

TABLE 23.4
USP Chapter <232> J-Values Compared with Limits of Quantitation for ICP-MS

Element	Concentration Limits for an Oral Drug with a Maximum Daily Dose of ≤10 g/day(μg/g)	J-Value with a Sample Dilution of 0.2 g/100 mL(μg/L)	~ICP-MS LOQ (IDL × 10) (μg/L)	Factor Improvement (J-Value/LOQ)
Cadmium	0.5	1	0.0007	1,430
Lead	0.5	1	0.0004	2,500
Arsenic	1.5	3	0.004	750
Mercury	3.0	6	0.01	600
Cobalt	5	10	0.0005	20,000
Vanadium	10	20	0.0005	40,000
Nickel	20	40	0.002	20,000
Thallium	0.8	1.6	0.0001	16,000
Gold	10	20	0.001	20,000
Palladium	10	20	0.0002	100,000
Iridium	10	20	0.001	20,000
Osmium	10	20	0.0005	40,000
Rhodium	10	20	0.0005	40,000
Ruthenium	10	20	0.002	10,000
Selenium	15	30	0.002	15,000
Silver	15	30	0.001	30,000
Platinum	10	20	0.001	20,000
Lithium	55	110	0.0005	220,000
Antimony	120	240	0.002	120,000
Barium	140	280	0.0005	560,000
Molybdenum	18	36	0.0007	51,430
Copper	130	260	0.003	86,670
Tin	600	1,200	0.003	400,000
Chromium	1,100	2,200	0.002	1,100,000

suited to the technique. However, the ETA technique is very time-consuming and labor-intensive, so it probably wouldn't be a practical solution in a high-throughput pharmaceutical production laboratory. The comparison between FAA and ETA is exemplified in Table 23.2.

Table 23.3 shows that axial-ICP-OES offers some possibilities for monitoring oral drugs because the vast majority of the improvement factors are higher than 1. These numbers could be further improved, especially for the heavy metals, by using a much higher sample weight in the sample preparation procedure without compromising the method. (Note: As most commercial ICP-OES instrumentation offers both axial and radial capability, it was felt that the axial performance was most appropriate for this comparison.)

TABLE 23.5
USP Chapter <2232> J-Values Compared with Limits of Quantitation for FAA, ETA, ICP-OES, and ICP-MS

Element	Concentration Limits for Dietary Supplement with a Maximum Daily Dose of ≤10 g/day(μg/g)	J-Value with a Sample Dilution of 2 g/0.2 g100 mL(μg/L)	FAA LOQ(μg/L)	ETA LOQ(μg/L)	Axial ICP-OES LOQ(μg/L)	ICP-MS LOQ(μg/L)	Factor Improvement(J-Value/LOQ)			
							FAA	ETA	ICP-OES	ICP-MS
Cadmium	0.5	10/1	7	0.02	0.02	0.0007	1.4	500	500	1430
Lead	1.0	20/2	140	0.5	10	0.0004	0.14	40	2	5000
Arsenic[b]	1.5	30/3	0.2[a]	0.5	10	0.004	150[a]	60	3	750
Mercury[b]	1.5	30/3	0.1[a]	5	10	0.01	300[a]	6	3	300

[a] These values are based on HGAA for arsenic and CVAA for mercury.
[b] In Chapter <2232>, As = inorganic and Hg = total.

However, it can be seen in Table 23.4 that ICP-MS shows significant improvement factors for all impurities, which are not offered by any other technique. Even for the four heavy metals, there appears to be ample improvement to monitor them with good accuracy and precision. The added benefit of using ICP-MS is that it would also be suitable for the other methods of pharmaceutical delivery, such as parenteral or inhalation, where the PDE levels are typically one or two orders of magnitude lower. Additionally, if arsenic or mercury levels were found to be higher than the PDE levels, it would be relatively straightforward to couple ICP-MS with high-performance liquid chromatography (HPLC) to monitor the speciated forms of these elements if required.

Table 23.5 shows the equivalent data for the four heavy metal contaminants in a dietary or herbal supplement using the PDEs defined in Chapter <2232>. It should be pointed out that the PDE limits for Pb and Hg are different from those in Chapter <232>. Also, the J-value and dilution factor column in this table is based on an optimum sample prep of 2 g/100 mL (for AA and ICP-OES) and 0.2 g/100 mL (for ICP-MS).

FINAL THOUGHTS

It is important to understand that there are many factors to consider when selecting a trace element technique most suited to the demands of your application. Sometimes, one technique stands out as being the clear choice, whereas other times, it is not quite so obvious. And as is true with many applications, more than one technique is often suitable. However, the current methodology described in the new USP chapters <232>, <2232>, and <233> presents unique challenges, not only from a perspective of performance capability, but also because of the validation protocols that have to be met in order to show the suitability of the technique to the analytical procedure being used. From a practical perspective, there is no question that to meet all the PDE limits in all pharmaceutical delivery methods, ICP-MS is probably the most appropriate technique. However, for oral delivery products, especially liquid medications or those that can be easily brought into solution with a suitable aqueous or organic solvent, axial-ICP-OES could offer a less costly approach. And if the sample workload requirements are not so demanding, ETA could provide a solution. On the other hand, with Chapter <2232>, only four elemental contaminants need to be monitored, so AA might be a viable option, particularly if the workload is not high. However, like the oral drug example, ICP-OES could also work, especially if larger sample weights and lower dilutions are used. But once again, ICP-MS has shown it has the detection limits and throughput capability to be the optimum technique of choice for these types of samples.

REFERENCES

1. R.J. Thomas. Choosing the right trace element technique: Do you know what to look for? *Today Chemist at Work*, October 1999.
2. R. J. Thomas. Emerging technology trends in atomic spectroscopy are solving real-world application problems. *Spectroscopy Magazine*, 29(3), 42–51, 2014.
3. D. Davis and H. Furukawa. Using XRF as an alternative technique to plasma spectrochemistry for the new USP and ICH directives on elemental impurities in pharmaceutical materials. *Spectroscopy Magazine*, 32(7), 12–17 2017.

24 Do You Know What It Costs to Run Your Atomic Spectroscopy Instrumentation?

Chapter 23 gave an overview of the performance differences between flame atomic absorption (FAA), graphite furnace atomic absorption (GFAA) (electrothermal atomization [ETA]), inductively coupled plasma optical emission spectrometry (ICP-OES), and inductively coupled plasma mass spectrometry (ICP-MS) for pharmaceutical applications, but selection of the optimum technique might also involve the cost of the instrumentation, particularly if funds are limited. This chapter looks at the running costs of the four atomic spectroscopy (AS) techniques.

For the purpose of this evaluation, let us make the assumption that the major operating costs associated with running AS instrumentation are the gases, electricity, and consumable supplies. For comparison purposes, the exercise will be based on a typical laboratory running its instrument for 2½ days (20 h) per week and 50 weeks a year (1000 h per year). This data is based on the cost of gases, electricity, and instrument consumables in the United States in 2017. They have been obtained from a number of publically available commercial sources, including suppliers of industrial and high-purity gases, independent utility companies, a number ICP-MS instrument vendors, and sample introduction and consumable suppliers). It's also important to emphasize that these costs might vary slightly based on the different vendor instrument designs and technologies being used.

GASES

FAA

Most FAA systems use acetylene (C_2H_2) as the combustion gas, and air or nitrous oxide (N_2O) as the oxidant. Air is usually generated by an air compressor, but the C_2H_2 and N_2O come in high-pressure cylinders. Normal atomic absorption–grade C_2H_2 cylinders contain 380 ft^3 (10,760 L) of gas. N_2O is purchased by weight and comes in cylinders containing 56 lb of gas, which is equivalent to 490 ft^3 (13,830 L). A cylinder of C_2H_2 costs approximately \$200, whereas a cylinder of N_2O costs about \$70. These prices have remained fairly stable over the past few years. Normal C_2H_2 gas flows in FAA are typically 2 L/min when air is the oxidant and 5 L/min when N_2O is the oxidant. N_2O gas flows are on the order of 10 L/min.

AIR–C_2H_2

Mixtures are used for the majority of elements, whereas an N_2O–C_2H_2 mixture has traditionally been used for the more refractory elements. So, for this costing exercise, we will assume that half the work is done using air–C_2H_2, and for the other half, N_2O–C_2H_2 is used. Therefore, a typical laboratory running the instrument for 1000 h per year will consume 16 cylinders of C_2H_2, which is equivalent to $3200 per year, and 22 cylinders of N_2O costing $1500, making a total of $4700.

ETA

The only gas that the ETA process used on a routine basis is high-purity argon, which costs about $100 for a 340 ft^3 (9630 L) cylinder. Typically, argon gas flows of up to 300 mL/min are required to keep an inert atmosphere in the graphite tube. At these flow rates, 540 h of use can be expected from one cylinder. Therefore, a typical laboratory running its instrument for 1000 h per year would consume almost two cylinders, costing $200.

ICP-OES AND ICP-MS

The consumption of gases in ICP-OES and ICP-MS is very similar. They both use a total of approximately 15–20 L/min (~1000 L/h) of gaseous argon (including plasma, nebulizer, auxiliary, and purge flows), which means a cylinder of argon (9630 L) would last only about 10 h. For this reason, most users install a Dewar vessel containing a liquid supply of argon. Liquid argon tanks come in a variety of different sizes, but a typical Dewar system used for ICP-OES or ICP-MS holds about 240 L of liquid gas, which is equivalent to 6300 ft^3 (178,000 L) of gaseous argon. (Note: The Dewar vessel can be bought outright, but is normally rented.) It costs about $350 to fill a 240 L Dewar vessel with liquid argon. At a typical argon flow rate of 17 L/min total gas flow, a full vessel would last for almost 175 h. Again, assuming a typical laboratory runs its instrument for 1000 h per year, this translates to six fills at approximately $350 each, which is equivalent to about $2100 per year. If cylinders were used, about 100 would be required, which would elevate the cost to almost $10,000 per year.

Note: When liquid argon is stored in a Dewar vessel, there is a natural bleed-off to the atmosphere when the gas reaches a certain pressure. For this reason, a bank of argon cylinders is probably the best option for laboratories that do not use their instruments on a regular basis. Some of the newer ICP-OES instruments operate at approximately 60%–70% argon consumption compared with older instruments. So this should be taken into consideration if this technology is being used.

Another added expense with ICP-MS is that if it is fitted with collision/reaction cell technology, the cost of the collision or reaction gas will have to be added to the running costs of the instrument. Fortunately, for most applications, the gas flow is usually less than 5 mL/min, but for the collision/reaction interface approach, typical gas flows are 100–150 mL/min. The most common collision/reaction gases used are hydrogen, helium, and ammonia. The cost of high-purity helium is on the order of \$400 for a 300 ft^3 (8500 L) cylinder, whereas that of a cylinder of hydrogen or ammonia is approximately \$250. One cylinder of either gas should be enough to last 1000 h at these kinds of flow rates. So, for this costing exercise, we will assume that the laboratory is running a collision/reaction cell or interface instrument, with an additional expense of \$650. If other collision/reaction gases are being used, these should be factored into the calculation.

It should also be pointed out that some collision/reaction cells require high-purity gases with extremely low impurity levels, because of the potential of the contaminants in the gas to create additional by-product ions. This can be achieved either by purchasing laboratory-grade gases and cleaning them up with a gas purification system (getter), or by purchasing ultra-high-purity gases directly from the gas supplier. If the latter option is chosen, you should be aware that ultra-high-purity helium (99.9999%) is approximately twice the price of laboratory-grade helium (99.99%), whereas ultra-high-purity hydrogen is approximately four times the cost of laboratory-grade hydrogen.

ELECTRICITY

Calculations for power consumption are based on the average cost of electricity, which is currently about \$0.10 per kilowatt hour (kWh) in the United States. The cost will vary depending on the location and demand, but it represents a good approximation for this exercise. So the following formula has been used for calculating the cost of electricity usage for each technique:

Cost per kWh (\$) × Power consumption (kW) × Annual usage (h)

FAA

The power in an FAA system is basically used for the hollow cathode lamps and the onboard microprocessor that controls functions like burner head position, lamp selection, photomultiplier tube voltage, and grating position. A typical instrument requires less than 1 kW of power. If it is used for 1000 h per year, it will draw less than 1000 kW of total power, which is ~\$100 per year.

ETA

A graphite furnace system uses considerably more power than an FAA system because a separate power unit is used to heat the graphite tube. In routine operation, there is a slow ramp heating of the tube for ~3 min until it reaches an atomization temperature of about 2700°C, requiring a maximum power of ~3 kW. This heating cycle, combined with the power requirements for the rest of the instrument, costs ~\$300 for a system that is run 1000 h per year.

ICP-OES AND ICP-MS

Both these techniques can be considered the same with regard to power requirements as the radio-frequency (RF) generators are of very similar design. Based on the voltage, magnitude of the electric current, and number of lines used, the majority of modern instruments draw about 5 kW of total power. This works out to be ~$500 for an instrument that is run 1000 h per year.

CONSUMABLES

Because of the fundamental differences between the four AS techniques, it is important to understand that there are considerable differences in the cost of consumables. In addition, the cost of the same component used in different techniques can vary significantly between different vendors and suppliers. So, the data has been taken from a number of different sources and averaged.

FAA

The major consumable supplies used in FAA are the hollow cathode lamps. Depending on usage, you should plan to replace three of them every year, at a cost of $300–$500 for a good-quality, single-element lamp. However, if a continuum source atomic absorption system is being used, there will not be a requirement to replace lamps on a regular basis. Other minor costs are nebulizer tubing and autosampler tubes. These are relatively inexpensive but should be planned for. The total cost of lamps, nebulizer tubing, and a sufficient supply of autosampler tubes should not exceed $1500–$2000 per year, based on 1000 h of instrument usage.

ETA

As long as the sample type is not too corrosive, a GFAA tube should last about 300 heating cycles (firings). Based on a normal heating program of 3 min per replicate, this represents 20 firings per hour. If the laboratory is running the instrument 1000 h per year, it will carry out a total of 20,000 firings and use 70 graphite tubes in the process. There are many designs of graphite tubes, but for this exercise, we will base the calculation on platform-based tubes that cost about $50 each when bought in bulk. If we add the cost of graphite contact cylinders, hollow cathode lamps, and a sufficient supply of autosampler cups, the total cost of consumables for a graphite furnace will be approximately $5000–$6000 per year.

ICP-OES

The main consumable supplies in ICP-OES are in the plasma torch and in the sample introduction area. The major consumable is the torch itself, which consists of two concentric quartz tubes and a sample injector made of either quartz or some ceramic material. In addition, a quartz bonnet normally protects the torch from the RF coil. There are many different demountable torch designs available, but they all cost about $600–$700 for a complete system. Depending on sample workload and matrices being analyzed, it is normal to go through a torch every 4–6 months. In addition to the torch, other parts that need to be replaced or at least need to have spares include the nebulizer, spray chamber, and sample capillary and pump tubing. When all these

TABLE 24.1
Annual Operating Costs ($US) for the Four AS Techniques for a Laboratory Running an Instrument 1000 h per Year (20 h per Week)

Technique	Gases ($)	Power ($)	Consumable Supplies ($)	Total ($)
FAA	4,700	100	1,750	6,550
ETA	200	300	5,500	6,000
ICP-OES	2,100[a]	500	3,100	5,700
ICP-MS	2,750[a,b]	500	10,000	13,250

[a] Using a liquid argon supply.
[b] Using a collision/reaction cell.

items are added together, the annual cost of consumables for ICP-OES is on the order of $3000–$3200.

ICP-MS

In addition to the plasma torch and sample introduction supplies, ICP-MS requires consumables that are situated inside the mass spectrometer. The first area is the interface region between the plasma and the mass spectrometer, which contains the sampler and skimmer cones. These are traditionally made of nickel, which is recommended for most matrices, or platinum for highly corrosive samples and organic matrices. A set of nickel cones costs $700–$1000, whereas a set of platinum cones costs about $3000–$4000. Two sets of nickel cones and perhaps one set of platinum cones would be required per year. The other major consumable in ICP-MS is the detector, which has a lifetime of approximately 1 year, and costs about $1200–$1800. When all these are added together with the torch, the sample introduction components, and the vacuum pump consumables, investing in ICP-MS supplies represents an annual cost of $9,000–$11,000.

The approximate annual cost of gases, power, and consumable supplies of the four AS techniques being operated for 1000 h per year is shown in Table 24.1.

COST PER SAMPLE

We can take the data given in Table 24.1 a step further and use these numbers to calculate the operating costs per individual sample, assuming that a laboratory is determining 10 analytes per sample. Let us now take a look at each technique to see how many samples can be analyzed, assuming the instrument runs 1000 h per year.

FAA

A duplicate analysis for a single analyte in FAA takes about 20 s. This is equivalent to 180 analytes per hour, or 180,000 analytes per year. For 10 analytes, this represents 18,000 samples per year. Based on an annual operating cost of $6550, this equates to $0.36 per sample.

ETA

A single analyte by ETA takes about 5–6 min for a duplicate analysis, which is equivalent to approximately 10 analytes per hour, or 10,000 analytes per year. For 10 analytes per sample, this represents 1000 samples per year. Based on an annual operating cost of $6000, this equates to $6.00 per sample.

ICP-OES

A duplicate ICP-OES analysis for as many analytes as you require takes about 3 min. So for 10 analytes, this is equivalent to 20 samples per hour, or 20,000 samples per year. Based on an annual operating cost of $5700, this equates to $0.30 per sample.

ICP-MS

ICP-MS also takes about 3 min to carry out a duplicate analysis for 10 analytes, which is equivalent to 20,000 samples per year. Based on an annual operating cost of $13,250, this equates to $0.66 per sample.

Operating costs for all four AS techniques for the determination of 10 analytes per sample are summarized in Table 24.2.

It must also be emphasized that this comparison does not take into account the detection limit requirements, but is based on instrument operating costs alone. These figures have been generated for a typical workload using what would be considered the average cost of gases, power, and consumables in the United States. Even though there will be geographical differences in the cost of these items in other parts of the world, the comparative costs should be very similar. Every laboratory's workload and analytical needs are unique, so this costing exercise should be treated with caution and only be used as a guideline for comparison purposes.

For a more accurate assessment of your lab's workload, the costing exercise should be carried out using your sample throughput and analyte requirements.

TABLE 24.2
Operating Costs for a Sample Requiring 10 Analytes, Based on the Instrument Being Used for 1000 h per Year

Technique	Operating Cost for 10 Analytes per Sample ($US)
FAA	0.36
ETA	6.50
ICP-OES[a]	0.29
ICP-MS[a,b]	0.66

[a] Using a liquid argon supply.
[b] Using a collision/reaction cell.

However, whatever the workload, it is a good exercise to show that there are running cost differences between the major AS techniques. If required, it can be taken a step further by also including the purchase price of the instrument, the cost of installing a clean room, the cost of sample preparation, and the salary of the operator. This would be a very useful exercise, as it would give a good approximation of the overall cost of analysis, and therefore it could be used as a guideline for calculating what a laboratory might charge for running samples on a commercial basis.

FINAL THOUGHTS

It can be seen from this evaluation that based on the annual operating costs, FAA, ETA, and ICP-OES are all very similar, with ICP-MS being approximately twice as expensive to operate. However, when the number of samples or analytes is taken into consideration, the picture changes quite dramatically. It is also important to remember that there are many criteria to consider when selecting a trace element technique. Operating costs are just one of them, and they should not prevent you from choosing an instrument if your analytical requirements change, such as the need for lower detection limits. But, if more than one of these techniques fulfills your analytical demands, then knowledge of the operating costs should help you make the right decision.

25 The Risk Assessment Approach

Risk assessment (noun): A systematic process of evaluating the potential risks that may be involved in a projected activity or undertaking.

The development of the International Conference on Harmonisation of Technical Requirements for Registration of Pharmaceuticals for Human Use (ICH) Q3D guideline and United States Pharmacopeia (USP) Chapter <232> is not a blanket regulatory demand for testing of pharmaceutical products but more a drive to ensure that patients are not exposed to unnecessary elemental impurities through the ignorance of their manufacturers. Modern analytical technology is capable of measuring specific elemental impurities at levels way below the thresholds where they may cause harm to humans, so we could simply test every product for almost every element in the periodic table. However, the regulatory requirement is for the drug producer to assess their manufacturing process for the potential introduction of elemental impurities and, where necessary, provide controls that may or may not include specific metals testing. The Food and Drug Administration (FDA) draft guidance on ICH Q3D (2016)[1] clearly indicates that the agency accepts that testing may not be needed if the risk assessment conclusion is that the inclusion of elemental impurities is a low risk. This chapter, written by Jon Sims, exemplifies the concept of a risk assessment approach as required for regulatory submissions of drug products.

ICH Q3D recommends that manufacturers conduct a product risk assessment by first identifying known and potential sources of elemental impurities. Manufacturers should consider all potential sources of elemental impurities, such as elements intentionally added, elements potentially present in the materials used to prepare the drug product, and elements potentially introduced from manufacturing equipment or container closure systems. Manufacturers should then evaluate each elemental impurity likely to be present in the drug product by determining the observed or predicted level of the impurity and comparing it with the established permissible daily exposure (PDE). If the risk assessment fails to show that an elemental impurity level is consistently less than the control threshold (defined as being 30% of the established PDE in the drug product), additional controls should be established to ensure that the elemental impurity level does not exceed the PDE in the drug product. These additional controls could be included as in-process controls or in the specifications of the drug product or components.

BENEFITS OF USING A RISK ASSESSMENT APPROACH

Rather than routinely testing drug products against a broad specification for elemental impurities, which may cause delays in product delivery, a correct use of the risk assessment process ensures that targeted and appropriate testing of materials is performed where control is needed, and this creates the possibility that the manufacturer will be able to

- Test for specific metals only
- Test occasional lots, for example, six batches per year
- Not require testing postapproval

The structured scientific-based and data-driven risk assessment ensures that the control strategy is appropriate and does not impact the product quality or patient safety.

ESSENTIAL READING

The ICH Q3D guideline provides in-depth coverage of both risk assessments and the toxicological safety for all the specified elements. It is supported by a comprehensive set of training materials, which are free to users via the ICH website (reference given later in the chapter). The FDA has published guidance on the implementation of ICH Q3D, which is also free to view from the FDA website. In addition, there have been several conferences held by the Product Quality Research Institute (PQRI) (United States), the FDA (United States), and Joint Pharmaceutical Analysis Group (JPAG) (United Kingdom), bringing together industry and regulatory staff in order to discuss effective approaches to compliance since the pharmaceutical industry will have thousands of products to submit data for in the near future. These conferences have produced valuable resources in the form of presentation slides and white papers that are available to members of the organizations, and therefore they are worth joining to access this kind of information.

PERFORMING THE RISK ASSESSMENT

The first thing you need before starting the risk assessment is the route of administration and dose range for your product since the systemic exposure of the human system to elemental impurities varies with the route of administration—oral, inhaled, topical, or parenteral. In addition, some metals exhibit higher toxicity in some forms of administration than others.

With this information, we can refer to Table 25.1 taken from ICH Q3D and determine what the maximum specification limits will be for our product, provided it has oral, parenteral, or inhaled administration. Further guidance is being developed by ICH to define PDEs for cutaneous and transdermal administration, which is expected to be finalized by mid-2018 and will be published as ICH Q3D R1. Since the intact skin serves as a barrier to absorption, it is possible that not all EIs in Q3D will require cutaneous and transdermal PDEs, streamlining the risk assessment process for these products. Pending publication of this revision, the best resource for other routes of administration is ICH Training Module 1 for ICH Q3D, available from the ICH website.

TABLE 25.1
Reproduction of Table A.2.1, "Permitted Daily Exposures for Elemental Impurities," in ICH Q3D Step 4 Document, December 2014

Element	Class	Oral PDE (µg/day)	Parenteral PDE (µg/day)	Inhalational PDE (µg/day)
Cd	1	5	2	2
Pb	1	5	5	5
As	1	15	15	2
Hg	1	30	3	1
Co	2A	50	5	3
V	2A	100	10	1
Ni	2A	200	20	5
Tl	2B	8	8	8
Au	2B	100	100	1
Pd	2B	100	10	1
Ir	2B	100	10	1
Os	2B	100	10	1
Rh	2B	100	10	1
Ru	2B	100	10	1
Se	2B	150	80	130
Ag	2B	150	10	7
Pt	2B	100	10	1
Li	3	550	250	25
Sb	3	1200	90	20
Ba	3	1400	700	300
Mo	3	3000	1500	10
Cu	3	3000	300	30
Sn	3	6000	600	60
Cr	3	11000	1100	3

From this PDE data, we can calculate the control threshold for any element that we risk assess (30% of the PDE), which was neatly explained by the FDA in Section F of its 2015 guidance document:

> If the total elemental impurity level from all sources in the drug product is expected to be consistently less than 30 percent of the PDE, then additional controls are not required, provided that the applicant has appropriately assessed the data and demonstrated adequate controls on elemental impurities.

Although this has been removed in the 2016 revision, it does capture the intent of having the control threshold concept. The term *additional controls* means testing and specifications. With knowledge of the route of administration, the information in Table 25.2, also from ICH Q3D, allows us to refine the number of elements we have to consider in our risk assessment since most elements not used in our process may be discounted.

TABLE 25.2
Reproduction of Table 5.1, "Elements to Be Considered in the Risk Assessment," in ICH Q3D Step 4 Document, December 2014

Element	Class	Intentionally Added (All Routes)	Not Intentionally Added		
			Oral	Parenteral	Inhalational
Cd	1	Yes	Yes	Yes	Yes
Pb	1	Yes	Yes	Yes	Yes
As	1	Yes	Yes	Yes	Yes
Hg	1	Yes	Yes	Yes	Yes
Co	2A	Yes	Yes	Yes	Yes
V	2A	Yes	Yes	Yes	Yes
Ni	2A	Yes	Yes	Yes	Yes
Tl	2B	Yes	No	No	No
Au	2B	Yes	No	No	No
Pd	2B	Yes	No	No	No
Ir	2B	Yes	No	No	No
Os	2B	Yes	No	No	No
Rh	2B	Yes	No	No	No
Ru	2B	Yes	No	No	No
Se	2B	Yes	No	No	No
Ag	2B	Yes	No	No	No
Pt	2B	Yes	No	No	No
Li	3	Yes	No	Yes	Yes
Sb	3	Yes	No	Yes	Yes
Ba	3	Yes	No	No	Yes
Mo	3	Yes	No	No	Yes
Cu	3	Yes	No	Yes	Yes
Sn	3	Yes	No	No	Yes
Cr	3	Yes	No	No	Yes

This is due to the very low risk of certain elements being present unexpectedly in our raw materials and process equipment due to their low abundance in natural sources. Class 1 and 2A metals must always be assessed irrespective of route of administration, but this does not mean they must be routinely tested for in an approved product; rather, data should be collected during the assessment phase to determine whether they are likely to occur in the finished product at levels at or near the PDE.

ICH Q3D provides a clear structure for companies to follow in designing their risk assessment process, which is summarized as a fishbone diagram, as shown in Figure 25.1.

With five likely routes for the introduction of elemental impurities, it is clear that if the inputs are known to be "clean" with respect to the relevant elemental impurity limits, then the drug product will be acceptable without testing for elemental impurities. The ICH has provided pragmatic guidance:

> The applicant's risk assessment can be facilitated with information about the potential elemental impurities provided by suppliers of drug substances, excipients, container

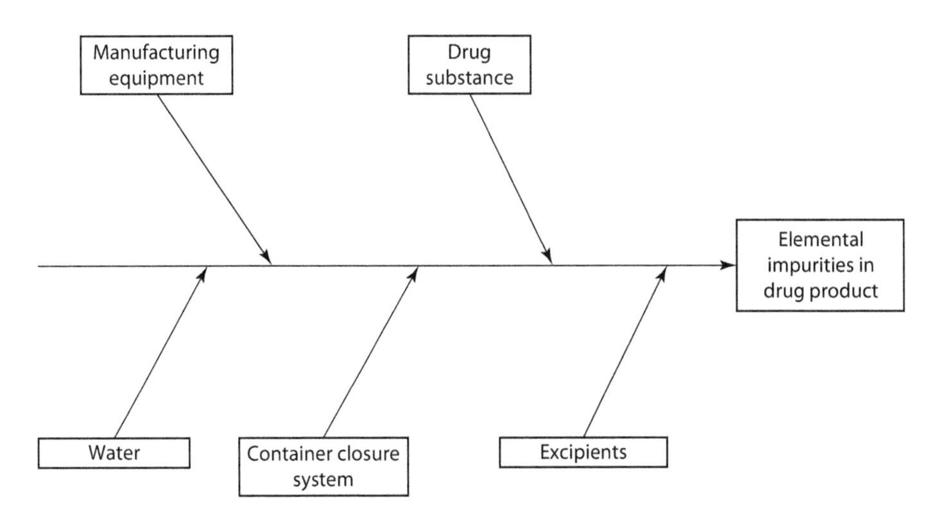

FIGURE 25.1 ICH Q3D risk assessment fishbone diagram.

closure systems, and manufacturing equipment. The data that support this risk assessment can come from a number of sources that include, but are not limited to:

- Prior knowledge;
- Published literature;
- Data generated from similar processes;
- Supplier information or data;
- Testing of the components of the drug product;
- Testing of the drug product.

In summary, testing is not the mandated requirement, but data is critical. This means that as a company becomes more familiar with the elemental impurity risk assessment process, it will be able to leverage information gained in the development of previous products, enabling it to streamline the process. Similarly, as information is published in the scientific literature, the reliance on analytical testing will be reduced.

If we consider the five sources of elemental impurities, it is clear that some are more likely to have issues than others, so in order of complexity, they are water, manufacturing equipment, container closure system, drug substance, and excipients.

WATER

From ICH Q3D, "the risk of inclusion of elemental impurities from water can be reduced by complying with compendial (e.g., European Pharmacopoeia, Japanese Pharmacopoeia, US Pharmacopeial Convention) water quality requirements, if purified water or water for injection is used in the process(es)." In practice, monitoring of water purification is a focus of FDA facility audits, and therefore, as long as no significant change to the quality of the water supply to the site occurs, it is unlikely that the purification process supporting the claim of compliance to pharmacopeial standards will fail to also control unwanted elemental impurities. Therefore, water can be used without testing for elemental impurities.

MANUFACTURING EQUIPMENT

The risk of inclusion of elemental impurities can be reduced through process understanding, equipment selection, equipment qualification, and good manufacturing practices (GMP) processes.

> The specific elemental impurities of concern should be assessed based on knowledge of the composition of the components of the manufacturing equipment. The assessment of this source of elemental impurities is one that can be utilized potentially for many drug products using similar process trains and processes. (Q3D)

It can be argued that a specific plant configuration could be qualified for elemental impurity leaching by testing the qualification batches produced as part of the new drug application (NDA) submission process, and that subsequent processes using that configuration will not need test data provided that the reagents are not significantly different, for example, pH 7 changed to pH 11.

CONTAINER CLOSURE SYSTEM

This term refers to all the packaging components potentially contacting the drug product. It is known that packaging components can leach impurities and manufacturers have to provide studies on extractables and leachables as part of the product registration.

ICH Q3D states the following:

> The probability of elemental leaching into solid dosage forms is minimal and does not require further consideration in the assessment. For liquid and semi-solid dosage forms there is a higher probability that elemental impurities could leach from the container closure system into the drug product during the shelf-life of the product.
>
> Studies to understand potential extractables and leachables from the final/actual container closure system should be performed. (Q3D)

Fortunately, we already have two very useful publications on elemental impurities that can be used in the risk assessment process,[2,3] and these studies by pharmaceutical industry experts cover a wider range of elements than those covered by ICH Q3D in more than 100 test articles. The information provided can be used in the elemental impurity risk assessment process by providing the identities of commonly reported elements and data to support probability estimates of those becoming elemental impurities in the drug product. Furthermore, recommendations are made related to establishing elements of potential product impact for individual materials.

DRUG SUBSTANCE

Drug substance manufacture uses catalysts (e.g., Pd) that are known potential contaminants and will have a control strategy in place as part of the process development, and so are considered low risk but may have input materials that have non-GMP precursors and therefore provide a high risk of introducing unexpected elemental impurities.

EXCIPIENTS

ICH Q3D explains that "elemental impurities are often associated with mined materials and excipients. The presence of these impurities can be variable, especially with respect to mined excipients, which can complicate the risk assessment. The variation should be considered when establishing the probability for inclusion in the drug product."

Excipients sourced from plants (e.g., cellulose), mined (e.g., talc), or from animals (e.g., lactose and gelatin) can potentially be contaminated through man-made pollution or natural sources, particularly with As, Cd, Hg, and Pb.

It is considered that highly refined excipients are low risk with respect to elemental impurities—examples are cellulose and lactose. However, data shows that elevated levels of Class 1 and 2A metals are commonly encountered in mined excipients, such as talc and calcium carbonate, so these are high risk.

A paper on elemental impurities in excipients was published in *Journal of Pharmaceutical Sciences*,[4] which reports the testing of 190 samples from 31 different excipients and 15 samples from 8 drug substances provided through the International Pharmaceutical Excipient Council of the Americas.

The Elemental Impurities Pharma Consortium is, at the time of this writing, working on a commercial database comprising analytical data on elemental impurities from more than 100 different materials (pharmaceutical excipients, dyes, etc.).

RISK ASSESSMENT FREQUENCY

A full risk assessment will need to have been completed as part of the regulatory submission for a new product filing and appropriate actions taken ready for manufacture. Once a product is approved, the manufacturer will need to update the risk assessment whenever changes are made to the process, for example, when a packaging change is made or a new raw material supplier is introduced to the supply chain. The updated risk assessment may require the introduction of additional testing or could justify reduction of testing.

The new risk assessment and any actions should be submitted as part of the annual report to the FDA as per the Chemistry Manufacturing and Controls (CMC) guidance issued in 2014.[5]

CASE STUDY

In order to understand the framework within which your risk assessment is taking place, a simple case study is presented for an Active Pharmaceutical Ingredient (API) and a formulated product. The suggestions presented here in conjunction with the ICH Q3D training materials should enable you to begin preparing good-quality risk assessments. The case study uses paracetamol (acetaminophen or Tylenol in the United States), as this is a generically available, high-dosage (10 g/day) over-the-counter medication.

DRUG SUBSTANCE

The manufacture of drug substances for pharmaceutical use requires the application of current good manufacturing practices (cGMP). Approval of a manufacturing process by regulatory agencies will be granted where cGMP compliance can be demonstrated. However, not all the process steps or inputs can be managed within the cGMP facility—at some point, a starting material or reagent will be brought in from a fine chemical source where lower standards of quality control apply than are expected for cGMP. Figure 25.2 shows the current standard industrial process for the manufacture of paracetamol (Hoechst Celanese synthesis)[6] from phenol (non-GMP starting material), which is referred to in this discussion.

The regulatory approval will reflect that these boundaries are accepted by the inspector as providing a low risk in terms of the molecular structure and related impurities. For elemental impurities, this is not yet the case and you must consider these non-GMP parts of your drug substance manufacture in terms of the risk of inclusion of unexpected elemental impurities. This does not necessarily mean test, but collect information and make a rational risk assessment. Therefore, consider the following:

- How is your non-GMP material likely to be manufactured, and do different routes offer more or less risk? Some will use catalysis, and the catalyst may vary between suppliers (see Class 2B). Phenol is the starting material for paracetamol manufacture and is produced from the extraction of coal tar or by synthesis from petroleum fractions using a number of different routes. No use of catalysis is likely, thus ruling out Class 2B.
- The number of chemical steps following the non-GMP material, which may eliminate the elemental impurity if present. Logically, a 10-step process is low risk compared with a 2-step one; similarly, specific steps may be highly likely to remove certain elements from organic chemistry reactions. Paracetamol manufactured by the most cost-effective route (Hoechst Celanese synthesis) has three steps from phenol without metal catalysis. Therefore, no Class 2B risk during GMP process but FDA may consider too few steps for elemental impurity wash out if phenol is contaminated.

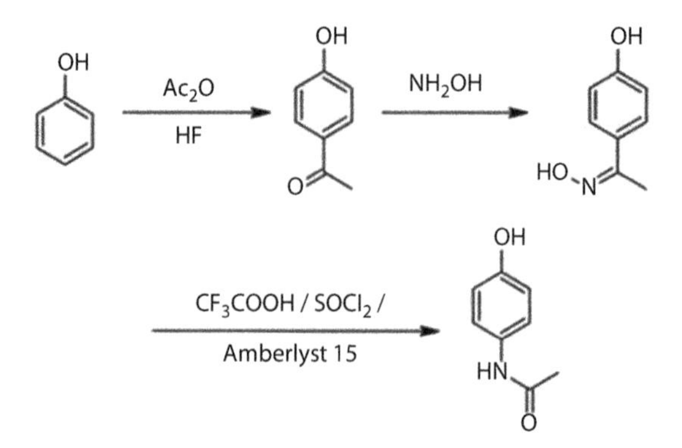

FIGURE 25.2: Hoechst Celanese synthesis for paracetamol.

- Solvents are typically produced using distillation and so may be considered low risk.
- Also, be aware of catalyst supports or resins, as these are not necessarily well characterized in terms of potential impurities and may well be applied in steps near the final drug substance preparation step. In the final paracetamol process step, the Hoechst Celanese synthesis uses Amberlyst 15, a macroreticular polystyrene-based ion exchange resin with a strongly acidic sulfonic group. It serves as an excellent source of strong acid. It can also be used several times. Is this an elemental impurity scavenger or a potential source of contamination?

If necessary, collect data from suppliers to identify where potential variability may arise for elemental impurity inclusion.

Drug Product

Formulations contain many components, some of which contribute very low percentages of the total weight (or volume) of the finished article, for example, ink written on tablets showing strength, or colors used in tablet coating. An example generic formulation for paracetamol tablets is shown in Table 25.3, and it can be seen that two of the ingredients (Nipasol and Nipagin) are present at a level that, at the maximum daily dose for paracetamol, could almost be replaced with cadmium or lead without the drug product failing the USP or ICH limit.

Quality control for these excipients, which included an identification test such as infrared or chromatography, would guarantee that no risk is associated with their use, and so no elemental impurity data would be needed in the risk assessment. Additionally, three cellulose-based excipients are used and, as stated above (under excipients), are considered low risk, and it can probably be concluded that no significant contribution of elemental impurities to the drug product is likely.

TABLE 25.3
Paracetamol (Tylenol or Acetaminophen) Formulation

Component	%w/w	Wt (mg)
Paracetamol (Active ingredient)	59	500
Maize Starch (Binder)	5	42.5
Croscarmellose (Disintegrant)	3	25.5
Nipagin/Methyl Paraben (Preservative)	0.03	0.255
Nipasol/Propyl Paraben (Preservative)	0.07	0.595
Microcrystalline Cellulose (Fillter)	20	170
Microcrystalline Cellulose 102 (Fillter)	11	94.15
Talc (antisticking)	0.5	4.25
Silicon Dioxide Colloidal (antiadherent)	1	8.5
Magnesium Stearate (Lubricant)	0.5	4.25

Test Data on a New Product

As part of the research and development effort during the development of a new drug product, it might be worth collecting data on a number of samples to confirm the conclusion of your risk assessment. There are valuable samples available from studies supporting regulatory submissions that are used for analytical testing, and therefore the addition of ICP screening to the normal protocols will be simple. The three business processes to consider for ICP screening are

1. Extractable and leachables testing, where the packaging components have not been seen before and the dose form is not an oral solid dose. See the "Container Closure System" section above.
2. Stability studies. Compare $T = 0$ and final time point. This data will validate the complete package for the conditions applied on storage with respect to temperature and humidity. This is also not necessary where solid dose formulations are in use.
3. Validation and qualification batches of product (or drug substance if required). These batches are at scale and use the suppliers and process equipment that commercial supplies will come from and will confirm the risk particularly for Class 2A from the manufacturing equipment.

Screening Method

For screening to support the risk assessment discussion, a fully validated method is not necessarily required, and in fact, a semiquantitative scan, or screening methodology or technique, could be applied as long as some indication of recovery and sensitivity is demonstrated, for example, by spiking a sample at the control threshold with the target elemental impurities.

Where an elemental impurity is expected at a significant level, then the data should be collected using a validated procedure, as this data will be used to establish, with the regulatory authority, an appropriate control strategy.

Documentation

At the time of writing, a number of scenarios exist for the implementation of elemental impurities risk assessments where the drug products are policed by the FDA:

- For new drug products submitted under an NDA or abbreviated new drug application (ANDA), manufacturers should include a summary of the risk assessment in any new application for which ICH Q3D or USP <232> or <233> applies. The P.2 section (Pharmaceutical Development) is suggested as a suitable location for this summary.[1]
- For existing products (approved prior to implementation of ICH Q3D or USP <232> or <233>), if any changes are made to reflect the requirements of elemental impurities testing, then a risk assessment summary should be included in any supplemental application or annual report.

- Where products are not approved under an NDA or ANDA, the risk assessment should be available at the manufacturing site for review by the FDA.

The training modules for ICH Q3D (available from ICH's website) contain four excellent example risk assessments featuring oral and parenteral drug products, and show clearly how to use the summary table format presented in Appendix 4 of the ICH Q3D Step 4 document.[7]

It has been made clear by representatives of many regulatory bodies that they will need a clear presentation of your risk assessment decision making—simply submitting data tables is not sufficient. The expression used is "tell the story," and do not assume that the inspector will be able to re-create your scientific decisions when looking at the tables. This is exemplified in the ICH training modules cited above.

REFERENCES

1. Center for Drug Evaluation and Research (CDER). Elemental impurities in drug products: Guidance for industry. Silver Spring, MD: CDER, 2016.
2. D. R. Jenke et al. Materials in manufacturing and packaging systems as sources of elemental impurities in packaged drug products: A literature review. *PDA Journal of Pharmaceutical Science and Technology*, 69, 1–48, 2015.
3. D. R. Jenke et al. A compilation of metals and trace elements extracted from materials relevant to pharmaceutical applications such as packaging systems and devices. *PDA Journal of Pharmaceutical Science and Technology*, 67 354–375, 2013.
4. G. Li, D. Schoneker, K. L. Ulman, J. J. Sturm, L. M. Thackery, J. F. Kauffman. Elemental impurities in pharmaceutical excipients. *Journal of Pharmaceutical Sciences*, 104(12), 4197–4206, 2015.
5. Food and Drug Administration (FDA). FDA guidance for industry: CMC postapproval manufacturing changes to be documented in annual reports. Silver Spring, MD: FDA, 2014.
6. Fritch et al. Production of Acetaminophen. U.S. Patent 5155273, filed 1990.
7. ICH Q3D training package modules 0–9. http://www.ich.org/products/guidelines/quality/article/quality-guidelines.html#3-5.

26 Regulatory Inspection Readiness

Providing data for regulatory submissions and product release documentation is only part of the challenge facing the modern pharmaceutical analytical laboratory. External inspections of facilities and procedures means that a lot of infrastructure is required to support the actual analytical operation covering people, equipment, facilities, and data management. In this chapter, prepared by Jon Sims, we look at the specifics involved in preparing your company for these challenges.

QUALITY SYSTEM

Activities performed in analytical chemistry laboratories within a pharmaceutical manufacturing facility are controlled by a range of policies and standard operating procedures (SOPs) covering safety and quality. Key areas of quality management that an inspector would expect to be covered by either corporate-level SOPs or local departmental SOPs are

- Change control
- Validation master plan (VMP)
- Corrective action and preventative action (CAPA)
- Out-of-specification investigations
- Failure investigations
- Audits
- Training records
- Data archiving
- Equipment calibration and maintenance
- Administration of software and user accounts
- Data backup and restoration

Therefore, it may be necessary to complete new SOPs specifically for the implementation of the inductively coupled plasma (ICP) instrumentation.

PEOPLE

It is necessary to be able to demonstrate that staff performing current good manufacturing practices (cGMP) analyses have the correct qualifications and up-to-date training in all aspects of the role they are undertaking in the laboratory. The key documents that are required are training records and role descriptions.

FACILITIES

The suitability of laboratories for ICP analysis relies on two things:

1. The supplies specified by the vendor for power, gases, ventilation, and access
2. The background elements leaching from the building materials. The ICP instrumentation is sensitive and excellent for measuring volatile elements in floor polishes, ceiling tiles, and other unexpected sources.

EQUIPMENT

It is not possible to introduce new analytical equipment into a laboratory for cGMP analyses without careful preparation and system validation. Regulatory inspections may focus on the equipment life cycle management process in place in the manufacturing company, which should be applied to all cGMP equipment. The best practice in equipment life cycle management is influenced by good automated manufacturing practice (GAMP), which is both a technical subcommittee of the International Society for Pharmaceutical Engineering (ISPE) and a set of guidelines for manufacturers and users of automated systems in the pharmaceutical industry. GAMP is aimed at creating compliance for bespoke automated systems but can be adapted to cover commercial off-the-shelf (COTS) systems, which includes all commercially available ICP instruments suitable for International Conference on Harmonisation of Technical Requirements for Registration of Pharmaceuticals for Human Use (ICH) Q3D and United States Pharmacopeia (USP) <232> testing. A definition of COTS equipment[1] is

- Exists *a priori*—not to be developed to purchaser specification
- Nontrivial install base (more than a few copies)
- Buyer has no access to source code
- Vendor controls development

Typically, for an ICP used for ICH or USP testing, the documentation available for inspection should include the following.

USER REQUIREMENTS SPECIFICATION

Before purchase, the desired features and performance of the equipment should have been documented, for example, PC control, Windows 7 or above, networked, capable of detecting all elements in USP <232>, sensitive to 1 ng/L, and 21 Code of Federal Regulations (CFR) Part 11 compliant.

FUNCTIONAL REQUIREMENTS SPECIFICATION

From the user requirements specification (URS), a functional specification is drawn up in which a detailed breakdown of the URS requirements is documented. For example, a desire to be 21 CFR compliant in the URS would be expanded to have entries in the functional requirements specification (FRS) for each of the key parts of 21 CFR 11 that relate to the instrument, such as 11.10(a)–(h) but not 11.10(i)–(k), which relate to training and laboratory compliance procedures.

DESIGN SPECIFICATION

The design specification (DS) would be prepared from the URS and FRS in cases where a bespoke instrument is being developed for a customer by an instrument manufacturer and is unlikely to be the case for GMP laboratories using ICP instruments. However, relevant DS information is readily available in the instrument manufacturer's literature as sales brochures, and preinstall documents will list PC specifications, mass and wavelength ranges, sensitivity, and other key performance criteria.

SUPPLIER ASSESSMENT

Once the URS is established, it is possible to screen suppliers before evaluating equipment that meet the URS. It is likely that many more features will be available than are in the URS, and that is not an issue since meeting the URS should mean your instrument will be appropriate for compliant testing of the manufactured products.

Once an equipment supplier can meet the URS, some consideration should be given to confirming that the supplier produces equipment with appropriate quality procedures (change controls, software life cycle, etc.) and has the capability to provide support to the installed instrument in a reasonable time frame. For example, an instrument manufacturer based in the United States may sell systems in a remote territory but have no local support infrastructure such that equipment repairs or preventative maintenance (PM) visits cause significant downtime. There may be no local alternative, but the decision making should be documented.

VALIDATION PLAN

A validation plan should be available that provides a review and linkage of all the activities, both planned and executed. Starting from the URS and a preinstall approval of the installation qualification/operation qualification (IQ/OQ) and performance qualification (PQ) documentation provided by the instrument supplier and referring to the quality system at the user location (SOPs for master validation plan, electronic records, calibration, maintenance, change control, etc.), traceability between URS and IQ/OQ test scripts should be established. Any functionality required by the URS that is not covered by an appropriate test script (where that URS requirement could be tested) will need a bespoke test script preparing and approving before qualification starts. An example of the structure of a traceability matrix for a COTS instrument is shown in Table 26.1.

It should be emphasized that responsibility for validation is on the user and not the vendor of the equipment.

TABLE 26.1
Traceability Matrix for COTS Instrument

User Requirement	Description	GxP (Y/N)	Functional Requirement	Design Specification	Test Script Reference
URS 1.0	Software runs on networked Windows 7 PC	N	FRS 2.1.1 and 2.1.2	Manufacturer specification April 2010	Manufacturer IQ part number 1234 section 2 prerequisites
URS 1.1	Software is 21 CFR 11 compliant	Y	FRS 1.1.1–1.1.25 inclusive	Manufacturer installation guide and OQ part number 9234 section 1	Manufacturer OQ section 4 tests 1–5 inclusive plus companytest_1
URS 2.0	Capable of detecting all elements in USP <232>	Y			

Note: GxP in a pharmaceutical business refers to either good manufacturing practice (GMP) or good laboratory practice (GLP).

IQ/OQ AND PQ

The IQ/OQ documentation should mainly be provided by the instrument vendor, but as stated in the "Validation Plan" section, care should be taken to ensure that the IQ/OQ fully tests the functionality relied on to demonstrate compliance with ICH and USP. Bespoke tests may be written by the user and/or the vendor, and undertaken by the end user, providing they have the correct training to operate the equipment. PQ testing that will be employed after maintenance visits should be designed to confirm that the equipment is suitable for its intended use, and therefore could be either some tests that are also in the OQ package or specific tests based on the routine methods used on the instrument. The relationship between the URS, FRS, DS, IQ, OQ, and PQ is shown schematically in Figure 26.1.

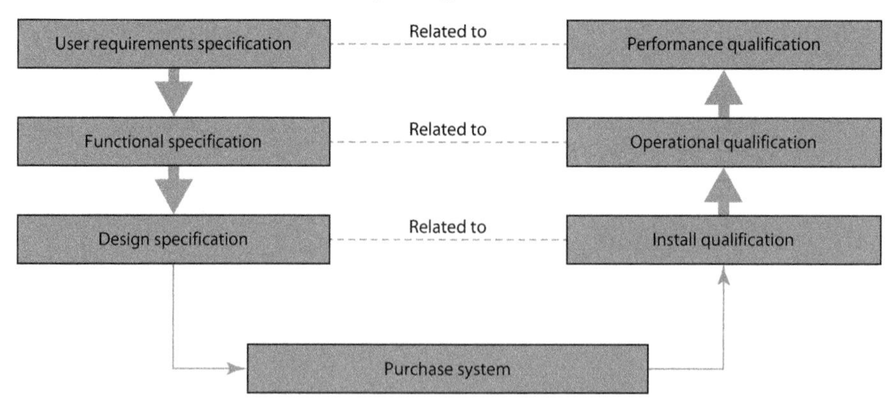

FIGURE 26.1 GAMP-derived workflow and relationships.

SYSTEM ADMINISTRATION

The system administrator is responsible for maintaining and setting up user accounts to ensure compliance with 21 CFR Part 11 and data integrity requirements by controlling access to the system and any external data. Typically, the administrator should be independent of the laboratory and have no operating rights on the instrumentation—it is also not good practice to have multiple accounts for an individual, for example, separate admin and user accounts. The administrator will also be responsible for system restoration following failures and reporting any incidents relating to the computer system.

PREVENTATIVE MAINTENANCE AND CALIBRATION

Ongoing PM should be scheduled and adhered to. Calibration and PQ testing should demonstrate system reliability in line with URS and confirm the validated state. Appropriate system suitability and calibration test limits should be in place.

TRAINING

User training must be documented and regularly reviewed in line with the corporate training SOP to ensure that data inspection during an audit can be traced to an individual with appropriate training on the instrument and software version in use. In particular, training should be clearly recorded after system changes, such as software updates or the addition of new accessories, and this may be simple in-house instruction or major vendor training, as appropriate.

CHANGE CONTROL MANAGEMENT

Change control management (CCM) is important in maintaining validated status; in particular, service engineers should be clear in what they can alter without permission, particularly software and firmware upgrades that may require requalification. All significant changes to the system should be considered with respect to the need for change control documentation before they are carried out and relevant system retesting planned for execution postchange.

PERIODIC REVIEW

Instrument validation status needs periodic review to ensure that GMP is being maintained. This includes confirming that all validation data is complete, correctly stored, and available for inspection. In addition, all change controls must have been appropriately completed and documented, and all changes to the instrument managed appropriately within the change control process. The interval between reviews should be specified in the VMP.

DATA MANAGEMENT

21 CFR Part 11

The most significant development of recent years in analytical chemistry is the move to electronic raw data and the associated risk of companies or individuals falsifying results by editing the data files produced by their laboratory instruments or data storage systems. The Food and Drug Administration (FDA) introduced the 21 CFR Part 11 regulations, and in a similar style, the European Medicines Agency (EMA) produced Annex 11 to cover the expectations of how manufacturers control their electronic records with both computer system protocols and procedural systems. Guidance on 21 CFR Part 11 is available on the FDA website.[2]

Most instrument vendors supporting pharmaceutical customers provide 21 CFR Part 11 tools within their software to address these needs, but it must be understood that vendor software will only provide data security if the PC operating system and user accounts are correctly configured as defined by the vendor; otherwise, the security is undermined by access from Windows programs such as Explorer.

Data Integrity

What is data integrity? The current FDA guidance[3] states:

> Data integrity refers to the completeness, consistency, and accuracy of data. Complete, consistent, and accurate data should be attributable, legible, contemporaneously recorded, original or a true copy, and accurate.

The focus on data integrity has been driven by a high number of incidents observed by FDA auditors where GMP data sets have had missing or altered files, incomplete audit trails, and lack of clarity on who carried out operations.

In particular, as noted in the section above on 21 CFR Part 11, incomplete control of permissions at a Windows level may allow the editing or renaming of data files such that failing measurements can be suppressed in place of passing results, or that one set of data can be copied and used to release multiple batches of material.

During the qualification process, sufficient testing should be carried out to confirm that data cannot be falsified or adulterated without evidence being captured through audit trails or other secure tracking mechanisms.

Data Backup and Recovery

Electronic records require secure storage and, when archived, must be complete, including raw data, methods, calibrations, and audit trails, such that the results can be recreated from that backup in the event that the original is lost. It is therefore essential that any backup or archive process is fully tested for reproducibility of data following restoration. Test scripts should be included in the IQ/OQ process for this purpose.

Computer System Validation

Computer system validation (CSV) establishes documented evidence providing confidence that a computer-controlled analytical instrument will consistently produce a result meeting the specifications of the system. This means that retesting a sample will produce the same value, within analytical variability of the technique, each time assuming the instrument is set up correctly. Similarly, reprocessing a data file or recalling data from backup will give the same numerical result each time, assuming the same processing method is used. This process covers hardware, vendor software, and PC + operating system. CSV will be met through completing the processes already covered in this chapter and will be captured in the validation documentation.

Decommissioning

Formal documentation on decommissioning equipment covers the following two key areas:

1. Safety - is the equipment a risk to those handling it after removal from the lab?
2. Regulatory -
   a. Identify the location of any cGMP raw data being retained by the organization.
   b. Procedure for accessing and maintaining retained cGMP data post decommissioning
   c. Confirming the date to which GMP compliance can be demonstrated.

Standard practice in the Pharmaceutical industry is to carry out a full PQ test set immediately prior to decommissioning. Where an instrument is decommissioned after a catastrophic failure the last date on which data can be claimed to be GMP compliant will typically be the last passing PQ test, although it may be possible to sufficiently demonstrate compliance from data acquired between PQ and failure due to the experimental design used in testing.

REFERENCES

1. V. Basili and B. Boehm. COTS based systems top 10 list. *Computer*, 34(5), 91–93, 2001.
2. Center for Drug Evaluation and Research. Electronic records; electronic signatures—Scope and application. Guidance for Industry Part 11. August 2003.
3. Food and Drug Administration. Data integrity and compliance with CGMP guidance for industry. Draft guidance. April 2016.

27 How to Select an ICP Mass Spectrometer *The Most Important Analytical Considerations*

When sample preparation and dilution factors are taken into consideration, ICP-MS is probably going to be the most suitable technique for carrying out the determination of elemental impurities in the majority of drug compounds and raw materials, because of its extremely low detection capability. Understanding the basic principles of ICP mass spectrometry is important but not absolutely essential to operate and use an instrument on a routine basis. However, understanding how these basic principles affect the performance of an instrument is a real benefit when evaluating the analytical capabilities of the technique. There are no bad commercial instruments on the market, and all are capable of generating high-quality data. However, they all have their own strengths and weaknesses. For that reason, the better informed you are going into an instrument evaluation, the better chance you have of selecting the right one for your organization. Having been involved in demonstrating ICP-MS equipment and running customer samples for more than 25 years, I know the mistakes that people make when they get into the evaluation process. So, in this chapter, I present a set of evaluation guidelines that hopefully will help you make the right decision.

OK, you have convinced your boss that ICP-MS is the ideal technique to meet the analytical demands of your pharmaceutical products and raw materials. Hopefully, the chapters on the fundamental principles have given you the basic knowledge and a good platform on which to go out and evaluate the marketplace. However, they do not really give you an insight into how to compare instrument designs, hardware components, and software features, which are of critical importance when you have to make a decision regarding which instrument to purchase. There are a number of high-quality commercial systems available in the marketplace, which look very similar and have very similar specifications, but how do you know which is the best one that fits your needs? This chapter, supported by the other chapters in the book, presents a set of evaluation guidelines to help you decide the most important analytical figures of merit for your application. You might not need to run all the tests, but experience has told me over the years that each one will give you valuable information about the performance of the instrument, depending on your evaluation objectives.

EVALUATION OBJECTIVES

It is very important before you begin the selection process to decide what your analytical objectives are. This is particularly important if you are part of an evaluation committee. It is fine to have more than one objective, but it is essential that all the members of the group begin the evaluation process with the objectives clearly defined. For example, is detection limit (DL) performance an important objective for your application, or is it more important to have an instrument that is easy to use? If the instrument is being used on a routine basis, maybe good reliability is also very critical. On the other hand, if the instrument is being used to generate revenue, perhaps sample throughput and cost of analysis are of greater importance. Every laboratory's scenario is unique, so it is important to prioritize before you begin the evaluation process. So, as well as looking at instrument features and components, the comparison should also be made with your analytical objectives in mind. Let us take a look at the most common ones that are used in the selection process. They typically include the following:

- Analytical performance
- Usability aspects
- Reliability issues
- Financial considerations

Let us examine these in greater detail.

ANALYTICAL PERFORMANCE

Analytical performance can mean different things to different people. The major reason that the trace element community was attracted to ICP-MS more than 30 years ago was its extremely low multielement DLs. Other multielement techniques, such as ICP-OES, offered very high throughput but just could not get down to ultratrace levels of ICP-MS. Even though electrothermal atomization (ETA) offered much better detection capability than ICP-OES, it did not offer the sample throughput capability that many applications demanded. In addition, ETA was predominantly a single-element technique and so was impractical for carrying out rapid multielement analysis. These limitations quickly led to the commercialization and acceptance of ICP-MS as a tool for rapid ultratrace element analysis. However, there are certain areas where ICP-MS is known to have weaknesses. For example, dissolved solids for most sample matrices must be kept below 0.2%; otherwise, this can lead to serious drift problems and poor precision.

Polyatomic and isobaric interferences, even in simple acid matrices, can produce unexpected spectral overlaps, which will have a negative impact on your data. High-resolution instrumentation and collision/reaction cell and interface technology are helping to alleviate these spectral problems, but they also have their limitations. Depending on the types of samples being analyzed, matrix components can dramatically suppress analyte sensitivity and affect accuracy. These potential problems can all be reduced to a certain extent, but different instruments approach and compensate for these problem areas in different ways. With a novice, it is often a basic lack of understanding of how a particular instrument works that makes the selection process more complicated than it really should be. So, any information that can help you prepare for the evaluation will put you in a much stronger position.

It should be emphasized that these evaluation guidelines are based on my personal experience and should be used in conjunction with other material in the open literature that has presented broad guidelines to compare figures of merit for commercial instrumentation.[1–3] In addition, you should talk with colleagues in the pharmaceutical or dietary supplements industries, who might have carried out an evaluation or are using a particular instrument for the analysis. For example, if they have gone through a lengthy evaluation process, they can give you valuable pointers or even suggest an instrument that is better suited to your needs. Finally, before we begin, it is strongly suggested that you narrow the actual evaluation to two, or maybe three, commercial products. By carrying out some preevaluation research, you will have a better understanding of what ICP-MS technology or instrument to focus on. For example, if funds are limited and you are purchasing ICP-MS for the very first time to carry out high-throughput environmental testing, it is probably more cost-effective to focus on single-quadrupole technology. On the other hand, if you are investing in a second system to enhance the capabilities of your quadrupole instrument, it might be worth taking a look at magnetic sector or "triple-quadrupole" collision/reaction cell technology. Or, if fast multielement transient peak analysis is your major reason for investing in ICP-MS, time-of-flight (TOF) or the Mattauch–Herzog simultaneous sector technology should be given serious consideration. I would like to add one final note, although it is not strictly a technical issue. If you are prepared to forego an instrument demonstration or do not need any samples run, you will be in a much stronger position to negotiate price with the instrument vendor. You should keep that in mind before you decide to get involved in a lengthy selection process.

So, let us begin by looking at the most important aspects of instrument performance. Depending on the application, the major performance issues that need to be addressed include the following:

- Detection capability
- Precision and signal stability
- Accuracy
- Dynamic range
- Interference reduction
- Sample throughput
- Transient signal capability

Detection Capability

Detection capability is a term used to assess the overall detection performance of an ICP mass spectrometer. There are a number of different ways of looking at detection capability, including instrument detection limit (IDL), elemental sensitivity, background signal, and background equivalent concentration (BEC). Of these four criteria, the IDL is generally thought to be the most accurate way of assessing instrument detection capability. It is often referred to as signal-to-background noise, and for a 99% confidence level is typically defined as three times the standard deviation (SD) of n replicates (n = ~10) of the sample blank and is calculated in the following manner:

$$\text{IDL} = \frac{3 \times \text{Standard deviation of background signd}}{\text{Analyteintensity} - \text{backgroung signd}} \times \text{analyte concentration}$$

However, there are slight variations of both the definition and calculation of IDLs, so it is important to understand how different manufacturers quote their DLs if a comparison is to be made. They are usually run in single-element mode, using extremely long integration times (5–10 s) to achieve the highest-quality data. So, when comparing DLs of different instruments, it is important to know the measurement protocol used.

A more realistic way of calculating analyte DL performance in your sample matrices is to use the method detection limit (MDL). The MDL is broadly defined as the minimum concentration of analyte that can be determined from zero with 99% confidence. MDLs are calculated in a manner similar to that of IDLs, except that the test solution is taken through the entire sample preparation procedure before the analyte concentration is measured multiple times. This difference between MDL and IDL is exemplified in Environmental Protection Agency (EPA) Method 200.8, where a sample solution at two to five times the estimated IDL is taken through all the preparation steps and analyzed. The MDL is then calculated in the following manner:

$$MDL = t \times S$$

where t = Student's t value for a 95% confidence level and specifies an SD estimate with $n - 1$ degrees of freedom (t = 3.14 for seven replicates), and S = the SD of the replicate analyses.

Both IDL and MDL are very useful to understand the capability of ICP-MS. However, whatever method is used to compare DLs of different manufacturers' instrumentation, it is essential to carry out the test using realistic measurement times that reflect your analytical situation. For example, if you are determining a group of elements across the mass range in a digested rock sample, it is important to know how much the sample matrix suppresses the analyte sensitivity, because the DL of each analyte will be impacted by the amount of suppression across the mass range. On the other hand, if you are carrying out high-throughput multielement analysis of drinking or wastewater samples, you probably need to be using relatively short integration times (1–2 s per analyte) to achieve the desired sample throughput. Or if you are dealing with a laser ablation or flow injection transient peak that lasts 10–20 s, it is absolutely critical that you understand the impact that the time has on DLs compared with a continuous signal generated with a conventional nebulizer. (In fact, analysis time and DLs are very closely related to each other and will be discussed later on in this chapter.) In other words, when comparing IDLs, it is absolutely critical that the tests represent your real-world analytical situation.

Elemental sensitivity is also a useful assessment of instrument performance, but it should be viewed with caution. It is usually a measurement of background-corrected intensity at a defined mass and is typically specified as counts per second (cps) per concentration (ppb or ppm) of a midmass element, such as $^{103}Rh^+$ or $^{115}In^+$. However, unlike DL, raw intensity usually does not tell you anything about the intensity of the background or the level of the background noise. It should be emphasized that instrument sensitivity can be enhanced by optimization of operating parameters, such as radio-frequency (RF) power, nebulizer gas flows, torch-sampling position, interface pressure, and sampler or skimmer cone geometry, but usually comes at the expense of other performance criteria, including oxide levels, matrix tolerance, or background intensity. So, be very cautious when you see an extremely high-sensitivity specification, because there is a strong probability that the oxide or background specs might also be high. For this reason, it is unlikely there will be an improvement in DL unless the increase in sensitivity comes with no compromise in the background level. It is also important to understand the difference between background and background noise when comparing specifications (the background noise is a measure of the stability of the background and is defined as the square root of the background signal). Most modern quadrupole instruments today specify 150 million–200 million cps/ppm rhodium ($^{103}Rh^+$) or indium ($^{115}In^+$) and <1–2 cps of background (usually at 220 amu), whereas magnetic sector instrument sensitivity specifications are typically 10–20 times the higher background, and 10 times the lower background.

Another figure of merit that is being used more routinely nowadays is BEC. BEC is defined as the intensity of the background at the analyte mass, expressed as an apparent concentration, and is typically calculated in the following manner:

$$\text{BEC} = \frac{3 \times \text{Standard deviation of background signd}}{\text{Analyteintensity} - \text{backgroung signd}} \times \text{analyte concentration}$$

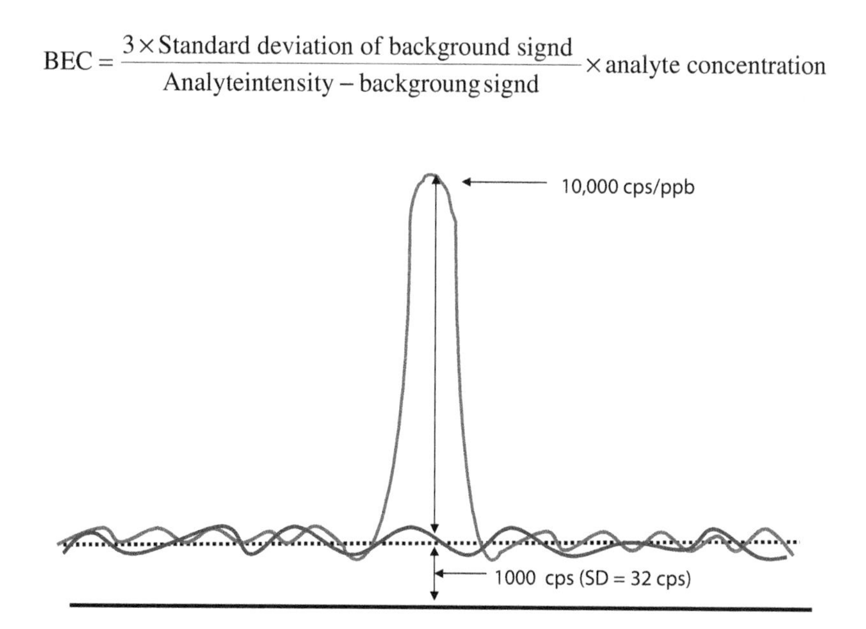

FIGURE 27.1 DL is calculated using the noise of the background, whereas BEC is calculated using the intensity of the background.

It is considered more of a realistic assessment of instrument performance in real-world sample matrices (especially if the analyte mass sits on a high background), because it gives an indication of the level of the background—defined as a concentration value. DLs alone can sometimes be misleading because they are influenced by the number of readings taken, integration time, cleanliness of the blank, and at what mass the background is measured—and are rarely achievable in a real-world situation. Figure 27.1 emphasizes the difference between DL and BEC.

In this example, 1 ppb of an analyte produces a signal of 10,000 cps and a background of 1000 cps. Based on the calculations defined earlier, the BEC is equal to 0.11 ppb because it is expressing the background intensity as a concentration value. On the other hand, the DL is 10 times lower because it is using the SD of the background (i.e., the noise) in the calculation. For this reason, BECs are particularly useful when it comes to comparing the detection capabilities of techniques such as cool plasma and collision/reaction cell technology, because they give you a very good indication of how efficient the background reduction process is.

It is also important to remember that peak measurement protocol will also have an impact on detection capability. As mentioned in Chapter 14, there are basically two approaches to measuring an isotopic signal in ICP-MS. There is the multichannel scanning approach, which uses a continuous smooth ramp of 1–20 channels per mass across the peak profile, and the peak-hopping approach, where the mass analyzer power supply is driven to a discrete position on the peak and allowed to settle, and a measurement is taken for a fixed period of time. This is usually at the peak maximum, but it can be as many points as the operator selects. This process is simplistically shown in Figure 27.2.

The scanning approach is best for accumulating spectral and peak shape information when doing mass calibration and resolution scans. It is traditionally used as a classical method development tool to find out what elements are present in the sample and to assess spectral interferences on the masses of interest. However, when the best possible DLs are required, it is clear that the peak-hopping approach is best.

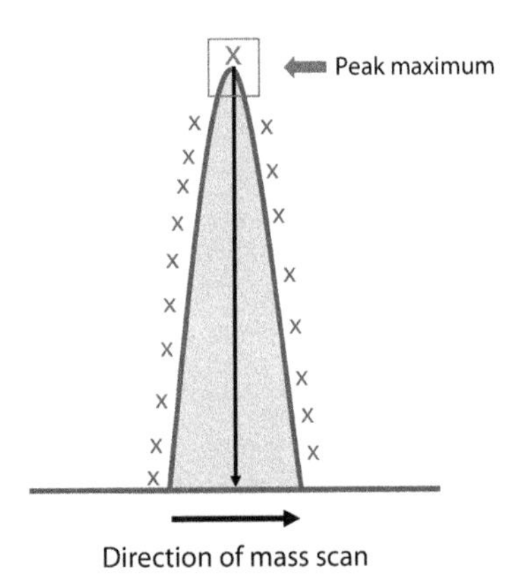

FIGURE 27.2 There are typically two approaches to peak quantitation—peak hopping (usually at peak maximum) and multichannel scanning (across the full width of the peak).

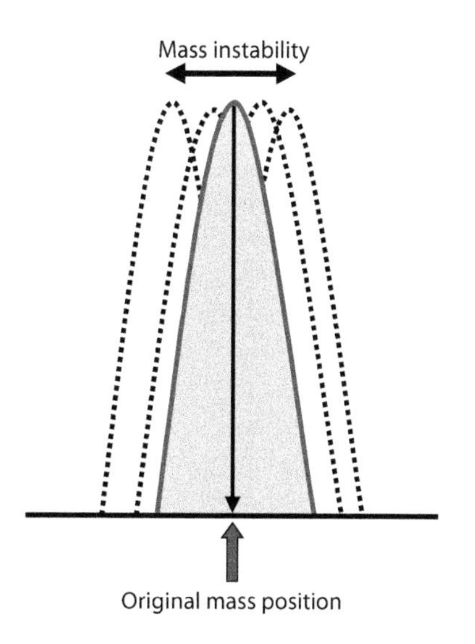

FIGURE 27.3 Good mass stability is critical for single-point, peak-hopping quantitation.

It is important to understand that to get the full benefit of peak hopping, the best DLs are achieved when single-point peak hopping at the peak maximum is chosen. It is well accepted that measuring the signal at the peak maximum will always give the best signal-to-background noise for a given integration time, and there is no benefit in spreading your available integration time over more than one measurement point per mass.[4] Instruments that use more than one point per peak for quantitation are sacrificing measurement time on the sides of the peak, where the signal-to-noise ratio is worse. However, the ability of the mass analyzer to repeatedly scan to the same mass position every time during a multielement run is of paramount importance for peak hopping. If multiple points per peak are recommended, this is a strong indication that the spectrometer has poor mass calibration stability, because it cannot guarantee that it will always find the peak maximum with just one point. Mass calibration specification, which is normally defined as a shift in peak position (in amu) over an 8 h period, is a good indication of mass stability. However, it is not always the best way to compare systems, because peak algorithms using multiple points are often used to calculate the peak position. A more accurate way is to assess the short-term and long-term mass stability by looking at relative peak positions over time. The short-term stability can be determined by aspirating a multielement solution containing four elements (across the mass range) and recording the spectral profiles using multichannel ramp scanning of 20 points per peak. Now repeat the multielement scan 10 times and record the peak position of every individual scan. Calculate the average and relative standard deviation (RSD) of the scan positions. The long-term mass stability can then be determined by repeating the test 8 h later to see how far the peaks have moved. It is important, of course, that the mass calibration procedure not be carried out during this time. Figure 27.3 shows what might happen to the peak position over time, if the analyzer's mass stability is poor.

Precision

Short- and long-term precision specifications are usually a good indication of how stable an instrument is (see Chapter 14). Short-term precision is typically specified as percent RSD of 10 replicates of 1–10 ppb of three elements across the mass range using 2–3 s integration times, whereas long-term precision is a similar test, but normally carried out every 5–10 min over a 4–8 h period. Typical short-term precision, assuming an instrument warm-up time of 30–40 min, should be approximately 1%–3%, whereas long-term precision should be on the order of 3%–5%—both determined without using internal standards. However, it should be emphasized that under these measurement protocols, it is unlikely you will see a big difference in the performance between different instruments in simple aqueous standards. A more accurate reflection of the stability of an instrument is to carry out the tests using a typical matrix that would be run in your laboratory at the concentrations you expect. It is also important that stability should be measured without the use of an internal standard. This will enable you to evaluate the instrument drift characteristics, without any type of signal correction method being applied.

It is recognized that the major source of drift and imprecision in ICP-MS, particularly with real-world samples, is associated with either the sample introduction area, design of the interface, or ion optics system. Some of the most common problems encountered are as follows:

- Pulsations and fluctuations in the peristaltic pump (if one is used), leading to increased signal noise
- Blockage of the nebulizer over time, resulting in signal drift—especially if the nebulizer does not have a tolerance for high dissolved solids
- Poor drainage, producing pressure changes in the spray chamber and resulting in spikes in the signal
- Buildup of solids in the sample injector, producing signal drift
- Changes in the electrical characteristics of the plasma, generating a secondary discharge and increasing ion energies
- Blockage of the sampler and skimmer cone orifice with sample material, causing instability
- Erosion of the sampler and skimmer cone orifice with high-concentration acids
- Coating of the ion optics with matrix components, resulting in slight changes in the electrical characteristics the of ion lens system

These can all be somewhat problematic, depending on the types of samples being analyzed. However, the most common and potentially serious problem with real-world matrices is the deposition of sample material on the interface cones and the ion optics over time. It does not impact short-term precision that much, because careful selection of internal standards matched to the analyte masses can compensate for slight instability problems. However, sample material, particularly matrix components found in environmental and geochemical samples, can have a dramatic effect on long-term stability. The problem is exaggerated even more in a high-throughput laboratory, because poor stability will necessitate more regular recalibration and might even require some samples to be rerun if quality control (QC) standards fall outside certain limits. There is no question that if an instrument has poor drift characteristics, it will take much longer to run an autosampler tray full of samples, and in the long term, this will result in much higher argon consumption.

For these reasons, it is critical that when short- and long-term precision are evaluated, you know all the potential sources of imprecision and drift. It is therefore important that you choose a matrix that is representative of your pharmaceutical samples, whether raw materials, an active pharmaceutical ingredient, or finished drug products, and will genuinely test the instrument out.

Whatever matrices are chosen, it must be emphasized that for the stability test to be meaningful, no internal standards should be used, the sample should contain less than 0.2% total dissolved solids, and the representative elements should be at a reasonably high concentration (1–10 ppb) and be spread across the mass range. In addition, no recalibration should be carried out for the length of the test, which should reflect your real-world situation.[5] For example, if you plan to run your instrument in a high-throughput environment, you might want to carry out an 8 h or even an overnight (12–16 h) stability test. If you are not interested in such long runs, a 2–4 h stability test will probably suffice. However, just remember, plan the test beforehand and make sure you know how to evaluate the vast amount of data that will be generated. It will be hard work, but I guarantee it will be worth it in order to fully understand the short- and long-term drift characteristics of the instruments you are evaluating.

One word of caution. If a syringe or piston pump is being used to deliver the sample, it will almost certainly give you better precision than a peristaltic pump. So make sure you are comparing like with like. It is almost pointless comparing the stability of an instrument that uses a peristaltic pump with that of an instrument that uses a syringe pump. If you are unsure, make sure that the same pumping system is being used on both instruments.

Isotope Ratio Precision

An important aspect of ICP-MS is its ability to carry out fast isotope ratio precision data, particularly if multiple isotopes are being used for quantitation. With this technique, two different isotopes of the same element are continuously measured over a fixed period of time. The ratio of the signal of one isotope to that of the other isotope is taken, and the precision of the ratios is then calculated. Analysts interested in isotope ratios are usually looking for the ultimate in precision. The optimum way to achieve this to get the best counting statistics would be to carry out the measurement simultaneously with a multicollector magnetic sector instrument, a TOF ICP-MS system, or one of the new Mattauch–Herzog simultaneous sector instruments. However, a quadrupole mass spectrometer is a rapid sequential system, so the two isotopes are never measured at exactly the same moment. This means that the measurement protocol must be optimized to get the best precision. As discussed earlier, the best and most efficient use of measurement time is to carry out single-point peak hopping between the two isotopes. In addition, it is also beneficial to be able to vary the total measurement time of each isotope, depending on their relative abundance. The ability to optimize the dwell time and the number of sweeps of the mass analyzer ensures that the maximum amount of time is being spent on the top of each individual peak where the signal-to-noise ratio is at its highest.[6]

It is also critical to optimize the efficiency cycle of the measurement. With every sequential mass analyzer, there is an overhead time called a *settling time* to allow the power supply to settle before taking a measurement. This time is often called non-analytical time, because it does not contribute to the quality of the analytical signal. The only time that contributes to the analytical signal is the *dwell time*, or the time that is actually spent measuring the peak. The measurement efficiency cycle (MEC) is a ratio of the dwell time to the total analytical time (which includes settling time) and is expressed as follows:

$$\text{MEC}(\%) = \frac{\text{No. sweeps} \times \text{dwell time} \times 100}{\left\{\text{No. sweeps} \times (\text{dwell time} + \text{settling time})\right\}}$$

It is therefore obvious that to get the best precision over a fixed period of time, the settling time must be kept to an absolute minimum. The dwell time and the number of sweeps are operator selectable, but the settling time is usually fixed because it is a function of the mass analyzer electronics. For this reason, it is important to know what the settling time of the mass spectrometer is when carrying out peak hopping. Remember, a shorter settling time is more desirable because it will increase the MED and improve the quality of the analytical signal.[7]

In addition, if isotope ratios are being determined on vastly different concentrations of major and minor isotopes using the extended dynamic range of the system, it is important to know the settling time of the detector electronics. This settling time will affect the detector's ability to detect the analog and pulse signals (or in dynamic attenuation mode with a pulse-only Extended Dynamic Range [EDR] system) when switching between measurement of the major and minor isotopes, which could have a serious impact on the accuracy and precision of the isotope ratio. So, for that reason, no matter how the higher concentrations are handled, shorter settling times are more desirable, so that the switching or attenuation can be carried out as quickly as possible.

This is shown in Figure 27.4, which shows a spectral scan of $^{63}Cu^+$ and $^{65}Cu^+$ using an automated pulse or analog EDR detection system. The natural abundance of these two isotopes is 69.17% and 30.83%, respectively. However, the ratio of these isotopes has been artificially altered to be 0.39% for $^{63}Cu^+$ and 99.61% for $^{65}Cu^+$. The intensity of ^{63}Cu is about 70,000 cps, which requires pulse counting, whereas the intensity of the $^{65}Cu^+$ is about 10 million cps, which necessitates analog counting. There is no question that the counting circuitry would miss many of the ions and generate erroneous concentration data if the switching between pulse and analog modes was not fast enough.

So, when evaluating isotopic ratio precision with a scanning device like a quadrupole, it is important that the measurement protocol and peak quantitation procedure are optimized. Isotope precision specifications are a good indication regarding what the instrument is capable of, but once again, these will be defined in aqueous-type standards, using relatively short total measurement times (typically 5 min). For that reason, if the test is to be meaningful, it should be optimized to reflect your

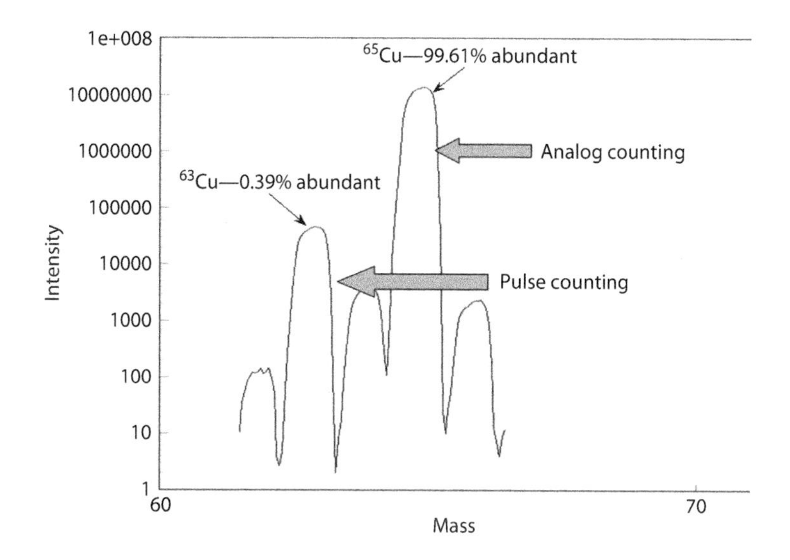

FIGURE 27.4 The detector electronics must be able to switch fast enough to detect isotope ratios that require both pulse and analog counting modes. (Copyright © 2013, all rights reserved, PerkinElmer Inc.)

real-world analytical situation. That is why if very high precision (low RSDs) is a requirement of the analysis, then quadrupole ICP-MS is probably not the optimum technique to use.

Accuracy

Accuracy is a very difficult aspect of instrument performance to evaluate because it often reflects the skill of the person developing the method and analyzing the samples, instead of the capabilities of the instrument itself. If handled correctly, it is a very useful exercise to go through, particularly if you can get hold of reference material (ideally of similar matrices to your own) whose values are well defined. However, when attempting to compare the accuracy of different instruments, it is essential that you prepare every sample yourself, including the calibration standards, blanks, unknown samples, QC standards, or certified reference material (CRM). I suggest that you make up enough of each solution to give to each vendor for analysis. By doing this, you eliminate the uncertainty and errors associated with different people making up different solutions. It then becomes more of an assessment of the capability of the instrument, including its sample introduction system, interface region, ion optics, mass analyzer, detector, and measurement circuitry, to handle the unknown samples, minimize interferences, and get the correct results.

A word of caution should be expressed at this point. Having worked in the field of ICP-MS for more than 30 years, I know that the experience of the person developing the method, running the samples, and performing the demonstration has a direct impact on the quality of the data generated in ICP-MS. There is no question in my mind that the analyst with the most application expertise has a much better chance of getting the right answer than someone who is either inexperienced or is not familiar with a particular type of sample. I think it is quite valid to compare the abilities of the application specialists because this might be the person who is supporting you. However, if you want to assess the capabilities of the instrument alone, it is essential to take the skill of the operator out of the equation. This is not as straightforward as it sounds, but I have found that the best way to "level the playing field" is to send some of your sample matrices to each vendor before the actual demonstration. This allows the application person to spend time developing the method and become familiar with the samples. You can certainly hold back on your CRM or QC standards until you get to the demonstration, but at least it gives each vendor some uninterrupted time with your samples. This also allows you to spend most of the time at the demonstration evaluating the instrument, assessing hardware components, comparing features, and getting a good look at the software. It is my opinion that most instruments on the market should get the right answer—at least for the majority of routine applications. So, even though the accuracies of different instruments should be compared, it is more important to understand how the result was generated, especially when it comes to the analysis of very difficult samples. This is especially true with a triple-quad ICP-MS system fitted with a collision/reaction cell. Method development with these kinds of instruments, whether it's carried out by a skilled operator or by using sophisticated decision-making software, is critical to achieving high-quality data.

Dynamic Range

When ICP-MS was first commercialized, it was primarily used to determine very low analyte concentrations. As a result, detection systems were only asked to measure concentration levels up to approximately five orders of magnitude. However, as the demand for greater flexibility grew, such systems were called on to extend their dynamic range to determine higher and higher concentrations. Today, the majority of commercial systems come standard with detectors that can measure signals up to 10 orders of magnitude.

As mentioned in Chapter 13, there are subtle differences between how various detectors and detection systems achieve this, so it is important to understand how different instruments extend the dynamic range. The majority of quadrupole-based systems on the market extend the dynamic range by using a discrete dynode detector operated in either pulse-only mode or a combination of pulse and analog mode. When evaluating this feature, it is important to know whether this is done in one or two scans because it will have an impact on the types of samples you can analyze. The different approaches have been described earlier, but it is worth briefly going through them again:

- *Two-scan approach*: Basically, two types of two-scan or prescan approaches have been used to extend the dynamic range. In the first one, a survey or prescan is used to determine what masses are at high concentrations and what masses are at trace levels. Then, the second scan actually measures the signals by switching rapidly between analog and pulse counting. In the second two-scan approach, the detector is first run in the analog mode to measure the high signals and then rescanned in pulse-counting mode to measure the trace levels.

- *One-scan approach*: This approach is used to measure both the high levels and trace concentrations simultaneously in one scan. This is typically achieved by measuring the ion flux as an analog signal at some midpoint on the detector. When more than a threshold number of ions are detected, the ions are processed through the analog circuitry. When fewer than a threshold number of ions are detected, the ions cascade through the rest of the detector and are measured as a pulse signal in the conventional way.
- *Using pulse-only mode*: The most recent development in extending the dynamic range is to use the pulse-only signal. This is achieved by monitoring the ion flux at one of the first few dynodes of the detector (before extensive electron multiplication has taken place) and then attenuating the signal by applying a control voltage. Electron pulses passed by the attenuation section are then amplified to yield pulse heights that are typical in normal pulse-counting applications. Under normal circumstances, this approach requires only one scan, but if the samples are complete unknowns, dynamic attenuation might need to be performed, where an additional premeasurement time is built into the settling time to determine the optimum detector attenuation for the selected dwell times used.

The methods that use a prescan or premeasurement time work very well, but they do have certain limitations for some applications, compared with the one-scan approach. Some of these include the following:

- The additional scan or measurement time means it will use more of the sample. Ordinarily, this will not pose a problem, but if sample volume is limited to a few hundred microliters, it might be an issue.
- If concentrations of analytes are vastly different, the measurement circuitry reaction time of a prescan system might struggle to switch quickly enough between high- and low-concentration elements. This is not such a major problem, unless the measurement circuitry has to switch rapidly between consecutive masses in a multielement run, or there are large differences in the concentrations of two isotopes of the same element when carrying out ratio studies. In both these situations, there is a possibility that the detection system will miss counting some of the ions and produce erroneous data.
- The other advantage of the one-scan approach is that more time can be spent measuring the peaks of interest in a transient peak generated by a flow injection or laser ablation system. With a detector that uses two scans or a prescan approach, much of the time will be used just to characterize the sample. It is exaggerated even more with a transient peak, especially if the analyst has no prior knowledge of elemental concentrations in the sample.

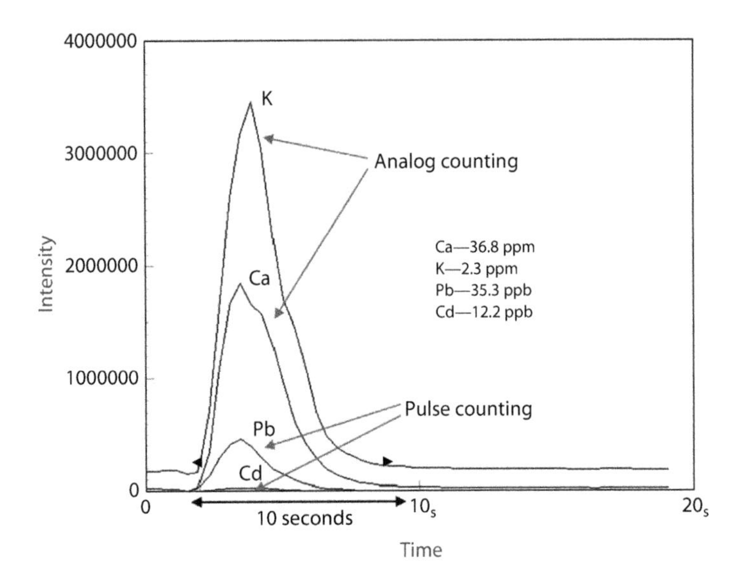

FIGURE 27.5 A one-scan approach to extending the dynamic range is more advantageous for handling a fast transient flow injection peak, in this example, generated by NIST 1643C. (From E. R. Denoyer, Q. H. Lu, *Atomic Spectroscopy,* **14**[6], 162–169, 1993.)

This final point is exemplified in Figure 27.5, which shows the measurement of a flow injection peak of NIST 1643C potable water CRM, using an automated simultaneous pulse and analog EDR system. It can be seen that the K and Ca are at parts per million levels, which requires the use of the analog counting circuitry, whereas the Pb and Cd are at parts per billion levels, which requires pulse counting.[8] This would not be such a difficult analysis for a detector, except that the transient peak has only lasted 10 s. This means that to get the highest-quality data, you want to be spending all the available time quantifying the peak. In other words, you cannot afford the luxury of doing a premeasurement, especially if you have no prior knowledge of the analyte concentrations.

For these reasons, it is important to understand how the detector handles high concentrations to evaluate them correctly. If you are truly interested in using ICP-MS to determine higher concentrations, you should check out the linearity of different masses across the mass range by measuring low-parts-per-trillion (~10 ppt), low-parts-per-billion (~10 ppb), and high-parts-per-million (~100 ppm) levels. Do not be hesitant to analyze a standard reference material (SRM) sample, such as one of the NIST 1643 series of drinking water reference standards, which has both high (ppm) and low (ppb) levels. Finally, if you know there are large concentration differences between the same analytes, make sure the detector is able to determine them with good accuracy and precision. On the other hand, if your instrument is only going to be used to carry out ultratrace analysis, it probably is not worth spending the time to evaluate the capability of the extended dynamic range feature.

However, it should be strongly emphasized that irrespective of which extended range technology is used, if low and high concentrations of the same analyte are expected in a suite of samples, it is unrealistic to think you can accurately quantitate down at the low end and at the top end of the linear range with the same calibration graph. If you want to achieve accurate and precise data at or near the limit of quantitation, you must run a set of appropriate calibration standards to cover your low-level samples. In addition, if you are expecting high and low concentrations in the same suite of samples, you have to be absolutely sure that a high-concentration sample has been thoroughly washed out from the spray chamber or nebulizer system, before a low-level sample is introduced. For this reason, caution must be taken when setting up the method with an autosampler, because if the read delay and integration times are not optimized for a suite of samples, erroneous results can be generated, which might necessitate a rerun under the manual supervision of the instrument operator.

Interference Reduction

As mentioned in Chapter 16, there are two major types of interferences that have to be compensated for: spectral and matrix (space charge and physical). Although most instruments approach the principles of interference reduction in a similar way, the practical aspect of compensating for them will be different, depending on the differences in hardware components and instrument design. Let us look at interference reduction in greater detail and compare the different approaches used.

Reducing Spectral Interferences

The majority of spectral interferences seen in ICP-MS are produced by either the sample matrix, solvent, plasma gas, or various combinations of them. If the interference is caused by the sample, the best approach might be to remove the matrix by some kind of ion exchange column. However, this can be cumbersome and time-consuming to do on a routine basis. If the interference is caused by solvent ions, simply desolvating the sample will have a positive effect on reducing the interference. For that reason, systems that come standard with chilled spray chambers to remove much of the solvent usually generate less sample-based oxide-, hydroxide-, and hydride-induced spectral interferences. There are alternative ways to reduce these types of interferences, but cooling the spray chamber or using a membrane desolvation system can be a very effective way of reducing the intensity of the solvent-based ionic species.

Spectral interferences are an unfortunate reality in ICP-MS, and it is now generally accepted that instead of trying to reduce or minimize them, the best way is to resolve the problem using high-resolution technology, such as a double-focusing magnetic sector mass analyzer.[9] Even though they are not considered ideal for a routine, high-throughput laboratory, they offer the ultimate in resolving power and have found a niche in applications that require ultratrace detection and a high degree of flexibility for the analysis of complex sample matrices. If you use a quadrupole-based instrument and are looking to purchase a second system to enhance the flexibility of your laboratory, it might be worth taking a serious look at magnetic sector technology. The full benefits of this type of mass analyzer for ICP-MS have been described in Chapter 10.

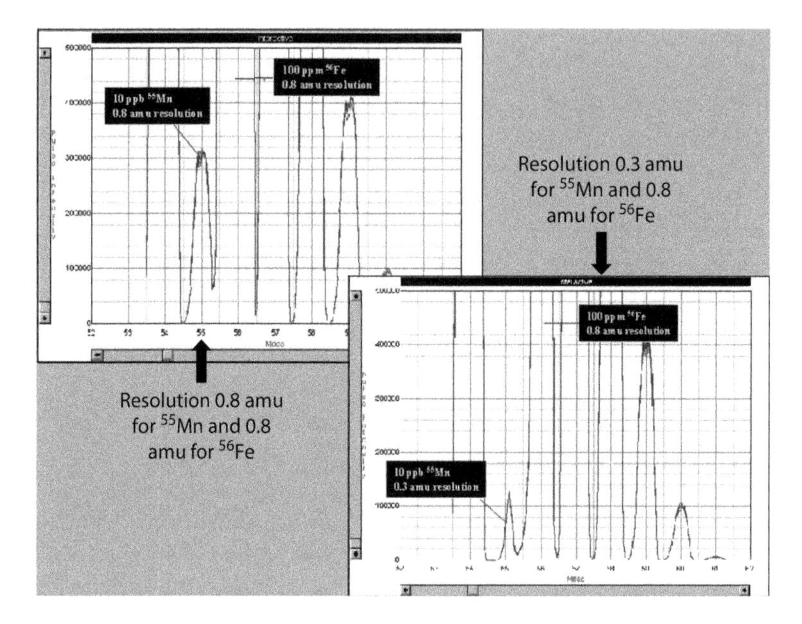

FIGURE 27.6 A resolution setting of 0.3 amu will improve the DL for $^{55}Mn^{+}$ in the presence of high concentrations of $^{56}Fe^{+}$. (Copyright © 2013, all rights reserved, PerkinElmer Inc.)

Let us now turn our attention to the different approaches used to reduce spectral interferences using quadrupole-based technology. Each approach should be evaluated on the basis of its suitability for the demands of your particular application.

Resolution Improvement

As described in Chapter 9, there are two very important performance specifications of a quadrupole—resolution and abundance sensitivity.[10] Although they both define the ability of a quadrupole to separate an analyte peak from a spectral interference, they are measured differently. Resolution reflects the shape of the top of the peak and is normally defined as the width of a peak at 10% of its height. Most instruments on the market have similar resolution specifications of 0.3–3.0 amu and typically use a nominal setting of 0.7–1.0 amu for all masses in a multielement run. For this reason, it is unlikely you will see any measurable difference when you make your comparison.

However, some systems allow you to change resolution settings on the fly on individual masses during a multielement analysis. Under normal analytical scenarios, this is rarely required, but at times it can be advantageous to improve the resolution for an analyte mass, particularly if it is close to a large interference and there is no other mass or isotope available for quantitation. This can be seen in Figure 27.6, which shows a spectral scan of 10 ppb $^{55}Mn^{+}$, which is monoisotopic, and 100 ppm $^{56}Fe^{+}$. The left-hand spectra show the scan using a resolution setting of 0.8 amu for both $^{55}Mn^{+}$ and $^{56}Fe^{+}$, whereas the right-hand spectra show the same scan, but using a resolution setting of 0.3 amu for $^{55}Mn^{+}$ and 0.8 amu for $^{56}Fe^{+}$. Even though the $^{55}Mn^{+}$

peak intensity is about three times lower at 0.3 amu resolution, the background from the tail of the large $^{56}Fe^+$ is about seven times less, which translates into a fivefold improvement in the $^{55}Mn^+$ DL at a resolution of 0.3 amu, compared with 0.8 amu.

Higher Abundance Sensitivity Specifications

The second important specification of a mass analyzer is abundance sensitivity, which is a measure of the width of a peak at its base. It is defined as the signal contribution of the tail of a peak at one mass lower and one mass higher than the analyte peak, and generally speaking, the lower the specification, the better the performance of the mass analyzer. The abundance sensitivity of a quadrupole is determined by a combination of factors, including the shape, diameter, and length of the rods; the frequency of the quadrupole power supply; and the slope of the applied RF and direct current (DC) voltages. Even though there are differences between designs of quadrupoles in commercial ICP-MS systems, there appears to be very little difference in their practical performance.

When comparing abundance sensitivity, it is important to understand what the numbers mean. The trajectory of an ion through the analyzer means that the shape of the peak at one mass lower than the mass M, that is, $(M - 1)$, is slightly different from the other side of the peak at one mass higher, that is, $(M + 1)$. For this reason, the abundance sensitivity specification for all quadrupoles is always worse on the low-mass side (–) than on the high-mass side (+), and is typically 1×10^{-6} at $M - 1$ and 1×10^{-7} at $M + 1$. In other words, an interfering peak of 1 million cps at $M - 1$ would produce a background of 1 cps at M, whereas it would take an interference of 10 million cps at $M + 1$ to produce a background of 1 cps at M. In theory, hyperbolic rods will demonstrate better abundance sensitivity than round ones, as will a quadrupole with longer rods and a power supply with higher frequency. However, you have to evaluate whether this produces any tangible benefits when it comes to the analysis of your real-world samples.

Use of Cool Plasma Technology

All the instruments on the market can be set up to operate under cool or cold plasma to achieve very low DLs for elements such as K, Ca, and Fe. Cool plasma conditions are achieved when the temperature of the plasma is cooled sufficiently low enough to reduce the formation of argon-induced polyatomic species.[11] This is typically achieved with a decrease in the RF power, an increase in the nebulizer gas flow, and sometimes a change in the sampling position of the plasma torch. Under these conditions, the formation of species such as $^{40}Ar^+$, $^{38}ArH^+$, and $^{40}Ar^{16}O^+$ is dramatically reduced, which allows the determination of low levels of $^{40}Ca^+$, $^{39}K^+$, and $^{56}Fe^+$, respectively.[12]

Under normal hot plasma conditions (typically RF power of 1200–1600 W and a nebulizer gas flow of 0.8–1.0 L/min), these isotopes would not be available for quantitation because of the argon-based interferences. Under cool plasma conditions (typically RF power of 600–800 W and a nebulizer gas flow of 1.2–1.6 L/min), the most sensitive isotopes can be used, offering low-parts-per-trillion detection in aqueous matrices. However, not all instruments offer the same level of cool plasma performance, so if these elements are important to you, it is critical to understand what kind of detection capability is achievable. A simple way to test cool plasma performance is to look at the BEC for iron at mass 56 with respect to cobalt at mass 59. This enables the background at mass 56 to be compared with a surrogate element, such as Co, which has a similar ionization potential as Fe, without actually introducing Fe into the system and contributing to the ArO^+ background signal. When carrying out this test, it is important to use the cleanest deionized water to guarantee that there is no Fe in the blank. First, measure the background in counts per second at mass 56 aspirating deionized water. Then, record the analyte intensity of a 1 ppb Co solution at mass 59. The ArO^+ BEC can be calculated as follows:

$$\mathrm{BEC}(\mathrm{ArO}^+) = \frac{\text{Intensity of deionized water background at mass } 56 \times 1 \text{ ppb}}{\text{Intensity of ppb Co at mass } 59 - \text{background at mass } 56}$$

The ArO^+ BEC at mass 56 will be a good indication of the DL for $^{56}Fe^+$ under cool plasma conditions. The BEC value will typically be about an order of magnitude greater than the DL.

Although most instruments offer cool plasma capability, there are subtle differences in the way it is implemented. It is therefore very important to evaluate the ease of setup and how easy it is to switch from cool to normal plasma conditions and back in an automated multielement run. Also, remember that there will be an equilibrium time in switching from normal to cool plasma conditions. Make sure you know what this is, because an equivalent read delay will have to be built into the method, which could be an issue if speed of analysis is important to you. If in doubt, set up a test to determine the equilibrium time by carrying out a short stability run while switching back and forth between normal and cool plasma conditions.

It is also critical to be aware that the electrical characteristics of a cool plasma are different from those of normal plasma. This means that unless there is a good grounding mechanism between the plasma and the RF coil, a secondary discharge can easily occur between the plasma and sampler cone. The result is an increased spread in kinetic energy of the ions entering the mass spectrometer, making them more difficult to control and steer through the ion optics into the mass analyzer. So, understand how this grounding mechanism is implemented and whether any hardware changes need to be made when going from cool to normal plasma conditions and vice versa (testing for a secondary discharge will be discussed later).

It should be noted that one of the disadvantages of the cool plasma approach is that cool plasma contains much less energy than a normal, high-temperature plasma. As a result, elemental sensitivity for the majority of elements is severely affected by the matrix, which basically precludes its use for the analysis of samples with a real matrix, unless the necessary steps are taken. This is shown in Figure 27.7, which shows cool plasma sensitivity for a selected group of elements in varying concentrations of nitric acid, and Figure 27.8, which shows the same group of elements under normal plasma conditions. It can be clearly seen that analyte sensitivity is dramatically reduced in a cool plasma as the acid concentration is increased, whereas under normal plasma conditions, the sensitivity for most of the elements varies only slightly with increasing acid concentration.[13]

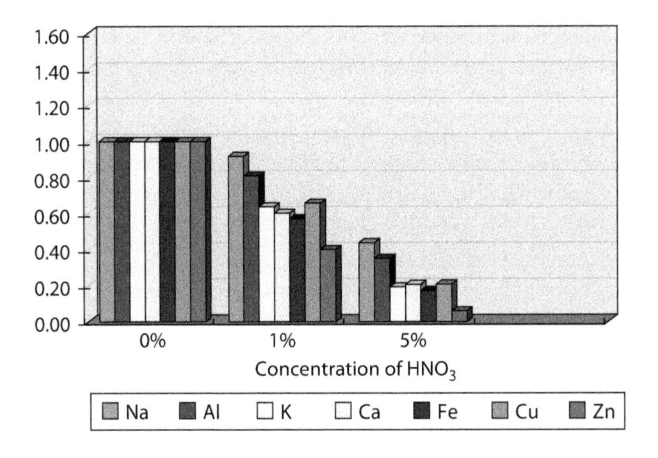

FIGURE 27.7 Sensitivity for a selected group of elements in varying concentrations of nitric acid, using cool plasma conditions (RF power–800 W, nebulizer gas–1.5 L/min). (From J. M. Collard, K. Kawabata, Y. Kishi, R. Thomas, *Micro*, January 2002.)

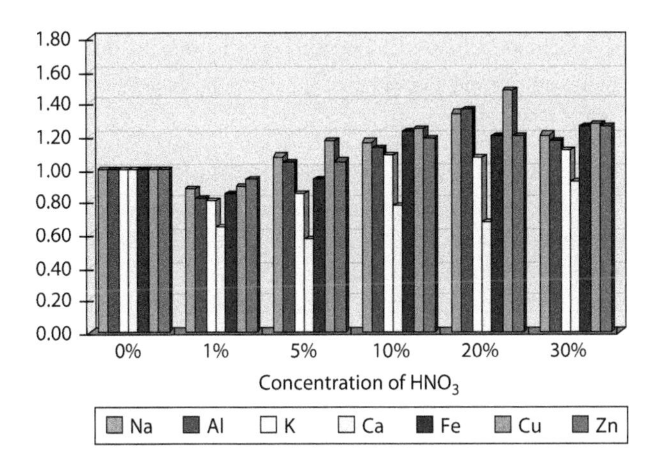

FIGURE 27.8 Sensitivity for a selected group of elements in varying concentrations of nitric acid, using normal plasma conditions (RF power–1600 W, nebulizer gas–1.0 L/min). (From J. M. Collard, K. Kawabata, Y. Kishi, R. Thomas, *Micro*, January 2002.)

In addition, because a cool plasma contains much less energy than a normal plasma, chemical matrices and acids with a high boiling point are often difficult to decompose in the plasma, which has the potential to cause corrosion problems on the interface of the mass spectrometer. This is the inherent weakness of the cool plasma approach—instrument performance is highly dependent on the sample being analyzed. As a result, unless simple aqueous-type samples are being analyzed, cool plasma operation often requires the use of standard additions or matrix matching to achieve satisfactory results. Additionally, to obtain the best performance for a full suite of elements, a multielement analysis often necessitates the use of two sets of operating conditions—one run for the cool plasma elements and another for normal plasma elements—which can be both time- and sample-consuming.

In fact, these application limitations have led some vendors to reject the cool plasma approach in favor of collision/reaction cell technology. So, it could be that the cool plasma capability of an instrument may not be that important if the equivalent elements are superior using the collision/reaction cell option. However, you should proceed with caution in this area, because on the current evidence, not all collision/reaction cell instruments offer the same kind of performance. For some instruments, cool plasma DLs are superior to the same group of elements determined in the collision cell mode. For that reason, an assessment of the suitability of using cool plasma conditions or collision/reaction cell technology for a particular application problem has to be made based on the vendor's recommendations.

For example, the recent development of a novel 34 MHz free-running designed RF generator using solid-state electronics has enhanced the capability of ICP-MS to analyze some real-world samples, particularly when using cool plasma conditions (refer to Chapter 16 on interference reduction). This new design, which is based on an air-cooled plasma load coil, allows the matching network electronics to rapidly respond to changes in the plasma impedance produced by different sampling conditions and sample matrices, while still maintaining low plasma potential at the interface region. This technology appears to offer some real benefits over traditional RF technology for some applications.

Using Collision/Reaction Cell and Interface Technology

Collision and reaction cells and interfaces are an available option on all quadrupole-based instruments today and are used to reduce the formation of harmful polyatomic spectral interferences, such as $^{38}ArH^+$, $^{40}Ar^+$, $^{40}Ar^{12}C^+$, $^{40}Ar^{16}O^+$, and $^{40}Ar_2^+$, to improve detection capability for elements such as K, Ca, Cr, Fe, and Se. However, when comparing systems, it is important to understand how the interference reduction is carried out, what types of collision/reaction gases are used, and how the collision/reaction cell or interface deals with the many complex side reactions that take place—reactions that can potentially generate brand-new interfering species and cause significant problems at other mass regions. The difference between collision cells and reaction cells and interfaces has been described in detail in Chapter 12. Two different approaches are used to reject these undesirable species. It can be done either by kinetic energy discrimination (KED) or by mass discrimination, depending on the type of multipole and the reaction gas used in the cell.

Unfortunately, the higher-order multipoles, such as hexapoles or octopoles, have less defined mass stability boundaries than lower-order multipoles, making them less than ideal to intercept these side reactions by mass discrimination. This means that some other mechanism has to be used to reject these unwanted species. The approach that has been traditionally used is to discriminate between them by kinetic energy. This is a well-accepted technique that is typically achieved by setting the collision cell potential slightly more negative than the mass filter potential. This means that the collision product ions generated in the cell, which have a lower kinetic energy as a result of the collision process, are rejected, whereas the analyte ions, which have a higher kinetic energy, are transmitted. This method works very well but restricts their use to inert gases, such as helium, and less reactive gases, such as hydrogen, because of the limitations of higher-order multipoles in efficiently controlling the multitude of side reactions.

However, the use of highly reactive gases such as ammonia and methane can lead to more side reactions and potentially more interferences unless the by-products from these side reactions are rejected. The way around this problem is to utilize a lower-order multipole, such as a quadrupole, inside the reaction/collision cell and use it as a mass discrimination device. The advantages of using a quadrupole are that the stability boundaries are much better defined than are those for a hexapole or an octapole, so it is relatively straightforward to operate the quadrupole inside the reaction cell as a mass or bandpass filter. Therefore, by careful optimization of the quadrupole electrical fields, unwanted reactions between the gas and the sample matrix or solvent, which could potentially lead to new interferences, are prevented. This means that every time an analyte and interfering ions enter the reaction cell, the bandpass of the quadrupole can be optimized for that specific problem and then changed on the fly for the next one.[14]

When assessing the capabilities of collision and reaction cells and interfaces, it is important to understand the level of interference rejection that is achievable, which will be reflected in the instrument's DL and BEC values for the particular analytes being determined. This has been described in greater detail in Chapter 12, but depending on the nature of interference being reduced, there will be differences between the collision/reaction cell methods, as well as with the collision/reaction interface approach. It is therefore critical to evaluate the capabilities of commercial instrumentation on the basis of your sample matrices and particular analytes of interest.

On the evidence published to date, it seems that the use of highly reactive gases appears to offer a more efficient way of reducing some interfering ion background levels because the optimum reaction gas can be selected to create the most favorable ion–molecule reaction conditions for each analyte. In other words, the choice and flow of the reaction gas can be optimized for each and every application problem.

The benefit of using highly reactive gases to reduce interferences has been confirmed by the recent development of the triple-quadrupole collision/reaction cell instruments, where an additional quadrupole is placed prior to the collision/reaction cell multipole and the analyzer quadrupole. This first quadrupole acts as a simple mass filter to allow only the analyte masses to enter the cell, while rejecting all other masses. With all nonanalyte, plasma, and sample matrix ions excluded from the cell, interference removal is then carried out using highly reactive gases in the collision/reaction cell. The analyte mass, free of the interference, is then passed into the analyzer quadrupole for separation and detection. The use of any kind of ion–molecule reaction chemistry is not as straightforward when it comes to developing methods, especially when new samples are encountered. In addition, if more than one reaction gas needs to be used, they might not be ideally suited for a high-sample-throughput environment because of the lengthy analysis times involved. However, I think it's fair to say that with the advent of intelligent, decision-making software routines, the choice of gases, cell conditions, and the overall method development process using a reaction cell has become relatively straightforward and user-friendly. This is true not only with a single quadrupole, but also using reaction chemistry with a triple-quad system.

On the other hand, the use of inert or low-reactivity gases and KED appears to offer a much simpler approach to reducing polyatomic spectral interferences. Normally, only one gas is used for a particular application problem, which is much better suited to routine analysis. It is possible to use other low gases, such as hydrogen, when helium does not work, but for the majority of elements, one gas is sufficient. For some applications, the collision gas is kept flowing all the time, even for elements that do not need a collision cell. However, its major analytical disadvantage is that its interference reduction capabilities are generally not as good as those of a system that uses highly reactive gases. Because more collisions are required with an inert gas to suppress the interfering ions, the analyte will also undergo more collisions, and as a result, fewer of the analyte ions will make it through the kinetic energy barrier at the exit of the cell. For that reason, DLs for the majority of the elements that benefit from a collision/reaction cell are generally poorer when using inert gases and KED than are those of a system that uses highly reactive gases and selective bandpass (mass) tuning.

However, it should be emphasized that when you are comparing systems, it should be done with your particular analytical problem in mind. In other words, evaluate the interference suppression capabilities of the different collision and reaction cell interface approaches by measuring BEC and DL performance for the suite of elements and sample matrices you are interested in. In other words, make sure it works for your application problem. This is even more important with the newer triple-quadrupole collision/reaction cell approach because it is complicated, and at this present moment in time, there are very few applications in the public domain.

Every laboratory's analytical scenario is different, so it is almost impossible to determine which approach is better for a particular application problem. If you are not pushing DLs but are looking for a simplified approach to running samples on a routine basis, then maybe a collision cell using KED best suits your needs. However, if your samples are spectrally more complex and you are looking for more performance and flexibility because your DL requirements are more challenging, then either the dynamic reaction cell using bandpass tuning or the triple quadrupole is probably the best way to go. Also, be aware that systems that use collisional mechanisms and KED will have to use either higher-purity gases or a gas purifier (getter) because of the potential for impurities in the gas generating unexpected reaction by-product ions, which could potentially interfere with other analyte ions (refer to Chapter 12).[15] Not only does this have the potential to affect the detection capability, but also ultra-high-purity gases are typically two to three times more expensive than industrial- or laboratory-grade gases. But at the end of the day, if you are investing in brand-new technology, it also depends on what kind of funds are available to solve a particular application problem.

Reduction of Matrix-Induced Interferences

As discussed in Chapter 16, there are three major sources of matrix-induced problems in ICP-MS. The first and simplest to overcome is often called a *sample transport* or *viscosity effect* and is a physical suppression of the analyte signal brought about by the matrix components. It is caused by the sample matrices' impact on droplet formation in the nebulizer or droplet size selection in the spray chamber. In some samples, it can also be caused by the variation in sample flow through the peristaltic pump. The second type of signal suppression is caused by the impact of the sample matrix on the ionization temperature of the plasma discharge. This typically occurs when different levels of matrix components or acids are aspirated into a cool or cold plasma. The ionization conditions in the low-temperature plasma are so fragile that higher concentrations of matrix components result in severe suppression of the analyte signal. The third major cause of matrix suppression is the result of poor transmission of ions through the ion optics owing to matrix-induced space charge effects.[16] This has the effect of defocusing the ion beam, which leads to poor sensitivity and DLs, especially when trace levels of low-mass elements are being determined in the presence of large concentrations of high-mass matrix elements. Unless an electrostatic compensation is made in the ion optic region, the high-mass element will dominate the ion beam, resulting in severe matrix suppression on the lighter ones. All these types of matrix interferences are compensated to varying degrees by the use of internal standardization, where the intensity of a spiked element that is not present in the sample is monitored in samples, standards, and blanks.

The single biggest difference in commercial instrumentation to focus the analyte ions into the mass analyzer is in the design of the ion lens system. Although they all basically do the same job of transporting the maximum number of analyte ions through the system, there have been many different ways of implementing this fundamental process, including the use of an extraction lens, multicomponent lens systems, a single-ion cylinder lens, right-angled reflectors, or multipole ion guide systems. First, it is important to know how many lens voltages have to be optimized. If a system has many lens components, it is probably going to be more complex to carry out optimization on a routine basis. In addition, the cleaning and maintenance of a multicomponent lens system might be a little more time-consuming. All these are possible concerns, especially in a routine environment where maybe the skill level of the operator is not so high.

However, the design of the ion-focusing system or the number of lens components used is not as important as its ability to handle real-world matrices.[17] Most lens systems can operate in a simple aqueous sample because there are relatively few matrix ions to suppress the analyte ions. The test of the ion optics comes when samples with a real matrix are encountered. When a large number of matrix ions are present in the system, they can physically "knock" the analyte ions out of the ion beam. This shows itself as a suppression of the analyte ions, which means that less analyte ions are transmitted to the detector in the presence of a matrix. For this reason, it is important to measure the degree of matrix suppression of the instrument being evaluated across the full mass range. The best way to do this is to choose three or four of your typical analyte elements spread across the mass range (e.g., $^{7}Li^{+}$, $^{63}Cu^{+}$, $^{103}Rh^{+}$, and $^{138}Ba^{+}$). Run a calibration of a 20 ppb multielement standard in 1% HNO_3. Then make up a synthetic sample of 20 ppb of the same elements in one of your typical matrices. Measure this sample against the original calibration.

The percentage matrix suppression at each mass can then be calculated as follows:

$$\frac{20\text{ ppb} - \text{Apparent concentration of 20 ppb analytes in your matrix}}{20\text{ ppb}} \times 100$$

There is a strong possibility that your own samples will not really test the matrix suppression performance of the instrument, particularly if they are simple aqueous-type samples. If this is the case and you really would like to understand the matrix capabilities of your instrument, then make up a synthetic sample of your analytes in 500 ppm of a high-mass element, such as thallium, lead, or uranium. For this test to be meaningful, you should tell the manufacturers to set up the ion optic voltages that are best suited for multielement analysis across the full mass range. If the ion optics are designed correctly for minimum matrix interferences, it should not matter whether it incorporates an extraction lens, uses a photon stop, or has an off-axis mass analyzer, or even whether it utilizes a single, multicomponent, or right-angled ion lens system.

It is also important to understand that an additional role of the ion optic system is to stop particulates and neutral species from making it through to the detector, which would increase the noise of the background signal. This will certainly impact the instrument's detection capability in the presence of complex matrices. Therefore, it is definitely worth carrying out a DL test in a difficult matrix, such as rock digests, soil samples, biological specimens, or metallurgical alloys, which tests the ability of the ion optics to transport the maximum number of analyte ions while rejecting the maximum number of matrix ions, neutral species, and particulates.

Another aspect of an instrument's matrix capability is its ability to aspirate lots of different types of samples, using both conventional nebulization and sampling accessories that generate a dry aerosol, such as laser ablation or a desolvation device. When changing sample types similar to this on a regular basis, parameters such as RF power, nebulizer gas flow, and sampling depth usually have to be changed. When this is done, there is an increased chance of altering the electrical characteristics of the plasma and producing a secondary discharge at the interface. All instruments should be able to handle this to some extent, but depending on how they compensate for the increase in plasma potential, parameters might need to be reoptimized because of the change in the spread of kinetic energy of the ions entering the mass spectrometer.[18] This may not be such a serious problem, but once again, it is important for you to be aware of this, especially if the instrument is running many different sample matrices on a routine basis.

Some of the repercussions of a secondary discharge, including increased doubly charged species, erosion of material from the skimmer cone, shorter lifetime of the sampler cone, a significantly different full-mass-range response curve with laser ablation, and the occurrence of two signal maximums when optimizing nebulizer gas flow, have been well reported in the literature.[19–21] On the other hand, systems that do not show signs of this phenomenon have reported an absence of these deleterious effects.[22]

A simple way of testing for the possibility of a secondary discharge is to aspirate one of your typical matrices containing approximately 1 ppb of a small group of elements across the mass range (such as $^{7}Li^{+}$, $^{115}In^{+}$, and $^{208}Pb^{+}$) and continuously monitor the signals while changing the nebulizer gas flow. In the absence of a secondary

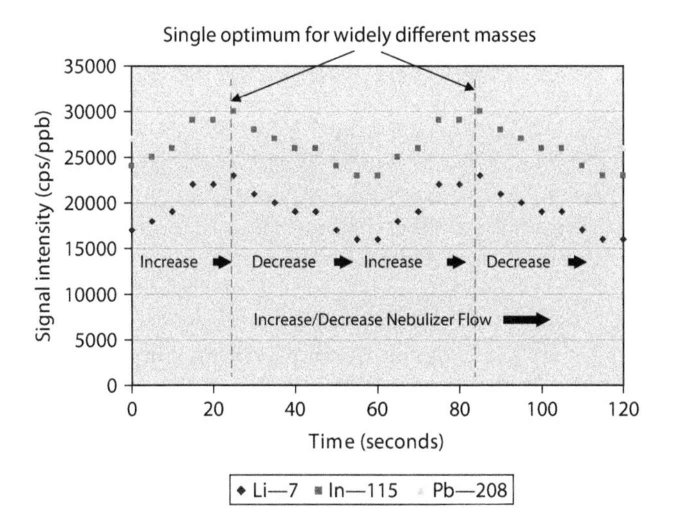

FIGURE 27.9 If the interface is grounded correctly, signals for 1 ppb $^{7}Li^{+}$, $^{115}In^{+}$, and $^{208}Pb^{+}$ should all track each other and have similar optimum values as the nebulizer gas flow is changed.

discharge, all three elements, which have widely different masses and ion energies, should track each other and have the same optimum nebulizer gas flow. This can be seen in Figure 27.9, which shows the signals for $^{7}Li^{+}$, $^{115}In^{+}$, and $^{208}Pb^{+}$ changing as the nebulizer gas flow is first increased and then decreased.

If the signals do not track each other or there is an erratic behavior in the signals, it could indicate that the normal kinetic energy of the ions has been altered by the change in the nebulizer gas flow. There are many reasons for this kind of behavior, but it could point to a possible secondary discharge at the interface, or that the RF coil–grounding mechanism is not working correctly.[23] It should be emphasized that Figure 27.9 is just a graphical representation of what the relative signals might look like and might not exactly be the same for all instruments.

Sample Throughput

In laboratories where high sample throughput is a requirement, the overall cost of analysis is a significant driving force determining what type of instrument is purchased. However, in a high-workload laboratory there sometimes has to be a compromise between the number of samples analyzed and the DL performance required. For example, if the laboratory wants to analyze as many samples as possible, relatively short integration times have to be used for the suite of elements being determined. On the other hand, if DL performance is the driving force, longer integration times need to be used, which will significantly impact the total number of samples that can be analyzed in a given time. This was described in detail in Chapter 12, but it is worth revisiting to understand the full implications of achieving high sample throughput.

It is generally accepted that for a fixed integration time, peak hopping will always give the best DLs. As discussed earlier, measurement time is a combination of time spent on the peak taking measurements (dwell time) and the time taken to settle (settling time) before the measurement is taken. The ratio of the dwell time to the overall measurement time is often called the *measurement efficiency*. The settling time, as we now know, does not contribute to the analytical signal but definitely contributes to the analysis time. This means that every time the quadrupole sweeps to a mass and sits on the mass for the selected dwell time, there is an associated settling time. The greater the number of points that have been selected to quantitate the mass, the longer the total settling time and the worse the overall measurement efficiency.

For example, let us take a scenario where 20 elements need to be determined in duplicate. For argument's sake, let us use an integration time of 1 s per mass, comprising 20 sweeps at 50 ms per sweep. The total integration time that contributes to the analytical signal and the DL is therefore 20 s per replicate. However, every time the analyzer is swept to a mass, the associated scanning and settling times must be added to the dwell time. The greater the number of points that are taken to quantify the peak, the greater the magnitude of the settling time that must be added. For this scenario, let us assume that three points per peak are being used to quantify the peaks. Let us also assume for this case that the quadrupole and detector have a settling time of 5 ms. This means that a 15 ms settling time will be associated with every sweep of each individual mass. So, for 20 sweeps of 20 masses, this is equivalent to 6 s of nonanalytical time every replicate, which translates into 12 s (plus 40 s of actual measurement time) for every duplicate analysis. This is equivalent to a $40/(12 + 40) \times 100\%$, or 77%, MEC. It does not take long to realize that the fewer the number of points taken per peak and the shorter the settling time, the better the measurement cycle. Just by reducing the number of points to one per peak and cutting the detector settling time by half, the nonanalytical time is reduced to 4 s, which is a $40/(2 + 40) \times 100\%$, or 95%, measurement efficiency per duplicate analysis. It is therefore very clear that the measurement protocol has a big impact on the speed of analysis and the number of samples that can be analyzed in a given time. So, if sample throughput is important, you should understand how peak quantitation is carried out on each instrument.

The other aspect of sample throughput is the time taken for the sample to be aspirated through the sample introduction system into the mass spectrometer, reach a steady-state signal, and then be washed out when the analysis is complete. The wash-in and washout characteristics of the instrument will most definitely impact its sample throughput capabilities. Therefore, it is important for you to know what these times are for the system you are evaluating. You should also be aware that if the instrument uses a computer-controlled peristaltic pump to deliver the sample to the nebulizer and spray chamber, it can be speeded up to reduce the wash-in and washout times. So, this should also be taken into account when evaluating the memory characteristics of the sample introduction system.

Therefore, if speed of analysis is important to your evaluation criteria, it is worth carrying out a sample throughput test. Choose a suite of elements that represents your analytical challenge. Assuming you are also interested in achieving good detection capability, let the manufacturer set the measurement protocol (integration time, dwell time, settling time, number of sweeps, points per peak, sample introduction wash-in and washout times, etc.) to get the best DLs. If you are interested in measuring high and low concentrations, also make sure that the extended dynamic range feature is implemented. Then, time how long it takes to achieve DL levels in duplicate from the time the sample probe goes into the sample to the time a result comes out on the screen or printer. If you have time, it might also be worth carrying out this test in an autosampler with a small number of your typical samples. It is important that the DL measurement protocol be used because factors such as integration times and washout times can be compromised to reduce the analysis time. All the measurement time issues discussed in this section and the memory characteristics of the sample introduction system will be fully evaluated with this kind of test. (Note: If high sample throughput is important, there are automated productivity enhancement systems on the market that by efficient delivery and washout of the sample are realizing a two- to threefold improvement in multielement analysis times— refer to Chapter 20 for details.)

Transient Signal Capability

The demands on an instrument to handle transient signals generated by sampling accessories, such as laser ablation, chromatography separation devices, flow injection, or electrothermal vaporization systems, are very different from those of conventional multielement analysis using solution nebulization. Because the duration of a sampling accessory signal is short compared with a continuous signal generated by a pneumatic nebulizer, it is critical to optimize the measurement time to achieve the best multielement signal-to-noise ratio in the sampling time available. This was addressed in greater detail in Chapter 20, but basically, the optimum design to capture the maximum amount of multielement data in a transient peak is to carry out the measurement in a simultaneous manner with a multicollector magnetic sector instrument, a TOF mass spectrometer, or the Mattauch–Herzog simultaneous detection sector instrument.

However, a scanning system such as a quadrupole instrument can achieve good performance on a transient peak if the measurement time is maximized to get the best multielement signal-to-noise ratio. Therefore, instruments that utilize short settling times are more advantageous, because they achieve a higher MEC. In addition, if the extended dynamic range is used to determine higher concentrations, the scanning and settling times of the detector will also have an impact on the quality of the signal. So, detectors that require two scans to characterize an unknown sample will use up valuable time in the quantitation process. For example, if a transient peak generated by a laser ablation device only lasts 10 s, a survey or prescan of 2 s will use up 20% of the available measurement time. This, of course, is a disadvantage when doing multielement analysis on a transient signal, especially if you have limited knowledge of the analyte concentration levels in your samples.

Single-Particle ICP-MS Transient Signals

Of all the applications involving transient peak measurements, single-particle ICP-MS is probably the most demanding, because the transient event only lasts a few milliseconds. Let's take a closer look at this rapidly emerging application area to better understand the measurement requirements. Single-particle ICP-MS XE "single-particle ICP-MS" is a new technique that has recently been developed for detecting and sizing metallic nanoparticles (NPs) at extremely low levels, in order to predict their environmental behavior. While this method is fairly new, it has shown a great deal of promise in several applications, including determining concentrations of NPs in complex matrices, such as wastewater effluents. The method involves introducing a liquid sample containing the NPs at a very dilute concentration into the ICP-MS. After the particles have been nebulized, ionized in the plasma, and separated by the mass analyzer, the resulting ions are detected and collected as time-resolved pulses. The number of pulses generated is directly related to the population of NPs in the sample, while the intensity of the pulses is related to the size (and mass) of the NPs.

In ICP-MS, dilute solutions of dissolved metals will produce relatively constant signals. If there are NPs of various sizes suspended in that solution, they will appear as pulses, which deviates from a steady-state continuous signal generated by the background of the dissolved metals. By using relatively short dwell times of a few milliseconds, the packets of pulses can be quantified as long as the pulses can be adequately resolved in the time domain. The pulse height is then compared with the signal intensity of dissolved metal in a set of calibration standards. This is represented in Figure 27.10, which shows Sample A containing dissolved metal in solution being analyzed and measured in the conventional way, and Sample B containing the same metal in the form of NPs being analyzed and measured as a transient pulse of ions. The integration (dwell time), which is the same in both examples, is shown as the dotted box. By subtracting the background produced by the dissolved analyte in the sample from the NP pulse height, and comparing the net signal intensity against the calibration standard, the concentration and therefore the size (and mass) distribution of the particle can be calculated using well-understood single-particle ICP-MS theory.

However, for this approach to work effectively at low concentrations, the speed of data acquisition and the response time of the ICP-MS quadrupole and detector electronics must be fast enough to capture the time-resolved NP pulses, which typically last only a few milliseconds or less. This is emphasized in Figure 27.11, which shows a real-world example of the time-resolved analysis of 30 nm silver NPs by single-particle ICP-MS. it can be seen that the silver NP pulse has been resolved with approximately 15 data points in < 1 ms. For this application, the ICP-MS should be capable of using dwell times shorter than the particle transient time, thus avoiding false signals generated from clusters of particles. In practice, this means using a dwell time of 10–100 μs so the pulse can be fully characterized.

For this kind of resolution, it is advantageous that the instrument measurement electronics are capable of very fast data acquisition rates, including dwell times that are as short as possible to capture the maximum number of data points within the transient event. It is also desirable that the quadrupole settling time is extremely short, so there is no wasted time waiting for the quadrupole power supply to stabilize. Ideally, it would beneficial if there was no settling time, so the quadrupole could just park itself on the mass of interest and just take measurements. Unfortunately, this is not typically a standard feature of the measurement protocol on most ICP-MS systems, because for multielement analysis, there has to be a built-in settling time

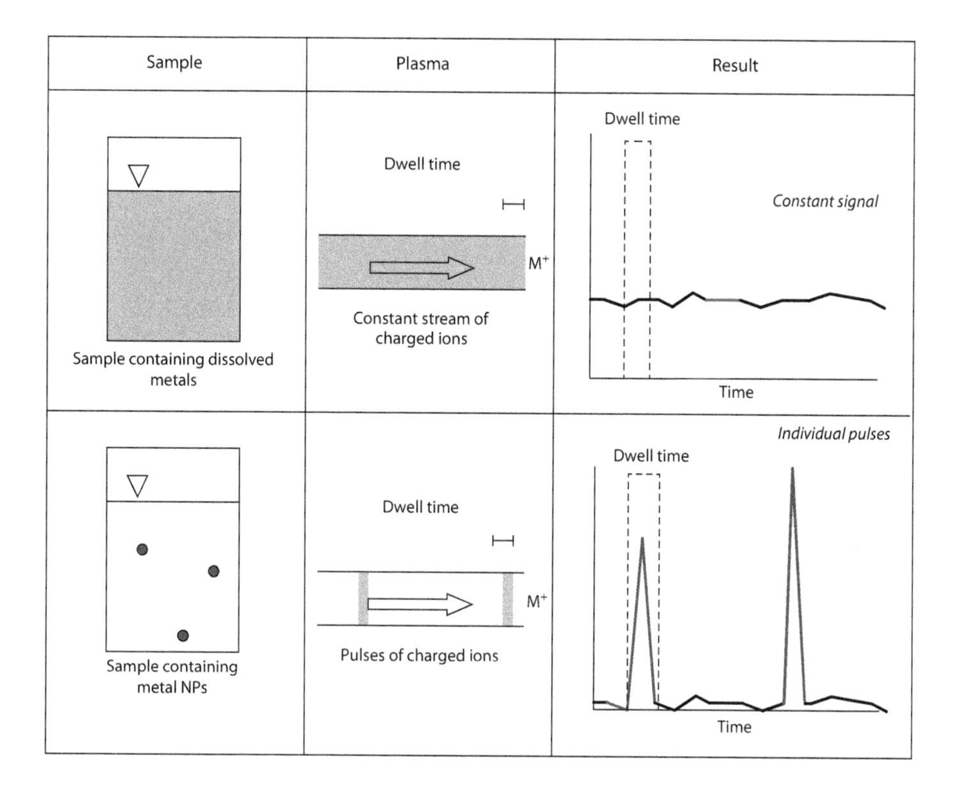

FIGURE 27.10 A comparison of a sample (A) containing dissolved metal in solution being analyzed and measured in the conventional way and another sample (B) containing the same metal in the form of nanoparticles being analyzed and measured as a transient pulses of ions. (Courtesy of the Colorado School of Mines.)

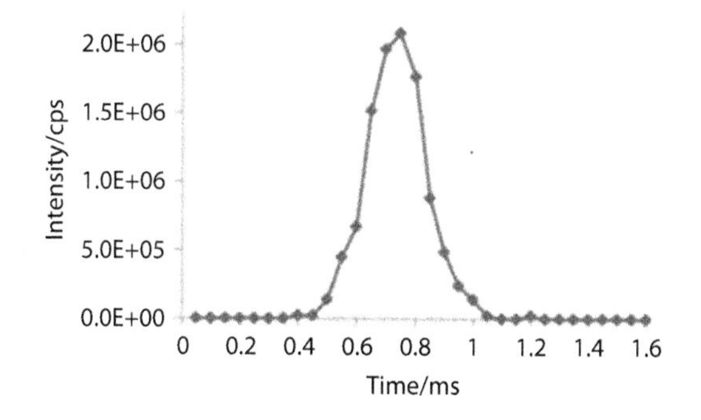

FIGURE 27.11 A time resolved analysis of 30 nm gold particles by sp-ICP-MS showing the pulse is fully characterized in less than 1 ms. (Copyright © 2013, all rights reserved, PerkinElmer Inc.)

as the quadrupole scans from one mass to the next. However, for optimum measurement conditions when one element is under investigation, there is no question that the ability to run a method with no settling time, very short dwell times, and fast data acquisition is extremely beneficial in order to increase the upper limit of the dynamic range when characterizing NPs using single-particle ICP-MS measurements.[24]

USABILITY ASPECTS

In most applications, analytical performance is a very important consideration when deciding what instrument to purchase. However, the vast majority of instruments being used today are operated by technician-level chemists. They usually have had some experience in the use of trace element techniques, such as atomic absorption (AA) or ICP-OES, but in no way could be considered experts in ICP-MS. Therefore, the usability aspects might be competing with analytical performance as the most important selection criteria, particularly if the application does not demand the ultimate in detection capability. Even though usability is in the eye of the user, there are some general issues that need to be addressed. They include, but are not limited to, the following:

- Ease of use
- Routine maintenance
- Compatibility with sampling accessories
- Installation requirements
- Technical support
- Training

Ease of Use

First, you need to determine the skill level of the operator who is going to run the instrument. If the operator is a PhD-type chemist, then maybe it is not critical that the instrument be easy to use. However, if the instrument is going to be used in a high-workload environment and possibly operated around the clock, such as in the pharmaceutical industry, there is a strong possibility that the operators will not be highly skilled. Therefore, you should be looking at how easy the software is to use and how similar it is to other trace element techniques that are used in your laboratory. This will definitely have an impact on the time it takes to get a person fully trained on the instrument. Another issue to consider is whether the person who runs the instrument on a routine basis is the same person who will be developing the methods. Correct method development is critical because it impacts the quality of your data, and therefore usually requires more expertise than just running routine methods. This is most definitely the case with collision/reaction cells and interfaces, especially if the method has never been done before. It can take a significant amount of time and effort to select the best gases, gas flows, and optimization of the cell parameters to maximize the reduction of interferences for certain analytes in a new sample matrix.

I am not going to get into different software features or operating systems, because it is a complicated criterion to evaluate and decisions tend to be made more on a personal preference or comfort level than on the actual functionality of ICP-MS software features. It is also a moving target, as instrument software is continually being modified and updated. However, there are differences in the way software feels. For example, if you have come from a mass spectrometry background, you are probably comfortable with fairly complex research-type software. Alternatively, if you have come from a trace element background and have used AA or ICP-OES, you are probably used to more routine software that is relatively easy to use. You will find that different vendors have come to ICP-MS from a variety of different analytical chemistry backgrounds, which is often reflected in the way they design their software. Depending on the way the instrument will be used, an appropriate amount of time should be spent looking at software features that are specific to your application needs. For example, if you are working in a high-throughput environmental laboratory, you might be interested in turnkey methods that are used to run a particular EPA methodology, such as Method 200.8 for the determination of elements in water and wastes. Alternatively, with more and instruments going into pharmaceutical production labs, look for customized methods for different pharmaceutical matrices. In addition, maybe you should also be looking very closely at all the features of the automated QC software, or if you do not have the time to export your data to an external spreadsheet to create reports, you might be more interested in software with comprehensive reporting capabilities.

In some highly regulated industries, the operating and reporting software needs to be compliant with a set of regulated standards and guidelines. For example, the pharmaceutical and food manufacturing industry is dictated by federal regulations set down by the Food and Drug Administration (FDA) in Title 21 CFR Part 11, which gives detailed requirements that computerized systems need to fulfill in order to allow electronic signatures and records in lieu of handwritten signatures on paper records. In summary, the regulations apply to validations for closed and open computerized systems, controlled access to the computerized system, content integrity, use of electronic signatures for authentication of electronic documents, audit trails for all records and signatures, and access to electronic records. It is therefore important that any ICP-MS that is used in a regulated environment has all the necessary software to be compliant (see Chapter 26 on inspection readiness for details).

Routine Maintenance

ICP mass spectrometers are complex pieces of equipment that, if not maintained correctly, have the potential to fail when you least expect them to. For that reason, a major aspect of instrument usability is how often routine maintenance has to carried out, especially if complex sample matrices are being analyzed. You must not lose sight of the fact that your samples are being aspirated into the sample introduction system and the resulting ions generated in the plasma are steered into the mass analyzer via the interface and ion optics. In other words, the sample, in one form or another, is in contact with many components inside the instrument. Even though modern instruments require much less routine maintenance than older-generation equipment, it is still essential to find out what components need to be changed and at what frequency to keep the instrument in good working order. Routine maintenance has been covered in great depth in Chapter 17, but you should be asking the vendor what needs to be changed or inspected on a regular basis and what type of maintenance should be done at daily, weekly, monthly, or yearly intervals. Some typical questions might include the following:

- If a peristaltic pump is being used to deliver the sample, how often should the tubing be changed?
- How often should the spray chamber drain system be checked?
- Can components be changed if a nebulizer gets damaged or blocked?
- How long does the plasma torch last?
- Are ceramic torches available, which tend to have a much longer lifetime?
- Can the torch sample injector be changed without discarding the torch?
- How is a neutral plasma maintained, and if an external shield or sleeves are used for grounding purposes, how often do they last?
- Is the RF generator solid state, or does it use a power amplifier (PA) tube? This is important because PA tubes are expensive consumable items that typically need replaced every 1–2 years.
- How often do you need to clean the interface cones, and what is involved in cleaning them and keeping the cone orifices free of deposits?
- How long do the cones last?
- Do you have a platinum cone trade-in service, and what is their trade-in value?
- Which type of pump is used on the interface, and if it is a rotary-type pump, how often should the oil be changed?
- What mechanism is used to keep the ion optics free of sample particulates or deposits?
- How often should the ion optics be cleaned?
- What is the cleaning procedure for the ion optics?
- Do the turbomolecular pumps require any maintenance?
- How long do the turbomolecular pumps typically last?
- Does the mass analyzer require any cleaning or maintenance?
- How long does the detector last, and how easy is it to change?
- What spare parts do you recommend to keep on hand? This can often indicate the components that are likely to fail most frequently.
- What maintenance needs to be carried out by a qualified service engineer, and how long does it take?

This is not an exhaustive list, but it should give you a good idea about what is involved in keeping an instrument in good working order. I also encourage you to talk to real-world users of the equipment to make sure you get their perspective of these maintenance issues. This is particularly important if you are investing in a brand-new instrument that hasn't been on the market long. There won't be a large number of users in the field, so it's important you talk to them to get their real-world perspective of the routine maintenance issues. For more information about how to maximize performance and reduce routine maintenance, check out the reference by Brennan and coworkers.[25]

Compatibility with Alternative Sampling Accessories

Alternative sample introduction techniques to enhance performance and productivity are becoming more necessary as ICP-MS is being utilized to analyze more complex sample types. Therefore, it is important to know if the sampling accessory is made by the ICP-MS instrument company or by a third-party vendor. Obviously, if it has been made by the same company, compatibility should not be an issue. However, if it is made by a third party, you will find that some sampling accessories work much better with some instruments than with others. It might be that the physical connection of coupling the accessory to the ICP-MS torch has been better thought out, or that the software "talks" to one system better than another. You should also understand how much routine maintenance a sampling accessory needs. The benefits of a rugged ICP-MS system, which requires very little maintenance, are negated if the sampling accessory needs to be cleaned every 5–10 samples. You should refer to Chapter 20 for more details on their suitability for pharmaceutical analysis, but if they are required, software and hardware compatibility should be one of your evaluation objectives. You should also read Chapter 21 if you are thinking of interfacing a high-performance liquid chromatography (HPLC) system to your ICP-MS for arsenic or mercury speciation studies.

Installation of Instrument

Installation of the instrument and where it is going to be located do not seem to be obvious evaluation objectives at first, but they could be important, particularly if space is limited. For example, is the instrument freestanding or bench mounted? Maybe you have a bench available but no floor space, or vice versa. It could be that the instrument requires a temperature-controlled room to ensure good stability and mass calibration. If this is the case, have you budgeted for this kind of expense? If the instrument is being used for ultratrace detection levels, does it need to go into a class 1, 10, or 100 clean room? If it does, what is the size of the room and do the roughing pumps need to be placed in another room? In other words, it is important to fully understand the installation requirements for each instrument being evaluated and where it will be located. Refer to Chapter 17 for more information on instrument installation.

Technical Support

Technical and application support is a very important consideration, especially if you have had no previous experience with ICP-MS. You want to know that you are not going to be left on your own after you have made the purchase. Therefore, it is important to know not only the level of expertise of the specialist who is supporting you, but also whether they are local to you or located in the manufacturer's corporate headquarters. In other words, can the vendor give technical help whenever you need it? Another important aspect related to application support is the availability of application literature. Is there a wide selection of material available for you to read, in the form of either Web-based application reports or references in the open literature, to help you develop your methods? Also, find out if there are active user or Internet-based discussion groups, because they will be an invaluable source of technical and application help. One such source of help in this area can be found on the PlasmaChem Listserver, a plasma spectrochemistry discussion group out of Syracuse University.[26]

TRAINING

Find out what kind of training course comes with the purchase of the instrument and how often it is run. Most instruments come with a 2- to 3-day training course for one person, but most vendors should be flexible regarding the number of people who can attend. Some manufacturers also offer application training, where they teach you how to optimize methods for major application areas. Historically, these customized application courses have included environmental, clinical, and semiconductor analysis. Today, all vendors are offering pharmaceutical training courses, as it is rapidly becoming the fastest-growing application segment for ICP-MS. Talk to other users in your field about the quality of the training they received when they purchased their instruments, and also ask them what they thought of the operator's manuals. You will often find that this is a good indication of how important a manufacturer thinks customer training is.

RELIABILITY ISSUES

To a certain degree, instrument reliability is impacted by routine maintenance issues and the types of samples being analyzed, but it is generally considered to be more of a reflection of the design of an instrument. Most manufacturers will guarantee a minimum percentage uptime for their instrument, but this number (which is typically ~95%) is almost meaningless unless you really understand how it is calculated. Even when you know how it is calculated, it is still difficult to make the comparison, but at least you should understand the implications if the vendor fails to deliver. Good instrument reliability is taken for granted nowadays, but this has not always been the case. When ICP-MS was first commercialized, the early instruments were a little unpredictable, to say the least, and quite prone to frequent breakdowns. However, as the technique became more mature, the quality of instrument components improved, and therefore the reliability improved. However, you should be aware that some components of the instrument are more problematic than others. This is particularly true when the design of an instrument is new or a model has had a major redesign. You will therefore find that in the life cycle of a newly designed instrument, the early years will be more susceptible to reliability problems than when the instrument is of an older design.

When we talk about instrument reliability, it is important to understand whether it is related to the samples being analyzed, the inexperience of the person operating the instrument, an unreliable component, or an inherent weakness in the design of the instrument. For example, how does the instrument handle highly corrosive chemicals, such as concentrated mineral acids? Some sample introduction systems and interfaces will be more rugged than others and require less maintenance in this area. On the other hand, if the operator is not aware of the dissolved solids limitation of the instrument, they might attempt to aspirate a sample, which will slowly block the interface cones, causing signal drift and, in the long term, possible instrument failure. Or, it could be something as unfortunate as a major component, such as the RF generator power amplifier tube, discrete dynode detector, or turbomolecular pump (which all have a finite lifetime) failing in the first year of use.

Service Support

Instrument reliability is very difficult to assess at the evaluation stage, so you have to look very carefully at the kind of service support offered by the manufacturer. For example, how close is a qualified support engineer to you, or what is the maximum amount of time you will have to wait to get a support engineer at your laboratory, or at least to call you back to discuss the problem? Ask the vendor if they have the capability for remote diagnostics, where a service engineer can remotely run the instrument or check the status of a component by "talking to" your system computer via a modem. Even if this approach does not fix the problem, at least the service engineer can come to your laboratory with a very good indication of what the problem could be.

You should know up front what a service visit is going to cost you, irrespective of what component has failed. Also, find out what routine maintenance jobs you can do and what requires an experienced service engineer. If it does require a service engineer, how long will it take them, because their time is not inexpensive. Most companies charge an hourly rate for a service engineer (which typically includes travel time as well), but if an overnight stay is required, fully understand what you are paying for (accommodations, meals, gas, etc.). Some companies might even charge for mileage between the service engineer's base and your laboratory. If you work in a commercial laboratory and cannot afford the instrument to be down for any length of time, find out what it is going to cost for 24/7 service coverage.

You can take a chance and just pay for each service visit, or you might want to budget for an annual preventative maintenance contract, where the service engineer checks out all the important instrumental components and systems frequently to make sure they are all working correctly. This might not be as critical if you work in an academic environment, where the instrument might be down for extended periods, but in my opinion, it is absolutely critical if you work in a commercial laboratory, which is using the instrument to generate revenue. Also, find out what is included in the contract, because some also cover the cost of consumables or replacement parts, whereas others just cover the service visits. These annual preventative maintenance contracts typically make up about 10%–15% of the cost of the instrument, but are well worth it if you do not have the expertise in-house, or if you just feel more comfortable having an insurance policy to cover instrument breakdowns.

Once again, talking to existing users will give you a very good perspective of the quality of the instrument and the service support offered by the manufacturer. There is no absolute guarantee that the instrument of choice is going to perform to your satisfaction 100% of the time, but if you work in a high-throughput, routine laboratory, make sure it will be down for the minimum amount of time. In other words, fully understand what it is going to cost you to maximize the uptime of all the instruments being evaluated.

FINANCIAL CONSIDERATIONS

The financial side of choosing an ICP mass spectrometer can often dominate the selection process. You may or may not have budgeted quite enough money to buy a top-of-the-line instrument, or perhaps you had originally planned to buy another lower-cost trace element technique, or you could be using funds left over at the end of your financial year. All these scenarios dictate how much money you have available and what kind of instrument you can purchase. In my experience, you should proceed with caution in this kind of situation, because if only one manufacturer is willing to do a deal with you, there is no real need to carry out the evaluation process. Therefore, you should budget at least 12 months before you are going to make a purchase and add another 10%–15% for inflation and any unforeseen price increases. In other words, if you want to get the right instrument for your application, never let price be the overriding factor in your decision. Always be wary of the vendor who will undercut everyone else to get your business. There could be a very good reason why they are doing this; for example, the instrument is being discontinued for a new model, or it could be having some reliability problems that are affecting its sales.

This is not to say that price is unimportant, but what might appear to be the most expensive instrument to purchase might be the least expensive to run. Therefore, you must never forget the cost of ownership in the overall financial analysis of your purchase. So, by all means compare the price of the instrument, computer, and any accessories you buy, but also factor in the cost of consumables, gases, and electricity based on your usage. Maybe instrument consumables from one vendor are much less expensive than those from another vendor. This is particularly the case with interface cones and plasma torches. Or maybe the purity of collision/reaction gases is more critical with one cell-based instrument than another. For example, there is a factor of four difference between the cost of high-purity (99.999%) hydrogen gas and ultra-high-purity (99.9999%) grade. Supplies of high-purity helium gas are diminishing, so the price is increasing quite significantly, and the gas is becoming much more difficult to obtain from your gas supplier.

So, be diligent when you compare prices. Look at the overall picture, and not just the cost of the instrument. Also, be aware of differences in sample throughput. Perhaps you can analyze more samples with one instrument because its measurement protocol is faster or it does not need recalibrating as often (less drift). It therefore follows that if you can get through your daily allocation of samples much faster with one instrument than another, then your argon consumption will be reduced.

Another aspect that should be taken into consideration is the salary of the operator. Even though you might think that this is a constant, irrespective of the instrument, you must assess the expertise required to run it. For example, if you are thinking of purchasing more complex technology, such as a magnetic sector instrument or a triple-quad collision/reaction cell instrument for a research-type application, the operator needs to be of a much higher skill level than, say, someone who is being asked to run a routine application with a quadrupole-based instrument. As a result, the salary of that person will probably be higher.

Finally, if one instrument has to be installed in a temperature-controlled, air-conditioned environment for stability purposes, the cost of preparing or building this kind of specialized room must be taken into consideration when doing your financial analysis. In other words, when comparing systems, never automatically reject the most expensive instrument. You will find that over the 10 years that you own the instrument, the cost of doing analysis and the overall cost of ownership are the more important evaluation criteria.

THE EVALUATION PROCESS: A SUMMARY

As mentioned earlier in this chapter, it is not my intention to compare instrument designs and features, but to give you some general guidelines as to what are the most important evaluation criteria. Besides being a framework for your evaluation process, these guidelines should also be used in conjunction with the other chapters in this book and the cited referenced information.

However, if you want to find the best instrument for your application needs, be prepared to spend a few months evaluating the marketplace. Do not forget to prioritize your objectives and give each of them a weighting factor based on their degree of importance for the types of samples you analyze. Be careful to take the evaluation in the direction you want to go and not where the vendor wants to take it. In other words, it is important to compare apples with apples and not to be talked into comparing an apple with an orange that looks like an apple! However, be prepared that there might not be a clear-cut winner at the end of the evaluation. If this is the case, then decide what aspects of the evaluation are most important and ask the manufacturer to put them in writing. Some vendors might be hesitant to do this, especially if it is guaranteeing instrument performance with your samples.

Talk to as many users in your field as you possibly can—not only ones given to you as references by the vendor, but also ones chosen by yourself. This will give you a very good indication of the real-world capabilities of the instrument, which can often be overlooked at a demonstration. You might find from talking to "typical" users that it becomes obvious which instrument to purchase. If that is the case and your organization allows it, ask the vendor what your options are if you do not have samples to run and you do not want a demonstration. I guarantee you will be in a much better position to negotiate a lower price.

Never forget that it is a very competitive marketplace, and your business is extremely important to each of the ICP-MS manufacturers. Hopefully, this book, and particularly this chapter, has not only helped you understand the fundamentals of the technique a little better but also given you some thoughts and ideas on how to find the best instrument for your needs. Refer to Chapter 29 for details on how to contact instrument vendors and consumables and accessories companies. And if you are still confused, I teach a half-day short course at the Pittsburgh conference every year: "How to Evaluate ICP-MS: The Most Important Analytical Considerations."

We talk about all these issues in more detail. But if I don't see you there, good luck with your evaluation!

REFERENCES

1. K. Nottingham. ICP-MS: It's elemental. *Analytical Chemistry*, 35A–38A, 2004.
2. Royal Society of Chemistry. Report by the Analytical Methods Committee: Evaluation of analytical instrumentation—Part X: Inductively coupled plasma mass spectrometers. *Analyst*, 122, 393–408, 1997.
3. A. Montasser, ed. Analytical figures of merit for ICP-MS. In *Inductively Coupled Plasma Mass Spectrometry: An Introduction to ICP Spectrometries for Elemental Analysis*. New York: Wiley-VCH, 1998, pp. 16–28.
4. E. R. Denoyer. *Atomic Spectroscopy*, 13(3), 93–98, 1992.
5. M. A. Thomsen. *Atomic Spectroscopy*, 21(5), 45–51, 2000.
6. L. Halicz, Y. Erel, and A. Veron. *Atomic Spectroscopy*, 17(5), 186–189, 1996.
7. R. Thomas. *Spectroscopy*, 17(7), 44–48, 2002.
8. E. R. Denoyer and Q. H. Lu. *Atomic Spectroscopy*, 14(6), 162–169, 1993.
9. R. Hutton, A. Walsh, D. Milton, and J. Cantle. *ChemSA*, 17, 213–215, 1991.
10. P. H. Dawson, ed. *Quadrupole Mass Spectrometry and its Applications*. Amsterdam: Elsevier, 1976. Reissued by AIP Press, Woodbury, NY, 1995.
11. S. J. Jiang, R. S. Houk, and M. A. Stevens. *Analytical Chemistry*, 60, 217, 1988.
12. K. Sakata and K. Kawabata. *Spectrochimica Acta*, 49B, 1027, 1994.
13. J. M. Collard, K. Kawabata, Y. Kishi, and R. Thomas. *Micro*, January 2002.
14. S. D. Tanner and V. I. Baranov. *Atomic Spectroscopy*, 20(2), 45–52, 1999.
15. B. Hattendorf and D. Günther. *Journal of Analytical Atomic Spectrometry*, 19, 600–606, 2004.
16. S. D. Tanner, D. J. Douglas, and J. B. French. *Applied Spectroscopy*, 48, 1373, 1994.
17. E. R. Denoyer, D. Jacques, E. Debrah, and S. D. Tanner. *Atomic Spectroscopy*, 16(1), 1, 1995.
18. R. C. Hutton and A. N. Eaton. *Journal of Analytical Atomic Spectrometry*, 5, 595, 1987.
19. A. L. Gray and A. Date. *Analyst*, 106, 1255, 1981.
20. E. J. Wyse, D. W. Koppenal, M. R. Smith, and D. R. Fisher. Presented at 18th FACSS Meeting, Anaheim, CA, October 1991, paper 409.
21. W. G. Diegor and H. P. Longerich. *Atomic Spectroscopy*, 21(3), 111, 2000.
22. D. J. Douglas and J. B. French. *Spectrochimica Acta*, 41B(3), 197, 1986.
23. E. R. Denoyer. *Atomic Spectroscopy*, 12, 215–224, 1991.
24. E. Heithmar and S. Pergantis. Characterizing concentrations and size distributions of metal-containing nanoparticles in waste water. EPA Report APM 272. Washington, DC: Environmental Protection Agency.
25. R. Brennan, G. Dulude, and R. Thomas. Approaches to maximize performance and reduce the frequency of routine maintenance in ICP-MS. *Spectroscopy Magazine*, 30(10), 12–25, 2015.
26. PlasmaChem Listserver. A discussion group for plasma spectrochemists worldwide. Syracuse, NY: Syracuse University. http://www.lsoft.com/scripts/wl.exe?SL1=PLASMACHEM-L&H=LISTSERV.SYR.EDU.

28 Plasma Spectrochemistry Glossary of Terms

In all my years of working with inductively coupled plasma (ICP) spectrochemistry, I have never come across any written material that included a basic dictionary of terms, primarily aimed at someone new to the technique. When I first became involved in the inductively coupled plasma optical emission spectrometry (ICP-OES) and inductively coupled plasma mass spectrometry (ICP-MS) techniques, most of the literature I read tended to give complicated descriptions of instrument components and explanations of fundamental principles that more often than not sailed over my head. It was not until I became more familiar with the technique that I began to get a better understanding of the complex jargon used in technical journals and presentations at scientific conferences. So, when I wrote my second ICP-MS textbook, I knew that a glossary of terms was an absolute necessity. In this book, we've expanded the glossary to also include ICP-OES terms, to cover the chapter compiled by Maura Rury. Even though the glossary is not exhaustive, it contains explanations and definitions of the most common ICP-OES and ICP-MS words, expressions, and terms used in this book. It should mainly be used as a quick reference guide. If you want more information about the subject matter, you should use the index to find a more detailed explanation of the topic in the appropriate book chapter. (Note: It does not include commercial names used by any of the instrument vendors.)

INDUCTIVELY COUPLED PLASMA MASS SPECTROMETRY GLOSSARY

A

AA: An abbreviation for **atomic absorption**.

abundance sensitivity: A way of assessing the ability of a mass separation device, such as a quadrupole, to identify and measure a small analyte peak adjacent to a much larger interfering peak. An abundance sensitivity specification is a combination of two measurements. The first is expressed as the ratio of the intensity of the peak at 1 atomic mass unit (amu) below the analyte peak to the intensity of the analyte peak, and the second is the ratio of the peak intensity 1 amu above the analyte mass to the intensity of the analyte peak. Because of the motion of the ion through the mass filter, the abundance sensitivity specification of a mass filtering device is always worse on the low-mass side than on the high-mass side.

active film multipliers: Another name for **discrete dynode multipliers**, which are used to detect, measure, and convert ions into electrical pulses in ICP-MS. *Also refer to* **channel electron multiplier** *and* **discrete dynode detector**.

addition calibration: A method of calibration in ICP-MS using standard additions. All samples are assumed to have a similar matrix, so spiking is only carried out on one representative sample and not the entire batch of samples, as per conventional standard additions used in graphite furnace **atomic absorption** analysis.

AE: An abbreviation for **atomic emission**.

aerosol: The result of breaking up a liquid sample into small droplets by the nebulization process in the sample introduction system. *Also refer to* **nebulizer** *and* **sample introduction system**.

aerosol dilution: A way of introducing a flow of argon gas between the nebulizer and the torch, which has the effect of reducing the sample's solvent loading on the plasma, so it can tolerate much higher total dissolved solids.

AF4: An abbreviation for asymmetrical flow field-flow fractionation.

alkylated metals: A metal complex containing an alkyl group. Typically detected by coupling liquid chromatography with ICP-MS. *Also refer to* **speciation analysis**.

alpha-counting spectrometry: A particle-counting technique that uses the measurement of the radioactive decay of alpha particles. *Also refer to* **particle-counting techniques**.

alternative sample introduction accessories: Alternative ways of introducing samples into an ICP mass spectrometer other than conventional nebulization. Also known as **alternative sample introduction devices**. Often used to describe desolvation techniques or laser ablation.

analog counting: A way of measuring high signals by changing the gain or voltage of the detector. *Also refer to* **pulse counting**.

Argon: The gas used to generate the plasma in an ICP.

argon-based interferences: A polyatomic spectral interference generated by argon ions combining with ions from the matrix, solvent, or any elements present in the sample.

array detectors: An ion detector based on solid-state, direct charge arrays, similar to charge injection device (CID) and charge-coupled device (CCD) technology used in ICP optical emission. By projecting all the separated ions from a mass separation device onto a two-dimensional array, these detectors can view the entire mass spectrum simultaneously. Used with the Mattauch–Herzog sector technology.

ashing: A sample preparation technique that involves heating the sample (typically in a muffle furnace) until the volatile material is driven off and an ash-like substance is left.

asymmetrical flow field-flow fractionation (AF4): Field-flow fractionation (FFF) is a single-phase chromatographic separation technique, where separation is achieved within a very thin channel, against which a perpendicular force field is applied. One of the most common forms of FFF is AF4, where the field is generated by a cross-flow applied perpendicular to the channel. Coupled with ICP-MS for the characterization of nanoparticles.

atom: A unit of matter. The smallest part of an element having all the characteristics of that element and consisting of a dense, central, positively charged nucleus surrounded by orbiting electrons. The entire structure has an approximate diameter of 10^{-8} cm and characteristically remains undivided in chemical reactions except for limited removal, transfer, or exchange of certain electrons.

atom-counting techniques: A generic name given to techniques that use atom or ion counting to carry out elemental quantitation. Some common ones, besides ICP-MS, include secondary ionization mass spectrometry (SIMS), thermal ionization mass spectrometry (TIMS), accelerator mass spectrometry (AMS), and fission track analysis (FTA).*Also refer to* **ionizing radiation counting techniques**.

atomic absorption (AA): An analytical technique for the measurement of trace elements that uses the principle of generating free atoms (of the element of interest) in a flame or electrothermal atomizer (ETA) and measuring the amount of light absorbed from a wavelength-specific light source, such as a hollow cathode lamp (HCL) or electrode discharge lamp (EDL).

atomic emission (AE): A trace element analytical technique that uses the principle of exciting atoms in a high-temperature source, such as a plasma discharge, and measuring the amount of light the atoms emit when electrons fall back down to a ground (stable) state.

atomic mass or weight: The average mass or weight of an atom of an element, usually expressed relative to the mass of carbon 12, which is assigned 12 atomic mass units.

atomic number: The number of protons in an atomic nucleus.

atomic structure: Describes the structural makeup of an atom. *Also refer to* **neutron**, **proton**, *and* **electron**.

attenuation (of the detector): Reduces the amplitude of the electrical signal generated by the detector, with little or no distortion. Usually carried out by applying a control voltage to extend the dynamic range of the detector. *Also refer to* **extended dynamic range**.

autocalibration: A way of carrying out calibration with an automated in-line sample delivery system.

autodilution: A way of carrying out automatic in-line dilution of large numbers of samples with no manual intervention by the operator.

autosampler: A device to automatically introduce large numbers of samples into the ICP-MS system with no manual intervention by the operator.

axial view: An ICP-OES system in which the plasma torch is positioned horizontally (end-on) to the optical system as opposed to the conventional vertical (radial) configuration. It is generally accepted that viewing the end of the plasma improves emission intensity by a factor of approximately 5- to 10-fold.

B

background equivalent concentration (BEC): Defined as the apparent concentration of the background signal based on the sensitivity of the element at a specified mass. The lower the BEC value, the more easily a signal generated by an element can be discerned from the background. Many analysts believe BEC is a more accurate indicator of the performance of an ICP-MS system than detection limit, especially when making comparisons of background reduction techniques, such as cool plasma or collision/reaction cell and interface technology.

background noise: The square root of the intensity of the blank in counts per second (cps) anywhere of analytical interest on the mass range. Detection limit (DL) is a ratio of the analyte signal to the background noise at the analyte mass. Background noise as an instrumental specification is usually measured at mass 220 amu (where there are no spectral features), while aspirating deionized water. *Also refer to* **background signal**, **instrument background noise**, *and* **detection limit**.

background signal: The signal intensity of the blank in counts per second (cps) anywhere of analytical interest on the mass range. Detection limit (DL) is a ratio of the analyte signal to the noise of the background at the analyte mass. Background as an instrumental specification is usually measured at mass 220 amu (where there are no spectral features), while aspirating deionized water. *Also refer to* **background noise**, **instrument background signal**, *and* **detection limit**.

bandpass tuning and filtering: A mechanism used in a dynamic reaction cell (DRC) to reject the by-products generated through secondary reactions utilizing the principle of mass discrimination. Achieved by optimizing the electrical fields of the reaction cell multipole (typically a quadrupole) to allow transmission of the analyte ion, while rejecting the polyatomic interfering ion.

BEC: An abbreviation for **background equivalent concentration**.

by-product ions: Ionic species formed as a result of secondary reactions that take place in a collision/reaction cell. *Also refer to* **secondary (side) reactions**.

C

calibration: A plot, function, or equation generated using calibration standards and a blank, which describes the relationship between the concentration of an element and the signal intensity produced at the analyte mass of interest. Once determined, this relationship can be used to determine the analyte concentration in an unknown sample.

calibration standard: A reference solution containing accurate and known concentrations of analytes for the purpose of generating a calibration curve or plot.

capacitive coupling: An undesired electrostatic (or capacitive) coupling between the voltage on the load coil and the plasma discharge, which produces a potential difference of a few hundred volts. This creates an electrical discharge or arcing between the plasma and sampler cone of the interface, commonly known as a "secondary discharge" or "pinch effect."

capillary electrophoresis (CE): *Refer to* **capillary zone electrophoresis**.

capillary zone electrophoresis (CZE): A chromatographic separation technique used to separate ionic species according to their charge and frictional forces. In traditional electrophoresis, electrically charged analytes move in a conductive liquid medium under the influence of an electric field. In capillary (zone) electrophoresis, species are separated based on their size-to-charge ratio inside a small capillary filled with an electrolyte. Its applicability to ICP-MS is mainly in the field of separation and detection of large biomolecules.

CCD: An abbreviation for charge-coupled device.

CE: An abbreviation for **capillary electrophoresis.** *Refer to* **capillary zone electrophoresis**.

cell: In ICP-MS terminology, a cell usually refers to a collision/reaction cell.

ceramic torch: A plasma torch where either (or all) the sample injector, inner tube, or outer tube is made of a ceramic material. Typically has a longer lifetime than a traditional quartz torch.

certified reference material (CRM): Well-established reference matrix that comes with certified values and associated statistical data that have been analyzed by other complementary techniques. Its purpose is to check the validity of an analytical method, including sample preparation, instrument methodology, and calibration routines, to achieve sample results that are as accurate and precise as possible and can be defended when subjected to intense scrutiny.

channel electron multiplier (CEM): A detector used in ICP-MS to convert ions into electrical pulses using the principle of multiplication of electrons via a potential gradient inside a sealed tube.

Channeltron®: Another name for a channel electron multiplier detector.

charge-coupled device (CCD): A type of solid-state detector technology for converting photons into an electrical signal. Typically applied to ICP-OES.

charge injection device (CID): A type of solid-state detector technology for converting photons into an electrical signal. Typically applied to ICP-OES.

charge transfer reaction: Sometimes referred to as "charge exchange." This is one of the ion–molecule reaction mechanisms that take place in a collision/reaction cell. It involves the transfer of a positive charge from the interfering ion to the reaction gas molecule, forming a neutral atom that is not seen by the mass analyzer. An example of this kind of reaction is

$$H_2 + {}^{40}Ar^+ = Ar + H_2^+$$

chemical modification: The process of chemically modifying the sample in electrothermal vaporization (ETV) ICP-MS work to separate the analyte from the matrix. *Also refer to* **chemical modifier** *and* **electrothermal vaporization**.

chemical modification: The process of chemically modifying the sample in electrothermal vaporization (ETV) ICP-MS work to separate the analyte from the matrix. *Also refer to* **chemical modifier** *and* **electrothermal vaporization**.

chemical modifier: A chemical or substance that is added to the sample in an electrothermal vaporizer to change the volatility of the analyte or matrix. Typically added at the ashing stage of the heating program to separate the vaporization of the analyte from the potential interferences of the matrix components. *Also refer to* **electrothermal vaporization**.

chromatographic separation device: Any device that separates analyte species according to their retention times or mobility through a stationary phase. When coupled with an ICP-MS system, it is used for the separation, detection, and quantitation of speciated forms of trace elements. Examples include liquid, ion, gas, size exclusion, and capillary electrophoresis chromatography. *Also refer to* **speciation analysis**.

chromatography terminology (as applied to trace element speciation): The following are some of the most important terms used in the chapter on trace element speciation. For easy access, they are contained in one section and not distributed throughout the glossary.

- **buffer:** A mobile-phase solution that is resistant to extreme pH changes, even with additions of small amounts of acids or bases.
- **chromatogram:** The graphical output of the chromatographic separation. It is usually a plot of peak intensity of the separated species over time.
- **column:** The main component of the chromatographic separation. It is typically a tube containing the stationary-phase material that separates the species and an eluent that elutes the species off the column.
- **counterions:** The mobile phase contains a large number of ions that have a charge opposite of that of the surface-bound ions. These are known as counterions, which establish equilibrium with the stationary phase.
- **dead volume:** Usually refers to the volume of the mobile phase between the point of injection and the detector that is accessible to the sample species, minus the volume of mobile phase that is contained in any union or connecting tubing.
- **gradient elution:** Involves variation of the mobile-phase composition over time through a number of steps, such as changing the organic content, altering the pH, changing the concentration of the buffer, or using a completely different buffer.
- **ion exchange:** A technique in which separation is based on the exchange of ions (anions or cations) between the mobile phase and the ionic sites on a stationary phase bound to a support material in the column.
- **ion pairing:** A type of separation that typically uses a reverse-phase column in conjunction with a special type of chemical in the mobile phase called an "ion-pairing reagent." *Also refer to* **reverse phase**.
- **isochratic elution:** An elution of the analytes or species using the same solvent throughout the analysis.
- **mobile phase:** A combination of the sample or species being separated or analyzed and the solvent that moves the sample through the column.
- **retention time:** The time taken for a particular analyte or species to be separated and pass through the column to the detector.

reverse phase: A type of separation that is typically combined with ion pairing, and essentially means that the column's stationary phase is less polar and more organic than the mobile-phase solvents.

stationary phase: A solid material, such as silica or a polymer, that is set in place and packed into the column for the chromatographic separation to take place.

clean room: The general description given to a dedicated room for the sample preparation and analysis of ultrapure materials. Usually associated with a number that describes the number of particulates per cubic foot of air (e.g., a class 100 clean room will contain 100 particles/ft^3 of air). It is commonly accepted that the semiconductor industry has the most stringent demands, which necessitates the use of class 10 and sometimes class 1 clean rooms.

cluster ions: Ions that are formed by two or more molecular ions combining together in a collision/reaction cell to form molecular clusters.

CMOS: An abbreviation for complementary metal oxide semiconductor. A technology used in the direct charge array detector, which is utilized in the Mattauch–Herzog simultaneous sector instrument.

cold plasma technology: Cool or cold plasma technology uses low-temperature plasma to minimize the formation of certain argon-based polyatomic species. Under normal plasma conditions (approximately 1000 W radio-frequency (RF) power and 1.0 L/min nebulizer gas flow), argon ions combine with matrix and solvent components to generate problematic spectral interferences, such as $^{38}ArH^+$, $^{40}Ar^+$, and $^{40}Ar^{16}O^+$, which impact the detection limits of a small number of elements, including K, Ca, and Fe. By using cool plasma conditions (approximately 600 W RF power and 1.6 L/min nebulizer gas flow), the ionization conditions in the plasma are changed so that many of these interferences are dramatically reduced and detection limits are improved.

cold vapor atomic absorption (CVAA): An analytical approach to determine low levels of mercury by generating mercuric vapor in a quartz cell and measuring the number of mercury atoms produced, using the principle of atomic absorption. *Also refer to* **hydride generation atomic absorption**.

collision cell: Specifically, a cell that predominantly uses the principle of collisional fragmentation to break apart polyatomic interfering ions generated in the plasma discharge. Collision cells typically utilize higher-order multipoles (such as hexapoles or octopoles) with inert or low-reactive gases (such as helium and hydrogen) to first stimulate ion–molecule collisions, and then kinetic energy discrimination to reject any undesirable by-product ionic species formed.

collision-induced dissociation (CID): A basic principle, first used for the study of organic molecules using tandem mass spectrometry, that relies on using a nonreactive gas in a collision cell to stimulate ion–molecule collisions. The more collision-induced daughter species that are generated, the better the chance of identifying the structure of the parent molecule.

collision/reaction cell (CRC) technology: A generic term applied to collision and reaction cells that use the principle of ion–molecule collisions and reactions to cleanse the ion beam of problematic polyatomic spectral interferences before they enter the mass analyzer. Both collision and reaction cells are positioned in the mass spectrometer vacuum chamber after the ion optics but prior to the mass analyzer. *Also refer to* **collision cell** *and* **reaction cell**.

collision/reaction interface (CRI) technology: A collision/reaction mechanism approach, which instead of using a pressurized cell, injects a gas directly into the interface between the sampler and skimmer cones. The injection of the collision/reaction gas into this region of the ion beam produces high collision frequency between the argon gas and the injected gas molecules. This has the effect of removing argon-based polyatomic interferences before they are extracted into the ion optics.

collisional damping: A mechanism that describes the temporal broadening of ion packets in a quadrupole-based dynamic reaction cell to dampen out fluctuations in ion energy. By optimizing cell conditions such as gas pressure, the radio-frequency (RF) stability boundary (q parameter), entrance and exit lens potentials, and cell rod offsets, it has been shown that fluctuation in ion energies can be dampened sufficiently to carry out isotope ratio precision measurements near their statistical limit.

collisional focusing: The mechanism of focusing ions toward the center of the ion beam in a collision/reaction cell. By using a neutral collision gas of lower molecular weight than the analyte, the analyte ions will lose kinetic energy and migrate toward the axis as a result of the collisions with the gaseous molecules. Therefore, the number of ions exiting the cell and reaching the detector will increase. *Also refer to* **collision cell** *and* **reaction cell.**

collisional fragmentation: The mechanism of breaking apart (fragmenting) a polyatomic interfering ion in a collision/reaction cell using collisions with a gaseous molecule. The predominant mechanism used in a collision cell, as opposed to a reaction cell. *Also refer to* **collision cell** *and* **reaction cell**.

collisional mechanisms: The mechanisms by which the interfering ion is reduced or minimized to allow the determination of the analyte ion. The most common collisional mechanisms seen in collision/reaction cells include collisional focusing, dissociation, and fragmentation, whereas the major reaction mechanisms include exothermic and endothermic associations, charge transfer, molecular associations, and proton transfer.

collisional retardation: A mechanism in a collision/reaction cell where the gas atoms or molecules undergo multiple collisions with the polyatomic interfering ion, retarding or lowering its kinetic energy. Because the interfering ion has a larger cross-sectional area than the analyte ion, it undergoes more collisions and, as a result, can be separated or discriminated from the analyte ion based on their kinetic energy differences.

concentric nebulizer: A nebulizer that uses two narrow concentric capillary tubes (one inside the other) to aspirate a liquid into the ICP-MS spray chamber. Argon gas is usually passed through the outer tube, which creates a Venturi effect, and as a result, the liquid is sucked up through the inner capillary tube.

cones: *Refer to* **interface cones**.

cooled spray chamber: A spray chamber that is cooled in order to reduce the amount of solvent entering the plasma discharge. Used for a variety of reasons, including reducing oxide species, minimizing solvent-based spectral interferences, and allowing the trouble-free aspiration of organic solvents.

cool plasma technology: *Refer to* **cold plasma technology**.

correction equation: A mathematical approach used to compensate for isobaric and polyatomic spectral overlaps. It works on the principle of measuring the intensity of the interfering species at another mass, which is ideally free of any interference. A correction is then applied by knowing the ratio of the intensity of the interfering species at the analyte mass to its intensity at the alternate mass.

counts per second (cps): Units of signal intensity used in ICP-MS. Number of detector electronic pulses counted per second.

cps: An abbreviation for **counts per second**.

CRC: An abbreviation for **collision/reaction cell technology**.

CRI: An abbreviation for **collision/reaction interface technology**.

CRM: An abbreviation for certified reference material.

cross-calibration: A calibration method that is used to correlate both pulse (low-level) and analog (high-level) signals in a dual-mode detector. This is possible because the analog and pulse outputs can be defined in identical terms (of incoming pulse counts per second) based on knowing the voltage at the first analog stage, the output current, and a conversion factor defined by the detection circuitry electronics. By carrying out a cross-calibration across the mass range, a dual-mode detector is capable of achieving approximately eight to nine orders of dynamic range in one simultaneous scan.

cross-flow nebulizer: A nebulizer that is designed for samples that contain a heavier matrix or small amounts of undissolved solids. In this design, the argon gas flow is directed at right angles to the tip of a capillary tube through which the sample is drawn up with a peristaltic pump.

CVAA: An abbreviation for **cold vapor atomic absorption**.

cyclonic spray chamber: A spray chamber that operates using the principle of centrifugal force. Droplets are discriminated according to their size by means of a vortex produced by the tangential flow of the sample aerosol and argon gas inside the spray chamber. Smaller droplets are carried with the gas stream into the ICP-MS, while the larger droplets impinge on the walls and fall out through the drain.

cylinder lens: A type of lens component used in the ion optics.

CZE: An abbreviation for **capillary zone electrophoresis**.

D

data-quality objectives: A term used to describe the quality goals of the analytical result. Typically achieved by optimizing the measurement protocol to achieve the desired accuracy, precision, or sample throughput required for the analysis.

DCD: An abbreviation for direct charge detector.

dead time correction: Sometimes ions hit the detector too fast for the measurement circuitry to handle in an efficient manner. This is caused by ions arriving at the detector during the output pulse of the preceding ion and not being detected by the counting system. This "dead time," as it is known, is a fundamental limitation of the multiplier detector and is typically 30–50 ns, depending on the detection system. A compensation or "dead time correction" has to be made in the measurement circuitry in order to count the maximum number of ions hitting the detector.

Debye length: The distance over which ions exert an electrostatic influence on one another as they move from the interface region into the ion optics. In the ion-sampling process, this distance is small compared with the orifice diameter of the sampler or skimmer cone. As a result, there is little electrical interaction between the ion beam and the cones, and relatively little interaction between the individual ions within the ion beam. In this way, the compositional integrity of the ion beam is maintained throughout the interface region.

desolvating microconcentric nebulizer: A microconcentric nebulizer that uses some type of desolvation system to remove the sample solvent. *Also refer to* **desolvation device** *and* **membrane desolvation**.

desolvating spray chamber: A general name given to a spray chamber that removes or reduces the amount of solvent from a sample using the principle of desolvation. Some of the approaches that are typically used include conventional water cooling, heating with cooling condensers, Peltier (thermoelectric) cooling, and membrane-based desolvation techniques.

desolvation device: A general name given to a device that removes or reduces the amount of solvent from a sample using the principle of desolvation. Some of the approaches that are typically used include conventional water cooling, heating and condensing units, Peltier (thermoelectric) cooling, or membrane-based desolvation techniques.

detection capability: A generic term used to assess the overall detection performance of an ICP mass spectrometer. There are a number of different ways of evaluating detection capability, including instrument detection limit (IDL), method detection limit (MDL), element sensitivity, and background equivalent concentration (BEC).

detection limit: Most often refers to the instrument detection limit (IDL) and is typically defined as a ratio of the analyte signal to the noise of the background at a particular mass. For a 99% confidence level, it is usually calculated as three times the standard deviation (SD) of 10 replicates (measurements) of the sample blank expressed as concentration units.

detector: A generic name used for a device that converts ions into electrical pulses in ICP-MS.

detector dead time: *Refer to* **dead time correction**.

devitrification: Crystalline breakdown of glass or quartz by a combination of chemical attack and elevated temperatures, typically associated with the plasma torch.

digital counting: Refers to the process of counting the number of pulses generated by the conversion of ions into an electrical signal by the detector measurement circuitry.

DIHEN: An abbreviation for **direct injection high-efficiency nebulizer**.

DIN: An abbreviation for **direct injection nebulizer**.

direct charge detector (DCD): A detector technology used to convert photons into an electric current. A type of complementary metal oxide semiconductor (CMOS) array ion detector used in the Mattauch–Herzog simultaneous sector instrument.

direct injection high-efficiency nebulizer (DIHEN): A more recent refinement of the direct injection nebulizer (DIN), which appears to have overcome many of the limitations of the original design.

direct injection nebulizer (DIN): A nebulizer that injects a liquid sample under high pressure directly into the base of the plasma torch. The benefit of this approach is that no spray chamber is required, which means that an extremely small volume of sample can be introduced directly into the ICP-MS with virtually no carryover or memory effects from the previous sample.

discrete dynode detector (DDD): The most common type of detector used in ICP-MS. As ions emerge from the quadrupole rods onto the detector, they strike the first dynode, liberating secondary electrons. The electron optic design of the dynode produces acceleration of these secondary electrons to the next dynode, where they generate more electrons. This process is repeated at each dynode, generating a pulse of electrons that are finally captured by the multiplier anode. *Also refer to* **active film multipliers**.

double-focusing magnetic sector mass spectrometer (analyzer): A mass spectrometer that uses a very powerful magnet combined with an electrostatic analyzer (ESA) to produce a system with very high resolving power. This approach, known as "double focusing," samples the ions from the plasma. The ions are accelerated in the plasma to a few kilovolts into the ion optic region before they enter the mass analyzer. The magnetic field, which is dispersive with respect to ion energy and mass, then focuses all the ions with diverging angles of motion from the entrance slit. The ESA, which is only dispersive with respect to ion energy, then focuses all the ions onto the exit slit, where the detector is positioned. If the energy dispersions of the magnet and ESA are equal in magnitude but opposite in direction, they will focus both ion angles (first focusing) and ion energies (second focusing) when combined together. *Also refer to* **electrostatic analyzer**.

double-pass spray chamber: A spray chamber that comprises an inner (central) tube inside the main body of the spray chamber. The smaller droplets are selected by directing the aerosol from the nebulizer into the central tube. The aerosol emerges from the tube, where the larger droplets fall out (because of gravity) through a drain tube at the rear of the spray chamber. The smaller droplets then travel back between the outer wall and the central tube into the sample injector of the plasma torch. The most common type of double-pass spray chamber is the Scott design.

doubly charged ion: A species that is formed when an ion is generated with a double positive charge as opposed to a normal single charge and produces an isotopic peak at half its mass. For example, the major isotope of barium at mass 138 amu also exhibits a doubly charged ion at mass 69 amu, which can potentially interfere with gallium at mass 69. Some elements, such as the rare earths, readily form doubly charged species, whereas others do not. Formation of doubly charged ions is also impacted by the ionization conditions (radio-frequency [RF] power, nebulizer gas flow, etc.) in the plasma discharge.

DRC: An abbreviation for **dynamic reaction cell.**

Droplet: Refers to individual particles (either small or large) that make up an aerosol generated by the nebulizer.

dry plasma: When a sample is introduced into the plasma that does not contain any liquid or solvent, such as laser ablation, electrothermal vaporization (ETV), or desolvation sample introduction systems.

duty cycle (%): Also known as the "measurement duty cycle." It refers to the actual peak measurement time and is expressed as a percentage of the overall integration time. It is calculated by dividing the total peak quantitation time (dwell time × number of sweeps × replicates × elements) by the total integration time ([dwell time + settling/scanning time] × number of sweeps × replicates × elements).

dwell time: The time spent sitting (dwelling) on top of the analytical peak (mass) and taking measurements.

dynamically scanned ion lens: A commercial ion optic approach to focus the maximum number of ions into the mass analyzer. In this design, the voltage is dynamically ramped on the fly in concert with the mass scan of the analyzer. The benefit is that the optimum lens voltage is placed on every mass in a multielement run to allow the maximum number of analyte ions through, while keeping the matrix ions down to an absolute minimum. This is typically used in conjunction with a grounded stop acting as a physical barrier to reduce particulates, neutral species, and photons from reaching the mass analyzer and detector.

dynamic reaction cell (DRC): A type of collision/reaction cell. Unlike a simple collision cell, a quadrupole is used instead of a hexapole or octopole. A highly reactive gas, such as ammonia or methane, is bled into the cell, which is a catalyst for ion–molecule chemistry to take place. By a number of different reaction mechanisms, the gaseous molecules react with the interfering ions to convert them into either an innocuous species different from the analyte mass or a harmless neutral species. The analyte mass then emerges from the DRC, free of its interference, and is steered into the analyzer quadrupole for conventional mass separation. Through careful optimization of the quadrupole electrical fields, unwanted reactions between the gas and the sample matrix or solvent, which could potentially lead to new interferences, are prevented. Therefore, every time an analyte and interfering ions enter the DRC, the bandpass of the quadrupole can be optimized for that specific problem and then changed on the fly for the next one.

E

EDR: An abbreviation for **extended dynamic range**, used in detector technology.

electrodynamic forces: Flow of the ion beam through the interface region, where the positively charged ions of varying mass-to-charge ratios exert no electrical influence on each other.

electron: A negatively charged fundamental particle orbiting the nucleus of an atom. It has a mass equal to 1/1836 of a proton's mass. Removal of an electron by excitation in the plasma discharge generates a positively charged ion.

electrostatic analyzer (ESA): An ion-focusing device (utilizing a series of electrostatic lens components) that varies the electric field to allow the passage of ions of certain energy. In ICP-MS, it is typically used in combination with a conventional electromagnet to focus ions based on their angular motion and their kinetic energy to produce very high resolving power. *Also refer to* **double-focusing magnetic sector mass spectrometer (analyzer)**.

electrothermal atomization (ETA): An atomic absorption (AA) analytical technique that uses a heated metal filament or graphite tube (in place of the normal flame) to generate ground-state analyte atoms. The sample is first injected into the filament or tube, which is heated up slowly to remove the matrix components. Further heating then generates ground-state atoms of the analyte, which absorb light of a particular wavelength from an element-specific, hollow cathode lamp source. The amount of light absorbed is measured by a monochromator (optical system) and detected by a photomultiplier or solid-state detector, which converts the photons into an electrical pulse. This absorbance signal is used to determine the concentration of that element in the sample. Typically used for parts per billion–level determinations.

electrothermal vaporization (ETV): A sample pretreatment technique used in ICP-MS. Based on the principle of electrothermal atomization (ETA) used in atomic absorption (AA), ETV is not used to generate ground-state atoms, but instead uses a carbon furnace (tube) or metal filament to thermally separate the analytes from the matrix components and then sweep them into the ICP mass spectrometer for analysis. This is achieved by injecting a small amount of the sample into a graphite tube or onto a metal filament. After the sample is introduced, drying, charring, and vaporization are achieved by slowly heating the graphite tube or metal filament. The sample material is vaporized into a flowing stream of carrier gas, which passes through the furnace or over the filament during the heating cycle. The analyte vapor recondenses in the carrier gas and is then swept into the plasma for ionization.

elemental fractionation: A term used in laser ablation. It is typically defined as the variation in intensity of a particular element over time compared with the total amount of dry aerosol generated by the sample. It is generally sample and element specific, but there is evidence to suggest that the shorter-wavelength excimer lasers exhibit better elemental fractionation characteristics than the longer-wavelength Nd:YAG design because they produce smaller particles that are easier to volatilize.

endothermic reaction: In thermodynamics, this describes a chemical reaction that absorbs energy in the form of heat. In ICP-MS, it generally refers to an ion–molecule reaction in a collision/reaction cell that is not allowed to proceed because the ionization potential of the analyte ion is significantly less than that of the reaction gas molecule. *Also refer to* **exothermic reaction**.

engineered nanomaterial (ENM): This is a man-made material made of particles with <100 nm diameter that can be made to exhibit greater physical strength, enhanced magnetic properties, conduction of heat or electricity, greater chemical reactivity, or size-dependent optical properties. An example of an ENM is silver nanoparticles, which are added to detergents as a bactericide.

ENM: An abbreviation for engineered nanomaterials.

ESA: An abbreviation for **electrostatic analyzer**.

ETA: An abbreviation for **electrothermal atomization**.

ETV: An abbreviation for **electrothermal vaporization**.

excimer laser: A gas-filled laser in which a very short electrical pulse excites a mixture containing a halogen, such as fluorine, and a rare gas, such as argon or krypton. It produces a brief, intense pulse of ultraviolet light. The output of an excimer laser is used for writing patterns on semiconductor chips because the short wavelength can write very fine lines. In ICP-MS, the most common excimer laser used is ArF at 193 nm, and it is typically used to ablate material with a very small size, such as inclusions on the surface of a geological sample.

exothermic reaction: In thermodynamics, this describes a chemical reaction that releases energy in the form of heat. In ICP-MS, it generally refers to an ion–molecule reaction in a collision/reaction cell that is spontaneous because the ionization potential of the interfering ion is much greater than that of the reaction gas molecule. *Also refer to* **endothermic reaction**.

extended dynamic range (EDR): An approach used in ICP-MS to extend the linear dynamic range of the detector from five orders of magnitude up to eight or nine orders of magnitude. *Also refer to* **discrete dynode detector** *and* **Faraday cup detector**.

external standardization: The normal mode of calibration used in ICP-MS by comparing the analyte intensity of unknown samples with the intensity of known calibration or reference standards.

extraction lens: An ion lens used to electrostatically extract the ions out of the interface region.

F

FAA: An abbreviation for **flame atomic absorption**.

Faraday collector: Another name for a Faraday cup detector.

Faraday cup detector: A simple metal electrode detector used to measure high ion counts. When the ion beam hits the metal electrode, it will be charged, whereas the ions are neutralized. The electrode is then discharged to measure a small current equivalent to the number of discharged ions. By measuring the ion current on the metal part of the circuit, the number of ions in the circuit can be determined. Unfortunately, with this approach, there is no control over the applied voltage (gain). So, it can only be used for high ion counts, and therefore is not suitable for ultratrace determinations.

field-flow fractionation (FFF): A single-phase chromatographic separation technique, where separation is achieved within a very thin channel, against which a perpendicular force field is applied. One of the most common forms of FFF is asymmetrical flow FFF (AF4), where the field is generated by a cross-flow applied perpendicular to the channel. Coupled with ICP-MS for the characterization of nanoparticles.

FFF: An abbreviation for field-flow fractionation.

FGDW: An abbreviation for flue gas desulfurization wastewater.

FIA: An abbreviation for **flow injection analysis**.

flame atomic absorption (FAA): An atomic absorption analytical technique that uses a flame (usually air–acetylene or nitrous oxide–acetylene) to generate ground-state atoms. The sample solution is aspirated into the flame via a nebulizer and a spray chamber. The ground-state atoms of the sample absorb light of a particular wavelength from an element-specific, hollow cathode lamp source. The amount of light absorbed is measured by a monochromator (optical system) and detected by a photomultiplier or solid-state detector, which converts the photons into an electrical pulse. This absorbance signal is used to determine the concentration of the element in the sample. Typically used for parts per million–level determinations.

flatapole: A quadrupole with rods that have flat corners. Used in a particular commercial design of collision/reaction cell.

flight tube: A generic name given to the housing that contains a series of optical components that focus ions onto the detector of a time-of-flight (TOF) mass analyzer. There are basically two different kinds of flight tubes that are used in commercial TOF mass analyzers. One is the orthogonal design, where the flight tube is positioned at right angles to the sampled ion beam, and the other is the axial design, where the flight tube is in the same axis as the ion beam. In both designs, all ions are sampled through the interface region, but instead of being focused into the mass filter in the conventional sequential way, packets (groups) of ions are electrostatically injected into the flight tube at exactly the same time.

flow injection analysis (FIA): A powerful front-end sampling accessory for ICP-MS that can be used for preparation, pretreatment, and delivery of the sample. It involves the introduction of a discrete sample aliquot into a flowing carrier stream. Using a series of automated pumps and valves, procedures can be carried out online to physically or chemically change the sample or analyte before introduction into the mass spectrometer for detection.

flue gas desulfurization wastewater (FGDW): This is one of the most widely used technologies for removing pollutants, such as sulfur dioxide, from flue gas emissions produced by coal-fired power plants. Sometimes called the limestone-forced oxidation scrubbing system, but more commonly known as flue gas desulfurization (FGD), this process employs gas scrubbers to spray limestone slurry over the flue gas to convert gaseous sulfur dioxide to calcium sulfate.

Fractogram: The separated particles that exit the outlet port of a field-flow fractionation device into the detection system (e.g., UV/Vis or ICP-MS) are displayed as a temporal signal called a fractogram (similar to a chromatogram in chromatographic separation).

fringe rods: A set of four short rods operated in the radio-frequency (RF)-only mode, positioned at the entrance of a quadrupole mass analyzer. Their function is to minimize the effect of the fringing fields at the entrance of a quadrupole mass analyzer and thus improve the efficiency of transmission of ions into the mass analyzer. They are usually straight, but it has been suggested that curved fringe rods might reduce background levels.

fusion mixture: A compound or mixture added to solid samples as an aid to get them into solution. Fusion mixtures are usually alkaline salts (e.g., lithium metaborate and sodium carbonate) that are mixed with the sample (in powdered form) and heated in a muffle furnace to create a chemical or thermal reaction between the sample and the salt. The fused mixture is then dissolved in a weak mineral acid to get the analytes into solution.

G

gamma-counting spectrometry: A particle-counting technique that uses the measurement of the radioactive decay of gamma particles. *Also refer to* **particle-counting techniques**.

gas dynamics: In ICP-MS, the flow and velocity of the plasma gas through the interface region. It dictates that the composition of the ion beam immediately behind the sampler cone be the same as the composition in front of the cone because the expansion of the gas at this stage is not controlled by electrodynamics. This happens because the distance over which ions exert influence on one another (the Debye length) is small compared with the orifice diameter of the sampler or skimmer cone. Consequently, there is little electrical interaction between the ion beam and the cone and relatively little interaction between the individual ions in the beam. In this way, gas dynamics ensures that the compositional integrity of the ion beam is maintained throughout the interface region.

getter (gas purifier): A device that "cleans up" inorganic and organic contaminants in pure gases. The getter usually refers to a metal that oxidizes quickly and, when heated to a high temperature (usually by means of radio-frequency [RF] induction), evaporates and absorbs or reacts with any residual impurities in the gas.

GFAA: An abbreviation for **graphite furnace atomic absorption**.

graphite furnace atomic absorption (GFAA): An electrothermal atomization (ETA) analytical technique that specifically uses a graphite tube (in place of the normal flame) to generate ground-state analyte atoms. The sample is first injected into the tube, which is heated up slowly to remove the matrix components. Further heating then generates ground-state atoms of the analyte, which absorb light of a particular wavelength from an element-specific, hollow cathode lamp source. The amount of light absorbed is measured by a monochromator (optical system) and detected by a photomultiplier or solid-state detector, which converts the photons into an electrical pulse. This absorbance signal is used to determine the concentration of that element in the sample. Typically used for parts per billion–level determinations. *Also refer to* **electrothermal atomization**.

grounding mechanism: A way of eliminating the secondary discharge (pinch effect) produced by capacitive (radio-frequency [RF]) coupling of the load coil to the plasma. This undesired coupling between the RF voltage on the load coil and the plasma discharge produces a potential difference of a few hundred volts, which creates an electrical discharge (arcing) between the plasma and sampler cone of the interface. This mechanism varies with different instrument designs, but basically involves grounding the load coil to make sure the interface region is maintained at zero potential.

H

half-life: The time required for half the atoms of a given amount of a radioactive substance to disintegrate. This principle is used in particle-counting measuring techniques.

heating zones: The zones that describe the different temperature regions within a plasma discharge, where the sample passes through. The most common zones include the preheating zone (PHZ), where the sample is desolvated; the initial radiation zone (IRZ), where the sample is broken down into its molecular form; and the normal analytical zone (NAZ), where the sample is first atomized and then ionized.

HEN: An abbreviation for **high-efficiency nebulizer**.

hexapole: A multipole containing six rods, used in collision/reaction cell technology.

HGAA: An abbreviation for **hydride generation atomic absorption**.

high-efficiency nebulizer (HEN): A generic name given to a nebulizer that is very efficient, with very little wastage. Usually used to describe direct injection or micro-concentric-designed systems, which deliver all or a very high percentage of the sample aerosol into the plasma discharge.

high-resolution mass analyzer: A generic name given to a mass spectrometer with very high resolving power. Commercial designs are usually based on the double-focusing magnetic sector design.

high-sensitivity interface (HSI): HSIs are offered as an option with most commercial ICP-MS systems. They all work slightly differently but share similar components. By using a slightly different cone geometry, higher vacuum at the interface, one or more extraction lenses, or modified ion optic design, they offer up to 10 times the sensitivity of a traditional interface. However, their limitations are that background levels are often elevated, particularly when analyzing samples with a heavy matrix. Therefore, they are more suited for the analysis of clean solutions.

high-solids nebulizers: Nebulizers that are used to aspirate higher concentrations of dissolved solids into the ICP-MS. The most common types used are the Babbington, V-groove, and cone-spray designs. Not widely used for ICP-MS because of the dissolved solids limitations of the technique, but sometimes used with flow injection sample introduction techniques.

hollow ion mirror: A more recent development in ion-focusing optics. The ion mirror, which has a hollow center, creates a parabolic electrostatic field to reflect and refocus the ion beam at right angles to the ion source. This allows photons, neutrals, and solid particles to pass through it, while allowing ions to be reflected at right angles into the mass analyzer. The major benefit of this design is the highly efficient way the ions are refocused, offering extremely high sensitivity and low background across the mass range.

homogenized sample beam: The laser beam in an excimer laser, which produces a much flatter beam profile and more precise control of the ablation process.

hydride generation atomic absorption (HGAA): A very sensitive analytical technique for determining trace levels of volatile elements, such as As, Bi, Sb, Se, and Te. Generation of the elemental hydride is carried out in a closed vessel by the addition of a reducing agent, such as sodium borohydride, to the acidic sample. The resulting gaseous hydride is swept into a special heated quartz cell (in place of the traditional flame burner head), where atomization occurs. Atomic absorption quantitation is then carried out in the conventional way, by comparing the absorbance of unknown samples against known calibration or reference standards.

hydrogen atom transfer: An ion–molecule reaction mechanism in a collision/reaction cell where a hydrogen atom is transferred to the interfering ion, which is converted to an ion at one mass higher.

hyperbolic fields: The four rods that make up a quadrupole are usually cylindrical or elliptical in shape. The electrical fields produced by these rods are typically hyperbolic in shape.

hyper skimmer cone: The name for an additional cone used in one commercial ICP-MS interface design, used in order to tighten the ion beam entering the ion optics.

I

ICP: An abbreviation for **inductively coupled plasma**.

ICP-OES: An abbreviation for **inductively coupled plasma optical emission spectrometry**.

IDL: An abbreviation for instrument detection limit.

impact bead (nebulizer): A type of spray chamber more commonly used in atomic absorption spectrometers. The aerosol from the nebulizer is directed onto a spherical bead, where the impact breaks the sample into large and small droplets. The large droplets fall out due to gravitational force, and the smaller droplets are directed by the nebulizer gas flow into the atomization, excitation, or ionization source.

inductively coupled plasma (ICP): The high-temperature source used to generate ions in ICP-MS. It is formed when a tangential (spiral) flow of argon gas is directed between the outer and middle tubes of a quartz torch. A load coil (usually copper) surrounds the top end of the torch and is connected to a radio-frequency (RF) generator. When RF power (typically 750–1500 W) is applied to the load coil, an alternating current oscillates within the coil at a rate corresponding to the frequency of the generator. The RF oscillation of the current in the coil creates an intense electromagnetic field in the area at the top of the torch. With argon gas flowing through the torch, a high-voltage spark is applied to the gas, causing some electrons to be stripped from their argon atoms. These electrons, which are caught up and accelerated in the magnetic field, then collide with other argon atoms, stripping off still more electrons. This collision-induced ionization of the argon continues in a chain reaction, breaking down the gas into argon atoms, argon ions, and electrons, forming what is known as an ICP discharge at the open end of the plasma torch.

inductively coupled plasma optical emission spectrometry (ICP-OES): A multi-element technique that uses an inductively coupled plasma to excite ground-state atoms to the point where they emit wavelength-specific photons of light, characteristic of a particular element. The number of photons produced at an element-specific wavelength is measured using high-resolving optical components to separate the analyte wavelengths and a photon-sensitive detection system to measure the intensity of the emission signal produced. This emission signal is directly related to the concentration of that element in the sample. Commercial instrumentation comes in two configurations: a traditional radial view, where the plasma is vertical and is viewed from the side (side-on viewing), and an axial view, where the plasma is positioned horizontally and is viewed from the end (end-on viewing).

infrared (IR) lasers: Laser ablation systems that operate in the IR region of the electromagnetic spectrum, such as the Nd:YAG laser, which has its primary wavelength at 1064 nm.

instrument background noise: Square root of the spectral background of the instrument (in cps), usually measured at mass 220 amu, where there are no spectral features. *Also refer to* **background signal**, **background noise**, *and* **detection limit**.

instrument background signal: Spectral background of the instrument (in cps), usually measured at mass 220 amu, where there are no spectral features. *Also refer to* **background signal**, **background noise**, *and* **detection limit**.

instrument detection limit (IDL): A way of assessing an instrument's detection capability. It is often referred to as signal-to-background noise, and for a 99% confidence level, it is typically defined as three times the standard deviation (SD) of 10 replicates of the sample blank.

integration time: The total time spent measuring an analyte mass (peak). Comprising the time spent dwelling (sitting) on the peak multiplied by the number of points used for peak quantitation multiplied by the number of scans used in the measurement protocol. *Also refer to* **duty cycle, measurement duty cycle, peak measurement protocol, settling time**, *and* **dwell time**.

interface: The plasma discharge is coupled to the mass spectrometer via the interface. The interface region comprises a water-cooled metal housing containing the sampler cone and the skimmer cone, which directs the ion beam from the central channel of the plasma into the ion optic region.

interface cones: Refer to the sampler and skimmer cones housed in the interface region. *Also refer to* **interface** *and* **interface region.**

interface pressure: The pressure between the sampler cone and skimmer cone. This region is maintained at a pressure of approximately 1–2 torr by a mechanical roughing pump.

interface region: A region comprising a water-cooled metal housing containing the sampler cone and the skimmer cone, which directs and focuses the ion beam from the central channel of the plasma into the ion optic region.

interferences: A generic term given to a nonanalyte component that enhances or suppresses the signal intensity of the analyte mass. The most common interferences in ICP-MS are spectral, matrix, or sample transport in nature.

internal standardization (IS): A quantitation technique used to correct for changes in analyte sensitivity caused by variations in the concentration and type of matrix components found in the sample. An internal standard is a nonanalyte isotope that is added to the blank solution, standards, and samples before analysis. It is typical to add three or four internal standard elements to the samples to cover all the analyte elements of interest across the mass range. The software adjusts the analyte concentration in the unknown samples by comparing the intensity values of the internal standard elements in the unknown sample to those in the calibration standards. Because ICP-MS is prone to many matrix- and sample transport–based interferences, internal standardization is considered necessary to analyze most sample types.

ion: An electrically charged atom or group of atoms formed by the loss or gain of one or more electrons. A cation (positively charged ion) is created by the loss of an electron, and an anion (negatively charged ion) is created by the gain of an electron. The valency of an ion is equal to the number of electrons lost or gained and is indicated by a plus sign for cations and a minus sign for anions. ICP-MS typically involves the detection and measurement of positively charged ions generated in a plasma discharge.

ion chromatography (IC): A chromatographic separation technique used for the determination of anionic species, such as nitrates, chlorides, and sulfates. When coupled with ICP-MS, it becomes a very sensitive hyphenated technique for the determination of a wide variety of elemental ionic species.

ion energy: In ICP-MS, it refers to the kinetic energy of the ion, in electron volts (eV). It is a function of both the mass and velocity of the ion ($KE = \frac{1}{2}mV^2$). It is generally accepted that the spread of kinetic energies of all the ions in the ion beam entering the mass spectrometer must be on the order of a few electron volts to be efficiently focused by the ion optics and resolved by the mass analyzer.

ion energy spread: The variation in kinetic energy of all the ions in the ion beam emerging from the ionization source (plasma discharge). It is generally accepted that this variation (spread) of kinetic energies must be on the order of a few electron volts to be efficiently focused by the ion optics and resolved by the mass analyzer.

ion flow: The flow of ions from the interface region through the ion optics into the mass analyzer.

ion-focusing guide: An alternative name for the **ion optics**.

ion-focusing system: An alternative name for the **ion optics**.

ion formation: The transfer of energy from the plasma discharge to the sample aerosol to form an ion, by traveling through the different heating zones in the plasma, where the sample is first dried, vaporized, and atomized, and then finally converted to an ion.

ionization source: In ICP-MS, the ionization source is the plasma discharge, which reaches temperatures of up to 10,000 K to ionize the liquid sample.

ionizing radiation counting techniques: Particle-counting techniques, such as alpha, gamma, and scintillation counters, that are used to measure the isotopic composition of radioactive materials. However, the limitation of particle-counting techniques is that the half-life of the analyte isotope has a significant impact on the method detection limit. This implies that they are better suited for the determination of short-lived radioisotopes, because meaningful data can be obtained in a realistic amount of time. They have also been successfully applied to the quantitation of long-lived radionuclides, but unfortunately require a combination of extremely long counting times and large amounts of sample to achieve low levels of quantitation.

ion kinetic energy: *Refer to* **ion energy**.

ion lens: Often referred to as a single-lens component in the ion optic system. *Also refer to* **ion optics**.

ion lens voltages: The voltages put on one or more lens components in the ion optic system to electrostatically steer the ion beam into the mass analyzer. *Also refer to* **ion optics**.

ion mirror: A more recent development in ion-focusing optics. With this design, a parabolic electrostatic field is created with a hollow ion mirror to reflect and refocus the ion beam at right angles to the ion source. The ion mirror is an electrostatically charged ring, which is hollow in the center. This allows photons, neutrals, and solid particles to pass through it, while allowing ions to be reflected at right angles into the mass analyzer. The major benefit of this design is the highly efficient way the ions are refocused, offering extremely high sensitivity across the mass range, with very little compromise in oxide performance. In addition, there is very little contamination of the ion optics because a vacuum pump sits behind the ion mirror to immediately remove these particles before they have a chance to penetrate further into the mass spectrometer.

ion–molecule chemistry: A chemical reaction between the analyte or interfering ion and molecules of the reaction gas in a collision/reaction cell. A reactive gas, such as hydrogen, ammonia, oxygen, methane, or gas mixtures, is bled into the cell, which is a catalyst for ion–molecule chemistry to take place. By a number of different reaction mechanisms, the gaseous molecules react with the interfering ions to convert them into either an innocuous species different from the analyte mass or a harmless neutral species. The analyte mass then emerges from the cell free of its interference and is steered into the analyzer quadrupole for conventional mass separation. In some cases, the chemistry can take place between the gaseous molecule and the analyte to form a new analyte ion free of the interfering species.

ion optics: Comprises one or more electrostatically charged lens components that are positioned immediately after the skimmer cone. They are made up of a series of metallic plates, barrels, or cylinders, which have a voltage placed on them. The function of the ion optic system is to take ions after they emerge from the interface region and steer them into the mass analyzer. Another function of the ion optics is to reject the nonionic species, such as particulates, neutral species, and photons, and prevent them from reaching the detector. Depending on the design, this is achieved by using some kind of physical barrier, positioning the mass analyzer off axis relative to the ion beam, or electrostatically bending the ions by 90° into the mass analyzer.

ion packet: A "slice of ions" that is sampled from the ion beam in a time-of-flight (TOF) mass analyzer. In the TOF design, all ions are sampled through the interface cones, but instead of being focused into the mass filter in the conventional way, packets (groups) of ions are electrostatically injected into the flight tube at exactly the same time. Whether the orthogonal (right-angle) or axial (straight-on) approach is used, an accelerating potential is applied to the continuous ion beam. The ion beam is then "chopped" by using a pulsed voltage supply to provide repetitive voltage "slices" at a frequency of a few kilohertz. The "sliced" packets of ions are then allowed to "drift" into the flight tube, where the individual ions are temporally resolved according to their differing velocities.

ion repulsion: The degree to which positively charged ions repel each other as they enter the ion optics. The generation of a positively charged ion beam is the first stage in the charge separation process. Unfortunately, the net positive charge of the ion beam means that there is now a natural tendency for the ions to repel one another. If nothing is done to compensate for this repulsion, ions of higher mass-to-charge ratio will dominate the center of the ion beam and force the lighter ions to the outside. The degree of loss will depend on the kinetic energy of the ions—ions with high kinetic energy (high-mass elements) will be transmitted in preference to ions with medium (midmass elements) or low kinetic energy (low-mass elements).

isobar (or isobaric): Used in the context of atomic principles, it refers to two or more atoms with the same atomic mass (same number of neutrons) but different atomic number (different number of protons). *Also refer to* **isobaric interferences**.

isobaric interferences: The word *isobaric* is used in the context of atomic principles, and refers to two or more atoms with the same atomic mass but different atomic number. In ICP-MS, they are a classification of spectrally induced interferences produced mainly by different isotopes of other elements in the sample, creating spectral interferences at the same mass as the analyte.

Isotope: A different form of an element having the same number of protons in the nucleus (i.e., same atomic number) but a different number of neutrons (i.e., different atomic mass). There are 275 isotopes of the 81 stable elements in the periodic table, in addition to more than 800 radioactive isotopes. Isotopes of a single element possess very similar properties.

isotope dilution: An absolute means of quantitation in ICP-MS based on altering the natural abundance of two isotopes of an element by adding a known amount of one of the isotopes. The principle works by spiking the sample solution with a known weight of an enriched stable isotope. By knowing the natural abundance of the two isotopes being measured, the abundance of the spiked enriched isotope, the weight of the spike, and the weight of the sample, it is possible to determine the original trace element concentration. It is considered one of the most accurate and precise quantitation techniques for elemental analysis by ICP-MS.

isotope ratio: The ability of ICP-MS to determine individual isotopes makes it suitable for an isotopic measurement technique called "isotope ratio analysis." The ratio of two or more isotopes in a sample can be used to generate very useful information, such as an indication of the age of a geological formation, a better understanding of animal metabolism, and the identification of sources of environmental contamination. Similar to isotope dilution, isotope ratio analysis uses the principle of measuring the exact ratio of two isotopes of an element in the sample. With this approach, the isotope of interest is typically compared with a reference isotope of the same element, but can also be referenced to an isotope of another element.

isotope ratio precision: The reproducibility or precision of measurement of isotope ratios is very critical for some applications. For the highest-quality isotopic ratio precision measurements, it is generally acknowledged that either magnetic sector or time-of-flight (TOF) instrumentation offers the best approach over quadrupole ICP-MS.

isotopic abundance: The percentage abundance of an isotope compared with the element's total abundance in nature. *Also refer to* **natural abundance** *and* **relative abundance of natural isotopes**.

K

KE: An abbreviation for **kinetic energy**.

KED: An abbreviation for **kinetic energy discrimination**.

kinetic energy (KE): The energy possessed by a moving body due to its motion. It is equal to one-half the mass of the body times the square of its speed (velocity): $KE = \frac{1}{2}mV^2$. For KE as applied to moving ions, *refer to* **ion energy**.

kinetic energy discrimination (KED): In collision/reaction cell technology, it is one way to separate the newly formed by-product ions from the analyte ions. It is typically achieved by setting the collision cell potential (voltage) slightly more negative than the mass filter potential. This means that the collision by-product ions generated in the cell, which have a lower kinetic energy as a result of the collision process, are rejected, whereas the analyte ions, which have a higher kinetic energy, are transmitted to the mass analyzer.

L

laser ablation: A sample preparation technique that uses a high-powered laser beam to vaporize the surface of a solid sample and sweep it directly into the ICP-MS system for analysis. It is mainly used for samples that are extremely difficult to get into solution or for samples that require the analysis of small spots or inclusions on the surface.

laser absorption: The "coupling" efficiency of the sample with the laser beam in laser ablation work. The more light the sample absorbs, the more efficient the ablation process becomes. It is generally accepted that the shorter-wavelength excimer lasers have better absorption characteristics than the longer-wavelength infrared laser systems for ultraviolet-transparent or opaque materials, such as calcites, fluorites, and silicates, and as a result generate a smaller particle size and higher flow of ablated material.

laser fluence: A term used to describe the power density of a laser beam in laser ablation studies. It is defined as the laser pulse energy per focal spot area, measured in joules per square centimeter. It is related to laser irradiance, which is the ratio of the fluence to the width of the laser pulse.

laser irradiance: A term used to describe the power density of a laser beam in laser ablation studies. Laser irradiance is the ratio of the laser pulse energy per focal spot area (i.e., fluence) to the width of the laser pulse. *Also refer to* **laser fluence**.

laser sampling: *Refer to* **laser ablation**.

laser vaporization: *Refer to* **laser ablation**.

laser wavelength: The primary wavelength of the optical components used in the design of a laser ablation system.

linear plane array detectors: Solid-state detector technology used to measure the mass spectrum in a simultaneous manner (recently commercialized in the Mattauch–Herzog magnetic sector instrument).

load coil: Another name for the radio-frequency (RF) coil used to generate a plasma discharge. *Also refer to* **RF generator**.

low-mass cutoff: This is a variation on bandpass filtering in a collision/reaction cell, which uses slightly different control of the filtering process. By operating the cell in the radio-frequency (RF)-only mode, the quadrupole's stability boundaries can be tuned to cutoff low masses where the majority of the interferences occur.

low-temperature plasma: An alternative name for cool or cold plasma.

M

magnetic field: A region around a magnet, an electric current, or a moving charged particle that is characterized by the existence of a detectable magnetic force at every point in the region and by the existence of magnetic poles. In ICP-MS, it usually refers to the magnetic field around the radio-frequency (RF) coil of the plasma discharge or the magnetic field produced by a quadrupole or an electromagnet.

magnetic sector mass analyzer: A design of mass spectrometer used in ICP-MS to generate very high resolving power as a way of reducing spectral interferences. Commercial designs typically utilize a very powerful magnet combined with an electrostatic analyzer (ESA). In this approach, known as the double-focusing design, the ions from the plasma are sampled. In the plasma, the ions are accelerated to a few kilovolts into the ion optic region before they enter the mass analyzer. The magnetic field, which is dispersive with respect to ion energy and mass, then focuses all the ions with diverging angles of motion from the entrance slit. The ESA, which is only dispersive with respect to ion energy, then focuses all the ions onto the exit slit, where the detector is positioned. If the energy dispersions of the magnet and ESA are equal in magnitude but opposite in direction, they will focus both ion angles (first focusing) and ion energies (second focusing) when combined together. *Also refer to* **electrostatic analyzer**.

mass analyzer: The part of the mass spectrometer where the separation of ions (based on their mass-to-charge ratio) takes place. In ICP-MS, the most common type of mass analyzers are quadrupole, magnetic sector, and time-of-flight (TOF) systems.

mass calibration: The ability of the mass spectrometer to repeatedly scan to the same mass position every time during a multielement analysis. Instrument manufacturers typically quote a mass calibration stability specification for their design of mass analyzer based on the drift or movement of the peak (in atomic mass units) position over a fixed period of time (usually 8 h).

mass calibration stability: *Refer to* **mass calibration**.

mass discrimination: Sometimes called "mass bias." In ICP-MS, it occurs when a higher-concentration isotope is suppressing the signal of the lower-concentration isotope, producing a biased result. The effect is not so obvious if the concentrations of the isotopes in the sample are similar, but can be quite significant if the concentrations of the two isotopes are vastly different. If that is the case, it is recommended to run a standard of known isotopic composition to compensate for the effects of the suppression.

mass filter: Another name for a mass analyzer.

mass filtering discrimination: A way of discriminating between analyte ions and the unwanted by-product interference ions generated in a collision/reaction cell.

mass resolution: A measure of a mass analyzer's ability to separate an analyte peak from a spectral interference. The resolution of a quadrupole is nominally 1 amu and is traditionally defined as the width of a peak at 10% of its height.

mass scanning: The process of electronically scanning the mass separation device to the peak of interest and taking analytical measurements. Basically, two approaches are used: single-point peak hopping, in which a measurement is typically taken at the peak maximum, and the multipoint-scanning approach, in which a number of measurements are taken across the full width of the peak. *Also refer to* **ramp scanning**, **integration time**, **dwell time**, **settling time**, **peak measurement protocol**, *and* **peak hopping**.

mass separation: The process of separating the analyte ions from the nonanalyte, matrix, solvent, and interfering ions with the mass analyzer.

mass separation device: Another name for a mass analyzer.

mass-shift mode: A mode used with a triple-quadruple collision/reaction cell instrument. In this configuration, Q1 and Q2 are set to different masses. Similar to the on-mass mode, Q1 is set to the precursor ion mass (analyte and on-mass polyatomic interfering ions), controlling the ions that enter the octopole collision/reaction cell. However, in the mass-shift mode, Q2 is then set to the mass of a target reaction product ion containing the original analyte. Mass-shift mode is typically used when the analyte ion is reactive, while the interfering ions are unreactive with a particular collision/reaction cell gas.

mass spectrometer: The mass spectrometer section of an ICP-MS system is generally considered to be everything in the vacuum chamber from the interface region to the detector, including the interface cones, ion optics, mass analyzer, detector, and vacuum pumps.

matching network (RF): The matching network of the radio-frequency (RF) generator compensates for changes in impedance (a material's resistance to the flow of an electric current) produced by the sample's matrix components or differences in solvent volatility. In crystal-controlled generators, this is usually done with mechanically driven servo-type capacitors. With free-running generators, the matching network is based on electronic tuning of small changes in the RF brought about by the sample, solvent, or matrix components.

mathematical correction equations: Used to compensate or correct for spectral interference in ICP-MS. Similar to interelement corrections (IECs) used in ICP-OES, they work on the principle of measuring the intensity of the interfering isotope or interfering species at another mass, which is ideally free of any interferences. A correction is then applied, depending on the ratio of the intensity of the interfering species at the analyte mass to its intensity at the alternate mass.

Mathieu stability plot: A graphical representation of the stability of an ion as it passes through the rods of a multipole mass separation device. It is a function of the ratio of the radio-frequency (RF) to the direct current (DC) placed on each pair of rods. A plot of these ratios of multiple ions traveling through the multipole shows which ions are stable and make it through the rods to the detector and which ions are unstable and get ejected from the multipole. The most well-defined stability boundaries are obtained with a quadrupole and become more diffuse with higher-order multipoles, such as hexapoles and octopoles.

matrix interferences: There are basically three types of matrix-induced interferences. The first, and simplest to overcome, is often called a "sample transport or viscosity effect" and is a physical suppression of the analyte signal brought on by the level of dissolved solids or acid concentration in the sample. The second type of matrix suppression is caused when the sample matrix affects the ionization conditions of the plasma discharge, which results in varying amounts of signal suppression, depending on the concentration of the matrix components. The third type of matrix interference is often called "space charge matrix suppression." This occurs mainly when low-mass analytes are being determined in the presence of larger concentrations of high-mass matrix components. It has the effect of defocusing the ion beam, and unless any compensation is made, the high-mass matrix element will dominate the ion beam, pushing the lighter elements out of the way, leading to low sensitivity and poor detection limits. The classical way to compensate for matrix interferences is to use internal standardization.

matrix separation: Usually refers to some kind of chromatographic column technology to remove the matrix components from the sample before it is introduced into the ICP-MS system.

Mattauch–Herzog magnetic sector design: One of the earliest designs of double-focusing magnetic sector mass spectrometers. In this design, which was named after the German scientists who invented it, two or more ions of different mass-to-charge ratios are deflected in opposite directions in the electrostatic and magnetic fields. The divergent monoenergetic ion beams are then brought together along the same focal plane. Recently commercialized using a simultaneous-based direct charge array detector.

MDL: An abbreviation for method detection limit.

measurement duty cycle: Also known as the duty cycle, it refers to a percentage of actual quantitation time compared with total integration time. It is calculated by dividing the total quantitation time (dwell time × number of sweeps × replicates × elements) by the total integration time ([dwell time + scanning plus settling time] × number of sweeps × replicates × elements).

membrane desolvation: Can be used with any sample introduction technique to remove solvent vapors. However, it is typically used with an ultrasonic or microconcentric nebulizer to remove the solvent from a liquid sample. In this design, the sample aerosol enters the membrane desolvator, where the solvent vapor passes through the walls of a tubular microporous polytetrafluoroethylene (PTFE) or Nafion membrane. A flow of argon gas removes the volatile vapor from the exterior of the membrane, while the analyte aerosol remains inside the tube and is carried into the plasma for ionization.

method detection limit (MDL): The MDL is broadly defined as the minimum concentration of analyte that can be determined from zero with 99% confidence. MDLs are calculated in a manner similar to that for instrument detection limits (IDLs), except that the test solution is taken through the entire sample preparation procedure before the analyte concentration is measured multiple times.

microconcentric nebulizer: Based on the concentric nebulizer design, but operates at much lower flow rates. Conventional nebulizers have a sample uptake rate of about 1 mL/min with an argon gas pressure of 1 L/min, whereas microconcentric nebulizers typically run at less than 0.1 mL/min and typically operate at much higher gas pressure to accommodate the lower sample flow rates.

microflow nebulizer: A generic name for nebulizers that operate at much lower flow rates than conventional concentric or cross-flow designs. *Also refer to* **microconcentric nebulizer**.

microporous membrane: A tubular membrane made of an organic microporous material, such as Teflon or Nafion, used in membrane desolvation. The sample aerosol enters the desolvation system, where the solvent vapor passes through the walls of the tubular membrane. A flow of argon gas then removes the volatile vapor from the exterior of the membrane, while the analyte aerosol remains inside the tube and is carried into the plasma for ionization.

microsampling: A generic name given to any front-end sampling device in atomic spectrometry that can be used for the preparation, pretreatment, and delivery of the sample to the spectrometric analyzer. The most common type of microsampling device used in ICP-MS is the flow injection technique, which involves the introduction of a discrete sample aliquot into a flowing carrier stream. Using a series of automated pumps and valves, procedures can be carried out online to physically or chemically change the sample or analyte, before introduction into the mass spectrometer for detection.

microwave digestion: A method of digesting difficult-to-dissolve solid samples using microwave technology. Typically, a dissolution reagent such as a concentrated mineral acid is added to the sample in a closed acid-resistant vessel contained in a specially designed microwave oven. By optimizing the current, temperature, and pressure settings, difficult samples can be dissolved in a relatively short time compared with traditional hot–plate sample digestion techniques.

microwave dissolution: An alternative name for **microwave digestion**.

microwave-induced plasma (MIP): The most basic form of electrodeless plasma discharge. In this device, microwave energy (typically 100–200 W) is supplied to the plasma gas from an excitation cavity around a glass or quartz tube. The plasma discharge in the form of a ring is generated inside the tube. Unfortunately, even though the discharge achieves a very high power density, the high excitation temperatures only exist along a central filament. The bulk of the MIP never goes above 2000–3000 K, which means it is prone to very severe matrix effects. In addition, it is easily extinguished when aspirating liquid samples, so it has found a niche as a detection system for gas chromatography.

MIP: An abbreviation for **microwave-induced plasma**.

molecular association reaction: An ion–molecule reaction mechanism in a collision/reaction cell, where an interfering ion associates with a neutral species (atom or molecule) to form a molecular ion.

molecular cluster ions: Species that are formed by two or more molecular ions combining together in a reaction cell to form molecular clusters.

molecular spectral interferences: Another name for polyatomic spectral interferences, which are typically generated in the plasma by the combination of two or more atomic ions. They are caused by a variety of factors, but are usually associated with the argon plasma or nebulizer gas used, matrix components in the solvent or sample, other elements in the sample, or entrained oxygen or nitrogen from the surrounding air.

monodisperse particulates: Nanoparticles, which are predominantly one size.

M/S mode: A mode used in a triple-quadrupole collision reaction cell where the first quadrupole acts as a simple ion guide allowing all ions through to the collision/reaction cell, similar to a traditional single-quad ICP-MS system that uses a collision cell.

M/S M/S mode: A mode used in a triple-quadrupole collision reaction cell where the first quadrupole is operated with a 1 amu fixed bandpass window, allowing only the target ions to enter the collision/reaction cell. This process can be implemented in two different ways: either the on-mass or mass-shift modes.

multichannel analyzer: The data acquisition system that stores and counts the ions as they strike the detector. As the ions emerge from the end of the quadrupole rods, they are converted into electrical pulses by the detector and stored by the multichannel analyzer. This multichannel data acquisition system typically has 20 channels per mass, and as the electrical pulses are counted in each channel, a profile of the mass is built up over the 20 channels, corresponding to the spectral peaks of the analyte masses being determined.

multichannel data acquisition: The process of storing and counting ions in ICP-MS. *Also refer to* **multichannel analyzer**.

multicomponent ion lens: An ion lens system consisting of several lens components, all of which have a specific role to play in the transmission of the analyte ions into the mass filter. To achieve the desired analyte specificity, the voltage can be optimized on every ion lens and is usually combined with an off-axis mass analyzer to reject unwanted photons and neutral species.

multipole: The generic name given to a mass filter that isolates an ion of interest by applying direct current (DC) or radio-frequency (RF) currents to pairs of rods. The most common type of mass analyzer multipole used in ICP-MS is the quadrupole (four rods). However, other higher orders of multipoles used in collision/reaction cell technology include hexapoles (six rods) and octopoles (eight rods).

m/z: Another way of expressing mass-to-charge ratio.

N

nanomaterials: Nanomaterials can occur in nature, such as clay minerals and humic acids; they can be incidentally produced by human activity, such as diesel emissions or welding fumes, or they can be specifically engineered to exhibit unique optical, electrical, physical, or chemical characteristics. Also refer to engineered nanomaterials.

nanometrology: The measurement and characterization of nanoparticles.

nanoparticles (NPs): Particles that are released from engineered nanomaterials when they enter the environment.

natural abundance: The natural amount of an isotope occurring in nature. *Also refer to* **isotopic abundance**.

natural isotopes: Different isotopic forms of an element that occur naturally on or beneath the earth's crust.

Nd:YAG laser: Nd:YAG is an acronym for neodymium-doped yttrium aluminum garnet, a compound that is used as the lasing medium for certain solid-state lasers. In this design, the YAG host is typically doped with around 1% neodymium by weight. Nd:YAG lasers are optically pumped using a flashlamp or laser diodes and emit light with a wavelength of 1064 nm in the infrared region. However, for many applications, the infrared light is frequency doubled, tripled, quadrupled, or quintupled by using additional optical components to generate output wavelengths in the visible and ultraviolet regions. Typical wavelengths used for laser ablation/ICP-MS work include 532 nm (doubled), 266 nm (quadrupled), and 213 nm (quintupled). Pulsed Nd:YAG lasers are usually operated in the so-called "Q-switching" mode, where an optical switch is inserted in the laser cavity, waiting for a maximum population inversion in the neodymium ions before it opens. Then the light wave can run through the cavity, depopulating the excited laser medium at maximum population inversion. In this Q-switched mode, output powers of 20 MW and pulse durations of less than 10 ns are achieved.

nebulizer: The component of the sample introduction system that takes the liquid sample and pneumatically breaks it down into an aerosol using the pressure created by a flow of argon gas. The concentric and cross-flow designs are the most common in ICP-MS.

neutral species: Species generated in the plasma torch that have no positive or negative charge associated with them. If they are not eliminated, they can find their way into the detector and produce elevated background levels.

neutron: A fundamental particle that is neutral in charge, found in the nucleus of an atom. It has a mass equal to that of a proton. The number of neutrons in the atomic nucleus defines the isotopic composition of that element.

Nier–Johnson magnetic sector design: Nier–Johnson double-focusing magnetic sector instrumentation is the technology that all modern magnetic sector instrumentation is based on. Named after the scientists who developed it, Nier–Johnson geometry comes in two different designs, the "standard" and "reverse" Nier–Johnson geometry. Both these designs, which use the same basic principles, consist of two analyzers: a traditional electromagnet analyzer and an electrostatic analyzer (ESA). In the standard (sometimes called "forward") design, the ESA is positioned before the magnet, and in the reverse design, it is positioned after the magnet.

ninety-degree (90°) ion lens: A design of ion optics that bends the ion beam 90°.

NP: *Refer to* nanoparticles.

O

octopole: A multipole mass filtering device containing eight rods. In ICP-MS, octopoles are typically used in collision/reaction cell technology.

off-axis ion lens: An ion lens system that is not on the same axis as the mass analyzer. Designed to stop particulates, neutral species, and photons from hitting the detector.

on-mass mode: A mode used with the triple-quadrupole collision/reaction cell. In this configuration, Q1 and Q2 are both set to the target mass. Q1 allows only the precursor ion mass to enter the cell (analyte and on-mass polyatomic interfering ions). The octopole collision/reaction cell then separates the analyte ion from the interferences using the reaction chemistry of a reactive gas, while Q2 measures the analyte ion at the target mass after the on-mass interferences have been removed by reactions in the cell.

oxide ions: Polyatomic ions that are formed between oxygen and other elemental components in the plasma gas, sample matrix, or solvent. They are generally not desirable because they can cause spectral overlaps on the analyte ions. Oxide formation is typically worse in the cooler zones of the plasma and, as a result, can be reduced by optimizing the radio-frequency (RF) power, nebulizer gas flow, and sampling position.

P

parabolic field: Shape of the magnetic fields produced by a quadrupole.

particle-counting techniques: Include alpha, gamma, and scintillation counters that are used to measure the isotopic composition of radioactive materials. However, the limitation of particle-counting techniques is that the half-life of the analyte isotope has a significant impact on the method's detection limit. This means that to get meaningful data in a realistic amount of time, they are better suited for the determination of short-lived radioisotopes. They have been successfully applied to the quantitation of long-lived radionuclides, but unfortunately require a combination of extremely long counting times and large amounts of sample to achieve low levels of quantitation.

peak hopping: A quantitation approach in which the quadrupole power supply is driven to a discrete position on the analyte mass (normally the maximum point) and allowed to settle (settling time), and then a measurement is taken for a fixed amount of time (dwell time). The integration time for that peak is the dwell time multiplied by the number of scans (scan time). Multielement peak quantitation involves peak hopping to every mass in the multielement run. *Also refer to* **measurement duty cycle** *and* **peak measurement protocol**.

peak integration: The process of integrating an analytical peak (mass). *Also refer to* **integration time**, **peak measurement protocol**, *and* **measurement duty cycle**.

peak measurement protocol: The protocol of scanning the quadrupole and measuring a peak in ICP-MS. In multielement analysis, the quadrupole is scanned to the first mass. The electronics are allowed to settle (settling time) and left to dwell for a fixed period of time at one or multiple points on the peak (dwell time), and signal intensity measurements are taken (based on the dwell time). The quadrupole is then scanned to the next mass and the measurement protocol repeated. The complete multielement measurement cycle (sweep) is repeated as many times as is needed to make up the total integration per peak and the number of required replicate measurements per sample analysis.

peak quantitation: The process of quantifying the peak in ICP-MS using calibration standards. *Also refer to* **peak hopping**, **peak integration**, *and* **peak measurement protocol**.

Peltier cooler: A thermoelectric cooler using the principle of generating a cold environment by creating a temperature gradient between two different materials. It uses electrical energy via a solid-state heat pump to transfer heat from a material on one side of the device to a different material on the other side, thus producing a temperature gradient across the device (similar to a household air conditioning system).

peristaltic pump: A small pump in the sample introduction system that contains a set of minirollers (typically 12) all rotating at the same speed. The constant motion and pressure of the rollers on the pump tubing feeds the sample through to the nebulizer. Peristaltic pumps are usually used with cross-flow nebulizers.

photon stop: A grounded metal disk in the ion lens system that is used as a physical barrier to stop particulate matter, neutral species, and photons from getting to the detector.

physical interferences: An alternative term used to describe sample transport- or viscosity-based suppression interferences.

pinch effect: An effect caused by an undesired electrostatic (capacitive) coupling between the voltage on the load coil and the plasma discharge, which produces a potential difference of a few hundred volts. This capacitive coupling is commonly referred to as the pinch effect and shows itself as a secondary discharge (arcing) in the region where the plasma is in contact with the sampler cone.

piston pump: A pump using a small piston or syringe to introduce the sample to the nebulizer (used instead of a peristaltic pump). They typically produce a more stable signal.

plasma discharge: Another name for an inductively coupled plasma (ICP).

plasma source: Refers to the radio-frequency (RF) hardware components that create the plasma discharge, including the RF generator, matching network, plasma torch, and argon gas pneumatics.

plasma torch: Another name for the quartz torch that is used to generate the plasma discharge. The plasma torch consists of three concentric tubes: an outer tube, middle tube, and sample injector. The torch can either be one piece, where all three tubes are connected, or have a demountable design, in which the tubes and the sample injector are separate. The gas (usually argon) that is used to form the plasma (plasma gas) is passed between the outer and middle tubes at a flow rate of 12–17 L/min. A second gas flow (auxiliary gas) passes between the middle tube and the sample injector at 1 L/min, and is used to change the position of the base of the plasma relative to the tube and the injector. A third gas flow (nebulizer gas), also at 1 L/min, brings the sample, in the form of a fine-droplet aerosol, from the sample introduction system and physically punches a channel through the center of the plasma. The sample injector is often made from other materials besides quartz, such as alumina, platinum, and sapphire, if highly corrosive materials need to be analyzed.

polyatomic spectral interferences: Another name for molecular-based spectral interferences, which are typically generated in the plasma by the combination of two or more atomic ions. They are caused by a variety of factors, but are usually associated with the argon plasma or nebulizer gas used, matrix components in the solvent or sample, other elements in the sample, or entrained oxygen or nitrogen from the surrounding air.

polydisperse particulates: Nanoparticles that are many different sizes and dimensions.

precursor ion: Usually refers to a polyatomic or isobaric interfering ion that is formed in the plasma, as opposed to a product (or by-product) ion that is formed in the collision/reaction cell.

product ion: Usually refers to a product (or by-product) ion that is formed in the collision/reaction cell, as opposed to a precursor interfering ion (polyatomic or isobaric) that is formed in the plasma.

proton: A stable, positively charged fundamental particle that shares the atomic nucleus with a neutron. It has a mass 1836 times that of the electron.

proton transfer: A reaction mechanism in a collision/reaction cell in which the interfering polyatomic species gives up a proton, which is then transferred to the reaction gas molecule to form a neutral atom.

pulse counting: Refers to the conventional mode of counting ions with the detector measurement circuitry. Depending on the type of detection system that is used, an ion emerges from the quadrupole and strikes the ion-sensitive surface (discrete dynode, Channeltron, etc.) of the detector to generate electrons. These electrons move down the detector and generate more secondary electrons. This process is repeated at each stage of the detector, producing a pulse of electrons that is finally captured by the detector's collecting and counting circuitry.

Q

QID An abbreviation for quadrupole ion deflector.

quadrupole: The most common type of mass separation device used in commercial ICP-MS systems. It consists of four cylindrical or hyperbolic metallic rods of the same length (15–20 cm) and diameter (approximately 1 cm). The rods are typically made of stainless steel or molybdenum and sometimes coated with a ceramic coating for corrosion resistance. A quadrupole operates by placing both a direct current (DC) field and a time-dependent alternating current (AC) of radio frequency (RF) 2–3 MHz on opposite pairs of the four rods. By selecting the optimum AC/DC ratio on each pair of rods, ions of a selected mass are then allowed to pass through the rods to the detector, while the others are unstable and ejected from the quadrupole.

quadrupole ion deflector (QID): A commercial ion optics design that bends the ion beam at right angles.

quadrupole power supply: Another name for the electronic components that control the radio-frequency (RF) and direct current (DC) voltages to change the mass filtering characteristics.

quadrupole scan rate: Scan rates of commercial quadrupole mass analyzers are on the order of 2500 amu/s. The quadrupole scan rate and the slope at which the radio-frequency (RF) and direct current (DC) voltages of the quadrupole power supply are scanned will determine the desired resolution setting. A steeper slope translates to higher resolution, whereas a shallower slope means poorer resolution.

quadrupole stability regions: The region of the Mathieu stability plot where the trajectory of an ion is stable and makes it through to the end of the quadrupole rods. All commercial ICP-MS systems that utilize quadrupole technology as the mass separation device operate in the first stability region, where resolving power is typically on the order of 500–600. If the quadrupole is operated in the second or third stability regions, resolving powers of 4000 and 9000, respectively, have been achieved. However, improving resolution using this approach has resulted in a significant loss of signal and higher background levels.

quantitative methods: The different kinds of quantitative analyses available in ICP-MS, which include traditional quantitative analysis (using external calibration, standard additions, or addition calibration), semiquantitative routines (semiquant), isotope dilution (ID) methods, isotope ratio (IR) measurements, and classical internal standardization (IS).

quartz torch: The standard plasma torch used in ICP-MS. *Also refer to* **plasma torch**.

R

radioactive isotope: Sometimes known as "radioisotope," a radioactive isotope is a natural or artificially created isotope of an element having an unstable nucleus that decays, emitting alpha, beta, or gamma rays until stability is reached. The stable end product is typically a nonradioactive isotope of another element.

radio-frequency (RF) generator: The power supply used to create the plasma discharge. Hardware includes the RF generator, matching network, plasma torch, and argon gas pneumatics.

ramp scanning: One of the two approaches for quantifying a peak in ICP-MS (peak hopping being the other). In the multichannel ramp-scanning approach, a continuous smooth ramp of $1 - n$ channels (where n is typically 20) per mass is made across the peak profile. Mainly used for accumulating spectral and peak shape information when doing mass scans. It is normally used for doing mass calibration and resolution checks and as a classical qualitative method development tool to find what elements are present in the sample and to assess their spectral implications on the masses of interest. Full-peak ramp scanning is not normally used for doing rapid quantitative analysis, because valuable analytical time is wasted taking data on the wings and valleys of the peak where the signal-to-noise ratio is poorest. For this kind of work, peak hopping is normally chosen.

reaction cell: A collision/reaction cell that specifically uses ion–molecule reactions to eliminate the spectral interference. Often used to describe a dynamic reaction cell (DRC).

reaction mechanism: The mechanism by which the interfering ion is reduced or minimized to allow the determination of the analyte ion. The most common collisional mechanisms seen in collision/reaction cells include collisional focusing, dissociation, and fragmentation, whereas the major reaction mechanisms include exothermic/endothermic associations, charge transfer, molecular associations, and proton transfer.

reactive gases: In ICP-MS, the term refers to gases such as hydrogen, ammonia, oxygen, methane, or those used to stimulate ion–molecule reactions in a collision/reaction cell (CRC) or collision/reaction interface (CRI).

relative abundance of natural isotopes: The isotopic composition expressed as a percentage of the total abundance of that element found in nature.

resolution: A measure of the ability of a mass analyzer to separate an analyte peak from a spectral interference. The resolution of a quadrupole is nominally 1 amu and is traditionally defined as the width of a peak at 10% of its height.

resolving power: Although resolving power and resolution are both a measure of a mass analyzer's ability to separate an analyte peak from a spectral interference, the term *resolving power* is normally associated with magnetic sector technology and is represented by the equation $R = m/\Delta m$, where m is the nominal mass at which the peak occurs and Δm is the mass difference between two resolved peaks. The resolving power of commercial double-focusing magnetic sector mass analyzers is on the order of 1,000–10,000, depending on the resolution setting chosen.

response tables: The intensity values for known concentrations of every elemental isotope stored in the instrument's calibration software. When semiquanitative analysis is carried out, the signal intensity of an unknown sample is compared against the stored response tables. By correcting for common spectral interferences and applying heuristic, knowledge-driven routines in combination with numerical calculations, a positive or negative confirmation can be made for each element present in the sample.

reverse Nier–Johnson double-focusing magnetic sector instrumentation: The technology that all modern magnetic sector instrumentation is based on. Named after the scientists who developed it, Nier–Johnson geometry comes in two different designs, the "standard" and "reverse" Nier–Johnson geometry. Both these designs, which use the same basic principles, consist of two analyzers: a traditional electromagnet analyzer and an electrostatic analyzer (ESA). In the standard (sometimes called "forward") design, the ESA is positioned before the magnet, and in the reverse design, it is positioned after the magnet.

RF generator: An alternative name for radio-frequency generator.

right-angled ion lens design: A recent development in ion-focusing optics, which utilizes a parabolic ion mirror to bend and refocus the ion beam at right angles to the ion source. The ion mirror incorporates a hollow structure that allows photons, neutrals, and solid particles to pass through it, while allowing ions to be deflected at right angles into the mass analyzer.

roughing pump: Traditional mechanical roughing or oil-based pumps are used in ICP-MS to pump the interface region down to approximately 1–2 torr and also to back up the turbomolecular pump used in the ion optics region of the mass spectrometer.

ruby laser: Ruby laser systems operate at 694 nm in the visible region of the electromagnetic spectrum.

S

S/B: An abbreviation for **signal-to-background ratio**.

sample aerosol: *Refer to* **aerosol**.

sample digestion: The process of digesting a sample by traditional hot plate, fusion techniques, or microwave technology to get the matrix and analytes into solution.

sample dissolution: The process of dissolving a sample by traditional hot-plate, fusion techniques, or microwave technology to get the matrix and analytes into solution.

sample injector: The central tube of the plasma torch that carries the sample aerosol mixed with the nebulizer gas. It can be a fixed part of the quartz torch, or it can be separate (demountable) and be made from other materials, such as alumina, platinum, and sapphire, for the analysis of highly corrosive materials.

sample introduction system: The part of the instrument that takes the liquid sample and puts it into the plasma torch as a fine-droplet aerosol. It comprises a nebulizer to generate the aerosol and a spray chamber to reject the larger droplets and allow only the smaller droplets into the plasma discharge.

sample preparation: The entire process of preparing the sample for aspiration into the ICP mass spectrometer.

sampler cone: A part of the mass spectrometer interface region, where the ion beam from the plasma discharge first enters. The sampler cone, which is the first cone of the interface, is typically made of nickel or platinum and contains a small orifice of approximately 0.8–1.2 mm diameter, depending on the design. The sampler cone is much more pointed than the skimmer cone.

sample throughput: The rate at which samples can be analyzed.

sample transport interferences: A term used to describe a physical suppression of the analyte signal caused by matrix components in the sample. It is more exaggerated with samples having high levels of dissolved solids, because they are transported less efficiently through the sample introduction system than aqueous-type samples. *Also refer to* **physical interferences**.

sampling accessories: Customized sample introduction techniques optimized for a particular application problem or sample type. The most common types used today include the following: laser ablation/sampling (LA/S), flow injection analysis (FIA), electrothermal vaporization (ETV), desolvation systems, direct injection nebulizers (DINs), and chromatography separation techniques.

scan time: The mass analyzer scan time is the time it takes to scan from one isotope to the next.

Scott spray chamber: A sealed spray chamber with an inner tube inside a larger tube. The sample aerosol from the nebulizer is first directed into the inner tube. The aerosol then travels the length of the inner tube, where the larger droplets fall out by gravity into a drain tube and the smaller droplets return between the inner and outer tube, where they eventually exit into the sample injector of the plasma torch.

secondary discharge: Another term used for the **pinch effect**.

secondary (side) reactions: Reactions that occur in a collision/reaction cell that are not a part of the main interference reduction mechanism. If not anticipated and compensated for, secondary reactions can lead to erroneous results.

semiquant: An abbreviated name used to describe **semiquantitative analysis**.

semiquantitative analysis: A method for assessing the approximate concentration of up to 70 elements in an unknown sample. It is based on comparing the intensity of a small group of elements against known response tables stored in the instrument's calibration software. By correcting for common spectral interferences and applying heuristic, knowledge-driven routines in combination with numerical calculations, a positive or negative confirmation can be made for each element present in the sample.

settling time: The time taken for the mass analyzer electronics to settle before a peak intensity measurement is taken for the operator-selected dwell time. The dwell time can usually be selected on an individual mass basis, but the settling time is normally fixed because it is a function of the mass analyzer and detector electronics.

shadow stop: A grounded metal disk that stops particulate matter, neutral species, and photons from getting to the detector. It is considered a part of the ion optics and is sometimes called a photon stop.

side reactions: Reactions that occur in a collision/reaction cell that are not a part of the main interference reduction mechanism. If not anticipated and compensated for, secondary reactions can lead to erroneous results.

signal-to-background ratio (S/B): The ratio of the signal intensity of an analyte to its background level at a particular mass. When considering the noise of the background signal (standard deviation of the signal), it is typically used as an assessment of the detection limit for that element. *Also refer to* **detection limit**, **background signal**, *and* **background noise**.

single-particle ICP-MS: A technique used to characterize nanoparticles. The method involves introducing nanoparticle (NP)-containing samples, at very dilute concentration, into the ICP-MS and collecting time-resolved data.

single-point peak hopping: A quantitation in which the quadrupole power supply is driven to a discrete position on the analyte mass (normally the maximum point) and allowed to settle (settling time), and a measurement is taken for a fixed amount of time (dwell time). The integration time for that peak is the dwell time multiplied by the number of scans (scan time). Multielement peak quantitation involves peak hopping to every mass in the multielement run. *Also refer to* **measurement duty cycle** *and* **peak measurement protocol**.

skimmer cone: A part of the mass spectrometer interface region where the ion beam from the plasma discharge first enters. The skimmer cone, which is the second cone of the interface, is typically made of nickel or platinum and contains a small orifice of approximately 0.5–0.8 mm diameter, depending on the design. The skimmer cone is much less pointed than the sampler cone.

solvent-based interferences: Spectral interferences derived from an elemental ion in the solvent (e.g., water or acid) and combining with another ion from either the sample matrix or plasma gas (argon) to produce a polyatomic ion that interferes with the analyte mass.

SOP: An abbreviation for standard operating procedure.

space charge effect: A type of matrix-induced interference that produces a suppression of the analyte signal. This occurs mainly when low-mass analytes are being determined in the presence of larger concentrations of high-mass matrix components. It has the effect of defocusing the ion beam, and without compensation, the high-mass matrix element will dominate the ion beam, pushing the lighter elements out of the way, leading to low sensitivity and poor detection limits. The classical way to compensate for a space charge matrix interference is to use an internal standard of a mass similar to that of the analyte.

speciation analysis: In ICP-MS, it is the study and quantification of different species or forms of an element using a chromatographic separation device coupled to an ICP mass spectrometer. In this configuration, the instrument becomes a very sensitive detector for trace element speciation studies when coupled with high-performance liquid chromatography (HPLC), ion chromatography (IC), gas chromatography (GC), or capillary electrophoresis (CE). In these hybrid techniques, element species are separated on the basis of their chromatograph retention and mobility times and then eluted or passed into the ICP mass spectrometer for detection. The intensity of the eluted peaks is then displayed for each isotopic mass of interest in the time domain.

spectral interferences: A generic name given to interferences that produce a spectral overlap at or near the analyte mass of interest. In ICP-MS, there are two main types of spectral interference that have to be taken into account. Polyatomic spectral interferences (or molecular-based spectral interferences) are typically generated in the plasma by the combination of two or more atomic ions. They are caused by a variety of factors, but are usually associated with the argon plasma or nebulizer gas used, matrix components in the solvent or sample, other elements in the sample, or entrained oxygen or nitrogen from the surrounding air. The other type is an isobaric spectral interference, which is caused by different isotopes of other elements in the sample creating spectral interferences at the same mass as the analyte.

SP-ICP-MS: An abbreviation for single-particle ICP-MS.

spray chamber: The component of the sample introduction system that takes the aerosol generated by the nebulizer and rejects the larger droplets for the more desirable smaller droplets.

SRM: An abbreviation for **standard reference material**.

stability: The ability of a measuring device to consistently replicate a measurement. In ICP-MS, it usually refers to the capability of the instrument to reproduce the signal intensity of the calibration standards over a fixed period of time without the use of internal standardization. Short-term stability is generally defined as the precision (as percent relative standard deviation [% RSD]) of 10 replicates of a single or multielement solution, whereas long-term stability is defined as the precision (as % RSD) of a fixed number of measurements over a 4–8 h time period of a single or multielement solution. However, stability in mass spectrometry can *also refer to* mass calibration stability, which is the ability of the mass spectrometer to repeatedly scan to the same mass position every time during a multielement analysis.

stability boundaries and regions: The radio-frequency (RF) and direct current (DC) boundaries of the Mathieu stability plot where an ion is stable as it passes through a quadrupole mass filtering device. *Also refer to* **Mathieu stability plot**.

standard additions: A method of calibration that provides an effective way to minimize sample-specific matrix effects by spiking samples with known concentrations of analytes. In standard addition calibration, the intensity of a blank solution is first measured. Next, the sample solution is "spiked" with known concentrations of each element to be determined. The instrument measures the response for the spiked samples and creates a calibration curve for each element for which a spike has been added. The calibration curve is a plot of the blank subtracted intensity of each spiked element against its concentration value. After creating the calibration curve, the unspiked sample solutions are then analyzed and compared with the calibration curve. Depending on the slope of the calibration curve and where it intercepts the x-axis, the instrument software determines the unspiked concentration of the analytes in the unknown samples.

standardization methods: The different types of calibration routines available in ICP-MS, including quantitative analysis (external calibration and standard additions), semiquantitative analysis, isotope dilution, isotope ratio, and internal standardization methods.

standard reference material (SRM): Well-established reference matrix that comes with certified values and associated statistical data that have been analyzed by other complementary techniques. Their purpose is to check the validity of an analytical method, including sample preparation, instrument methodology, and calibration routines, to achieve sample results that are as accurate and precise as possible and can be defended under intense scrutiny.

syringe pump: *Refer to* piston pump.

T

thermoelectric cooling device: Better known as a Peltier cooler, it generates a cold environment by creating a temperature gradient between two different materials. It uses electrical energy via a solid-state heat pump to transfer heat from a material on one side of the device to a different material on the other side, thus producing a temperature gradient across the device (similar to a household air conditioner).

thermoelectric flowmeter: A device to measure the liquid flow through a nebulizer to check for any blockages or breakages.

time-of-flight mass spectrometry (TOFMS): A mass spectrometry technique based on the principle that the kinetic energy (KE) of an ion is directly proportional to its mass (m) and velocity (V), which can be represented by the equation $KE = \frac{1}{2}mV^2$. Therefore, if a population of ions with different masses is given the same KE by an accelerating voltage (U), the velocities of the ions will all be different, depending on their masses. This principle is then used to separate ions of different mass-to-charge (m/z) ratio in the time (t) domain, over a fixed flight path distance (D), represented by the equation $m/z = 2Ut^2/D^2$. The simultaneous nature of sampling ions in TOF offers distinct advantages over traditional scanning (sequential) quadrupole technology for ICP-MS applications, where large amounts of data need to be captured in a short amount of time, such as the multielement analysis of transient peaks (laser ablation, flow injection, etc.).

time-of-flight mass spectrometry (axial design): There are basically two different sampling approaches that are used in commercial time-of-flight (TOF) mass analyzers: the axial and orthogonal designs. In the axial design, the flight tube is in the same axis as the ion beam, whereas in the orthogonal design, the flight tube is positioned at right angles to the sampled ion beam. The axial approach applies an accelerating potential in the same axis as the incoming ion beam as it enters the extraction region. Because the ions are in the same plane as the detector, the beam has to be modulated using an electrode grid to repel the "gated" packet of ions into the flight tube. This kind of modulation generates an ion packet that is long and thin in cross section (in the horizontal plane), which is then resolved in the time domain according to the different ionic masses. *Also refer to* **time-of-flight mass spectrometry**.

time-of-flight mass spectrometry (orthogonal design): There are basically two different sampling approaches that are used in commercial time-of-flight (TOF) mass analyzers, the axial and orthogonal designs. In the axial design, the flight tube is in the same axis as the ion beam, whereas in the orthogonal design, the flight tube is positioned at right angles to the sampled ion beam.

With the orthogonal approach, an accelerating potential is applied at right angles to the continuous ion beam from the plasma source. The ion beam is then "chopped" by using a pulsed voltage supply coupled to the orthogonal accelerator to provide repetitive voltage "slices" at a frequency of a few kilohertz. The "sliced" packets of ions, which are typically tall and thin in cross section (in the vertical plane), are then allowed to "drift" into the flight tube, where the ions are temporally resolved according to their differing velocities. *Also refer to* **time-of-flight mass spectrometry**.

TOFMS: An abbreviation for **time-of-flight mass spectrometry**.

torch design: Refers to the different kinds of commercially available torch designs.

trace metal speciation studies: *Refer to* **speciation analysis**.

transient signal (peak): A signal that lasts for a finite amount of time, compared with a continuous signal that lasts for as long as the sample is being aspirated. Transient peaks are typically generated by alternative sampling devices, such as laser ablation, flow injection, or chromatographic separation systems, where discrete amounts of sample are introduced into the ICP mass spectrometer.

triple-cone interface: A commercial design of an ICP-MS interface that uses three cones. *Also refer to* hyper skimmer cone.

triple-quadrupole collision/reaction cell: A commercial collision/reaction cell design that has an additional quadrupole prior to the collision/reaction cell multipole and the analyzer quadrupole. This first quadrupole acts as a simple mass filter to allow only the analyte masses to enter the cell, while rejecting all other masses. With all nonanalyte, plasma, and sample matrix ions excluded from the cell, sensitivity and interference removal efficiency is significantly improved compared with traditional collision/reaction cell technology coupled with a single-quadrupole mass analyzer.

turbomolecular pump (turbo pump): A type of vacuum pump used to maintain a high vacuum in the ion optics and mass analyzer regions of the ICP mass spectrometer. These pumps work on the principle that gas molecules can be given momentum in a desired direction by repeated collision with a moving solid surface. In a turbo pump, a rapidly spinning turbine rotor strikes gas (argon) molecules from the inlet of the pump toward the exhaust, creating and maintaining a vacuum. In the case of ICP-MS, two pumps are normally used, a large pump for the ion optic region, which creates a vacuum of approximately 10^{-3} torr, and another small pump for the mass analyzer region, which generates a vacuum of 10^{-6} torr. However, some designs use a twin-throated turbo pump, in which one powerful pump is used with two outlets, one for the ion optics and one for the mass analyzer region.

twin-throated turbomolecular pump: In some designs of ICP mass spectrometer, a single twin-throated turbo pump is used instead of two separate pumps. In this design, one powerful pump is used with two outlets, one for the ion optics and one for the mass analyzer region.

U

ultrasonic nebulizer (USN): A type of desolvating nebulizer that generates an extremely fine-droplet aerosol for introduction into the ICP mass spectrometer. The principle of aerosol generation using this approach is based on a sample being pumped onto a quartz plate of a piezoelectric transducer. An electrical energy of 1–2 MHz is coupled to the transducer, which causes it to vibrate at high frequency. These vibrations disperse the sample into a fine-droplet aerosol, which is carried in a stream of argon. With a conventional USN, the aerosol is passed through a heating tube and a cooling chamber, where most of the sample solvent is removed as a condensate before it enters the plasma. However, commercial USNs are also available with membrane desolvation systems.

ultraviolet (UV) laser: A generic name given to a laser ablation system that works in the UV region of the electromagnetic spectrum. The three most common wavelengths used in commercial equipment are all UV lasers. They include the 266 nm (frequency-quadrupled) Nd:YAG laser, the 213 nm (frequency-quintupled) Nd:YAG laser, and the 193 nm ArF excimer laser system. *Also refer to* **excimer laser** *and* **Nd:YAG laser**.

universal cell: The name applied to a commercial collision/reaction cell that offers the capability of either a collision cell using inert gases and KED or a dynamic reaction cell using highly reactive gases.

USN: An abbreviation for **ultrasonic nebulizer**.

V

vacuum chamber: The region of the mass spectrometer that is under negative pressure created by a combination of roughing and turbomolecular pumps. As the ion beam moves from the plasma, which is at atmospheric pressure (760 torr), it enters the interface region between the sampler and skimmer cone (1–2 torr) before it is focused through the ion optic vacuum chamber region (10^{-3} torr) and eventually goes through the mass analyzer vacuum chamber (at 10^{-6} torr). *Also refer to* **turbomolecular pump**.

vacuum gauge: Used to measure the pressure in the different vacuum chambers of the mass spectrometer.

vacuum pump: A number of vacuum pumps are used to create the vacuum in an ICP mass spectrometer. Two roughing pumps are used, one for the interface region and another to back up the first turbomolecular pump of the ion optic region. Also, two turbomolecular pumps (or in some designs, one twin-throated pump) are used, one for the ion optics and another for the main mass analyzer region. *Also refer to* **roughing pump**, **turbomolecular pump**, *and* **twin-throated turbomolecular pump**.

vapor phase decomposition (VPD): A technique used to dissolve and collect trace metal impurities on the surface of a silicon wafer, by rolling a few hundred microliters of hydrofluoric acid (HF) over the surface.

visible laser: A laser that operates in the visible region of the electromagnetic spectrum. An example is the ruby laser, which operates at 694 nm.

W

wavelength: A name commonly used to identify an emission line in nanometers used in ICP-OES. Also used to describe a type of laser ablation system (e.g., a ruby laser operating at a 694 nm wavelength) used to couple to an ICP-MS to analyze solid samples directly.

wet plasma: When a liquid sample is introduced or aspirated into the plasma, the plasma is called a "wet plasma."

X

***x*-position:** Refers to the alignment of the plasma torch in the lateral (sideways) position. Typically carried out to maximize sensitivity or to optimize sampling conditions for cool plasma use.

x-ray: A part of the electromagnetic spectrum that has a wavelength in the range of 0.01–10 nm.

x-ray fluorescence: An analytical technique that uses the emission of characteristic "secondary" (or fluorescent) x-rays from a material that has been excited by bombarding with high-energy x-rays or gamma rays. The phenomenon is widely used for elemental analysis and chemical analysis, particularly in the study of solid materials, such as metals, glass, ceramics, soils, and rocks.

Y

***y*-position:** Refers to the alignment of the plasma torch in the longitudinal (vertical) position. Typically carried out to maximize sensitivity or to optimize sampling conditions for cool plasma use.

Z

***z*-position:** Refers to the alignment of the plasma torch in relation to the distance from the interface cone (in and out position). Typically carried out to maximize sensitivity or to optimize sampling conditions for cool plasma use.

INDUCTIVELY COUPLED PLASMA OPTICAL EMISSION SPECTROMETRY GLOSSARY

A

aerosol: A group of small-diameter liquid droplets that are formed when the nebulizer breaks up a continuously pumped solution.

argon: One of the noble gases in the periodic table. This gas is used to generate a plasma in ICP-based instruments.

array detector: A group of photoactive elements, arranged in a linear or two-dimensional configuration, which is used to measure photons from a plasma. Arrays can consist of a variety of detectors, including photodiodes, charge-coupled device (CCD) image sensors, and charge injection device (CID) image sensors.

atomic emission (AE): This is sometimes used to refer to the technique inductively coupled plasma atomic emission spectroscopy (ICP-AES), and refers to the emission from atoms in a sample. This term has been replaced with the term *optical emission*, which is more accurate. *Refer to* optical emission.

axial view: This describes the measurement of light down the central channel of the plasma, which includes emission from the entire length of the plasma. Currently technology allows axial measurements to be made, regardless of whether the torch is mounted vertically or horizontally in the instrument.

B

background equivalent concentration (BEC): For a given analyte, this is the concentration of that analyte that produces a net emission signal equal to the background signal.

background noise: The uncertainty affiliated with measuring a blank solution. This value is often calculated by taking the square root of the background signal.

background signal: The intensity of the signal emitted when a blank solution is measured.

C

ceramic torch: A torch for an ICP-OES that is made from a ceramic material that makes the torch more robust and typically increases its lifetime. The outer, intermediate, and injector tubes can all be made out of ceramic.

charge-coupled device (CCD): A type of solid-state detector that can convert photons into photoelectric signals. This type of detector is used in many commercially available ICP-OES instruments.

charge injection device (CID): A type of solid-state detector that can convert photons into photoelectric signals. This type of detector is used in many commercially available ICP-OES instruments.

chemical interferences: These occur when the standards behave differently from the samples as they enter the plasma. These types of interferences typically result from changes in temperature within the plasma. They can be overcome by utilizing ionization suppressants or using a radially configured plasma.

complementary metal oxide semiconductor (CMOS): This is a type of technology that is used to construct circuits that are used in many applications, including imaging sensors for detectors in ICP-OES instruments.

concentric nebulizer: A nebulizer design that uses a stream of high-velocity argon gas to create an area of low pressure at its tip, causing a constant flow of solution to break apart into small droplets as it leaves the nebulizer. This nebulizer can be fed via a pumped solution, or it can operate via free aspiration.

cooled spray chamber: A spray chamber that is housed inside a cooled chamber that lowers the temperature inside the spray chamber and reduces the amount of aerosol that gets transported from the spray chamber to the plasma.

cyclonic spray chamber: A spray chamber design that is commonly used in ICP-OES instruments. As the name implies, this spray chamber accepts aerosol from the nebulizer and forms it into a virtual cyclone before transferring the smallest droplets into the plasma. This spray chamber is manufactured with or without a center tube, which acts as an additional impact surface for the aerosol.

D

desolvating nebulizer: A nebulizer that desolvates (removes a significant amount of the solvent from) the aerosol prior to its passage into the plasma. This system utilizes a heated or cooled spray chamber and a solvent membrane to remove either aqueous or organic solvent from the nebulized aerosol.

detection limit (DL): Sometimes referred to as the LOD, the detection limit is the lowest concentration of an analyte that can be distinguished from a solution that does not contain that analyte (i.e., a blank solution).

detector: A device used to convert photons from the instrument's plasma to a digital signal that can be read out by a microprocessor.

devitrification: A chemical and temperature-related process that breaks down the structure of a glass or quartz component (typically a torch).

direct injection high-efficiency nebulizer (DIHEN): This is a modification of the original direct injection nebulizer, which incorporates some design improvements.

direct injection nebulizer (DIN): A nebulizer design that injects sample solution directly into the base of the instrument's plasma, eliminating the need for a spray chamber. This nebulizer maximizes the transport efficiency of the aerosol and minimizes solution waste and memory effects between samples.

double-pass spray chamber: A spray chamber design.

droplet: A small drop of solution. During the process of introducing samples into the instrument's plasma, samples must be converted from continuous solutions to a group of small-diameter droplets.

H

high-efficiency nebulizer (HEN): A term used to describe nebulizers that have been designed to convert a significant percentage of the pumped solution into a usable aerosol. Very little solution is sent to waste.

high-solids nebulizers: These nebulizers are designed with the intent to create a usable aerosol from solutions with high levels of dissolved solids. These include V-groove and Babington nebulizer designs.

I

inductively coupled plasma (ICP): A body of gas ions that are formed and maintained via the interaction between radio-frequency and magnetic fields.

This type of plasma is used as an atomization or ionization source in inductively coupled plasma optical emission spectrometry.

inductively coupled plasma optical emission spectrometry (ICP-OES): An elemental analysis technique that uses an inductively coupled plasma to desolvate, vaporize, atomize, and excite liquid samples. The resulting emitted light is collected, measured, and used to quantify the elements present in the original sample.

injector: Also known as the sample injector, this is the center tube in the plasma torch, and it is used to transport the sample aerosol and inject it into the center of the plasma.

instrument detection limit (IDL): This is often used synonymously with the limit of detection (LOD) and refers to the signal-to-background noise capability of the instrument for a given analyte. This is often calculated in accordance with the International Union of Pure and Applied Chemistry (IUPAC) definition as three times the standard deviation of 10 replicate measurements of a blank solution.

integration time: The amount of time the instrument allows light to pass to the detector to collect emission from the plasma.

interferences: In plasma-based techniques, this is a broad term that refers to something that hinders the instrument's ability to precisely and accurately quantify elemental impurities in a sample. Interferences can be chemical, physical, or spectral in nature and sometimes involve more than one correction technique to ensure quality results.

internal standardization: A technique used to overcome some physical and chemical interferences in ICP-based techniques. This technique involves the addition of an analyte to all blanks, standards, and samples that will behave in a manner similar to that of the analytes being measured. Sample results are reported as a ratio between the analyte and the internal standard, which corrects for small changes in intensity that are due to physical or chemical changes in the solutions. These small changes, if left uncorrected, result in errors in the reported results.

L

load coil: A term that refers to the radio-frequency (RF) coil in a plasma-based instrument. This coil maintains the RF signal being generated by the instrument's RF generator, which helps sustain the plasma.

M

method detection limit (MDL): A conservative estimate of the instrument's detection limits. The MDL can be calculated in a number of different ways; however, it is the minimum concentration of an analyte that can be distinguished from the absence of the analyte and quantified with a high degree of confidence (95%–99%).

microconcentric nebulizer: Based on the concentric design, this nebulizer operates at significantly lower solution flow rates than the standard concentric nebulizer.

N

nebulizer: One of the components in the sample introduction system, the nebulizer takes a continuous solution and breaks it up into a body of small-diameter droplets (aerosol).

P

Peltier cooler: A thermoelectric cooler that is used to cool the spray chamber and maintain a specific cooled temperature during a sample run.
peristaltic pump: A type of pump that uses rollers to positively displace sample solutions, pushing them through flexible tubing. These pumps are used to move solution through the sample uptake and drain tubing that is used on many commercially available ICP-OES instruments.
physical interferences: Occur when the nebulization and/or transport efficiency of the standards differs from that of the samples. Physical interferences are usually overcome by utilizing internal standardization and preparing the calibration standards in a matrix that matches that of the samples.

Q

quartz torch: A torch for an ICP-OES that is made from quartz. The outer, intermediate, and injector tubes can all be made out of quartz.

R

radial view: This describes the measurement of light perpendicular to the direction of the plasma.
radio-frequency (RF) generator: The RF generator typically consists of a direct current (DC) power supply, an RF generator board, and a number of other electrical components that work together to generate an RF field that is necessary for maintaining the plasma in an ICP-based instrument.

S

sample introduction system: This refers to the components that transport a sample solution into the instrument's plasma, while converting it from a continuous solution to a fine aerosol. The components that make up the sample introduction system include the nebulizer, spray chamber, and torch, as well as the pump and affiliated pump tubing.
Scott spray chamber: This is a type of double-pass spray chamber.

signal-to-background ratio (S/B or SBR): For a given analyte, this is the intensity ratio of the signal and the background at a particular concentration. When the S/B is calculated in a blank solution, this is often used to represent the detection limit for that analyte.

spray chamber: One of the components in the sample introduction system, the spray chamber takes the aerosol from the nebulizer and filters out the larger-diameter droplets such that only small-diameter droplets are transferred to the plasma. The larger droplets are sent directly to waste.

T

torch: One of the components in the sample introduction system, the torch houses the plasma and receives the sample aerosol that was generated and transported by the nebulizer and spray chamber.

torch injector: Also known as the injector or the sample injector, this is the center tube in the plasma torch, and it is used to transport the sample aerosol and inject it into the center of the plasma.

U

ultrasonic nebulizer (USN): A nebulizer design that uses high-frequency vibrations to convert a solution into an aerosol with small-diameter droplets. Unlike conventional nebulizers, an USN generates a higher percentage of smaller droplets, which improves the transport of solution to the plasma and increases the analyte signal. Most commercially available USNs use a heated or cooled chamber and/or a membrane desolvation system to remove excess solvent from the aerosol prior to being transported to the plasma.

V

viewing height: This refers to where the emitted light is being collected from the plasma when making measurements in radial view mode. This viewing position is usually in reference to the height above the top turn of the load (radio-frequency [RF]) coil.

W

wavelength: In ICP-OES, wavelength-specific photons are emitted from excited atoms and ions. and measured by the instrument's detector.

29 Useful Contact Information

The final chapter of the book is dedicated to providing you with useful contact information related to atomic spectroscopy (AS). It includes contact details for manufacturers of inductively coupled plasma optical emission spectrometry (ICP-OES) and inductively coupled plasma mass spectrometry (ICP-MS) instrumentation, instrument consumables, sample introduction components, and alternative sources of sampling accessories, together with suppliers of laboratory chemicals, calibration standards, certified reference materials, high-purity gases, deionized water systems, and clean-room equipment. I have also included information about the major scientific conferences, professional societies, publishing houses, and Internet discussion groups, and the most popular ICP-related journals. It is sorted alphabetically by category. However, some vendors sell many different products, so they are listed under the category represented by their major product line. Please note that contact information for users in North America is given, even for vendors whose corporate headquarters are outside the United States. Please visit the websites given for support issues or ordering information in other parts of the world.

Certified Reference Materials and Calibration Standards

National Research Council (NRC) of Canada
1500 Montreal Road
Ottawa, Ontario, K1A 0R9, Canada
Phone: 613-993-9391
Fax: 613-952-8239
www.nrc-cnrc.gc.ca

NIST
100 Bureau Drive, Stop 200
Gaithersburg, MD 20899
Phone: 301-975-6776
Fax: 301-975-2183
www.nist.gov

VHG Labs
276 Abby Road
Manchester, NH 03103
Phone: 603-622-7660
Fax: 603-622-5180
www.vhglabs.com

High Purity Standards
P.O. Box 41727
Charleston, SC 29423
Phone: 843-767-7900
Fax: 843-767-7906
www.highpuritystandards.com

Inorganic Ventures Inc.
300 Technology Drive
Christiansburg, VA 24073
Phone: 540-585-3030
Fax: 540-585-3012
www.inorganicventures.com

SPEX CertiPrep
203 Norcross Avenue
Metuchen, NJ 08840
Phone: 800-522-7739
Fax: 732-603-9647
www.spexcertiprep.com

SCP Science
21800 Clark Graham
Baie D'urfe, H9X 4B6, Canada
Phone: 800-361-6820
Fax: 514-457-4499
www.scpscience.com

Conostan Standards
A Division of SCP Science
348 Route 11
Champlain, NY 12919
Phone: 800-361-6820
Fax: 800-253-5549
www.conostan.com

LGC Standards USA
276 Abby Road
Manchester, NH 03103
Phone: 603-206-0799
Toll-free: 1-850-LGC-USA1
www.lgcgroup.com

United States Pharmacopeial Convention (USP)
12601 Twinbrook Parkway Rockville, MD
20852-1790
Phone: 800-227-8772
www.usp.org/reference-standards

Chromatographic Separation Equipment

Agilent Technologies
2850 Centerville Road
Wilmington, DE 19808
Phone: 302-633-8264
Fax: 302-633-8916
www.chem.agilent.com

Dionex Corporation
A Thermo Company
1228 Titan Way
P.O. Box 3603
Sunnyvale, CA 94088
Phone: 408-737-0700
Fax: 408-730-9403
www.dionex.com

PerkinElmer Life and Analytical Sciences
710 Bridgeport Avenue
Shelton, CT 06484
Phone: 800-762-4000
Fax: 203-944-4914
www.perkinelmer.com

Shimadzu Scientific Instruments (North America)
7102 Riverwood Drive
Columbia, MD 21046
Phone: 410-381-1227
Fax: 410-381-122
http://www.ssi.shimadzu.com/

Thermo Fisher Scientific
81 Wyman Street
Waltham, MA 02454
Phone: 781-622-1000
Fax: 781-622-1207
www.thermoscientific.com

Waters Corporation
34 Maple Street
Milford, MA 01757
Phone: 508-478-2000 Fax: 508-872-1990
www.waters.com

Clean Rooms and Equipment

Microzone Corp.
86 Harry Douglas Drive
Ottawa, Ontario, K2S 2C7, Canada
Phone: 613-831-8318
Fax: 613-831-8321
www.microzone.com

Clestra Hauserman
259 Veterans Lane, Suite 201
Doylestown, PA 18901
Phone: 267-880-3700
Fax: 267-880-3705
www.clestra-cleanroom.com

Consumables: ICP-MS Detectors

SGE Inc.
2007 Kramer Lane
Austin, TX 78758
Phone: 800-945-6154
Fax: 512-836-9159
www.sge.com

Spectron Inc.
1601 Eastman Avenue, Suite 205
Ventura, CA 93003
Phone: 805-642-0400
Fax: 805-642-0300
www.spectronus.com

Consumables: Sample Introduction and Interface Components

Burgener Research Inc.
1680-2 Lakeshore Road W
Mississauga, Ontario, L5J 1J5, Canada
Phone: 905-823-3535
Fax: 905-823-2717
www.burgenerresearch.com

CPI International
5580 Skylane Boulevard
Santa Rosa, CA 95403
Phone: 800-878-7654
Fax: 707-545-7901
www.cpiinternational.com

Elemental Scientific Inc. (ESI)
2440 Cumming Street
Omaha, NE 68131
Phone: 402-991-7800
Fax: 402-997-7799
www.elementalscientific.com

Glass Expansion Pty
4 Barlows Landing Road, Unit #2
Pocasset, MA 02559
Phone: 505-563-1800
Fax: 505-563-1802
www.geicp.com

Meinhard Glass Products
An Elemental Scientific Company
700 Corporate Circle, Suite A
Golden, CO 80401
Phone: 303-277-9776
Fax: 303-216-2649
www.meinhard.com

Precision Glassblowing
14775 E. Hindsdale Avenue
Centennial, CO 80112
Phone: 303-693-7329
Fax: 303-699-6815
www.precisionglassblowing.com

SCP Science
21800 Clark Graham
Baie D'urfe, H9X 4B6, Canada
Phone: 800-361-6820
Fax: 514-457-4499
www.scpscience.com

Spectron Inc.
1601 Eastman Avenue, Suite 205
Ventura, CA 93003
Phone: 805-642-0400
Fax: 805-642-0300
www.spectronus.com

Expositions and Conferences

Eastern Analytical
P.O. Box 633
Montchanin, DE 19710
Phone: 610-485-4633
Fax: 610-485-9467
www.eas.org

FACSS (SCIX)
1201 Don Diego Avenue
Santa Fe, NM 87505
Phone: 505-820-1648
Fax: 505-989-1073
www.facss.org

Pittsburgh Conference and Exposition
300 Penn Center Boulevard, Suite 332
Pittsburgh, PA 15235
Phone: 412-825-3220
Fax: 412-825-3224
www.pittcon.org

Plasma Winter Conference
c/o Dr. Ramon Barnes ICP Information Newsletter
P.O. Box 666 Hadley MA 01035-0666
Phone: 239-674-9430
Fax: 239-674-9431
http://icpinformation.org/

High-Purity Gases

Air Liquide Specialty Gases LLC
Corporate Headquarters
6141 Easton Road
Plumsteadville, PA 18949
Phone: 215-766-8860
Fax: 215-766-2476
www.airliquide.com

Air Products and Chemicals
7201 Hamilton Boulevard
Allentown, PA 18195
Phone: 610-481-4911
Fax: 610-481-5900
www.airproducts.com

Praxair Inc.
Worldwide Headquarters
39 Old Ridgebury Road
Danbury, CT 06810
Phone: 800-772-9247
Fax: 716-879-2040
www.praxair.com

Scott Specialty Gases
A Division of Air Liquide
6141 Easton Road, P.O. Box 310
Plumsteadville, PA 18949
Phone: 215-766-8861
Fax: 215-766-2476
www.scottgas.com

High-Purity and Reagent Chemicals

Sigma-Aldrich Chemicals
Customer Support
P.O. Box 14508
St. Louis, MO 63178
Phone: 800-325-3010
Fax: 800-325-5052
www.sigmaaldrich.com

Eichrom Technologies LLC
1955 University Lane
Lisle, IL 60532
Phone: 630-963-0320
Fax: 630-963-1928
www.eichrom.com

Fisher Scientific Inc.
300 Industry Drive
Pittsburgh, PA 15275
Phone: 800-766-7000
Fax: 800-926-1166
www.fishersci.com

Mallinckrodt Baker
222 Red School Lane
Phillipsburg, NJ 08865
Phone: 908-859-9315
Fax: 908-859-9385
www.mbigloballabcatalog.com

ICP-MS Instrumentation: Magnetic Sector Technology

Thermo Fisher Scientific
81 Wyman Street
Waltham, MA 02454
Phone: 781-622-1000
Fax: 781-622-1207
www.thermoscientific.com

Nu Instruments
Sales and Support
8 Magnolia Street
Newburyport, MA 01950
Phone: 978-465-2484
Fax: 978-465-2484
www.nu-ins.com

SPECTRO Analytical Instruments Inc.
91 McKee Drive
Mahwah, NJ 07430
Phone: 201-642-3000
Fax: 201-642-3091
www.spectro.com

ICP-MS Instrumentation: Quadrupole Technology

Agilent Technologies
ICP-MS Systems
2850 Centerville Road
Wilmington, DE 19808
Phone: 302-633-8264
Fax: 302-633-8916
www.chem.agilent.com

Analytic Jena
100 Cummings Center, Suite 234-N
Beverly, MA, 01915
Phone: 781-376-9899
Fax: 781-376-9897
http://us.analytik-jena.com

PerkinElmer Inc.
710 Bridgeport Avenue
Shelton, CT 06484
Phone: 800-762-4000
Fax: 203-944-4914
www.perkinelmer.com

Shimadzu Scientific Instruments (North America)
7102 Riverwood Drive
Columbia, MD 21046
Phone: 410-381-1227
Fax: 410-381-122
http://www.ssi.shimadzu.com/

Thermo Fisher Scientific
81 Wyman Street
Waltham, MA 02454
Phone: 781-622-1000
Fax: 781-622-1207
www.thermoscientific.com

ICP-MS Instrumentation: Time-of-Flight Technology

GBC Scientific
151A North State Street
P.O. Box 339
Hampshire, IL 60140
Phone: 847-683-9870
Fax: 847-683-9871
www.gbcsci.com

TOFWERK, USA
2760 29th Street, Suite 1F
Boulder, CO 80301
Phone: 303-524-2205
http://www.tofwerk.com/contact-us/

ICP-OES/AS Instrumentation

Agilent Technologies
2850 Centerville Road
Wilmington, DE 19808
Phone: 302-633-8264
Fax: 302-633-8916
www.chem.agilent.com

Analytic Jena
100 Cummings Center, Suite 234-N
Beverly, MA, 01915
Phone: 781-376-9899
Fax: 781-376-9897
http://us.analytik-jena.com

Jobin Yvon/HORIBA Scientific
3880 Park Avenue
Edison, NJ 08820-3097
Phone: 732-494-8660
Fax: 732-549-5125
http://www.horiba.com/

PerkinElmer Inc.
710 Bridgeport Avenue
Shelton, CT 06484
Phone: 800-762-4000
Fax: 203-944-4914
www.perkinelmer.com

Shimadzu Scientific Instruments
7102 Riverwood Drive
Columbia, MD 21046
Phone: 410-381-1227
Fax: 410-381-122
http://www.ssi.shimadzu.com/

SPECTRO Analytical Instruments Inc.
91 McKee Drive
Mahwah, NJ 07430
Phone: 201-642-3000
Fax: 201-642-3091
www.spectro.com

Teledyne Leeman Labs
110 Lowell Road
Hudson, NH 03051
Phone: 603-886-8400
Fax: 603-886-4322
www.teledyneleemanlabs.com/

Thermo Fisher Scientific
81 Wyman Street
Waltham, MA 02454
Phone: 781-622-1000
Fax: 781-622-1207
www.thermoscientific.com

Internet Discussion Group

PLASMACHEM List Server
312 Heroy Geology Laboratory
University of Syracuse
Syracuse, NY 13244
Phone: 315-443-1261 (Michael Cheatham)
Fax: 315-443-3363
To subscribe, e-mail:
mmcheath@mailbox.syr.edu

European Virtual Institute for Speciation Analysis (EVISA)
A forum dedicated to trace element speciation analysis
http://www.speciation.net/

Journals and Magazines

Analytical Chemistry
1155 16th Street, NW
Washington, DC 20036
Phone: 202-872-4570
Fax: 202-872-4574
www.pubs.acs.org/ac

Applied Spectroscopy
201b Broadway Street
Frederick, MD 21701
Phone: 301-694-8122
Fax: 301-694-6860
www.s-a-s.org

International Labmate Ltd
Oak Court Business Center
Sandridge Park
Porters Wood, St. Albans
Hertfordshire, AL3 6PH, United Kingdom
Phone: 44-1727-840310
www.labmate-online.com

Journal of Analytical Atomic Spectrometry (JAAS)
Thomas Graham House
Science Park
Milton Rd.
Cambridge, CB4 4WF, United Kingdom
Phone: 44-1223-420066
Fax: 44-1223-420247
www.rsc.org

Pharmaceutical Technology
485 Route One South
Building F, Suite 210
Iselin, NJ 08830
Phone: 732-596-0276
Fax: 732-647-1235
http://www.pharmtech.com/

Spectroscopy Magazine
485 Route One South
Building F, First Floor
Iselin, NJ 08830
Phone: 732-596-0276
Fax: 732-225-0211
www.spectroscopyonline.com

Laser Ablation and Laser-Induced Breakdown Spectroscopy Equipment

Applied Spectra Inc.
46665 Fremont Boulevard
Fremont, CA 94538
Phone: 510-657-7679
http://appliedspectra.com/

Electro Scientific Industries (ESI)
13900 NW Science Park Drive
Portland, OR 97229-5497
Phone: 800-331-4708
Fax: 503-671-5551
https://www.esi.com

Teledyne Cetac Technologies
14306 Industrial Road
Omaha, NE 68144
Phone: 402-733-2829
Fax: 402-733-5292
http://www.teledynecetac.com/

TSI Inc.
500 Cardigan Road Shoreview,
Minneapolis, MN 55126
Phone: 651-483-0900
Fax: 651-490-3824
http://www.tsi.com

Microwave Digestion Equipment

CEM Corporation
3100 Smith Farm Road
Matthews, NC 62810
Phone: 800-726-3331
Fax: 704-821-5185
www.cem.com

Milestone Inc.
160 B Shelton Road
Monroe, CT 06468
Phone: 203-925-4240
Fax: 203-925-4241
www.milestonesci.com

SCP Science
21800 Clark Graham
Baie D'urfe, H9X 4B6, Canada
Phone: 800-361-6820
Fax: 514-457-4499
www.scpscience.com

Anton Paar USA Inc.
10215 Timber Ridge Drive
Ashland, VA 23005
Phone: 804-5501051
Fax: 804-5501051
www.info.us@anton-paar.com

PerkinElmer Inc.
710 Bridgeport Avenue
Shelton, CT 06484
Phone: 800-762-4000
Fax: 203-944-4914
www.perkinelmer.com

Performance and Productivity Enhancement Technology

Teledyne Cetac Technologies
14306 Industrial Road
Omaha, NE 68144
Phone: 402-733-2829
Fax: 402-733-5292
www.cetac.com

Elemental Scientific Inc. (ESI)
1500 N 24th Street
Omaha, NE 68110
Phone: 402-991-7800
Fax: 402-997-7799
www.elementalscientific.com

Glass Expansion Pty
4 Barlows Landing Road, Unit #2
Pocasset, MA 02559
Phone: 505-563-1800
Fax: 505-563-1802
www.geicp.com

Professional Societies, Regulatory Agencies, and Pharmacopeias

American Chemical Society (ACS)
1155 16th Street NW
Washington, DC 20036
Phone: 800-227-5558
Fax: 202-872-4615
www.pubs.acs.org

American Society for Mass Spectrometry (ASMS)
2019 Galisteo Street, Bldg. I-1
Santa Fe, NM 87505
Phone: 505-989-4517
Fax: 505-989-1073
www.asms.org

American Society for Testing Materials (ASTM)
100 Barr Harbor Drive
West Conshohocken, PA 19428
Phone: 610-832-9605
Fax: 610-834-3642
www.astm.org

International Society for Pharmaceutical Engineering (ISPE)
7200 Wisconsin Avenue, Suite 305
Bethesda, MD 20814
Phone: 301-364-9201
Fax: 240-204-6024
http://www.ispe.org/

Society for Applied Spectroscopy (SAS)
5320 Spectrum Drive, Suite C
Frederick, MD 21703
Phone: 301-694-8122
Fax: 301-694-6860
www.s-a-s.org

United States Pharmacopeial Convention (USP)
12601 Twinbrook Parkway
Rockville, MD 20852-1790
Phone: 800-227-8772
www.usp.org/

U.S. Food and Drug Administration (FDA)
10903 New Hampshire Avenue
Silver Spring, MD 20993
Phone: 888-463-6332)
https://www.fda.gov/

European Pharmacopeia (Ph. Eur.)
https://www.edqm.eu/en/european-pharmacopoeia-9th-edition

Japanese Pharmacopeia (JP)
http://jpdb.nihs.go.jp/jp14e/

International Convention for Harmonisation (ICH)
http://www.ich.org/home.html

European Medicines Agency (EMA)
http://www.ema.europa.eu/ema

Publishers

CRC Press/Taylor and Francis
6000 Broken Sound Parkway NW, Suite 300
Boca Raton, FL 33487
Phone: 561-994-0555
Fax: 561-989-9732
www.crcpress.com

Elsevier Science Publishing
Journals Customer Service
3251 Riverport Lane
Maryland Heights, MO 63043
Phone: 800-222-9570
www.elsevier.com

John Wiley & Sons
111 River Street
Hoboken, NJ 07030-5774
Phone: 201-748-6000
Fax: 201-748-6088
www.wiley.com

American Laboratory
395 Oyster Point Boulevard,
#321 South San Francisco, CA 94080
Phone: 650-234-5600
http://www.americanlaboratory.com

Sample Preparation Equipment: Cryogenic Grinding, Fusion Systems, and Pelletizing Equipment

SPEX SamplePrep
203 Norcross Avenue
Metuchen, NJ 08840
Phone: 800-522-7739
Fax: 732-603-9647
www.spexsampleprep.com

SCP Science
21800 Clark Graham
Baie D'urfe, H9X 4B6, Canada
Phone: 800-361-6820
Fax: 514-457-4499
www.scpscience.com

Vacuum Pumps and Components

Oerlikon Leybold Vacuum (USA) Inc.
5700 Mellon Road
Export, PA 15632
Phone: 800-764-5369
Fax: 724-733-1217
www.leyboldvacuum.com

Agilent Technologies
Vacuum Products Division
121 Hartwell Avenue
Lexington, MA 02421
Phone: 781-861-7200
Fax: 781-860-5405
www.chem.agilent.com/en-US/ProductsServices/Instruments-Systems/Vacuum-Technologies/Pages/default.aspx

XRF Equipment

Shimadzu Scientific Instruments (North America)
7102 Riverwood Drive
Columbia, MD 21046
Phone: 410-381-1227
Fax: 410-381-122
http://www.ssi.shimadzu.com/

SPECTRO Analytical Instruments Inc.
91 McKee Drive
Mahwah, NJ 07430
Phone: 201-642-3000
Fax: 201-642-3091
www.spectro.com

Bruker Daltonics
Chemical Analysis Group
3500 West Warren Avenue
Fremont, CA 94538
Phone: 510-683-4300
Fax: 510-687-1217
www.bdal.com/chemicalanalysis

Applied Rigaku Technologies
9825 Spectrum Drive, Bldg. 4, Suite 475
Austin, TX 78717
Phone: 512-225-1796
Fax: 512-225-1797
https://www.rigakuedxrf.com

Index